W0079580

Molecular Electro-Optics

Electro-Optic Properties of Macromolecules
and Colloids in Solution

NATO ADVANCED STUDY INSTITUTES SERIES

A series of edited volumes comprising multifaceted studies of contemporary scientific issues by some of the best scientific minds in the world, assembled in cooperation with NATO Scientific Affairs Division.

Series B. Physics

Recent Volumes in this Series

This series is published by an international board of publishers in conjunction with NATO Scientific Affairs Division

A Life Sciences	Plenum Publishing Corporation	
B Physics	London and New York	
C Mathematical and	D. Reidel Publishing Company	
Physical Sciences	Dordrecht, Boston and London	
D Behavioral and	Sijthoff & Noordhoff International	
Social Sciences	Publishers	
E Applied Sciences	Alphen aan den Rijn, The Netherlands, and	
	Germantown, U.S.A.	

Molecular Electro-Optics

Electro-Optic Properties of Macromolecules and Colloids in Solution

Edited by
Sonja Krause

Rensselaer Polytechnic Institute
Troy, New York

PLENUM PRESS • **NEW YORK AND LONDON**
Published in cooperation with NATO Scientific Affairs Division

Library of Congress Cataloging in Publication Data

NATO Advanced Study Institute on Molecular Electro-optics (1980: Rensselaer Polytechnic Institute) Molecular electro-optics.

(NATO advanced study institutes series. Series B, Physics; v. 64)
"Proceedings of a NATO Advanced Study Institute on Molecular Electro-optics, held July 14-24, 1980, at Rensselaer Polytechnic Institute, Troy, New York" – Verso of t. p.
Bibliography: p.
Includes index.
1. Macromolecules–Electric properties–Congresses. 2. Macromolecules–Optical properties –Congresses. 3. Colloids–Electric properties–Congresses. 4. Colloids–Optical properties– Congresses. I. Krause, Sonja. II. Title. III. Series.
QD381.9.E38N22 1980   547.7'04572   81-1314
ISBN-13: 978-1-4684-3916-8   e-ISBN-13: 978-1-4684-3914-4   AACR2
DOI: 10.1007/978-1-4684-3914-4

Proceedings of a NATO Advanced Study Institute on
Molecular Electro-Optics, held July 14–24, 1980,
at Rensselaer Polytechnic Institute, Troy, New York

© 1981 Plenum Press, New York
Softcover reprint of the hardcover 1st edition 1981
A Division of Plenum Publishing Corporation
233 Spring Street, New York, N.Y. 10013

All rights reserved

No part of this book may be reproduced, stored in a retrieval system, or transmitted, in any form or by any means, electronic, mechanical, photocopying, microfilming, recording, or otherwise, without written permission from the publisher

PREFACE

The Advanced Study Institute on Molecular Electro-Optics was held on the campus of the Rensselaer Polytechnic Institute, Troy, New York, USA, from July 14 through July 24, 1980. This Advanced Study Institute was attended by sixteen invited lecturers and by forty-eight other participants.

The present volume contains the texts of all of the invited lectures presented at the Institute. Although these lectures were supplemented by many animated discussions and by numerous short contributed papers, it was not possible to include these in the present volume.

Molecular electro-optics is a difficult subject for research because it incorporates areas of theoretical physics such as electromagnetic theory and hydrodynamics of rotational diffusion, experimental physics such as lasers, optics, electric pulsers, and data collection via analog to digital converters and signal averagers, and physical chemistry of macromolecules and colloids in solution (colloid science, biophysical chemistry, double layer polarization). This volume includes chapters on all of these subjects as well as introductions to magnets-optics and to electrophoretic light scattering.

The Advanced Study Institute was sponsored mainly by the North Atlantic Treaty Organization whose financial support made this meeting possible. Additional financial aid was supplied by the National Institutes of Health of the USA through their Fogarty International Center and the National Institute for Arthritis, Metabolism, and Digestive Diseases. Industrial contributers consisted of the General Electric Company, Cober Electronics, and Malvern Scientific Corporation.

I would like to thank all the people whose assistance during the planning and operation of this Institute and during the preparation of this volume was indispensable.

Beth McGraw, who served as our Meeting Secretary and worked
tirelessly both during the meeting and during the preparation
of this Volume,

Jeff Boyer, a senior student in Chemistry who was our
meeting projectionist and who helped our visitors feel
at home,

Krystyna Szumilin, my postdoctoral associate, and
Nancy Letko, my graduate student, who worked hard during
the meeting with the various necessary tasks,

Robert Metzger, Manager of Special Events at Rensselaer
Polytechnic Institute who made everything flow smoothly,

Victoria R. Lee, who typed the major portion of this volume,
and Mary K. Nolan, Elizabeth R. Schoonmaker, Susan J. Mangione,
and Patricia S. Connell who typed the rest,

Walter W. Goodwin, my husband, who helped keep me on an even
keel before an during the meeting and during preparation
of this book,

all the lecturers and participants who made the meeting a
stimulating and friendly one.

Sonja Krause
Director of the Institute

CONTENTS

A HISTORY OF MOLECULAR ELECTRO-OPTICS

Chester T. O'Konski

Department of Chemistry
University of California
Berkeley, CA

PHENOMENOLOGICAL DESCRIPTION AND EQUATIONS OF THE KERR ELECTRO-OPTIC EFFECT

Before describing Kerr's discovery of the first electro-optic effect to be detected, let us consider the physical principles of the phenomenon. Figure 1 illustrates the electric double refraction, or birefringence, and introduces conventional terminology. A plane polarized light beam enters a sample placed between plane parallel electrodes connected to an external voltage source. The direction of polarization of the light beam is specified by the direction of its electric vector, E_i, which is normally oriented at 45° with respect to the direction of the applied electric field, as illustrated in the view along the direction of the light beam in Fig. 1(a). The vector of the incident light, E_i, may be decomposed into two components, one parallel to the direction of the applied electric field, designated $E_{||}$, and the other, $E_{\perp}$, perpendicular to the direction of the applied electric field. These two vectors are in phase as the electromagnetic radiation enters the medium. Under the influence of the external field the refractive indexes of the medium change and become unequal for the parallel and the perpendicular components; the sample behaves like a uniaxial crystal with its unique optic axis in the direction of the applied field. The two component rays traverse the medium at different velocities, and emerge out of phase, as illustrated in Fig. 1(b). The resultant, E_e, produces a vector which describes an ellipse.

In the absence of an electric field, the light coming out of the Kerr cell is plane polarized and is extinguished by the crossed analyzer, but in the presence of the field, the emergent light has

1

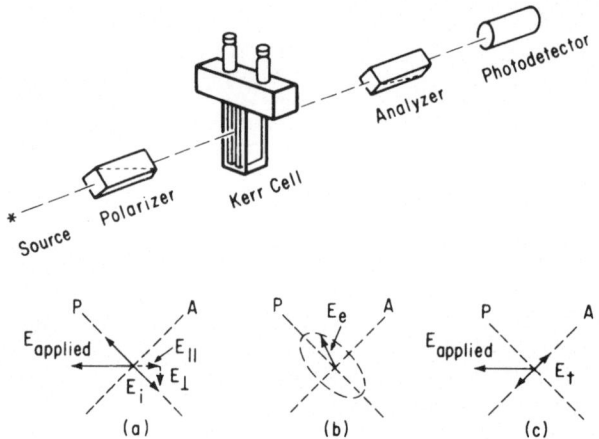

Fig. 1. Apparatus for Kerr electro-optic effect (electric bire-
         fringence) experiments. Below it are shown the states of
         polarization of light: (a) after passing through the
         polarizer, (b) after passing through the cell, and (c)
         after passing through the analyzer. E_i is the electric
         vector of the plane-polarized light incident upon the
         sample, and it can be decomposed into two in-phase compo-
         nents, $E_{||}$ and $E_\perp$, parallel and perpendicular, respectively,
         to the applied electric field. E_e is the vector of the
         elliptically polarized light emerging from the birefringent
         solution in the cell, and E_t is the component of E_e trans-
         mitted by the analyzer.

an electric field component perpendicular to that of the incident
polarized light, so that a portion of the light is transmitted by
the analyzer, as shown by the vector E_t in Fig. 1(c). The intensity of the transmitted light, relative to that of the incident
light, is directly related to the degree of optical anisotropy produced by the applied electric field.

    The optical path length of the cell, for the parallel comp-
nent of the light, may be expressed by

$$N_{\|} = \frac{L n_{\|}}{\lambda_0} , \tag{1}$$

where $N_{\|}$ is the number of wavelengths of light in the cell of length L, $n_{\|}$ is the refractive index for the parallel light component, and λ_0 is the wavelength of light in vacuum. A similar equation applies to the perpendicular component. It follows that the phase retardation, or the optical pathlength difference in radians, δ, is given by

$$\delta = 2\pi(N_{\|} - N_{\perp}) = \frac{2\pi L \ \Delta n}{\lambda_0} \tag{2}$$

Here $\Delta n = n_{\|} - n_{\perp}$, and it is called the double refraction or the birefringence of the medium. The value of δ, and thus Δn, can be obtained from measurements of the increase in the intensity of light produced by application of the electric field to the specimen between crossed polarizer and analyzer oriented at 45° with respect to the applied field.

In most substances at low electric field intensities, the birefringence at a specified wavelength varies with the square of the electric field, i.e.,

$$\Delta n = n_{\|} - n_{\perp} = KnE^2 \tag{3}$$

where E is the applied electric field intensity, and n is the mean refractive index. This is known as the Kerr law, and K is called the Kerr constant.

The specific Kerr constant of a substance in solution or suspension is a measure of the electric and optical properties of the molecular kinetic units; it is given by

$$K_{sp} = \frac{K}{C_v} \tag{4}$$

where C_v is the volume fraction of the solute. In using Eq. (3) for a solute, the contribution of other components to Δn is first subtracted out.

KERR'S WORK

The discovery of the first electro-optic effect was reported in 1875 by the Reverend John Kerr, Mathematical Lecturer in the Free-Church Training College of Glasgow, in an article entitled "A New Relation Between Electricity and Light." He mentions Maxwell's famous "Treatise on Electricity and Magnetism," first published in 1873 [1], which marked the achievement of the mathematical theory of electromagnetic radiation. But Kerr's chief inspiration apparently was Faraday's persistent idea of, and unsuccessful searches for, electro-optic phenomena. In the opening paragraph of his paper [2], Kerr writes:

> The thought which led me to the following inquiry was briefly this: -- that if a transparent and optically isotropic insulator were subjected properly to intense electrostatic force, it should act no longer as an isotropic body upon light sent through it. Faraday was often occupied with expectations of this kind; and he has mentioned in his memoir on the Magnetization of Light, and elsewhere in his "Researches," how he experimented in this very direction, upon electrolytes as well as dielectrics, at different times and in many ways, but always without success. As far as I remember, I have not read or heard of an attempt in this field by any other naturalist. I proceed to offer a few notes of some recent experiments of my own. The investigation is not so complete as I should wish it to be; but it has been carried forward as far as my limited time and means would allow. At present I confine myself to solid dielectrics, reserving the case of liquids for a second paper.

Kerr then goes on to describe his experiments on plate glass "electrified" with the aid of a Ruhmkorff induction apparatus which gave a spark of 20 to 25 cm in air. He observed the glass plate between two crossed Nicol prisms. He noticed that when the electric field was turned on, with its direction 45° from that of the electric vector of the polarized beam, light began to appear through the glass in about 2 sec, it brightened continuously for about 30 sec, and then disappeared gradually to zero after the electric field was removed. Several experiments were described. Finding variables which affected the phenomenon was of interest in seeking to understand its origins. Observations were reported on clear amber resin and on quartz as well as on glass. He found that the "electrified" quartz and glass acted upon transmitted light as if they were compressed along the electric lines of force (positive birefringence), while the electrified resin acted as if it were stretched along the lines of force (negative birefringence).

Previous experiment on strain birefringence played an important part in the interpretation of these qualitative studies.

It is interesting to read Kerr's application of Faraday's views as to the constitution and function of dielectrics in his theoretical interpretation of the results:

> When the induction terminals were charged, the particles of the dielectric throughout the field are electrically polarized, and tend accordingly to arrange themselves end to end, and to cohere in files along the lines of force, just as iron filings do in a magnetic field. As far as this tendency of the polarized particles toward a file arrangement along the lines of force takes effect, there is a new molecular structure induced in the dielectric ... and ... we may assume ... that the effect of electric force is to superinduce a uniaxial structure upon the primitive structure.

Kerr considered some alternative possibilities. His last paragraph in that first article is as follows:

> I have made some experiments, and have had a good many reflections, bearing on other explanations (other than the one mentioned above) of the phenomena; and I think it not unlikely that strains due to the mutual actions of intensely charged shells of the dielectric, or strains due to the changes of temperature, may have something to do with the facts. But in the meantime I offer the preceding remarks as a sketch of what appears to me to be the only probable theory.

In his second paper [3] Kerr reported electric birefringence observations in various liquids, including carbon disulfide, benzene, paraffin oil, kerosene, turpentine (two samples of opposite optical rotations), olive oil, and castor oil. He noticed that the olive oil gave a negative birefringence, like the glass in his earlier experiments, whereas the other liquids displayed a positive birefringence, like resin. Optical compensators were already known at this time, but Kerr did not have one; therefore, he improvised one with a slip of glass held with the hands and subjected to tension which he knew produced positive birefringence, with the slow optic axis along the direction of stretch. He observed semiquantitatively that the birefringence varied many fold, being strongest in carbon disulfide and weakest in paraffin oil and kerosene. He also noted that the birefringence appeared and disappeared in all liquids essentially instantaneously with the application of the field.

He reiterated the idea that the particles of "dielectrified" bodies tend to arrange themselves in files along the lines of force, and suggested that the lines of electric force are lines of compression in one class of dielectrics and lines of extension in another class. He also stated that "the facts ... in whatever way interpreted ... give promise of some new insight into that interesting subject, the molecular mechanism of electric action." The last paragraph of his second paper is:

> I cannot conclude without expressing a hope, amounting almost to a belief, that the plate-cell charged with carbon disulfide will develop from the present crude beginning into a valuable physical instrument, a very delicate optical electrometer.

Thus, although his molecular explanation of the effect was partly erroneous, by postulating a "new molecular structure" and attributing the birefringence to alignment of the molecules into "files," in addition to an (implied) orientation phenomenon, his hope that his cell would become a valuable instrument was indeed prophetic.

He also noticed and reported the effect of the electric field upon the dust particles commonly seen in liquid samples. When the particles in benzene and carbon disulfide were numerous enough, they formed a chain between the electrodes. At the instant of discharge, the chains were sometimes seen to break up violently, and when the particles were so few as not to give chains, they appeared as a set of "sparkling points ... which dart hither and thither through the central parts of the electric field." Thus, both pearl-chain formation and the peculiar kinetic effects observed with small particles in high electric fields in liquids were observed in his pioneering experiments.

In two papers published in 1879 [4,5] Kerr extended his work by making observations on various liquids. With the aid of a "Government Fund" he improved his apparatus and began quantitative measurements of retardation by measuring the weights on the compensator plate. A Thomson electrometer was introduced to measure the applied electrostatic potentials. His optical observations were entirely visual, of course; the light source was a flickering paraffin candle! (Edison developed the first practical electric lamp in 1880.) Most liquids Kerr studied displayed a positive birefringence, and most of those were aromatic substances. Oils of vegetable and animal origin generally showed a negative birefringence. Kerr also observed and reported in detail quite a number of curious effects due to conductivity, spark discharges, inhomogeneities, and electrokinetic phenomena. Apparently these have never been followed up systematically and could be interesting to look at even today with the better apparatus and advanced theoretical understanding we now possess.

In 1880, unsatisfied with the quantitative nature of the previous measurements, Kerr set out to determine the relationship between birefringence and the applied electric field intensity: "In
the leisure of my last summer holidays I resumed the inquiry with
better means, carried it forward for some weeks with great care."
He now had a Jamin compensator [6], an "admirable instrument," and
used sunlight reflected from a bright cloud as well as the light
from a paraffin candle as sources. After a number of measurements
on carbon disulfide, he concluded that the optical retardation was
proportional to the square of the applied electric field intensity.
One can infer that he had made quantitative observations on other
liquids and was confident that this was a general law, although data
on other liquids were not reported systematically.

In 1882 Kerr published two more papers on qualitative observations of electric birefringence, calling attention to the interesting repeats and variations of his own experiments by Röntgen [7],
who had added new observations on water, sulfuric ether, and glycer-
ine. Kerr added more than a hundred new liquids to the previous 27
liquids on which he had reported earlier. Among the liquids measured were bromine, molten phosphorus and sulfur, some hydrocarbons,
molten paraffin, various alcohols and carboxylic acids, allyl ethers,
iodides, bromides, and fluorides [8]. The second paper in that year
[9] reported on various mercaptans, sulfides, esters of various carboxylic acids, nitrates and nitrites, acetone and some aldehydes,
some nitriles and amines, additional chlorides and bromides, and
miscellaneous oils and inorganic and organic compounds. These
observations were purely qualitative, although the relative strengths
of the effects were sometimes mentioned. He noted the resemblance
of most liquids to positive uniaxial crystals, and that they were
more numerous than the negatively refracting liquids. He remarked
that the electro-optic and chemical characters of compounds are
closely connected and that regular progressions sometimes taking
the form even of a change of sign, occur in homologous series. He
did not think in terms of the orientation of an optically anisotropic
molecule in the electric field; rather his explanations were in more
macroscopic terms as follows:

> Electro-optic double refraction is brought about
> by a special state of the dielectric, the state of
> essentially directional strain, which is concomitant
> and a condition of the maintenance of electric stress.
> ... There is, first, the directional transmission of
> electric force from limit to limit of the field; there
> is, secondly, the special dioptric action of the force-
> transmitting medium, an action precisely similar to
> that of glass directionally stressed by tensions or
> pressures. ... I insist upon the essentially direc-
> tional character of electric strain, in opposition to
> a theory of electro-optic action which has been advanced
> by Prof. Quincke.

Other electro-optic papers of Kerr include "Experiments on a Fundamental Question in Electro-Optics" and "Reduction of Relative Retardations to Absolute," both published in 1894 [10,11]. Kerr, who published his first electro-optic work at the age of 51, died at 83 in 1907, shortly before a mathematical theory of the Kerr effect based upon orientation of anisotropic molecules was published by Langevin [12,13].

A commemorative biography of J. Kerr was printed [13a] and has been reprinted [13b]; it provides an interesting view of the circumstances surrounding his discovery, some personal aspects of his life, and glimpses of the University setting at Glasgow and of responses of his contemporary scientists. Also mentioned is Kerr's discovery of the Kerr magneto-optic effect. The original electric birefringence cells are described and illustrated [13b] and his Royal Medal Citation (1898) is also reprinted [13b].

DISPUTATIONS AND VERIFICATION OF KERR

Apparently the first person to attempt to repeat Kerr's experiments with plate glass was Gordon in 1876 [14]. He got a negative result. In the same year, Lodge [15] mentioned in a brief footnote that he repeated Kerr's experiments with glass, "but not without failures sufficient to excite my admiration for the skill and patience involved in the discovery." Mackenzie [16] failed to reproduce Kerr's pehnomenon and thought it was produced by heat only. Röntgen [7] got negative results for glass too, and he decided that "accidental influences" were at work. Quincke [17] confirmed Kerr's results for glass, using flint glass, and also for carbon disulfide. He believed that the birefringence was a secondary effect arising from mechanical stress produced by the field; thus, he disagreed with Kerr's interpretation.

In 1882, following Kerr's two papers that year, Brongersma [18] also succeeded in repeating Kerr's experiments on glass, although he remarked about difficulties with strain double refraction produced by drilling the glass. Others also had reported problems with this in attempting to reproduce Kerr's work. Brongersma did not accept Kerr's assertion that the optical effects were due to a primary action of the electric field upon molecular alignment, remarking:

In my opinion, it is not sufficiently proved
that these phenomena cannot be of secondary order.
The motions of the molecules on their arrangement in
a limited portion of the plate (cell), may have for
their consequence a development of heat; and this...
may be the cause of the observed double refraction.

In 1880 Röntgen [7] confirmed the existence of the Kerr effect in liquids. This was further substantiated by Brongersma [18], who remarked:

> Kerr saw in these phenomena a confirmation of Faraday's theory respecting dielectrics. That great physicist already regarded it as probable that under the influence of electricity an isotropic body passes into the anisotropic state, so that it behaves like a double refracting crystal. He did not, however, succeed in confirming this by experiment.

Another "Kerr effect" is the Kerr magneto-optic effect, reported in 1877 [19]. In this work, Kerr measured the rotation of a beam of plane polarized light reflected from the surface of magnetized iron. He had expected to find the plane of polarization rotated in the process of reflection, and confirmed that this is so. This discovery was apparently considered more important than his earlier one by the physicists of that day [13b].

DEVELOPMENT OF THE THEORY OF THE KERR EFFECT

The first clear proposal that the Kerr effect may be due to molecular orientation appears to be that of Larmor [19a], who stated in 1897: "The double refraction induced in dielectrics in a strong electric field is possibly mainly due to molecular orientation, as also that arising from mechanical strain." A theory of the Kerr effect based upon a classical harmonic oscillator model was developed by Voigt [20] shortly after the discovery of the electron. He showed that in an electric field, which produces a quasi-elastic force, the oscillator will no longer be harmonic and that its frequency will be split into two characteristic values. The two absorptions will be parallel and perpendicular to the applied field and can produce an optical anisotropy in the medium. This proposal was critically examined in 1910 by Langevin [12,13] who was the first to set up equations for the energy of an electrically anisotropic molecule as a function of orientation, and to use the Boltzmann equation to calculate the probability of orientation as a function of angle (orientation distribution function). He followed through to obtain an equation for the Kerr constant, but he did not introduce a permanent dipole moment, and treated only the case of axial symmetry. By comparing the theories with experimental data for CS_2, he showed that the Voigt effect was much too small to explain the Kerr constant. Later, Landenburg and Kopfermann [21] were able to observe the Voigt effect in sodium vapor at low pressures. They detected both a shift of the absorption line and the double refraction of the vapor produced by an electric field.

The wavelength dependence of the Kerr constant was discussed by Havelock [22] on the basis of ideas first suggested by Larmor [19a]. Havelock proposed a deformable cavity idea for the Kerr effect and also considered the possibility of orientation of aniso- tropic molecules as giving rise to the Kerr effect.

Enderle [23] and Voigt [23a] extended Langevin's theory to the more general case, but did not include a permanent dipole moment or optical activity. Pockels [24] considered the theory of simulta- neously applied electric and magnetic fields, as well as separately applied fields, and developed equations for suspensions. This study related to experimental observations by Cotton and Mouton [25]. In 1912, Debye interpreted the temperature dependence of the dielectric constant of polar substances by deriving the expression for the mean electric moment from Boltzmann statistics [26], and near the same time, Thomson [27] gave the same result. These treatments resemble the calculation first carried out by Langevin [28] to find the mean magnetic moment of gas molecules having a permanent magnetic moment. A theory of the Kerr effect which included permanent dipole moment was given by Born [29]. These developments were later reviewed by Szivessy [30] and by Briegleb and Wolf [31]. Mallemann [32] extended the theory to optically active molecules. Debye [33] de- veloped a quantitative treatment of the theory of the Kerr effect for generally anisotropic dipolar molecules in a gas.

Brief reviews of magneto-and electro-optics were give by Wood [33a]. A comprehensive early review of electric and magnetic double refraction was presented by Beams in 1932 [34]. While the differences of interpretation among the early workers are not dis- cussed, a full account of experiments and theory is given. This includes an outline of the Langevin-Born theory for gases, its modification for dense fluids by Raman and Krishman [35], and a discussion of the relationship of the Kerr effect to light scatter- ing and molecular optical anisotropy. There is a discussion of the theory in Born's [36] book on optics; the experimental verification of the Kerr law for liquids and the temperature and wavelength de- pendences of the Kerr effect also are reviewed. The use of Kerr cells was discussed by Beams in 1930 [36a].

Several investigators have treated the quantum theory of the Kerr effect [36,37-40]. Van Vleck [41] outlined how the Kerr ef- fect can be calculated from perturbation theory. Serber [40] dis- cussed the theories of depolarization of scattered light and of the Kerr effect. The current status of theories of electric birefrin- gence in gases and liquids has been summarized [41a]. The quantum theory and calculation of relevant molecular properties have been given [41b]. The quantum theory of molecular absorption in electric fields also has been discussed [41c]. A summary of theories of the Kerr constant is presented elsewhere in this volume·[41d].

APPLICATIONS TO MOLECULAR STRUCTURE STUDIES

The relationship between the Kerr constant and the extent of
the depolarization of scattered light, which depends upon the optical anisotropy of a molecule, was first recognized by Gans [42].
The theoretical relationships between molecular optical properties
and light scattering were further developed by Debye [33] and Debye
and Sack [43]. During the 1930s a great deal of work was done on
the theory of optical anisotropy and molecular structure by Stuart
and others, and this has been summarized in the comprehensive review
articles of Stuart [44,45]. More recent reviews were given by
Partington [46] and by Le Fèvre and Le Fèvre [47]. In these reviews,
experimental methods of measurement of the Kerr effect in gases and
insulating liquids are outlined, and many results and their interpretations are presented. A concise summary, including representative results on gases and liquids has been presented in Molecular
Electro-Optics [150].

OTHER ELECTRO-OPTIC EFFECTS

The magnetic analog of the Kerr effect, double refraction produced by a magnetic field transverse to the direction of propagation of a light beam, also is primarily due to orientation of the
molecules. This is known as the Cotton-Mouton effect, discovered
in 1905 [48], and it is not a subject of this book. Another related phenomenon, the Maxwell effect, or double refraction produced
by flow, was first observed in Canada balsam in 1874 [49], a date
so close to Kerr's discovery that one wonders whether or not this
finding, not mentioned by Kerr, was in fact an important encouragement for Kerr to renew Faraday's quest (see "KERR'S WORK").

There are several optical effects besides birefringence which
are related to the orientation produced by an electric field. Since
optical absorptions of molecules are generally anisotropic, an electric field which causes a preferred orientation of the molecules
will cause the absorption to be different for plane-polarized light
vibrating along the direction of the orienting electric field, as
compared with the absorption perpendicular to the direction of the
orienting field. This is known as linear electric dichroism (or,
simply, electric dichroism) and it was observed in microcrystalline
and colloidal suspensions in the early days of electro-optic studies [25,50-52]. Bjornstahl [52a] described the effects of an electric field on the optical transmission of nematic liquid crystals.

For ordinary-sized molecules, electric dichroism is a weak
effect, but in the last 10 to 15 years there have been advances in
experimental methods and in theory which make it important. When
it was shown that macromolecules can be oriented very completely

in strong electric fields (around 10^3 to 10^4 V/cm) as the electric
birefringence method was extended into the saturation region [53],
it became evident that measurements of dichroism as a function of
wavelength would be useful to determine the orientation of chromophores in ordered macromolecular structures. Studies were begun
here on dyes and on a dye-polyelectrolyte complex by Bergmann in
1959, and shortly thereafter by Paulson on the triiodide amylose
complex. These were discussed in an introduction to electric dichroism [53a] and has become an important method for macromolecules
[53b].

Changes of the absorption of polarized light may occur purely
by orientation when an electric field is applied, or there may also
be changes in intrinsic molecular absorption coefficients due to
oscillator strength or frequency changes or both. The intrinsic
molecular optical changes may be referred to as electrochromism.
Very high fields ordinarily are required to produce a measurable
electrochromism in dyes [54], but a much stronger effect was found
in a dye-polyelectrolyte complex [55]. The latter was shown,
through examination of the time and wavelength dependences, to consist of a combination of orientation-produced electric dichroism
and of electrochromism. The electrochromism was strong and was
attributed to changes in dye-dye stacking interactions; hence, it
was named "electrometachromism" [55].

The absorption changes produced by electric fields have been
extensively studied by Lippert [56], Czekalla and coworkers [57],
Liptay and Czekalla [58], and Labhart [59]. They have been reviewed
by Liptay [41c,60].

Observations of the state of polarization of the fluorescence
of light emitted from molecules in a fluid which are irradiated with
a polarized source can give information about the rotational relaxation times relative to the fluorescence emission time. Czekalla
and coworkers [57,60a] showed that an electric field produces an
additional polarization. Weber [61] has calculated this effect,
using a model of rotationally diffusing rigid spheres, for weak
electric fields. Further developments are making this effect very
interesting for chemical and biological studies, particularly when
relaxation effects can be observed. Fluorescence intensity modulation by electric fields applied to "labeled" membranes has been reviewed [61a]. A further contribution appears in this volume [61b].

Electric field-induced changes in light scattering were
observed in the early days of electro-optic work [62-64]. Interest
in this effect has been heightened by the advent of modern light-
scattering studies of macromolecules [65-67a] and by increased
interest in large biopolymers and certain inorganic polyelectrolytes,
some of which can be oriented quite completely. The electric light-
scattering method [67b] provides some structural parameters for

macromolecules not available by other methods; thus, in spite of
sensitivity limitations, it is quite useful for certain problems
[67c].

Electric fields produce significant changes of optical rotation
for certain interesting macromolecules, e.g., helices [68,69], and
useful new molecular parameters can be obtained [69a]. A closely
related phenomenon is electric circular dichroism, which also has
been investigated [70].

When polarized intense optical radiation is passed through a
substance the optical electric field will tend to align the molecules when they are optically anisotropic. Buckingham [71] calculated the magnitude of this effect and suggested the use of intense
light flashes to observe it. The same phenomenon contributes to
self-focusing of intense laser pulses [72,72a]. It has been discussed recently in relation to stimulated scattering as well as the
laser field Kerr effect [73].

The Stark effect is a well-known electro-optic phenomenon. It
has been used extensively for modulation in microwave spectroscopy
[74]. Recently, the effect of electric fields on high-resolution
spectra has been shown to be a useful modulation method in high-
resolution optical spectroscopy, and the subject is being reinvestigated [75-79] in relation to both birefringence and absorption
changes [41a,41c].

A relatively new method for studying macromolecules in solution
is Doppler broadening of scattered laser radiation. This reveals
the dynamics of macromolecular motions [80]. When combined with
electrophoresis, a powerful method of electrophoretic analysis results. This is one of the most recent developments in electro-
optics and has been described [80a].

An electric field will change the Gibbs free energy of liquid
phases, and thus should affect the critical mixing temperature of
partially miscible liquids. Since a strong opalescence occurs near
the critical mixing temperature, there should be an observable
electro-optic effect. Debye and Kleboth [81] examined a binary
liquid system and found that a field of 45,000 V/cm produced a
change of $0.015^{\circ}C$ in the critical mixing temperature. The electro-
optic effect (modulation of the opalescence by an electric field)
apparently has not been studied.

ELECTRO-OPTIC STUDIES OF MACROMOLECULES

    A desire to apply optical techniques to the measurement of
biopolymers led me to consider the extension of light-scattering
methods in Berkeley, California in the fall of 1948. While adapting
and testing available equipment offered for this purpose by Bruno
Zimm, using tobacco mosaic virus (TMV), it occurred to me that it
would be intriguing to measure the anisotropy of the scattering of
light as a function of its polarization as the particles are oriented
by an electric field. Flow birefringence of TMV had already been
studied, and Zimm reminded me that observing birefringence between
crossed polarizers is a very sensitive technique for detecting
optical anisotropy. Having dealt with the problem of achieving
maximum sensitivity at the threshold of detectable light-scattering
signals in the presence of the shot noise of a photomultiplier [82],
it was obvious that the Kerr technique, in which on measures
changes in the intensity of the transmitted light beam, offers an
important advantage over electric field light scattering, in which
one measures the relatively weak scattered light. Thus, before any
scattering measurements, I decided to switch to the electric birefringence technique for studies of biopolymers in solution. Zimm
and I set up crude cells using platinum electrodes, and we applied
sinusoidal fields of varying frequency to TMV solutions. Our attention was called by Wendell Stanley to the earlier Kerr effect studies on solutions of TMV by Max Lauffer [83], who employed 60 Hz
sinusoidal fields and a polarizing microscope for average retardation measurements. Later I learned of the elegant early study of
colloids by the Kerr effect in sinusoidal fields at varying frequencies by Errera, Overbeek, and Sack [84], and still later, of a paper
by Kuhn and Kuhn [84a] on "dragging" birefringence of macromolecules
in an electric field. A stimulating review which clearly emphasized
the poor state of understanding of the Kerr effect in disperse systems had been given by Heller [85]. Lauffer's work on TMV showed
that the field and frequency dependences of the Kerr effect can be
very complex and thus difficult to understand when making only average light-intensity measurements in sinusoidal fields. Thus it
became clear, after exploratory measurements here that measuring
steady-state responses to electric fields would be important for
definitive studies.· However, macromolecule solutions are often
aqueous ones, and biological macromolecules are usually polyelectrolytes, so an applied electric field causes energy dissipation
and undesired heating as well as electrophoresis. To observe build-
up and decay, and to minimize the extraneous effects, I introduced
short orienting pulses, produced electronically. To further reduce
electrophoresis, they were given polarities which alternated [86].
Synchronized single sweeps were not generally available on oscil-
loscopes at that time, but in a few years the single-pulse method
was introduced [87].

Unknown to me during my initial studies, Benoit was pursuing a suggestion of Ch. Sadron (Sadron, private communication, 1972; Benoit, private communication, 1972) to extend the interpretation of flow birefringence studies of macromolecules by measuring birefringence in electric fields. In 1949 Benoit published a note on pulsed technique using a motor-driven interrupter [88]. In another note [89], he presented an equation for the birefringence and its time dependence. That note, his 1950 Ph.D. thesis, and my 1950 paper with Zimm, all representing work completed by 1949, included derivation of the equation for the birefringence relaxation time, i.e., τ = $1/6\Theta$, which relates the birefringence field-free relaxation time, τ , to the rotational diffusion constant, Θ , for an axially symmetric particle. Benoit studied deoxyribonucleic acid (DNA) and vanadium pentoxide sols. I studied TMV solutions. He derived equations for the build-up of the birefringence. Zimm and I used square-wave fields and looked for frequency dispersion to get clues as to the origin of the orienting torque [86]; later, with Haltner, I introduced [87] the reversing pulse technique. Polarization of the counterion atmosphere was proposed as an important orienting mechanism for polyelectrolytes, and a theory of the Kerr constant for conducting systems came later. DNA was shown to have a very strong electro-optic effect [90]. These were exciting developments as the power of the pulsed electro-optic technique was brought to bear in the study of macromolecular structures in solution.

Also unknown to us in 1948 was the 1947 contribution of Kaye and Devaney [91] on the observation of the Kerr effect relaxation in viscous liquids, using an electronic pulse generator and an oscilloscope. Theirs was apparently the first direct observation of electro-Optic relaxation with pulses. No doubt, had their interest been in macromolecules, they would have shared the experimental findings, the puzzlements, and the theoretical developments which began to unfold the next decade. But to Kerr himself, in his study of glass and to several others -- e.g., Raman and Sirkar [9]] and Kitchin and Mueller [93] -- we must give the credit for having observed, indirectly, much earlier, in viscous liquids, the effect of electro-optic relaxation. The last two studies were by anomalous dispersion of the Kerr effect.

Kerr's work on electric birefringence of glass had shown a relaxation time of the order of seconds [2]. In 1900 Abraham and Lemoine [94] observed that the disappearance of the electric birefringence of liquids occurs in less than 10^{-8} sec -- a remarkable experimental achievement in that day. Peterlin and Stuart [95], in their important monograph on flow and electric birefringence, had already treated the sine-wave dispersion behavior of macromolecules in terms of rigid models for macromolecules having dipole moments and anisotropic electric and optical polarizabilities.

A group of investigators in the USSR also began developing electro-optic methods for studying macromolecules in the late 1940s. Tolstoi and Feofilov [96] discussed the possibility of improving the signals from a Kerr cell, then started pulse experiments [97] and studied some colloids [98]. Thus, in the late 1940s there were workers in three countries beginning to exploit the Kerr electro-optic relaxation method for macromolecule studies with photoelectronic instrumentation.

During the 1950s and 1960s there were developments in theory and methods that relate to an understanding of the electric field orientation effects operating on macromolecules in solution. Some of the highlights will be briefly mentioned here. Developments of the past decade are discussed elsewhere in this volume.

In Berkeley, Haltner took up experimental studies of tobacco mosaic virus (TMV) in depth [99]. Several preparations were examined; a very homogeneous one was studied intensively [87]. Bipolar pulses, an exponential discharge and pulsed sine wave trains at various frequencies were used to investigate the mechanism of electric orientation [100]. Because of its large and rigid structure, the behavior of the TMV macromolecule divided into three distinct time domains: the orientational relaxation occurring around 200 Hz, a plateau of steady Kerr constant extending to around 10 KHz, and then a broad anomalous dispersion of the Kerr constant continuing into the MHz region [100]. The macromolecular orientation was found to have a relatively negligible contribution from permanent dipole moment, and the high frequency anomalous dispersion was electrolyte dependent. The magnitude of the electric field orientation was around 50-fold greater than could be explained by the usual electronic and atomic polarization effects in insulating dielectrics; this, together with the high frequency anomalous dispersion, unequivocally established the predominating influence of the polyelectrolyte character, and focussed attention upon contributions from mobile counterions, both those bound to macromolecule sites and those freer to move in the diffuse counterion atmosphere [100].

The importance of counterions has also been recognized from studies of the anisotropy of electric conductivity in polyelectrolyte solutions [101,102], and from dielectric dispersion studies of polyelectrolytes [103,104]. The important role of particle and medium conductivities in determining the frequency-dependent dielectric and conductivity behavior of polyelectrolytes and charged colloidal particles, and the effects of shape were treated in an extension of the Maxwell-Wagner equations which included contributions from mobile bound ions and the counterions of a thin diffuse ion atmosphere. This is a useful approximation for visualizing limiting behavior [105]. Conductivity effects in particle and medium were explicitly included in the formulation of the electric

free energy problem when the electric field deformation of electrically conducting droplets in conducting media was treated [106].
Later, an extended Maxwell-Wagner model for conducting systems, with
surface conductivity contributions added for the mobile bound ions
and diffuse layer counterions, was used to derive equations for the
Kerr constant of rigid, conducting dipolar polyelectrolytes [107].
This theory predicted values in reasonable agreement with the earlier
experiments [107,100]. Equations are given in a later chapter on
the theory of the Kerr constant [108]. Other approximations to
treat counterion contributions to polyion polarizabilities have
been introduced [109,103,104] and will be discussed in the contributed papers at this study institute.

During the early 1950s attention also was given to the possibility of protonic contributions to the ion-transport polarization
mechanism of macromolecules [110]. Kirkwood's approach was to consider the contributions of fluctuations of protonic distribution on
a protein to the static dielectric increments. A careful test of
this mechanism was made with isoionic sephadex - fractionated deion-
ized bovine serum albumin monomer, using dielectric dispersion and
electric birefringence relaxation measurements [111,112]. It was
concluded that the proton relaxation mechanism was too slow to provide an important kinetic pathway relevant to whole-molecule reorien-
tations of the "permanent" dipole moment [111]. The pioneering
protein solution studies of "Mme. Directeur Krause" [113,114] provided the basis for this study and many others which have since
appeared [115].

The case of the collagen macromolecule is historically interesting because it was the first time that large contributions from
dipole moment and induced moment were separated in high electric
field orientation saturation experiments on a biopolymer [116].
The saturation was strong, so the value of the optical anisotropy
factor was determined also. The theoretical basis for this methodology was laid in a paper on the theory of the orientation of
macromolecules into the region of saturation with illustrative data
on tobacco mosaic virus, a synthetic polypeptide (poly-λ-benzyl-L-
glutamate) in an organic solvent and an aqueous polyelectrolyte,
sodium polyethylenesulfonate [117]. Subsequent extension of the
equations to additional molecular parameters has generalized the
method [118,119,120].

The high field methods were applied to DNA [121]. At essentially complete saturation in solutions of calf thymus DNA, a dramatic effect occurred after orientation during pulses between 10
and 20 kv/cm; the negative birefringence fell off considerably before the field was removed. This was ascribed to a field induced
change of macromolecular structure. Field induced transitions were
suggested for membrane macromolecules of nerve as a mechanism of
ion transport control in axonal nerve impulse transmission [121].

Similar observations have been reported with polypeptides recently
[122,123] and with polynucleotide complexes [124]. "Field-jump"
experiments of Pörschke presented later in this volume relate to
these so-called "chemical relaxation" effects of electric fields
[125].

The Kerr effect of synthetic polymers in solution has received
considerable attention in a systematic set of studies by Le Fèvre
and coworkers [126-137].

A large body of data on the rotational diffusion coefficients
of macromolecules has been accumulating the past three decades. The
brownian diffusion problem for rigid macromolecules in solution
under a pulsed electric field was treated in the fifties [138,139].
Shortly after the introduction of the reversing field method [100],
the resulting birefringence signals were calculated [140]. Extensions of these studies have been reviewed [53b,120].

A number of investigations of the electric birefringence and
dichroism of DNA have appeared [141-145,121]. Work on this important set of somewhat flexible macromolecules has been reviewed [53b,
146] and is continuing.

Liquid crystal systems were among the early systems studied
[147,148] and their interesting and useful electro-optic behavior
has been reviewed [149].

The nineteen seventies have brought on-line computer systems
for data recording and processing, improved optical and electronic
detection systems, and some advances in pulser capabilities. Furthermore, there have been studies from established and from several
new laboratories on a wide variety of systems - polymers, both
natural and synthetic, colloidal dispersions, biomembranes, biological cells and larger biological structures, e.g., living muscle
fibers. These are discussed later in this institute.

In the nineteen eighties we may hope for better coordination
of instrumentation improvements aided by industrial initiatives to
place electro-optic measuring systems more readily into the laboratories of polymer chemists and biologists, and in medical laboratories and clinics as well. It is not too early to be planning for
a sixth* International Symposium in 1982.

---

*
We count one for the meeting at the University of Liege (1974);
two for the University of California, Berkeley, meeting at Asilomar
(1975); three for Brunel University (1978); four for the joint Am.
Chem. Soc./Japanese Chem. Soc. Symposium in Honolulu (1979); and
this one at Rensselaer Polytechnic Institute is the fifth.

REFERENCES

1. J. C. Maxwell, "A Treatise on Electricity and Magnetism,"
    Oxford (1873).
2. J. Kerr, Phil. Mag., 50(4):337 (1875).
3. J. Kerr, Phil. Mag., 50(4):416 (1875).
4. J. Kerr, Phil. Mag., 8(5):85 (1879).
5. J. Kerr, Phil. Mag., 8(5):229 (1879).
6. J. Kerr, Phil. Mag., 9(5):157 (1880).
7. W. C. Röntgen, Phil. Mag., 10(5):77 (1880).
8. J. Kerr, Phil. Mag., 13(5):153 (1882).
9. J. Kerr, Phil. Mag., 13(5):248 (1882).
10. J. Kerr, Phil. Mag., 37(5):380 (1894).
11. J. Kerr, Phil. Mag., 38(5):144 (1894).
12. P. Langevin, Radium 7:249 (1910).
13. P. Langevin, Compt. Rend., 151:475 (1910).
13a. Anon., Proc. Roy. Soc., 82A, Obituary Notices, pp. i-v (1909).
13b. C. T. O'Konski, Appendix A, Mementos of John Kerr, in: "Molec-
    ular Electro-Optics," Part 1, Theory and Methods, Marcel Dekker
    Inc., New York, pp. 515-523 (1976).
14. J. E. H. Gordon, Phil. Mag., 2(5):203 (1876).
15. O. J. Lodge, Phil. Mag., 2(5):353 (1876).
16. J. J. Mackenzie, Ann. Physik M. Chem., 238(2):356 (1877).
17. G. Quincke, Phil. Mag., 10:537 (1880); Ann. Physik 10:536 (1880).
18. H. Brongersma, Phil. Mag. 14(5):127 (1882).
19. J. Kerr, Phil. Mag., 3(5):321 (1877).
19a. J. Larmor, Phil. Trans., A190:232 (1897); Aether and Matter,
    1:351 (1900).
20. W. Voigt, Ann. Physik, 4:197 (1901).
21. R. Ladenburg, and H. Kopfermann, Ann. Physik, 78(4):659 (1925).
22. T. H. Havelock, Proc. Roy. Soc. (London), A80:28 (1907); Phys.
    Rev., 28:136 (1909); Proc. Roy. Soc. (London), A84:492 (1911).
23. A. Enderle, Dissertation, Freiburg (1912).
23a. W. Voigt, Nadir. Kgl. Ges. Wiss. Göttingen, 577-93 (1912);
    Chem. Abstr., 7:3914 (1913).
24. F. Pockels, Radium 10:152 (1913).
25. A. Cotton, and H. Mouton, J. Chim. Phys., 10:692 (1912).
26. P. Debye, Physik Z., 13:97 (1912).
27. J. J. Thomson, Phil. Mag., 28:757 (1914).
28. P. Langevin, J. Phys., 4(4):678 (1905); Ann. Chim. Phys., 5(8):
    70 (1905).
29. M. Born, Ann. Physik, 55:177 (1918).
30. C. Szivessy, Handb. Physik, 21:S724 (1929).
31. G. Briegleb and K. L. Wolf, Fortschr. Chem. Physik Physik.
    Chem., 21:1-58 (1931).
32. R. de Mallemann, Ann. Physik 2:21 (1924); Compt. Rend., 193:523
    (1931).
33. P. Debye, Handb. Radiol., 6:597, 760 (1925); "Polar Molecules,"
    Dover, New York (1929).

33a. R. W. Wood, "Physical Optics," 3rd Ed., Macmillan, New York,
      Chap. 21 and 22 (1934).
34.   J. W. Beams, Rev. Mod. Phys., 4:133 (1932).
35.   C. V. Raman, and K. S. Krishnan, Proc. Roy. Soc. (London),
      A117:1 (1927).
36.   M. Born, "Optik," Springer, Berlin (1933).
36a. J. W. Beams, Rev. Sci. Instr., 1:780 (1930).
37.   R. de L. Kronig, Z. Physik, 45:458 (1927); 47:702 (1928).
38.   M. Born, and P. Jordan, "Elementare Quantenmechanik," Springer,
      Berlin (1930).
39.   Th. Neugebauer, Z. Physik, 73:386 (1932); 660 (1933); 86:392
      (1933).
40.   R. Serber, Phys. Rev., 43(2):1003 (1933).
41.   J. H. van Vleck, "Theory of Electric and Magnetic Suscepti-
      bilities," Oxford University Press, London, p. 366, (1932).
41a. A. D. Buckingham, Ch. 2, Electric Birefringence in Gases and
      Liquids, in: "Molecular Electro-Optics, Part 1. Theory and
      Methods," C. T. O'Konski, ed., Marcel Dekker, Inc., New York,
      pp. 27-62 (1976).
41b. T. K. Ha, Ch. 14 Quantum Theory and Calculation of Electric
      Polarizability, in: "Molecular Electro-Optics, Part 1. Theory
      and Methods," C. T. O'Konski, ed., Marcel Dekker, Inc., New
      York, pp. 471-513 (1976).
41c. W. Liptay, Ch. 6 Optical Absorption in an Electric Field, in:
      "Molecular Electro-Optics, Part 1. Theory and Methods,"
      C. T. O'Konski, ed., Marcel Dekker, Inc., New York, pp. 207-
      242 (1976).
41d. C. T. O'Konski, this volume, "Theory of Kerr Constant."
42.   R. Gans, Ann. Physik, 63(4):97 (1921).
43.   P. Debye, and H. Sack, 6(ii):69, 179 (1934).
44.   H. A. Stuart, "Hand- und Jahrbuch der Chemischen Physik,"
      Vol. 10, Part 3, A. Eucken and K. S. Wolf, eds., (1939).
45.   H. A. Stuart, "Die Struktur des freien Molekuls," Springer,
      Berlin, Chap. 7, p. 315 (1952).
46.   J. R. Partington, "An Advanced Treatise on Physical Chemistry,"
      Vols. 4 & 5, Longmans, Green, London (1953 & 1954).
47.   C. G. Le Fevre and R. J. W. Le Fevre, Rev. Pure Appl. Chem.,
      25:261 (1955).
48.   A. Cotton, and H. Mouton, Compt. Rend., 141:317, 349 (1905).
49.   J. C. Maxwell, Proc. Roy. Soc. (London), 22:46 (1874); Ann.
      Phys., 151:151 (1874).
50.   G. Meslin, Compt. Rend., 136:888, 930 (1903); J. Phys., 7:856
      (1908).
51.   J. Chaudier, Compt. Rend., 137:248 (1903); Ann. Chim. Phys.,
      15:67 (1908); Compt. Rend., 149:202 (1909).
52.   P. Drapier, Compt. Rend., 157:1063 (1913).
52a. Y. Bjornstahl, Ann. Physik, 56:161 (1918).
53.   C. T. O'Konski, K. Yoshioka, and W. H. Orttung, J. Phys. Chem.,
      63:1558 (1959).

53a. C. M. Paulson, Jr., Ch. 7 Electric Dichroism of Macromolecules, in: "Molecular Electro-Optics, Part 1. Theory and Methods," C. T. O'Konski, ed., Marcel Dekker, Inc., New York, pp. 243-273 (1976).

53b. E. Fredericq, and C. Houssier, "Electric Birefringence and Electric Dichroism," Clarendon, Oxford (1973).

54.   J. Kumamoto, J. C. Powers, Jr., and W. R. Heller, J. Chem. Phys. 36:2893 (1962).

55.   C. T. O'Konski, and K. Bergmann, J. Chem. Phys., 37:1573 (1962).

56.   E. Lippert, Z. Electrochem., 61:962 (1957).

57.   J. Czekalla, Z. Electrochem., 64:1221 (1960); Chimia (Aarau), 15:26 (1961); J. Czekalla, W. Liptay, and J. O. Meyers, Ber. Bunsenges. Physik. Chem., 67:465 (1963); J. Czekalla, and G. Wick, Z. Electrochem., 65:727 (1961).

58.   W. Liptay, Z. Naturforsch, 20a:272 (1965); W. Liptay and J. Czekalla, Z. Naturforsch, 15a:1072 (1960); Z. Electrochem., 65:721 (1961).

59.   H. Labhart, Adv. Chem. Phys., 13:179 (1967); H. Labhart, Chimia (Aarau), 15:20 (1961); Helv. Chim. Acta, 44:457 (1961); Experientia, 22:65 (1966); H. Labhart and G. Wagniere, Helv. Chim. Acta, 46:1314 (1963).

60.   W. Liptay, Mod. Quantum Chem., 3:45 (1965).

60a. J. Czekalla, and K. O. Meyer, Z. Physik. Chem. (Frankfurt), 27:185 (1961).

61.   G. Weber, J. Chem. Phys., 43:521 (1965).

61a. R. D. Keynes, Ch. 21 Electro-Optics of Nerve Membranes, in: "Molecular Electro-Optics, Part 2: Applications to Biopolymers," C. T. O'Konski, ed , Marcel Dekker, Inc., New York, pp. 743-760 (1978).

61b. G. Weill, this volume, Chap. on Polarized Fluorescence.

62.   H. Siedentopf, Wiss. Mikrosk. M., 29:1 (1912).

63.   H. R. Kruyt, Kolloid-Z., 19:161 (1916).

64.   H. Freundlich, Z. Elektrochem. 22:27 (1916).

65.   P. Debye, J. Appl. Phys., 15:338 (1944).

66.   B. Zimm, J. Chem. Phys. 16:1099 (1948).

67.   K. A. Stacey, "Light Scattering in Physical Chemistry," Academic, New York (1956).

67a. M. B. Huglin, "Light Scattering from Polymer Solutions," Academic, New York (1972).

67b. B. R. Jennings, Ch. 8 Electric Field Light Scattering, in: "Molecular Electro-Optics, Part 1: Theory and Methods, C. T. O'Konski, ed., Marcel Dekker, Inc., New York pp. 275-319 (1976).

67c. B. R. Jennings, this volume, Chap. on Light Scattering.

68.   I. Tinoco, Jr., J. Phys. Chem., 60:1619 (1956).

69.   N. Go, J. Chem. Phys., 43:1275 (1965); J. Phys. Soc. Japan, 23:88 (1967).

69a. I. Tinoco, Jr., Ch. 10 Circular Dichroism and Optical Rotation in an Electric Field, in: "Molecular Electro-Optics, Part 1. Theory and Methods, C. T. O'Konski, ed., Marcel Dekker, Inc., New York, pp. 367-379 (1976).

70.   S. J. Hoffman, and R. Ullman, J. Polymer Sci., C31:205 (1970).

71.   A. D. Buckingham, Proc. Phys. Soc. (London), B69:344 (1956).

72.   S. Kielich, IEEE J. Quantum Electronics, QE-4:744 (1968).

72a. N. Bloembergen, Am. J. Phys., 35:989-1023 (1967).

73.   R. Y. Chiao, and J. Godine, Phys. Rev., 185:430 (1969).

74.   W. Gordy, W. V. Smith, and R. R. Trambarulo, "Microwave Spectroscopy," Wiley, New York (1953); C. H. Townes and A. L. Schawlow, "Microwave Spectroscopy," McGraw-Hill, New York (1955).

75.   D. A. Dows, and A. D. Buckingham, J. Mol. Spectry., 12:189 (1964).

76.   A. D. Buckingham, and D. A. Ramsey, J. Chem. Phys., 43:3721 (1965).

77.   R. D. Conrad, and D. A. Dows, J. Mol. Spectry., 32:276 (1969).

78.   D. A. Haner, and D. A. Dows, J. Mol. Spectry., 34:296 (1970).

79.   J. M. Brown, A. D. Buckingham, and D. A. Ramsay, Can. J. Phys., 49:914 (1970).

80.   R. Pecora, Ann. Rev. Biophys. Bioeng., 1:257 (1972).

80a. W. H. Flygare, S. L. Hartford, and B. R. Ware, Ch. 9 Electrophoretic Light Scattering, in: "Molecular Electro-Optics, Part 1: Theory and Methods," C. T. O'Konski, ed., Marcel Dekker, Inc., New York, pp. 321-366 (1978).

81.   P. Debye, and K. Kleboth, J. Chem. Phys., 42:3155 (1965).

82.   F. T. Gucker, Jr., C. T. O'Konski, H. B. Pickard, and J. N. Pitts, Jr., J. Am. Chem. Soc., 69:2422 (1947).

83.   M. A. Lauffer, J.-Am. Chem. Soc., 61:2412 (1939).

84.   J. Errera, J. Th. G. Overbeek, and H. Sack, J. Chim. Phys., 32:681 (1935).

84a. W. Kuhn, and H. Kuhn, Helv. Chim. Acta, 27:493 (1944).

85.   W. Heller, Rev. Mod. Phys., 14:390 (1942).

86.   C. T. O'Konski, and B. Zimm, Science, 111:113 (1950).

87.   C. T. O'Konski, and A. J. Haltner, J. Am. Chem. Soc., 78:3604 (1956).

88.   H. Benoit, Compt. Rend., 228:1716 (1949).

89.   H. Benoit, Compt. Rend., 229:30 (1949).

90.   H. Benoit, Ann. Phys., 6:561 (1951).

91.   W. Kaye, and R. Devaney, J. Appl. Phys., 18:912 (1947).

92.   C. V. Raman, and S. C. Sirkar, Nature, 121:794 (1928).

93.   D. W. Kitchin, and H. Mueller, Phys. Rev., 32:979 (1928).

94.   H. Abraham, and J. Lemoine, J. Phys., 9(3):262 (1900).

95.   A. Peterlin, and H. A. Stuart, "Hand- und Jahrbuch der Chemischen Physik," Vol. 8, Part 1B, A. Euken, and K. L. Wolf, eds., Akademische Verlagsges., Leipzig, pp. 1-115 (1943); (Reprint, University of Michigan Press, Ann Arbor, 1948).

96. N. A. Tolstoi, and P. P. Feofilov, Dokl. Akad. Nauk SSSR,
     60:219 (1948); N. A. Tolstoi, Dokl. Akad. Nauk SSSR, 59:1563
     (1948).
97. N. A. Tolstoi, and P. P. Feofilov, Zh. Eksperim, i Teor, Fiz.,
     19:421 (1949).
98. N. A. Tolstoi, and P. P. Feofilov, Dokl. Adad. Nauk SSSR,
     66:617 (1949).
99. A. G. Haltner, Ph.D. Thesis, University of California,
     Berkeley (1955).
100. C. T. O'Konski, and A. G. Haltner, J. Am. Chem. Soc., 79:
     5634-5649 (1957).
101. G. Schwarz, Z. Physik, 145:563-584 (1956).
102. M. Eigen, and G. Schwarz, J. Colloid Science 12:181-194 (1957).
103. M. Mandel, Mol. Phys., 4:489-496 (1961); M. Mandel, and F.
     van der Touw, Dielectric Properties of Polylectrolytes, in:
     "Polyelectrolytes, E. Sélegny, ed., D. Reidel, Dortrecht, The
     Netherlands (1974).
104. M. Mandel, this volume.
105. C. T. O'Konski, J. Chem. Phys., 23:1559 (1955); J. Phys. Chem.,
     64:605-619 (1950).
106. C. T. O'Konski, and F. E. Harris, J. Phys. Chem. 61:1172-1174
     (1957).
107. C. T. O'Konski, and S. Krause, J. Phys. Chem., 74:3243-3250
     (1970).
108. C. T. O'Konski, this volume, "Theory of the Kerr Constant."
109. G. Schwarz, Z. Physik Chem. [N.F.], 19:286-314 (1959).
110. J. G. Kirkwood, and J. B. Schumaker, Proc. Natl. Acad. Sci.
     USA, 38:855 (1952).
111. P. Moser, P. G. Squire, and C. T. O'Konski, J. Phys. Chem.,
     70:744-756 (1966).
112. P. G. Squire, P. Moser, and C. T. O'Konski, Biochemistry,
     7:4261-4272 (1968).
113. S. Krause, Ph.D. Thesis, Dept. of Chemistry, University of
     California, Berkeley, CA (1957).
114. S. Krause, and C. T. O'Konski, J. Am. Chem. Soc., 81:5082-
     5088 (1959); Biopolymers, 1:503-515 (1963).
115. K. Yoshioka, Ch. 17 Electro-Optics of Polypeptides and Pro-
     teins, in: "Molecular Electro-Optics, Part 2. Applications
     to Biopolymers, C. T. O'Konski, ed., Marcel Dekker, Inc., New
     York, pp. 601-643 (1978).
116. K. Yoshioka, and C. T. O'Konski, Biopolymers, 4:499-507 (1966);
     Abstracts of papers, 136th Meeting, Am. Chem. Soc., Atlantic
     City, p. 28S (1959).
117. C. T. O'Konski, K. Yoshioka, and W. H. Orttung, J. Phys. Chem.,
     63:1558-1565 (1959).
118. M. J. Shah, J. Phys. Chem., 67:2215 (1963).
119. D. N. Holcomb, and I. Tinoco, Jr., J. Phys. Chem., 67:2691-
     2698 (1963).

120. C. T. O'Konski, and S. Krause, Ch. 3 Electric Birefringence
      and Relaxation in Solutions of Rigid Macromolecules, in:
      "Molecular Electro-Optics, Part 1.   Theory and Methods,"
      C. T. O'Konski, ed., Marcel Dekker, Inc., New York, pp. 63-
      120 (1976).

121. C. T. O'Konski, and N. C. Stellwagen, Biophys. J., 5:607-613
      (1965).

122. K. Kikuchi, and K. Yoshioka, Biopolymers, 12:2667-2679 (1973);
      15:1669-1676 (1976).

123. M. Fujimori, K. Kikuchi, K. Yoshioka, and S. Kubota, Bio-
      polymers, 18:2005-2013 (1979); K. Yoshioka, M. Fujimori, and
      K. Kikuchi, Intl. J. Biol. Macromolecules, 2 (in press) (1980).

124. E. Neumann, and A. Katchalsky, Proc. Natl. Acad. Sci. USA,
      69:993-997 (1972); A. Revzin, and E. Neumann, Biophys. Chem.,
      2:144-150 (1974).

125. D. Pörschke, this volume, Chapter on Threshold Effects.

126. C. G. Le Fèvre, R. J. W. Le Fèvre, and G. M. Parkins, J. Chem.
      Soc., 1958:1468-1474 (1958).

127. C. G. Le Fèvre, R. J. W. Le Fèvre, and G. M. Parkins, J. Chem.
      Soc., 1960:1418-1419 (1960).

128. M. Aroney, R. J. W. Le Fèvre, and G. M. Parkins, J. Chem. Soc.,
      1960:2890-2895 (1960).

129. C. G. Le Fèvre, and R. J. W. Le Fèvre, in: "Technique of
      Organic Chemistry," A. Wiessberger, ed., 3rd Ed., Vol. I, Part
      III, pp. 2459-2496, Interscience Publishers, New York (1960).

130. R. J. W. Le Fèvre, and K. M. S. Sundaram, J. Chem. Soc.,
      1962:1494-1502 (1962).

131. R. J. W. Le Fèvre, and K. M. S. Sundaram, J. Chem. Soc., 1962:
      4003-4008 (1962).

132. R. J. W. Le Fèvre, and K. M. S. Sundaram, J. Chem. Soc., 1963:
      1880-1887 (1963).

133. R. J. W. Le Fèvre, A. Sundaram, and K. M. S. Sundaram, J.
      Chem. Soc., 1963:3180-3188 (1963).

134. R. J. W. Le Fèvre, and K. M. S. Sundaram, J. Chem. Soc.,
      1963:3188-3193 (1963).

135. R. J. W. Le Fèvre, and K. M. S. Sundaram, J. Chem. Soc., 1963:
      3547-3554 (1963).

136. R. J. W. Le Fèvre, and K. M. S. Sundaram, J. Chem. Soc., 1964:
      556-562 (1964).

137. R. J. W. Le Fèvre, and K. M. S. Sundaram, J. Chem. Soc., 1964:
      3518-3523 (1964).

138. H. Benoit, Ann. Phys., 6:561-609 (1951).

139. I. Tinoco, Jr., J. Am. Chem. Soc., 77:4486-4489 (1955).

140. I. Tinoco, Jr., and K. Yamaoka, J. Phys. Chem., 63:423-427
      (1959).

141. H. Benoit, J. Chim. Phys., 47:719-721 (1950); 48:612-614 (1951)

142. G. A. Dvorkin, Dokl. Akad. Nauk SSSR, 135:739-742 (1960); G.
      A. Dvorkin, Biofizika, 6:403-409 (1961).

143. G. A. Dvorkin, and V. I. Krinskii, Dokl. Akad. Nauk SSSR, 140:
      942-945 (1961); G. A. Dvorkin, and E. I. Golub, Biofizika, 8:
      301-307 (1963); E. I. Golub, G. A. Dvorkin, and V. G. Nazarenko,
      Biokhimiya, 28:1041-1046 (1963); [Biochemistry (USSR), 28:
      769-773 (1963)].
144. N. Ise, M. Eigen, and G. Schwarz, Biopolymers, 1:343-352 (1963).
145. C. Houssier, and E. Fredericq, Biochim. Biophys. Acta, 88:450-
      452 (1964).
146. N. C. Stellwagen, Ch. 18 Electro-Optics of Polynucleotides
      and Nucleic Acids, in: "Molecular Electro-Optics, Part 2.
      Applications to Biopolymers, C. T. O'Konski, ed., Marcel
      Dekker, Inc., pp. 645-683 (1978).
147. N. A. Tolstoi, Zh. Eksperim. i Teor. Fiz., 19:319-327 (1949);
      N. A. Tolstoi, and P. P. Feofilov, Zh. Eksperim. i Tero. Fiz.,
      19:421-440 (1949).
148. P. A. Winsor, J. Colloid Sci., 10:101-106 (1955).
149. T. G. Scheffer, and H. Gruler, Ch. 22 Electro-Optics of Liquid
      Crystals, in: "Molecular Electro-Optics, Part 1. Theory and
      Methods, C. T. O'Konski, ed., Marcel Dekker, Inc., New York,
      pp. 761-818 (1976).
150. C. T. O'Konski, ed., Molecular Electro-optics, Part 1,
      Theory and Methods, M. Dekker, Inc., New York (1976);
      Part 2 Applications to Biopolymers, M. Dekker, Inc. New
      York (1978). (22 Chapters by contributing authors).

INTRODUCTION TO MODERN ELECTRO-OPTICS

B. R. Jennings

Electro-Optics Group, Physics Department
Brunel University
Uxbridge, U.K.

ABSTRACT

The optical properties of many fluids and molecular solutions
change when these media are subjected to suitable electric fields.
The best known effect is the induction of electric birefringence
which has become increasingly popular as a means of characterizing
the optical anisotropy and the geometry of molecules and particles
in dilute solution or suspension. During the past fifteen years,
electrically induced changes have been reported for a wide variety
of other optical phenomena. These include the absorption, fluorescence, optical rotation, light scattered intensity and light scattered fluctuations for fluid media. As each optical phenomenon has
its origins in different optical and structural characteristics,
their measurement leads to complementary molecular data. This paper
constitutes a brief description of recent novel extensions in electro-
optical apparatus, methodology and optical effects. By illustrating
the methods with representative experimental data from a wide class
of materials, their utility for molecular characterization is indicated.

I. INTRODUCTION

In 1875 John Kerr[1] published his classical scientific paper in
the Philosophical Magazine on 'A New Relation Between Electricity
and Light', in which he demonstrated to the world the ability of an
apparently isotropic medium to act as a uniaxial, anisotropic system
when an electric field was applied across it. This phenomenon of
'electric birefringence' or the 'Electric Kerr Effect' has become so
well established through its use for the measurement of the velocity
of light, high speed photography, and as an optical modulator in laser

technology, that many scientific dictionaries still classify it as "the" electro-optical effect. Without detracting from the importance of the Kerr effect, such classification is misleading, for one of the more exciting aspects of molecular characterisation studies in chemical physics during the last twenty years has been the development and harnessing of a whole host of novel electro-optical phenomena.

Birefringence has its origin in an anisotropy of refractive indices associated with different molecular axes. In dilute solution, such molecules adopt a random array and the medium is not birefringent. Application of an external electric field induces order and the inherent refractive index anisotropy is manifest as birefringence. Alternative anisotropic optical and structural molecular characteristics may be manifest by observing other optical phenomena (Fig. 1). For example a molecule may be able to better absorb light polarized parallel to one molecular axis than another. This gives rise to dichroism for a highly ordered medium. Fluorescence is a two-fold process of absorption followed by re-emission of light at a different wavelength. The ability to absorb and re-emit light depends strongly upon the axis presented to a polarized light beam for a fluorescent molecule. An aligned system therefore fluoresces in a different manner to a medium consisting of random reorienting anisotropic molecules. Other optical phenomena, which are influenced by molecular ordering, are the intensity of scattered light, the fluctuations in the scattered intensity, the optical rotation and the circular dichroism of the medium. All of these phenomena have been measured as means of characterizing the structure and properties of various molecular systems.

Modern apparatus and methodology have also had a strong influence on the development of electro-optics. It is surprising that, some hundred years after the announcement of the Kerr Effect, experimenters[2,3] were still using d.c. electric fields as the orienting force[2,3], whilst closely related dielectric measurements were being made with alternating electric fields. The frequency dependence of electro-optical phenomena, obtained from data using applied alternating electric fields, is one of the more interesting recent developments. It has its culmination in some very exciting experiments using lasers. In 1956 Buckingham[4] suggested that the oscillating electric vector in a laser beam might itself induce electro-optical effects. During the past five years laser-induced electro-optical effects, which are sometimes called 'optico-optical' effects, have been reported for solutions and suspensions[5]. Apart from being of phenomenological interest, this extension has great potential for more detailed local molecular structural investigations.

This paper is essentially a survey of a number of novel electro-optical effects which have been harnessed during recent years for the characterization of molecules in fluid media. Where given,

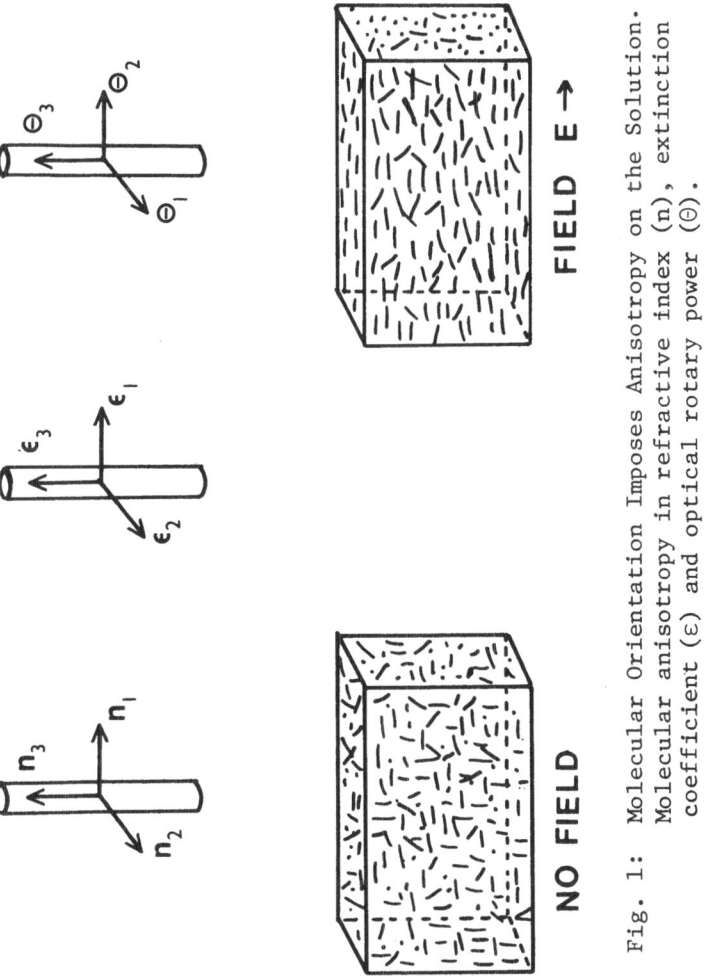

Fig. 1:   Molecular Orientation Imposes Anisotropy on the Solution.
Molecular anisotropy in refractive index (n), extinction
coefficient (ε) and optical rotary power (Θ).

theory for the various effects is predominantly for cylindrically
symmetric molecules which are rigid, uncharged and non-interacting.
The objective of the survey is to illustrate the variety, potential
and versatility of the electro-optical principle and to bring the
various methods to the attention of those scientists who are seeking
new methods for molecular characterization.

II.   ELECTRIC BIREFRINGENCE

Commonly known as the Kerr Effect, it owes its origin to anisotropy of the refractive indices associated with the major geometric
axes of the molecules. The effect can be manifest by placing the
test solution between a pair of electrodes in a suitable cell
(Fig. 2). A beam of well collimated light is passed centrally
through the sample, with the light initially linearly polarized at
45° azimuth to the electric field direction. The electric field
is applied as a potential difference across a pair of electrodes
spaced transversely to the light beam in the sample cell. Elliptically polarized light leaves the cell when the solution becomes
birefringent, as when the electric field is applied across the
sample. A quarter-wave plate (Q) and an analyzing polariser (A) are
set in parallel azimuth to each other.   These are 'crossed' with
the state of linear polarization transmitted through the initial
polarizer (P). The light penetrating such a system is conveniently
recorded using a photomultiplier and an oscilloscope (Fig. 3). A
low power laser may be used as the light source, but such a
component restricts the optical wavelengths available. For this
reason, arc sources are more commonly used. It is essential however that they be powered from a suitable smoothed electrical
supply.

The above mentioned optical arrangement generally results in a
photo multiplier output signal which is proportional to the square
of the birefringence (Δn) induced in the cell. This birefringence
is the difference in refractive indices both parallel ($n||$) and
perpendicular ($n\perp$) to the electric field direction. It is related
to the optical polarizability (g) anisotropy associated with the
solute molecules, their volume concentration (C_v) and the average
refractive index of the solution (n) by the equation [6]

$$\Delta n = (2\pi C_v/n)(g_3 - g_1)\, \Phi \qquad\qquad \dots (1)$$

In this expression the subscripts 3 and 1 indicate the relevant
parameters along the major (3) and a minor (1) axis of a cylindrical molecule. The orientation factor Φ describes the average
orientation of the molecules at any given time (t) or in any given
field of amplitude (E). If an alternating electric field is applied
then a root-mean-square value for E is used. The function Φ has

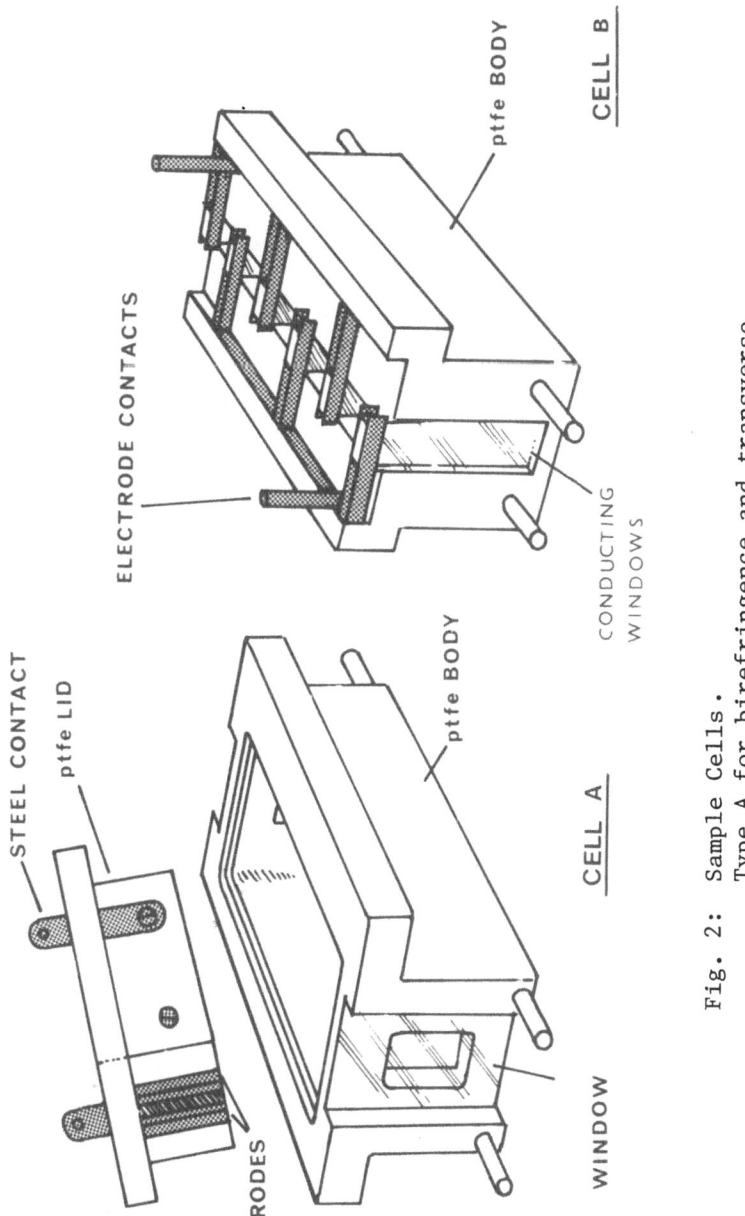

Fig. 2:   Sample Cells.
          Type A for birefringence and transverse
          dichroism measurements.
          Type B for electro-optical rotation and
          longitudinal dichroism measurements.

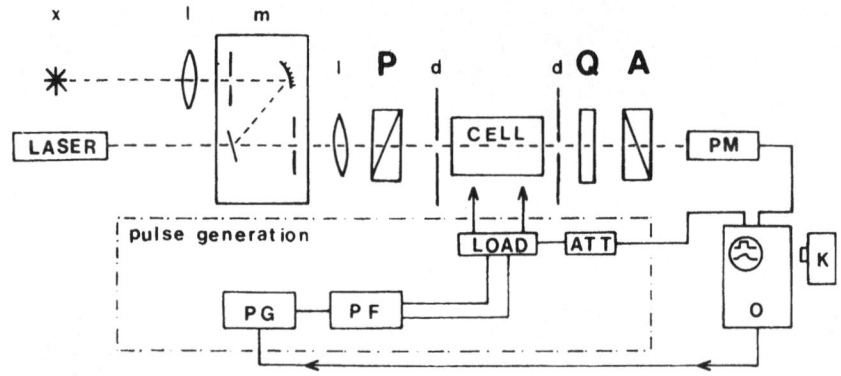

Fig. 3:   Schematic of Basic Apparatus.
          The arc (x), lenses (ℓ) and monochromator (m)
          provide an alternative source to the laser.
          Polarizer, P; λ/4 plate, Q; analyzer, A;
          diaphragms, d; PM, photomultiplier; PG and
          PF, pulse generator and former; ATT, attenuator;
          O, oscilloscope and K, camera.

been calculated and tabulated for a variety of situations. Under
conditions of low molecular orientation, as when the electrical
torque on the molecules is much less than the thermal energy (kT),
for rigid molecules oriented in a sinusoidal electric field of
angular frequency ω

$$\Phi = \left\{ \frac{(\mu_3^2 - \mu_1^2)}{(1 + 9\omega^2 \tau^2)k^2T^2} + \frac{(\alpha_3 - \alpha_1)}{kT} \right\} E^2/15 \qquad \ldots (2)$$

Here, the parameters μ_i and α_i are the magnitudes of the permanent
dipole moments and the electrical polarizabilities respectively, τ
is the molecular rotary relaxation time, k is the Boltzmann constant
and T the absolute temperature. It is convenient to express the
difference $(\alpha_3 - \alpha_1)$ as $\Delta\alpha$. The factor Φ does not depend upon the
optical property being measured; it simply describes the interaction
of the electric field with the electrical parameters of the molecule.
Hence this function reappears in theoretical descriptions of other
electro-optical phenomena. From equation (2) we note the following
factors. Firstly, by evaluating Φ from a particular experiment, its
quadratic dependence on E indicates the experimental fulfilment of
the condition for low degrees of molecular orientation. Under such
conditions one can isolate the electrical factor within the brace
of the equation. Secondly, only the permanent dipole contribution
is frequency dependent. Hence data recorded at two or more frequencies enable the discrete isolation of μ_i and α_i. Thirdly, the
parameter τ can be readily evaluated from the frequency dependence
of the data. In fact it is more convenient experimentally to
determine τ in another manner (see below). Once evaluated,
equations exist in the literature which relate τ to the length of
rigid rods[7], to the diameter of discs[8] or the axial ratio of
ellipsoids[8].

Two experimental advances have been made which have profoundly
influenced the methodology of electric birefringence experiments.
The first involves the use of short duration, high voltage, pulses
of electric field[9]. The light reaching the photomultiplier is then
transient in nature. When the field is applied the molecules align
against the viscous Brownian forces and reach an ultimate orientation equilibrium in the field. Termination of the field is
accompanied by molecular disorientation and the reversion of the
birefringence back to its pre-field value. The amplitude of the
response leads directly to the evaluation of the electrical
parameters of the molecules. Analysis of the decay rate is an
indicator of the rotary relaxation time τ of the particles and hence
of their geometric size. The second advance is in the development
of suitable photodetectors and output recorders which are capable
of following the rapid transient changes in the optical signal.
Initially the photomultiplier output was displayed on a suitable

oscilloscope and the transient trace recorded photographically and
analyzed at will. Currently, with the advent of fast transient
recorders, such responses can be digitized and analyzed electronically and immediately.

The advances in pulse generation technology have obviated the
need for measuring birefringence data as a function of frequency.
A simple procedure is now available. It involves the use of pulses
of alternating current as well as d.c. If a single shot d.c. pulse
is applied to the solution and a transient response recorded, then
the amplitude of the birefringence embraces contributions from both
μ_i and $\Delta\alpha$. A subsequent burst of the same r.m.s. field strength,
but of high frequency a.c. field, gives rise to a transient response
which responds to $\Delta\alpha$ alone. Hence from two rapid, successive pulses
all information needed for the evaluation of μ_i and $\Delta\alpha$ explicitly
can be obtained (Fig. 4). An additional advantage in using pulsed
fields is the rapid evaluation of τ from the decay of the optical
response following the field. For a monodisperse solution the
disorientation process follows the equation[10]

$$\Delta n = \Delta n_o \exp \left(- \frac{t}{\tau} \right) \qquad\qquad \ldots (3)$$

where Δn is the birefringence amplitude at any time t after the
termination of the electric pulse for which t = 0 and the birefringence has amplitude Δn_o. A rapid estimate of τ, and hence the particle size, can be made simply by observing the oscilloscope trace and
noting the time taken for the transient birefringence to decay to
approximately a third of its maximum value.

Electric birefringence studies have been conducted on an enormous number of materials. These have been reviewed often[2,3,11-13].
Recently, some interesting studies have been reported which demonstrate the suitability of the method for the study of complex molecular associations.

Proteoglycans are polysaccharide molecules found in the cartilage. They consist of a protein backbone with a radial distribution
of predominantly chondroitin sulphate side chains[14]. Typically, the
side chains are of 40 nm length (Fig. 5a). The proteoglycans
associate with hyaluronic acid. Such association is thought to be
of importance in arthritic processes[15]. The proposed model for the
association was as indicated in Fig. 5c. Transient electric birefringence data supported this[16] as seen in frames b and d of Fig.
5. The extended proteoglycan molecules gave a single, positive
transient which, with the addition of increasing amounts of
hyaluronic acid, gradually reduced in amplitude. This was
accompanied by the appearance of a negative birefringent component
having a much larger value of τ. The change of sign of the

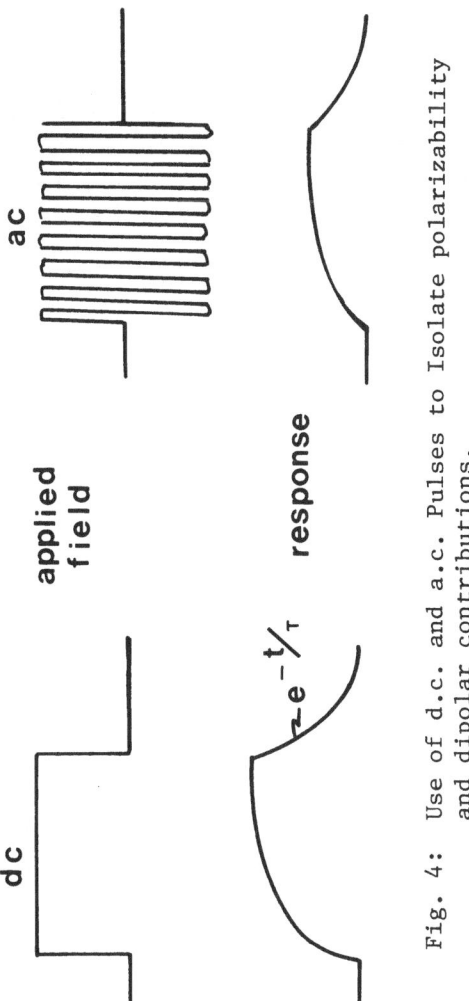

Fig. 4: Use of d.c. and a.c. Pulses to Isolate polarizability and dipolar contributions.

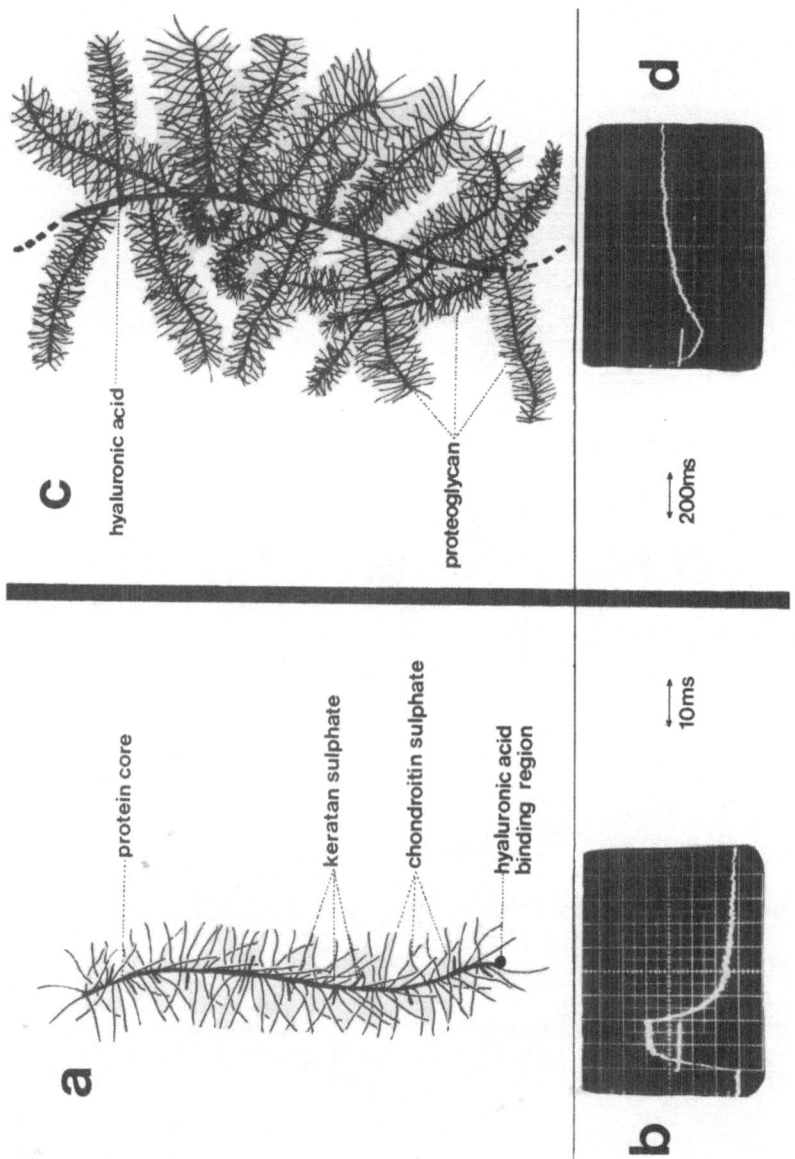

Figure 5. Proteoglycan Aggregation. Frames (a) and (c) are the models and (b) and (d) the electric birefringence responses. (Reproduced from ref. 16., courtesy of Wiley Interscience Inc.).

birefringence was consistent with the lateral grafting of the proteoglycan molecules on to the hyaluronic acid chains. The increased value of τ corresponded to a prolate ellipsoid with a major axis of the same length as the hyaluronic acid extended chains.

Another area of current growth is the estimation of particle size distributions for polydisperse samples. Various theoretical treatises have been published for the deconvolution of the multi-exponential transient decay into a distribution of relaxation times and hence of particle sizes [17-19]. In Fig. 6 the results of an experimental study on a suspension of sepiolite clay mineral are shown[20]. The analysis uses the transient decay for data obtained under low field and high field conditions and fits them to a mono-modal log-normal distribution function.

III.   OPTICAL KERR EFFECT

Electric birefringence is frequency dependent. With increasing frequency the birefringence amplitude passes through successive dispersion regions as the various electrical polarization mechanisms become unable to contribute to molecular orientation. At optical frequencies only the electronic contributions to the polarizability persist. Some twenty-four years ago it was realized that[4], with the advent of the laser, the electric vector in a light beam might itself give rise to the Kerr effect. During recent years the 'Optical Kerr Effect' has been realized in liquids[21], liquid crystals[22], macromolecular solutions and colloidal dispersions[5]. The principle is as follows. A conventional optical arrangement is assembled for the measurement of birefringence. This is essentially as described in the previous section. Instead of applying an electric field between metal electrodes across the sample, the optical array is arranged so that a light pulse from a high powered solid state laser traverses the solution along the same path as the birefringence probing beam. A convenient optical assembly is shown in Fig. 7. In general the effect is small. Furthermore, certain apparatus features have to be carefully considered. The probing light beam must be of different wavelength than the 'inducing' high powered light beam. Furthermore the probe must not be any multiple frequency of the inducing beam frequency. The pulse beam must be single moded in the TEM_{oo} mode, for only then is the electric vector unidirectional across the inducing laser pulse profile. Furthermore, the relative polarization states of the probe and inducing beam must be carefully adjusted to be at 45° relative azimuth upon entering the sample. Finally, and most important, the two beams must pass down exactly the same light path through the sample. In the Figure the mirrors m ensure this condition and bring the pulsed beam into and then out of the birefringence probe beam. Various optical filters ensure that the inducing beam does not reach the ultimate photodetector. In all other respects the principle is the

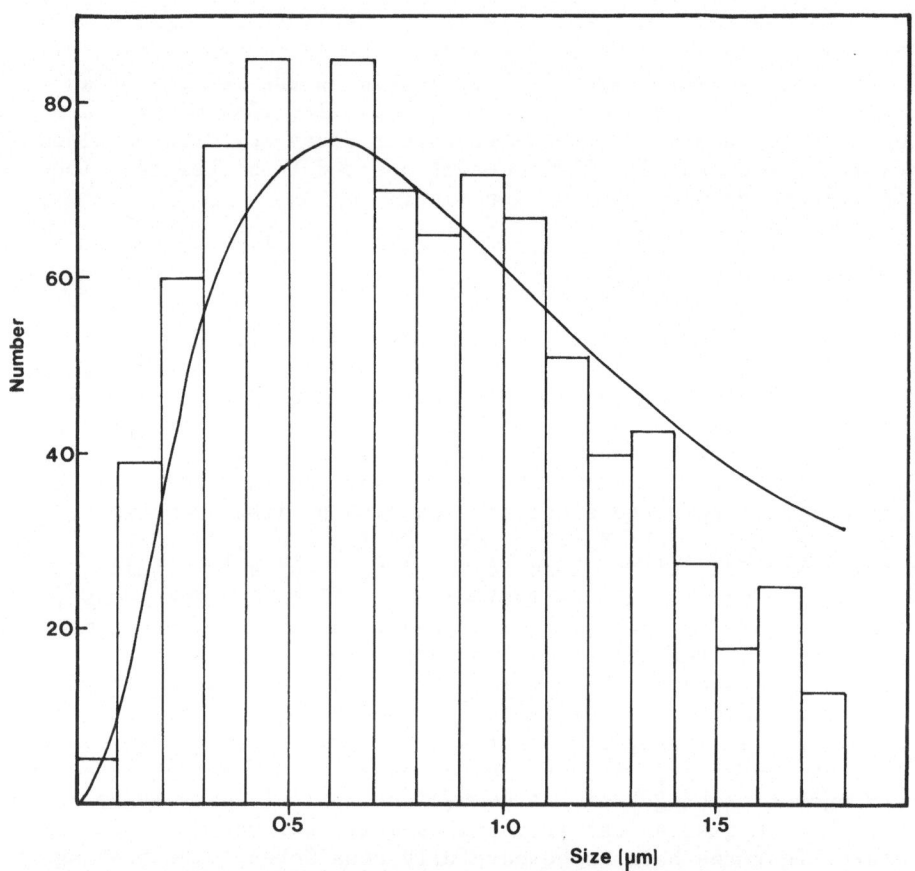

Fig. 6:   Log-normal Size Distribution from Transient
          Electric Birefringence Data (full line) Com-
          pared with Electron Microscopic Data.
          Results for a suspension of sepiolite rods.
          (From ref. 20, courtesy of Plenum Press Ltd.).

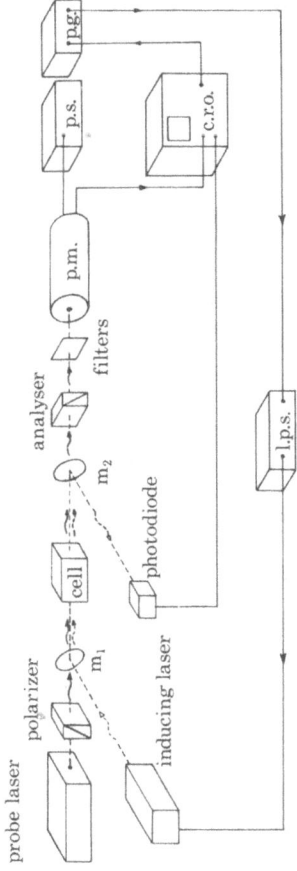

Fig. 7: Apparatus for Optical Kerr Effect Studies.
Components are dielectric mirrors, m_1, m_2;
p.m., photomultiplier; p.s., power supply;
l.p.s., laser power supply; p.g., pulse generator; c.s.o., oscilloscope.

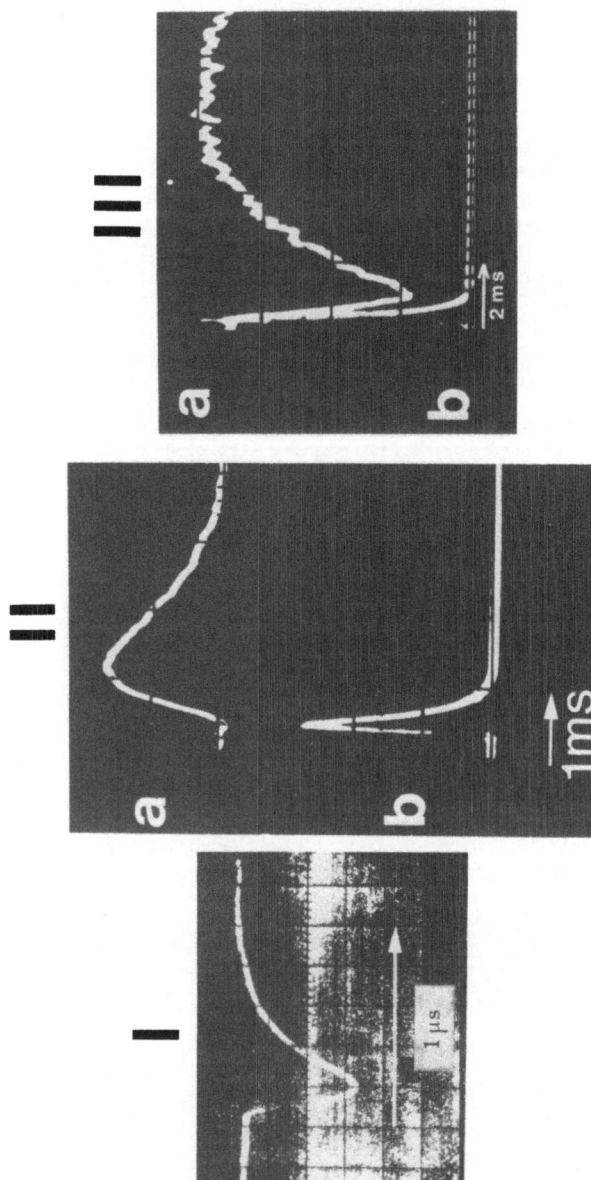

Figure 8. Optical Kerr Effect: I in MBA liquid crystal to a 'Q' switched YAG inducing pulse: II in Tobacco Rattle Virus Solution, andIII Sodium Bentonite sol using 'fixed-Q' YAG inducing pulses. (Data from refs. 5, 23; courtesy of the Royal Society and Macmillan Press).

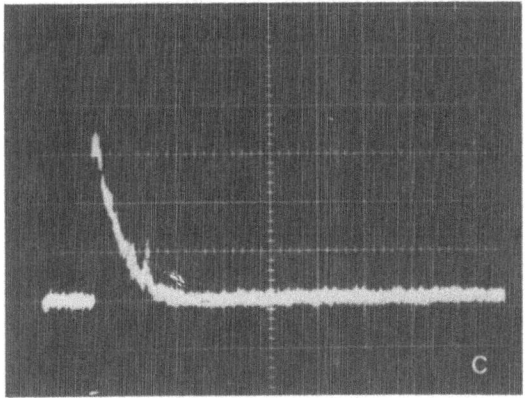

Fig. 9:   Laser Induced Scattered Intensity
          Changes in Bentonite Sols.
          A Nd-glass laser provided the in-
          ducing pulse of 5 joules in 1.5 ms,
          multimode; 40 ms per div. time scale.
          (Reproduced from ref. 24, courtesy of
          Plenum Press).

same as for electric birefringence.

Apart from its phenomenological interest, the optical Kerr effect is of special value for the following reasons. Firstly, the study of increasingly smaller molecules requires the generation of pulsed electric fields for electric birefringence measurements which are of ever increasing amplitude but reduced duration. The high powered, short duration, pulses one obtains by Q-switching lasers meet this requirement, and molecules of the order of tens of Angstroms in size should be measurable. Secondly, the discrete determination of the electronic contribution to the molecular polarizability should be possible. Thirdly, the fast laser pulses avoid some of the heating and conductivity problems associated with the application of longer duration electric pulses to ionic and conducting media.

A number of systems have been studied to date. In Fig. 8 data are shown for a liquid crystal, a virus and a colloid. It is convenient to mention in passing that laser fields can induce other optico-optical effects in large particulate suspensions, and optico-optical scattering effects are manifest in Fig. 9.

IV.   ELECTRIC LINEAR DICHROISM

The refractive index of a material is a complex property which can be expressed in the form $n(1 - i\kappa)$. We have considered how molecules can exhibit anisotropy in the real part of the refractive index through birefringence. Logically we might expect these molecules to have anisotropic imaginary components. The imaginary part of refractive index represents light energy loss and is thus related to the optical absorption. If a beam of incident intensity I_o passes through a solution of concentration C held in a cell of length ℓ, then the transmitted intensity I is given by

$$I = I_o \exp(-\varepsilon C \ell) \qquad \qquad \dots (4)$$

Here ε is the molecular extinction coefficient where $\varepsilon = 4\pi\kappa/c\lambda$ where c is the velocity of light. An isotropic molecule can be expected to exhibit various extinction coefficients ε_i in those regions where they are strongly absorbing. Here ε_i present extinction coefficients for light incident on the molecule parallel to the various molecular axes. A random array of molecules prior to the application of an electric field absorb light according to an average of the molecular extinction coefficients. Electric field alignment will therefore be accompanied by changes in the optical absorption as an incident polarized light beam is presented preferentially with the favored molecular axis. This change in adsorptive power is termed 'dichroism'. Hitherto it has been usual to

measure dichroic effects through two specific measurements. The
first involves light polarized parallel, and the second perpendicular, to the applied electric field vector. This can be conveniently undertaken in the apparatus of Fig. 3. The quarter-wave plate
and analyzer would be removed; the arc source would be preferred to
that of the laser supply, unless a dye laser is used, in order to
allow variable wavelength selectability. The two measurements would
then be made with a polarizer P, firstly parallel and then perpendicular, to the electric field direction across the sample.

For rigid, cylindrically symmetric molecules it has been shown
that a single, transient measurement is all that is needed for complete dichroism evaluation[25]. A special cell design is required in
which the field can be applied along the light beam. Such a 'longitudinal' experimental configuration means that the polarizer can
be dispensed with and measurements made in a modified simple spectro-
photometer[26]. This is because the electric field must always be
perpendicular to the electric vector at any instant in the light
beam owing to the transverse nature of light. Sample cells can be
constructed in which the optical windows are coated on their inner
surfaces with thin, conducting metal oxide films. The effect can
be enhanced through cell designs which incorporate many equivalent
series cells (Fig. 2,B).

If $\varepsilon_{||}$ and $\varepsilon_{\perp}$ be used for extinction coefficients observed
when the initial light beam polarization state is parallel or perpendicular to the applied field respectively, and if ε_L be used for
the longitudinal configuration, then in the electric field

$$\varepsilon_{\perp} = \varepsilon_L = \varepsilon_u - \frac{1}{3} (\varepsilon_3 - \varepsilon_1) \, \Phi \qquad \qquad \ldots \text{ (5)}$$

and

$$\varepsilon_{||} = \varepsilon_u + \frac{2}{3} (\varepsilon_3 - \varepsilon_1) \, \Phi \qquad \qquad \ldots \text{ (6)}$$

where $\varepsilon_u = (\varepsilon_3 + 2\varepsilon_1)/3$, and is the coefficient recorded in the
absence of any field. From equation (5) one sees that if Φ can be
evaluated from electric birefringence measurements, then ε_L and ε_u
are sufficient for complete evaluation of the molecular parameters
ε_3 and ε_1 at any particular wavelength. Alternatively, if electric
fields are available of sufficient amplitude, then at complete
molecular alignment Φ = unity and the parameters ε_1 can be realized
from electric dichroism measurements alone. It should be remembered
that by recording data in the transient manner, values of τ and
hence particle geometry, can be obtained simultaneously with the
extinction coefficient data.

Measurements on the dilute aqueous suspension of copper
phthalocyanine crystallities provide an illustrative example of the

method[27]. The crystallites are long and needle-like and are composed of layers of regularly stacked planar molecules in a continuous
zig-zag configuration[28]. The crystallite suspensions absorb strongly
with characteristic peaks in the red spectral region. All suspen-
sions demonstrated a dependence of $\varepsilon_{||}$ on field strength which was
linear in E^2 at low fields and became independent of E at high
fields. From such dependence, the parameters ε_1 and ε_3 were determined explicitly at 720 nm wavelength. From the data an electric
polarizability anisotropy of 8.5×10^{-25} Fm^2 per unit crystallite
length was obtained. The average particle length from τ was 130 nm.

The experiments were then repeated over the spectral range
$500 < \lambda$ (nm) < 750 and ε_1 and ε_3 determined. The wavelength dispersion of these parameters is shown in Fig. 10. One notices the
following factors. Firstly, at least three absorption peaks are
present. By analyzing the spectra in terms of composite Gaussian
contributions these peaks were found to be at 720, 640 and 578 nm.
Secondly, it is apparent that for $\lambda = 720$ and $\lambda = 640$, $\varepsilon_1 < \varepsilon_3$.
Hence at these wavelengths the absorption transition moment of the
crystallite is predominantly along the crystallite axis. This condition is reversed for the absorption occurring at 578 nm. From
these data it was not only possible to estimate the direction of the
transition moments within the crystallites, but also to relate them
to probable atomic origins within the planar constituent molecules.
The only alternative means of obtaining such molecular information
is to grow large crystals of the material and to study their optical
properties directly. Such large crystal growth is not generally
convenient.

Advantages of the electric dichroism method are thus as follows.
Colored solutions and suspensions can be studied and electrical and
geometrical parameters determined in addition to the extinction
coefficients at various wavelengths. Molecular structural characteristics can often be inferred.

V.   ELECTRICALLY INDUCED FLUORESCENCE EFFECTS

Fluorescence is a two-fold property which involves the absorption and subsequent re-emission of light along specific directions
or transition moments within an active fluorescent chemical group
(called a fluorophor). Absorption is most efficient for light
which is polarized parallel to the relevant absorption moment,
whilst the emitted light is polarized parallel to the
emission transition moment direction. Clearly both the amount of
absorption, and hence the intensity and state of polarization of the
emission, depend upon the orientation of the transition moments with
respect to the polarization direction of the incoming light beam.
The fluorescence from a collection of many identical fluorophors
will also depend upon their collective degree of alignment within

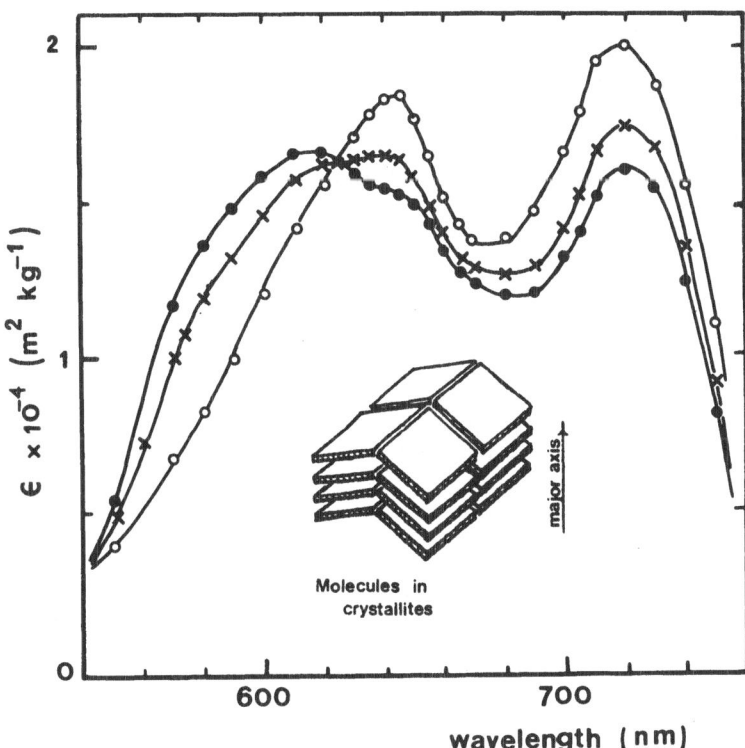

Fig. 10: Spectral Variation of the Extinction Coefficients
for Copper Phthalocyanine Particles in Water.
Open circles, filled circles and crosses represent
ε_3, ε_1 and ε_u respectively. (Data from ref. 27,
courtesy of Pergamon Press).

the medium.

Many macromolecules are inherently fluorescent and are composed of regularly repeated fluorophors within their structure. Alternatively they can be made fluorescent by binding fluorescent dye molecules to them. Should the fluorophors have a regular binding geometry relative to the parent molecule, then in a dilute solution, overall directional order would not be encountered owing to random spatial array within the solvent. Alignment in an external electric field induces molecular order and, provided the fluorophors are bound with high directional specificity, is accompanied by changes in the emitted fluoroescence intensity and polarization state. The phenomenon can be characterized through the measurement of four polarized components of the fluoroescence, namely V_v, V_h, H_v and H_h. Here the major letters V or H represent the polarization state of the incident light, whilst the subscript letters v, h represent the polarized states for vertically and horizontally polarized emitted light. These components and electrically induced changes in them (prefixed Δ) can be recorded with the apparatus represented in Fig. 11. High sensitivity is achieved by using incident laser light whose polarization state can be controlled by a Fresnel rhomb assembly. The two photodetectors record the two polarized components of fluorescence for one state of incident polarized light. Here again pulsed electric fields are applied and the transient changes in the polarized components recorded.

Following very early pioneer work by Czekalla[30] and Weber[31], Weill and co-workers[32,33] have developed the theory of the effect for rod molecules. They have shown that for small degrees of molecular orientation the following interchangeable equations hold

$$\frac{\Delta V_h - \Delta H_v}{V_v + 2V_h} = \frac{\Delta H_h + 2\Delta V_h}{V_v + 2V_h} = \frac{\Delta H_h + 2\Delta H_v}{V_v + 2V_h} = \frac{3}{2}(\cos^2\psi - \cos^2\psi')\Phi \quad ..(7)$$

Here the angles ψ and ψ' are those between the major rod axis and the absorption or emission transition moments respectively. In the special case where these moments lie in a plane perpendicular to the rod molecular axis, then the following simpler equation applies

$$\Delta V_v / V_v = -(2E^2/21)\left\{(\mu^2/k^2T^2) + (\Delta\alpha/kT)\right\} \quad ... (8)$$

In such cases the factor in the brace can be evaluated with relative ease. Usually however one cannot assume such symmetric geometry for the transition moments and the theory is correspondingly complicated. However, as a general principle, a combination of electric birefringence and electric fluorescence experiments leads to the evaluation of ψ and ψ' for rigid rod molecules.

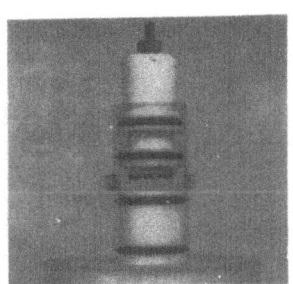

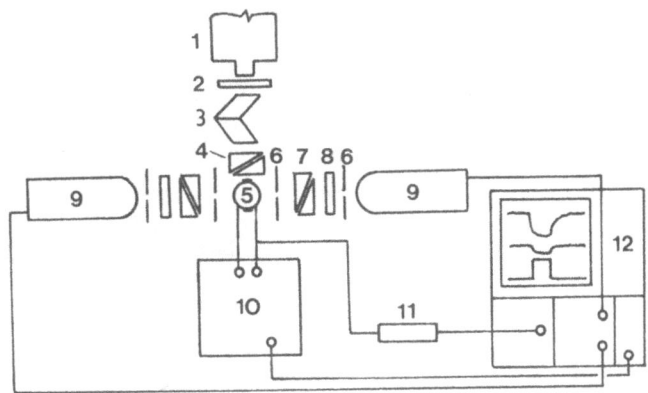

Fig. 11:   Transient Electro-Fluorescence Apparatus.
           1 - Argon laser, 2 - neutral density filter,
           3 - Fresnel rhomb, 4 and 7 - polarizers,
           5 - cell (also inset photograph), 6 - apertures,
           8 - cut-off filters, 9 - photomultiplier,
           10 - pulse generator, 11 - probe, 12 - oscillo-
           scope. (Reproduced from ref. 29, courtesy of
           Society of Photo-Optical Instrumentation
           Engineers).

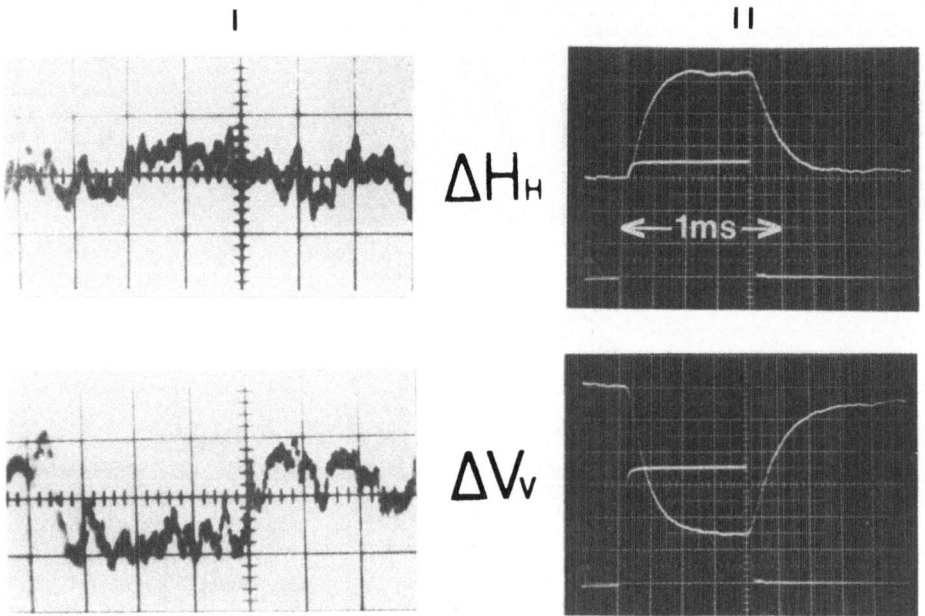

Fig. 12:   Electrically Induced Changes in Polarized Components
           of Fluorescence for DNA Solutions.
           I - Early results from ref. 32 on acridine orange
           sonicated DNA complex using a tungsten lamp source,
           and II - From ref. 34 on acridine orange native DNA
           complex using a laser source. (Data reproduced
           courtesy of Wiley Interscience and Institute of
           Physics).

In Fig. 12 transient changes in various polarized components of fluorescence are shown. For a dilute solution of DNA tagged with acridine orange the approximate evaluation of the binding geometry of the dye to the nucleic acid can be determined with simple reasoning. At any instant the total fluorescent intensity is equal to the sum of the components ($V_x + 2V_h$) for vertically polarized incident light. From the Figure ΔV_y is negative. This was also true for ΔV_h, indicating that the fluorescent intensity decreased with molecular alignment in the electric field. The cell electrodes applied the field in the vertical direction and thus aligned the macromolecules with their long axes increasingly parallel with both the field direction and the linear polarization state of the incident light. Reduction in the total fluorescence therefore indicates that the absorption moment does not favor such alignment and is thus predominantly perpendicular to the long DNA axis. Furthermore, the reduction in ΔV_v was greater than that for ΔH_v, indicating that the emission is preferentially associated with the horizontal rather than the vertical direction. Hence both absorption and emission transition moments are predominantly associated with perpendicular axes of the DNA structure. This is consistent with the dye molecules binding (or intercalating) the DNA molecules with the planar dye essentially parallel to the base pairs of the double helix.

A host of studies are current in the author's laboratory for studying the binding geometry of a number of dyes, chemotherapeutic agents and carcinogens to nucleic acids. In addition studies in Strasbourg by Weill and co-workers are aimed at the development of the theory of the method for evaluating complex binding geometries. The fluorescence method is one of the more useful and promising of the electro-optical techniques. It has already been applied to the study of nucleic acids[32-34], liquid crystals, polymers and colloids[29].

VI. ELECTRICAL-OPTICAL ROTATION

The ability of macromolecules to rotate the plane of polarization of a transmitted light beam is often used in biophysics as a measure of the concentration of molecular solutions. The specific rotation [Θ] is given by

$$[\Theta] = \Theta/C\ell \qquad \qquad \qquad \text{... (9)}$$

The optical rotatory power depends strongly on local molecular structure and is thus used as an empirical guide for conformation changes. What is not generally appreciated is that the phenomenon has its origin in specific electronic transitions. Thus optical rotation is another entity which has different values associated with various molecular axes. Fresnel[35] originally associated optical activity with helical symmetry in the structure. From this,

the well-known test of optically active molecules was devised, namely
that if the mirror image of the structure is not superimposable on
itself, then the molecules must be asymmetric and hence optically
active. A mirror image of a helix demonstrates that helices are
inherently optically active; as a right-handed helix appears left-
handed in a mirror. However an end-on mirror image is a circle.
Hence optical activity varies with molecular direction and one would
expect a change in the optical rotation upon alignment of such
molecules. Tinoco[36] first demonstrated this principle using a very
concentrated solution (some 10 per cent) of poly-benzyl-L-glutamate
in a helix promoting solvent. High voltage continuous electric
fields ($\simeq 10$ kV cm^{-1}) had to be applied. He also derived the theory
for the effect[37], namely

$$\Delta [\Theta] = \frac{2}{3} ([\Theta_3] - [\Theta_1]) \phi \qquad \qquad \dots (10)$$

where $[\Theta_i]$ indicates the rotary power associated with the various
molecular axes. In the absence of any aligning field

$$[\Theta] = \frac{1}{3} ([\Theta_3] + 2[\Theta_1]) \qquad \qquad \dots (11)$$

Here again electric fields are applied in the form of pulses and
transient changes detected in the optical rotation. From the pre-
field value $[\Theta]$ and the field-induced changes in this value, namely
$\Delta[\Theta]$, the parameters $[\Theta_i]$ can be evaluated. In addition, from the
transient data, the electrical properties and the sizes of the
molecules can be determined.

The apparatus of Fig. 3 can again be used. In this case the
quarterwave plate Q is removed and the analyzer and polarizer are
crossed. A special cell (type B, Fig. 2) must be used in which
the electric field is applied along the light path. This is essential to avoid birefringence phenomena. In addition, optical
rotation should be measured at a wavelength removed from any dichroic absorption band. As the optical rotation increases with
path length, so a multiple cell consisting essentially of a series
of cells must be used. Using suitable metal oxide coatings on the
inner faces, the field strength E can be maintained whilst the
optical path can be increased. After filling the cell the procedure is as follows. The optical rotation Θ is measured in the
absence of any electric field by offsetting P or A so as to restore
an initial zero transmission or 'crossed' condition. Upon application of a suitable pulsed electric field, a transient response
is detected by the photomultiplier as light penetrates the system.
This can be photographed or fed directly through a digitizer and
transient recorder. A typical response is shown in Fig. 13.

Transient optical rotation changes are difficult to obtain.

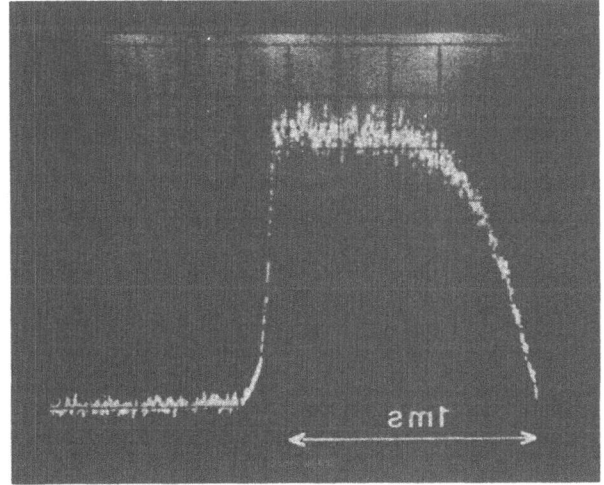

Fig. 13: Transient Change in the Optical Rotation of 1.8° for
          Polybenzyl-L-glutamate in 1,2-dichloroethane.
          E = 6.7 kV.cm^{-1}, λ = 630 nm. An argon pumped dye laser
          was used with a multiple series cell of 12 sections.
          (Reproduced from ref. 38 by courtesy of Taylor and
          Francis Ltd.).

The effect is very insensitive. However sensitivity increases with
the intensity of the light source and with the use of the multiple
cell principle. Using a dye laser as a tunable light source, and
electric fields of up to 7 kV cm^{-1} amplitude and of duration up to
10 ms, solutions of poly-benzyl-L-glutamate of some 10^{-3}g.ml^{-1} concentration have recently been studied[38]. From the transient changes,
values of $[\Theta_3]$ and $[\Theta_1]$ have been obtained over the wavelength range
340 < λ nm < 650. In addition a permanent dipole moment of μ =
2,970 Debye and τ = 30 μs were obtained. From τ a helix length
of 129 nm was found; combining these values leads to a value of
3.45 Debye for the dipole moment per monomer unit. This is extremely close to the value of 3.4 D for the α helix[39].

Despite its insensitivity the electrical-optical rotation
method has a number of endearing features. Already Θ is used as
an empirical measure of macromolecular conformation changes. Conventionally gross size and electrical parameters cannot be evaluated
from optical activity data. Using transient electric fields, the
changes $\Delta[\Theta]$ can be used to follow molecular conformation transitions and to evaluate size and electrical properties discretely
throughout the transitions. Such studies should prove very re-
warding for the study of helix-to-coil transitions of synthetic
and biopolymer systems.

VII. ELECTRIC LIGHT SCATTERING

Measurement of the scattered intensity from polymers and colloids has become a standard method for obtaining the relative
molecular mass (M) and radius of gyration (S) for molecules whose
major dimension is of the order of λ. Intramolecular interference
is encountered between the wavelets scattered from different
molecular regions. By recording the angular scattered intensity
the characteristics of the interference, and hence of the geometry
that gives rise to it, are evaluated. Measurements are made of the
scattered intensity (I) with observation angle Θ. Details of the
method of analysis through the Zimm[40] plot procedure are given in
a number of reviews[41-43].

The intermolecular interference is strongly dependent on
molecular orientation. The scattered intensity and its angular
dependence thus change significantly with molecular alignment.
Measurements of these changes lead to an alternative means of
evaluating the electrical parameters μ_i and α_i. Although restricted
to molecules of the order of λ in size, the method has a great
advantage in that it gives a wealth of molecular parameters from a
combination of both the pre-field and the in-field scattering data.

A number of commercial scattering photometers are available.
They can be readily converted for electro-optical measurements

essentially through the incorporation of suitable electrodes in the
scattering cell. Electric fields are applied and the intensity I
and the field induced changes ΔI are recorded at various angles Θ.
The theory is complicated as it varies with both the molecular shape
and the angle of observation. A typical equation for rigid molecules
at low degrees of orientation is as follows[44]

$$\Delta I/_I = (1 - 3\cos^2\Omega) \cdot Q \left\{ \frac{(\mu_3^2 - \mu_1^2)}{(1+9\omega^2\tau^2)k^2T^2} + \frac{(\alpha_3 - \alpha_1)}{kT} \right\} E^2 \quad \ldots(12)$$

Here Ω defines the electric field vector direction, whilst Q is independent of the electrical properties of the molecule but varies
with particle shape and Θ. All dependent parameters in Q can be
obtained from pre-field scattering measurements. The theory is
utilized as follows. In the absence of an electric field, I is
measured at various Θ and both M and S evaluated. Two consecutive
pulsed scattering transients are then obtained in response firstly
to a d.c. pulse and then to a pulse of a.c. field. From the amplitude of these two transients and the frequency dependence of the
terms in the brace of equation (12), $(\mu_3^2 - \mu_1^2)$ and $(\alpha_3 - \alpha_1)$ are
evaluated. Furthermore from the sign of scattered intensity changes
the predominance of μ_3 over μ_1 or α_3 over α_1 can be indicated.
Suffix 3 indicates the unique axis of rod, disc or ellipsoid.

Typical transient responses are shown in Fig. 14 for both d.c.
and a.c. pulsed fields when applied to a suspension of a synthetic
disc-like clay mineral[45]. Fields of 80 V cm^{-1} were of sufficient
amplitude. The effect of frequency on the magnitude of the response
is immediately apparent. From the complete study all of the following parameters were obtained on one and the same sample suspension. From the radius of gyration and on the assumption that
the discs were thin, a Z average diameter of 990 nm was obtained.
Two rotary relaxation times were seen to contribute predominantly
to the transient scattering decay phenomena, corresponding to
$\tau = 22$ and 50 ms. These were equivalent to discs of 740 nm and 980
nm respectively. Such close agreement in particle size is encouraging. From the signs of ΔI, it was seen that $\mu_3^2 > \mu_1^2$
and $\alpha_1 > \alpha_3$ with $(\mu_3^2-\mu_1^2)^{\frac{1}{2}} = 5.7 \times 10^{-25}$ Cm and $(\alpha_1-\alpha_3) = 2.5 \times 10^{-29}$Fm2.

The evaluation of M and S together with the electrical
properties affords a particularly useful means of monitoring particle geometry whilst the electrical properties change. Fig. 15
shows the high sensitivity of the polarizability factor of colloids
to the presence of surfactants. Such measurements are proving of
value in the study of the electric double layer, and hence the
stability, of colloidal suspensions.

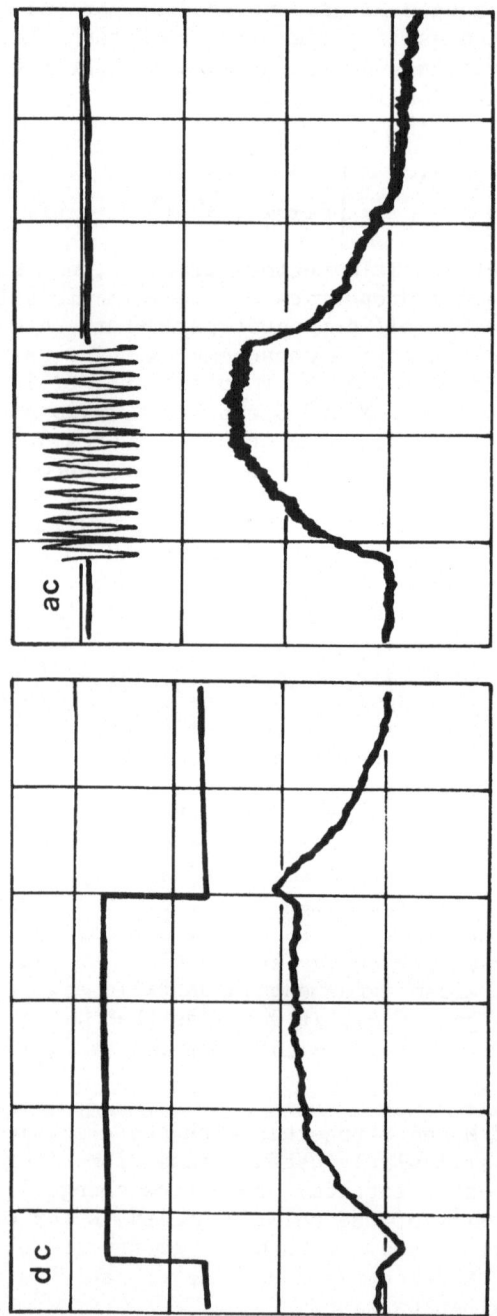

Fig. 14: Transient Scattering Effects from a Sol of Laponite Synthetic Clay.
Responses are to a d.c. and a 250 Hz a.c. pulse, each of 80 Vcm^{-1}
rms equivalent. Pulse durations of 100 ms and 70 ms. respectively.
(Data from ref. 45, courtesy of the Institute of Physics).

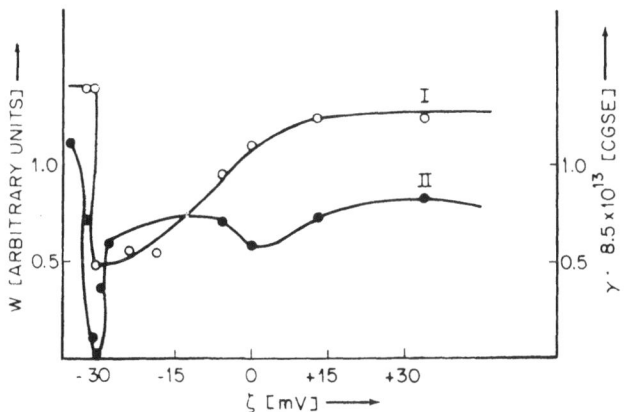

Fig. 15:   Variation of the Stability (W) and Polarizability
           (called γ here) with Zeta Potential (ζ) for a Sus-
           pension of Palygorskite Clay.
           The surfactant cetyl-pyridinium chloride had been
           added to vary ζ. (From ref. 46, courtesy of
           Academic Press Inc.).

VIII. ELECTROPHORETIC LIGHT SCATTERING

The light scattered from a dilute solution or suspension will exhibit short duration fluctuations due to the Brownian motion of the solute. If one can effectively 'count' the photons being received by a detector, then the fluctuations in the number received in a given time interval will be indicative of the molecular motion. By comparing the counts in neighboring intervals of time one can search for meaningful patterns in the molecular motions. This is currently done using correlation analysis of the fluctuations of the scattered intensity. The method goes under various names such as 'photon correlation spectroscopy' and 'fluctuation intensity spectroscopy'. A translatory, and sometimes an additional rotatory diffusion coefficient is evaluated from a correlation function $C(\tau)$ from which particle size can be determined[47].

The imposition of an electric field to a solution of charged molecules results in electrophoretic motion which can be readily detected through its influence on the scattered intensity fluctuations. The correlation function becomes oscillatory (Fig. 16) with a period (ΔT) which corresponds to the time taken by the particles to travel through a distance equivalent to the wavelength of the light. From the changes in the correlation function a measure of the electrophoretic mobility (u) is obtained directly. This mobility leads directly to the charge σ on the molecules. Electrophoretic light scattering therefore presents a method which complements existing electro-optical phenomena through the evaluation of σ. It is hoped that the method will be considerably developed in the next few years.

REFERENCES

1.   J. Kerr, Phil. Mag. 50:337 (1875).
2.   C. T. O'Konski, in 'Encyclopedia of Polymer Science & Technology' Vol. 9:551 (1969) Interscience, New York.
3.   K. Yoshioka and H. Watanabe, in 'Physical Principles and Techniques of Protein Chemistry', Academic Press, New York (1969).
4.   A. D. Buckingham, Proc. Phys. Soc. B69:344 (1956).
5.   B. R. Jennings and H. J. Coles, Nature 252:33 (1974).
6.   A. Peterlin and H. A. Stuart, 'Handbuch und Jahrbuch der Chemischen Physik,8(sec.1B), Becker & Erler, Leipzig (1943).
7.   S. Broersma, J. Chem. Phys. 32:1626 (1960).
8.   F. Perrin, J. Phys. Radium (Paris) 5:497 (1934).
9.   W. Kaye and R. Devaney, J. Appl. Phys. 18:913 (1947).
10. H. Benoit, Ann. Phys. (Paris), 6:561 (1951).
11. C. G. Le Fèvre and R. J. W. Le Fèvre, Rev. Pure Appl. Chem. 5:263 (1955).

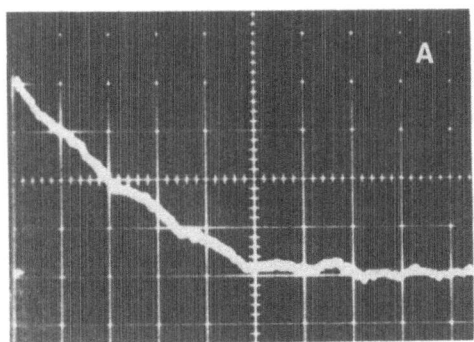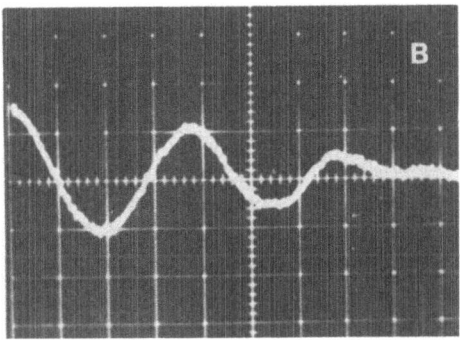

Fig. 16: Change in the Correlation Function for Bovine Serum
          Albumin Solutions Accompanying Electrophoretic Motion.
          (a) No electric field, $C = 5 \times 10^{-2}$ g.ml^{-1}; $\lambda = 515$
          nm, $\Theta = 4.5°$.
          (b) $E = 154$ Vcm^{-1}. (Reproduced from ref. 48, by
          courtesy of North Holland Publishers).

12. E. Fredericq and C. Houssier, 'Electric Dichroism and Electric
     Birefringence', Clarendon Press, Oxford (1973).
13. C. T. O'Konski (ed.), 'Molecular Electro Optics' Vol. 2, M.
     Dekker Inc., New York (1978).
14. H. Muir and T. E. Hardingham, in MTP International Review of
     Science, Series I, 5:105 (1975).
15. C. A. McDevitt and H. Muir, Ann. Rheum. Dis., Suppl. 2,34:137
     (1975).
16. A. R. Foweraker, M. Isles, B. R. Jennings, T. E. Hardingham,
     and H. Muir, Biopolymers 16:1367 (1977).
17. K. Tsuji, H. Watanabe and K. Yoshioka, Adv. Mol. Relxn. Proc.
     8:49 (1976).
18. H. J. Coles and G. Weill, Polymer 18:1235 (1977).
19. V. J. Morris, A. R. Foweraker and B. R. Jennings, Adv. Mol.
     Relaxn. & Intn. Proc. 12:65 (1978); 12:201 (1978); 12:211
     (1978).
20. A. R. Foweraker, V. J. Morris and B. R. Jennings, in 'Electro-
     Optics and Dielectrics of Macromolecules and Colloids', Ed.
     B. Jennings, Plenum Press (1979) p. 303.
21. G. Mayer and F. Gires, Compt. Rend. (Paris) 258:2039 (1964).
22. G. K. L. Wong and Y. R. Shen, Phys. Rev. Lett., 30:895 (1973).
23. B. R. Jennings and H. J. Coles, Proc. Royal Soc. Lond. A.
     348:525 (1976).
24. J. W. Parsons, R. L. Rowell and R. S. Farinato, in 'Electro-
     Optics and Dielectrics of Macromolecules and Colloids', Ed.
     B. Jennings, Plenum Press (1979) p. 385.
25. E. D. Bailey and B. R. Jennings, Applied Optics 11:527 (1972).
26. B. R. Jennings and E. D. Bailey, Nature 233:162 (1971).
27. A. R. Foweraker and B. R. Jennings, Spectrochim. Acta. 31A:1075
     (1975).
28. J. M. Robertson, J. Chem. Soc. 615 (1935).
29. P. J. Ridler and B. R. Jennings, SPIE Vol. 164 – 'Fourth
     European Electro-Optics Conference' (1979) p. 64.
30. J. Czekalla, Z. Electrochem. 64:1221 (1960).
31. G. Weber, J. Chem. Phys. 43:521 (1965)
32. G. Weill and C. Hornick, Biopolymers 10:2029 (1971).
33. G. Weill and J. Sturm, Biopolymers 14:2537 (1975).
34. P. J. Ridler and B. R. Jennings, J. Phys. E. (Sci. Instr.)
     10:558 (1977).
35. A. Fresnel, Oeuvres 2:479 (1868).
36. I. Tinoco, J. Amer. Chem. Soc. 81:1540 (1959).
37. I. Tinoco and W. G. Hammerle, J. Phys. Chem. 68:1619 (1956).
38. P. Hammond and B. R. Jennings, Mol. Phys. 39:1035 (1980).
39. A. Wada, J. Chem. Phys. 31:495 (1959).
40. B. H. Zimm, J. Chem. Phys. 16:1099 (1948).
41. K. A. Stacey, 'Light Scattering in Physical Chemistry',
     Butterworths, London (1956).
42. M. Kerker, 'The Scattering of Light and other Electromagnetic
     Radiation', Academic Press, New York (1969).

43. M. B. Huglin (ed.), 'Light Scattering from Polymer Solutions', Academic Press, New York (1972).
44. C. Wippler, J. Chim. Phys. 53:316 (1956).
45. J. F. Schweitzer and B. R. Jennings, J. Phys. D. (Appl. Phys.) 5:297 (1972).
46. S. P. Stoylov and I. Petkanchin, J. Coll. Int. Sci. 40:159 (1972).
47. B. Chu, 'Laser Light Scattering', Academic Press, New York (1974).
48. B. R. Ware and W. H. Flygare, Chem. Phys. Lett. 12:81 (1971).
49. B. R. Ware, Adv. Coll. Int. Sci. 4:1 (1974).

SMALL MOLECULES IN ELECTRIC AND OPTICAL FIELDS

A. D. Buckingham

University Chemical Laboratory
Lensfield Road
Cambridge CB2 IEW, United Kingdom

OPTICAL POLARIZABILITY

The optical properties of matter arise from the oscillations in the charge and current distributions induced by the incident electromagnetic field. For most purposes, the oscillations may be represented by the time-dependent electric dipole moment $\vec{\mu}(t)$ which arises from the linear polarization induced by the electric field $\vec{E}(t)$ of the light wave. The relationship between the vectors $\vec{\mu}(t)$ and $\vec{E}(t)$ may be written:

$$\mu_\alpha(t) = \sum_{\beta=x,y,z} \tilde{\alpha}_{\alpha\beta}\, E_\beta(t) \equiv \tilde{\alpha}_{\alpha\beta}\, E_\beta(t) \qquad (1)$$

where the polarizability $\tilde{\alpha}_{\alpha\beta}$ relates the α-component of $\vec{\mu}(t)$ to the β-component of $\vec{E}(t)$, and the subscripts α and β may represent any of the cartesian axes x,y,z. The polarizability $\tilde{\alpha}_{\alpha\beta}$ is a function of the frequency of the field and is complex at absorption frequencies, the real part giving the refractive index and the imaginary part the absorption.

The polarizability $\tilde{\alpha}_{\alpha\beta}$ is a second-rank tensor and in general it may have nine independent elements:

$$\tilde{\alpha}_{\alpha\beta} = \begin{bmatrix} \tilde{\alpha}_{xx} & \tilde{\alpha}_{xy} & \tilde{\alpha}_{xz} \\ \tilde{\alpha}_{yx} & \tilde{\alpha}_{yy} & \tilde{\alpha}_{yz} \\ \tilde{\alpha}_{zx} & \tilde{\alpha}_{zy} & \tilde{\alpha}_{zz} \end{bmatrix} \qquad (2)$$

Symmetry may reduce the number of independent elements, and in a system with cubic symmetry and without angular momentum (e.g. an Ar atom or a CH_4 molecule), there is only one independent component $\tilde{\alpha}_{xx} = \tilde{\alpha}_{yy} = \tilde{\alpha}_{zz} = \tilde{\alpha}$ and

$$\tilde{\alpha}_{\alpha\beta} = \tilde{\alpha}\delta_{\alpha\beta} \tag{3}$$

where $\delta_{\alpha\beta}$ is the Kronecker delta ($\delta_{\alpha\beta}$ is the unit tensor of the second rank). In general, the mean polarizability

$$\tilde{\alpha} = \frac{1}{3}\tilde{\alpha}_{\beta\beta} = \frac{1}{3}(\tilde{\alpha}_{xx} + \tilde{\alpha}_{yy} + \tilde{\alpha}_{zz}).$$

Quantum mechanical perturbation theory provides an expression for $\tilde{\alpha}_{\alpha\beta}$ for a molecule in the state n in terms of the transition dipole moments $<n|\vec{\mu}|j>$:[1]

$$\tilde{\alpha}_{\alpha\beta} = \sum_{j \neq n}\left[\frac{<n|\mu_\alpha|j> <j|\mu_\beta|n>}{\hbar(\omega_{jn} - \omega)} + \frac{<n|\mu_\beta|j> <j|\mu_\alpha|n>}{\hbar(\omega_{jn} + \omega)}\right] \tag{4}$$

$$= \alpha_{\alpha\beta} - i\alpha'_{\alpha\beta} \tag{5}$$

where ω is the angular frequency of the field E(t), that is, for light propagating in the z-direction,

$$E_\alpha(t) = E_\alpha^{(o)} \exp[-i\omega (t - zc^{-1})] \tag{6}$$

where c is the velocity of propagation. The energy of excitation from the molecular state n to the state j is $\hbar\omega_{jn}$.

On collecting the real and imaginary parts of $\tilde{\alpha}_{\alpha\beta}$ in (4), one obtains

$$\alpha_{\alpha\beta} = 2\hbar^{-1} \sum_{j \neq n} (\omega_{jn}^2 - \omega^2)^{-1} \omega_{jn} \text{Re}\{<n|\mu_\alpha|j> <j|\mu_\beta|n>\} = \alpha_{\beta\alpha} \tag{7}$$

$$\alpha'_{\alpha\beta} = -2\hbar^{-1} \omega \sum_{j \neq n} (\omega_{jn}^2 - \omega^2)^{-1} \text{Im}\{<n|\mu_\alpha|j> <j|\mu_\beta|n>\} = -\alpha'_{\beta\alpha} \tag{8}$$

where Re{x} and Im{x} are the real and imaginary parts of x. Equations (7) and (8) have singularities at the resonant frequency $\omega = \pm\omega_{jn}$, but in practice the non-zero linewidth of the absorption band prevents divergence. The factor $(\omega_{jn}^2 - \omega^2)^{-1}$ in (7) and (8) for frequencies near resonance is replaced by

$$Z(\omega, \omega_{jn}, \Gamma_{jn}) = f(\omega, \omega_{jn}, \Gamma_{jn}) + ig(\omega, \omega_{jn}, \Gamma_{jn}) \tag{9}$$

where Γ_{jn} is the width at half the maximum height of the absorption line-shape function g. For Wigner-Weisskopf line-shape, Z =

$(\omega^2_{jn} - \omega^2 - i\omega\Gamma_{jn})^{-1}$ and

$$f = \frac{\omega^2_{jn} - \omega^2}{(\omega^2_{jn} - \omega^2)^2 + \omega^2\Gamma^2_{jn}}, \qquad g = \frac{\omega\Gamma_{jn}}{(\omega^2_{jn} - \omega^2)^2 + \omega^2\Gamma^2_{jn}} \qquad (10)$$

The imaginary part of $\tilde{\alpha}_{\alpha\beta}$ leads to an induced dipole out of phase with the driving field and hence to absorption of the electromagnetic wave.[2]

If the wavefunctions n and j are necessarily complex, as in the presence of an external magnetostatic field or of electronic angular momentum (that is, if the molecule in the state n is not symmetric under time-reversal), there is an imaginary part of $\langle n|\mu_\alpha|j\rangle \langle j|\mu_\beta|n\rangle$ and the anti-symmetric contribution $\alpha'_{\alpha\beta}$ to the polarizability is non-zero.

The depolarization ratio ρ_0 of light elastically scattered in the y-direction from a beam linearly polarized in the x-direction and propagating in the z-direction is

$$\rho_0 = \frac{I_z}{I_x} = \frac{\langle \tilde{\alpha}^*_{zx} \tilde{\alpha}_{zx} \rangle}{\langle \tilde{\alpha}^*_{xx} \tilde{\alpha}_{xx} \rangle} = \frac{\tilde{\alpha}^*_{\alpha\beta} \tilde{\alpha}_{\gamma\delta} \langle k_\alpha k_\gamma i_\beta i_\delta \rangle}{\tilde{\alpha}^*_{\epsilon\phi} \tilde{\alpha}_{\xi\eta} \langle i_\epsilon i_\phi i_\xi i_\eta \rangle} \qquad (11)$$

where I_x and I_z are the polarized and depolarized intensities; $\vec{i}$, $\vec{j}$, $\vec{k}$ are unit vectors in the x, y, z directions, respectively, and the angular brackets $\langle \cdots \rangle$ denote statistical averages. If the rotational motion can be taken to be classical, then for any cartesian components $\alpha\beta\gamma\delta$,[3]

$$\langle i_\alpha i_\beta i_\gamma i_\delta \rangle = \frac{1}{15} (\delta_{\alpha\beta}\delta_{\gamma\delta} + \delta_{\alpha\gamma}\delta_{\beta\delta} + \delta_{\alpha\delta}\delta_{\beta\gamma})$$
$$\qquad (12)$$
$$\langle i_\alpha i_\beta k_\gamma k_\delta \rangle = \frac{1}{30} (4\delta_{\alpha\beta}\delta_{\gamma\delta} - \delta_{\alpha\gamma}\delta_{\beta\delta} - \delta_{\alpha\delta}\delta_{\beta\gamma})$$

and

$$\rho_0 = \frac{3\kappa^2 + 5\kappa'^2}{5 + 4\kappa^2} \qquad (13)$$

where

$$\kappa^2 = \frac{3\alpha^*_{\alpha\beta}\alpha_{\alpha\beta} - \alpha^*_{\alpha\alpha}\alpha_{\beta\beta}}{2\alpha^*_{\xi\xi}\alpha_{\eta\eta}} = \frac{\alpha^*_{\alpha\beta}\alpha_{\alpha\beta} - 3\alpha^*\alpha}{6\alpha^*\alpha} \qquad (14)$$

$$\kappa'^2 = \frac{3\alpha'_{\alpha\beta}{}^*\alpha'_{\alpha\beta}}{2\alpha^*_{\xi\xi}\alpha_{\eta\eta}} = \frac{\alpha'_{\alpha\beta}{}^*\alpha'_{\alpha\beta}}{6\alpha^*\alpha} \qquad (15)$$

The anisotropies κ and κ' cannot be separated experimentally by measuring ρ_0 at different angles. However, this separation can be achieved through the reversal coefficient $P(\theta)$ of Placzek[4], that is, the ratio of the intensities of left to right circularly polarized light scattered at an angle θ to the direction of propagation when the incident radiation is right-circular. For forward scattering $\theta = 0$, and for backward scattering $\theta = \pi$, and[5,6]

$$P(0) = [P(\pi)]^{-1} = \frac{6\kappa^2}{5 + \kappa^2 + 5\kappa'^2} \qquad (16)$$

Atomic sodium vapor depolarizes light at frequencies in the vicinity of the resonance lines through its anti-symmetric polarizability $\alpha'_{\alpha\beta}$.[7]

If the molecule is symmetric under time-reversal and has a three-fold or higher axis of symmetry, then there are two independent components of $\tilde{\alpha}_{\alpha\beta}$:

$$\alpha_{\alpha\beta} = \alpha\delta_{\alpha\beta} + \frac{2}{3}(\alpha_{\|} - \alpha)(\frac{3}{2}\ell_\alpha\ell_\beta - \frac{1}{2}\delta_{\alpha\beta}) \qquad (17)$$

where $\alpha_{\|}$ and $\alpha_{\perp}$ are the polarizabilities parallel and perpendicular to this axis which is in the direction of the unit vector $\vec{\ell}$; the mean polarizability α is $(1/3)(\alpha_{\|} + 2\alpha_{\perp})$.

From equation (7), we see that excited vibrational and electronic states with dipole-allowed transitions to the state n contribute to distortion of the molecule by $\vec{E}(t)$ that gives the induced dipole and hence $\alpha_{\alpha\beta}$. The contribution to $\alpha_{\alpha\beta}$ arising from excited vibrational levels of the ground electronic state has been called the "atomic polarizability" but a better name would be the "vibrational polarizability." It is present in all molecules which possess infrared activity and is due to the change in equilibrium structure due to the field $\vec{E}(t)$. For static fields ($\omega = 0$) the vibrational polarizability may be a significant fraction of the total, but at optical frequencies ($\omega \gg \omega_{vibration}$) it is negligible (and negative). Thus the scattering of light is due to the "electronic polarizability". However, an electrostatic field could modify the electronic polarizability by distorting the structure or by orienting the molecule, and this is considered in the next section on the Kerr effect.

From equation (4) or (7), it follows that static polarizabilities are necessarily positive for molecules in the ground state.

However, the energy $\hbar\omega_{jn}$ is negative if n is an excited state and j a state of lower energy, giving the possibility of a negative polarizability for an excited molecule. The second term in (4) is resonant at the emission frequency $\omega = \omega_{nj} = -\omega_{jn}$.

THE KERR EFFECT

The Kerr effect provides another important means of measuring anisotropy in $\alpha_{\alpha\beta}$. But before evaluating the Kerr constant of a gas, we need to consider the dipole induced by a strong electric field $\vec{E}(t)$. For a fixed orientation of the molecule, the dipole may be written as a power series in $\vec{E}$:

$$\mu_\alpha = \mu_\alpha^{(o)} + \alpha_{\alpha\beta}\,E_\beta + \frac{1}{2}\beta_{\alpha\beta\gamma}\,E_\beta E_\gamma + \frac{1}{6}\gamma_{\alpha\beta\gamma\delta}\,E_\beta E_\gamma E_\delta + \ldots \qquad (18)$$

$\mu_\alpha^{(o)}$ is the permanent electric dipole; $\beta_{\alpha\beta\gamma}$ and $\gamma_{\alpha\beta\gamma\delta}$ are the first and second hyperpolarizabilities, and they describe the non-linear polarization of the molecule by $\vec{E}$ and are responsible for frequency doubling and trebling, respectively. If the molecule is centrosymmetric $\mu_\alpha^{(o)}$ and $\beta_{\alpha\beta\gamma}$ must vanish. From (18), we can obtain the "differential polarizability" $\pi_{\alpha\beta}$ of a molecule in a field $\vec{E}$:

$$\pi_{\alpha\beta} = \frac{\partial\mu_\alpha}{\partial E_\beta} = \alpha_{\alpha\beta} + \beta_{\alpha\beta\gamma}E_\gamma + \frac{1}{2}\gamma_{\alpha\beta\gamma\delta}E_\gamma E_\delta + \ldots \qquad (19)$$

In the Kerr effect we are concerned with the anisotropy in the symmetric part of the differential polarizability (i.e. $(1/2)(\pi_{\alpha\beta} + \pi_{\beta\alpha})$) induced by a static electric field F_z:

$$<\pi_{zz} - \pi_{xx}> = <\alpha_{zz} - \alpha_{xx} + (\beta_{zzz} - \beta_{xxz})F_z + \frac{1}{2}(\gamma_{zzzz} - \gamma_{xxzz})F_z^2 + \cdots> \qquad (20)$$

The statistical averaging implied by the brackets $<\cdots>$ is for the molecule in the field F_z. The angular distribution function is:

$$\frac{\exp[-W(F_z,\Omega)/kT]}{\int\exp[-W(F_z,\Omega)/kT]d\Omega} = 1 + \frac{\mu_z^{(o)}F_z}{kT} + \frac{\left(\mu_z^{(o)}\right)^2 F_z^2}{2(kT)^2} + \frac{\alpha_{zz}^{(o)}F_z^2}{2kT} + \cdots \qquad (21)$$

where $\alpha_{\alpha\beta}^{(o)}$ is the static polarizability. Inserting (21) into (20) gives (neglecting terms in F_z^4):

$$<\pi_{zz} - \pi_{xx}> = F_z^2 [\frac{1}{2}\gamma_{\alpha\beta\gamma\delta} + \frac{\beta_{\alpha\beta\gamma}\mu_\delta^{(o)}}{kT} + \alpha_{\alpha\beta}(\frac{\mu_\gamma^{(o)}\mu_\delta^{(o)}}{2(kT)^2} + \frac{\alpha_{\gamma\delta}^{(o)}}{2kT})]x$$

$$<k_\alpha k_\beta k_\gamma k_\delta - i_\alpha i_\beta k_\gamma k_\delta> \qquad (22)$$

The averaging over molecular orientations may now be accomplished through equations (12), giving

$$\langle\pi_{zz}-\pi_{xx}\rangle = F_z^2\left[\frac{1}{3}\gamma+\frac{2\mu^{(o)}}{9kT}\beta+\frac{\alpha_{\alpha\beta}^{(o)}\alpha_{\alpha\beta}-3\alpha\alpha^{(o)}}{10kT}+(\alpha_{33}-\alpha)\frac{(\mu^{(o)})^2}{10(kT)^2}\right] \quad (23)$$

where α_{33} is the optical polarizability for the electric field in the direction of the permanent molecular dipole $\vec{\mu}^{(o)}$;

$$\beta = \frac{3}{10}\sum_{\alpha=1,2,3}(3\beta_{\alpha 3\alpha}-\beta_{\alpha\alpha 3}) \quad (24)$$

$$\gamma = \frac{1}{10}\sum_{\alpha,\beta=1,2,3}(3\gamma_{\alpha\beta\alpha\beta}-\gamma_{\alpha\alpha\beta\beta}) \quad (25)$$

For molecules with an axis of symmetry, (23) simplifies to:

$$\langle\pi_{zz}-\pi_{xx}\rangle = F_z^2\left[\frac{1}{3}\gamma+\frac{2\mu^{(o)}}{9kT}\beta+\frac{(\alpha_{\parallel}-\alpha_{\perp})}{15kT}\left(\alpha_{\parallel}^{(o)}-\alpha_{\perp}^{(o)}+\frac{(\mu^{(o)})^2}{kT}\right)\right] \quad (26)$$

The real part of (23) or (26) gives the birefringence and the imaginary part the electric dichroism.

If the molecules are non-rigid, the expressions (23) for the Kerr constant and (13) for the scattered light intensities must be averaged over the internal motion; this can influence the temperature dependence of $\langle\pi_{zz}-\pi_{xx}\rangle$ and ρ_0.

For molecules with cubic symmetry, $\mu^{(o)} = 0$, $\alpha_{\alpha\beta}$ is isotropic and given by (3), and (26) reduces to:

$$\langle\pi_{zz}-\pi_{xx}\rangle = \frac{1}{3}\gamma F_z^2 \quad (27)$$

Thus, at low gas densities the second hyperpolarizability γ is the sole contributor to the birefringence and dichroism in such cases and hence may be measured accurately[8]. For anisotropic molecules, γ is normally a minor contributor to the Kerr effect, as is $\mu^{(o)}\beta/kT$.[9] For dipolar molecules, the term in $(\alpha_{33}-\alpha)(\mu^{(o)})^2/(kT)^2$ is usually dominant.

In a field F_z, we may write $\langle\alpha_{zz}-\alpha_{xx}\rangle = \langle\alpha_{zz}-\frac{1}{2}(\alpha_{xx}+\alpha_{yy})\rangle =$

$\frac{3}{2}\langle\alpha_{zz}-\alpha\rangle$ and from (17), for molecules with a three-fold or higher axis of symmetry, $\langle\alpha_{zz}-\alpha_{xx}\rangle = (\alpha_{\parallel}-\alpha_{\perp})\langle P_2(\cos\theta)\rangle \quad (28)$

where $P_2(\cos\theta) = \frac{3}{2}\ell_z^2 - \frac{1}{2}$ is the second Legendre polynomial of the cosine of the angle between the molecular axis and the field.

In a field F_z

$$<P_1(\cos\theta)> = \frac{\mu^{(o)}F_z}{3kT} \tag{29}$$

$$<P_2(\cos\theta)> = \frac{F_z^2}{15kT}\left(\alpha_{\parallel}^{(o)} - \alpha_{\perp}^{(o)} + \frac{(\mu^{(o)})^2}{kT}\right) \tag{30}$$

For small dipolar molecules with $\mu^{(o)} \sim 1D$ in a field $F_z = 30kV/cm = 100esu$, $<P_1(\cos\theta)> \sim 10^{-3}$ and $<P_2(\cos\theta)> \sim 10^{-6}$. For a large molecule, $\mu^{(o)}$ may be much greater and it is possible to approach saturation when $<P_1(\cos\theta)> = <P_2(\cos\theta)> = 1$ and the orientation is independent of F_z.

For molecules with a small moment of inertia, such as H_2 or HF, the effects of quantization of the rotational motion tend to reduce $<P_1(\cos\theta)>$ and $<P_2(\cos\theta)>$. Thus, for H_2 at 300K, $<P_2(\cos\theta)>$ is reduced to 0.76 of its classical value given by (30).[10] However, these quantum effects are unimportant for all molecules except diatomic hydrides.

HYPERPOLARIZABILITY

Measurement of the first hyperpolarizability β are best carried out through second-harmonic generation in the presence of a strong electric field.[11,12] The effective non-linear polarizability giving second-harmonic generation through the dipole
$$\mu_\alpha(t) = (1/2)S_{\alpha\beta\gamma} E_\beta(t) E_\gamma(t)$$
oscillating at 2ω may be obtained from (18) and is

$$S_{\alpha\beta\gamma} = \beta_{\alpha\beta\gamma} + \gamma_{\alpha\beta\gamma\delta} F_\delta + \cdots = S_{\alpha\gamma\beta} \tag{31}$$

The first hyperpolarizability $\beta_{\alpha\beta\gamma}$ in (31) has resonances at $\omega = \omega_{jn}$ and at $2\omega = \omega_{jn}$ and, like $\alpha_{\alpha\beta}$ at optical frequencies, has an insignificant vibrational component; it therefore differs somewhat from $\beta_{\alpha\beta\gamma}$ in (18). In a static field F_z

$$<S_{zzz}> = \frac{F_z}{15}\left[2\gamma_{\alpha\alpha\beta\beta} + \gamma_{\alpha\beta\beta\alpha} + \frac{\mu^{(o)}}{kT}(2\beta_{\alpha\alpha3} + \beta_{3\alpha\alpha})\right] \tag{32}$$

$$<S_{xxz}> = \frac{F_z}{30}\left[3\gamma_{\alpha\alpha\beta\beta} - \gamma_{\alpha\beta\beta\alpha} + \frac{\mu^{(o)}}{kT}(3\beta_{\alpha\alpha3} - \beta_{3\alpha\alpha})\right] \tag{33}$$

so that measurements of second-harmonic intensities induced by F_z over a range of temperature provide information about β and γ.[12] In a dipolar species the contributions of β and γ to $<S>$ may be of the same order of magnitude, and of opposite sign.

OPTICAL ACTIVITY

Optical activity is normally studied through observations of
the rotation of the plane of polarization or through circular
dichroism. These effects result from interference of the forward
scattering with the incident wave. Another manifestation of optical activity is the differential scattering of right and left
circularly polarized light.[13]

To account for optical activity it is necessary either that the
anti-symmetric polarizability α'_{xy} be non-zero (as in a magnetostatic
field B_z) or that equation (1) be extended to allow for an oscillating dipole to be induced in a chiral molecule by the magnetic field
of the incident wave.[14] In the latter case, the oscillating electric
and magnetic dipole and electric quadrupole moments are[1]

$$\mu_\alpha(t) = \alpha_{\alpha\beta} \; E_\beta(t) + G'_{\alpha\beta} \dot{B}_\beta(t) \; \omega^{-1} + \tfrac{1}{3} A_{\alpha\beta\gamma} \nabla_\beta E_\gamma(t) + \cdots \quad (34)$$

$$m_\alpha(t) = -G'_{\beta\alpha} \; \dot{E}_\beta(t) \omega^{-1} + \cdots \quad\quad\quad\quad\quad (35)$$

$$\theta_{\alpha\beta}(t) = A_{\gamma\alpha\beta} \; E_\gamma(t) + \cdots \quad\quad\quad\quad\quad\quad (36)$$

The symmetry of a chiral molecule under time-reversal θ requires
these forms for the induced moments, since $\vec{\mu}(t)$, $\vec{\theta}(t)$, $\vec{E}(t)$ and
$\vec{\nabla E}(t)$ are symmetric, while the magnetic dipole $\vec{m}(t)$ and magnetic
field $\vec{B}(t)$ are antisymmetric under θ (so that the time derivative
$\dot{\vec{B}}(t)$ is symmetric). It is also helpful to enquire into the symmetry
properties of $\vec{\mu}$, $\vec{m}$ and $\vec{\theta}$ under parity P (space reversal). If the
coordinates of all particles in the molecule and in the source of
the fields are inverted through the origin, then $\vec{\mu}$, $\vec{E}$ and $\dot{\vec{E}}$ are
reversed while $\vec{m}$, $\vec{\theta}$, $\vec{B}(t)$ and $\vec{\nabla E}$ are unchanged. Hence, $\alpha_{\alpha\beta}$ is even
under both θ and P while $G'_{\alpha\beta}$ and $A_{\alpha\beta\gamma}$ are even under θ but
odd under P. Hence $G'_{\alpha\beta}$ and $A_{\alpha\beta\gamma}$ vanish for centrosymmetric molecules, and

$$G' = \frac{1}{3} \sum_{\alpha=1,2,3} G'_{\alpha\alpha}$$

changes sign on going from the d form of a molecule to the ℓ form.
Perturbation theory yields the equations:

$$G'_{\alpha\beta} = -2\hbar^{-1}\omega \sum_{j \neq n} Z(\omega, \; \omega_{jn}, \; \Gamma_{jn}) \; \text{Im}\{<n|\mu_\alpha|j> <j|m_\beta|n>\} \quad (37)$$

$$A_{\alpha\beta\gamma} = 2\hbar^{-1} \sum_{j\neq n} Z(\omega, \omega_{jn}, \Gamma_{jn})\omega_{jn} \ Re\{<n|\mu_\alpha|j><j|\Theta_{\beta\gamma}|n>\} \quad (38)$$

The optical rotation $\overset{\sim}{\phi}$ per unit path length is

$$\overset{\sim}{\phi} = -\frac{2\pi N\omega}{(4\pi\varepsilon_0)c} \ <\frac{1}{2}(\alpha'_{xy}-\alpha'_{yx}) -\frac{1}{c}[G'_{xx} + G'_{yy} + \frac{1}{3}\omega(A_{xyz} - A_{yxz})]> \quad (39)$$

where N is the number of molecules per unit volume. The circular dichroism is given by the imaginary part of (39). For an isotropic fluid:

$$\overset{\sim}{\phi} = -\frac{4\pi N\omega}{(4\pi\varepsilon_0)c^2} \ G' \quad (40)$$

Thus $A_{\alpha\beta\gamma}$ contributes to the components of the circular dichroism tensor but not to the trace.

DIFFERENTIAL SCATTERING

Chiral molecules differentially scatter right and left circularly polarized light, and measurements of the differential intensities provide information about the components of $G'_{\alpha\beta}$ and $A_{\alpha\beta\gamma}$.[13]

A magnetic field B_z in the direction of propagation discriminates between right and left circularly polarized light, as in magnetic circular dichroism[15] and in scattering[16]. So too does an electrostatic field F_x at right angles to the direction of propagation and scattering (see figure 1).[17]

To interpret the linear effect of F_x on the polarized scattering of light from a molecule such as methyl chloride, consider first the electric fields of right and left circularly polarized light traveling in the z-direction:

$$E_x^{\pm} = E^{(o)} \cos \omega(t - zc^{-1})$$
$$\quad (41)$$

$$E_y^{\pm} = \mp E^{(o)} \sin \omega(t - zc^{-1})$$

where the upper and lower signs refer to right and left, respectively. The field E_y induces a magnetic dipole $m_z(t)$ (see eqn. (35)):

$$m_z(t) = -G'_{yz} \ \dot{E}_y(t) \ \omega^{-1} = \pm G'_{yz}E^{(o)} \cos\omega(t-zc^{-1}) \quad (42)$$

and its radiation interferes with that from $\mu_x(t)$:

$$\mu_x(t) = \alpha_{xx} E_x(t) = \alpha_{xx} E^{(o)} \cos\omega(t-zc^{-1}) \quad (43)$$

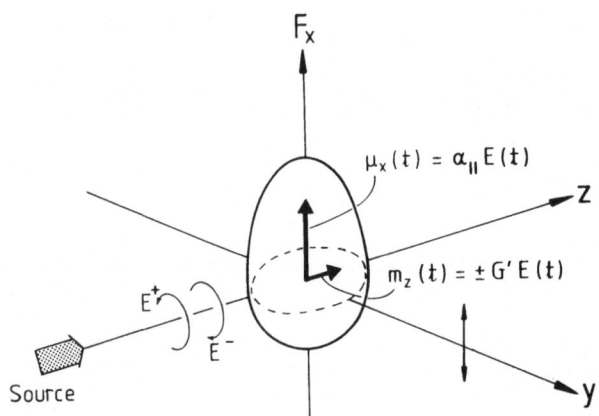

Figure 1. A dipolar molecule such as CH_3Cl oriented by the electric field F_x. The scattering from right (E^+) and left (E^-) circularly polarized light with x or z polarization is observed on the y axis.

giving opposite differential scattering in right and left circularly polarized light.

Differential depolarized scattering comes from the partial orientation of molecules by F_x. The scattering is due to

$$\mu_z(t) = \alpha_{zx} E_x(t) + \alpha_{zy} E_y(t) \tag{44}$$

$$m_x(t) = -G'_{yx} \dot{E}_y(t) \omega^{-1} = \pm G'_{yx} \dot{E}^{(o)} \cos\omega(t - zc^{-1}) \tag{45}$$

For a molecule such as CH_3Cl with its two independent polarizabilities $\alpha_{\|}$ and $\alpha_{\perp}$ and its single $G'_{\alpha\beta}$, namely $G'_{12} = -G'_{21}$ where the 3-axis is in the direction of the unit vector ℓ from the Cl to the C nucleus, (44) and (45) become

$$\mu_z(t) = (\alpha_{\|} - \alpha_{\perp}) E^{(o)} [\ell_x \ell_z \cos\omega(t - zc^{-1}) \mp \ell_y \ell_z \sin\omega(t-zc^{-1})] \tag{46}$$

$$m_x(t) = \mp G'_{12} \ell_z E^{(o)} \cos\omega(t - zc^{-1}) \tag{47}$$

The differential depolarized scattering $\Delta I_z^{\pm}$ results from interference in the radiation from the magnetic dipole and the first part
of $\mu_z(t)$ in (46). In the field F_x

$$<\ell_x\ell_z^{\ 2}> = \frac{\mu^{(o)}F_x}{kT} \qquad <\ell_x^{\ 2}\ell_z^{\ 2}> = \frac{\mu^{(o)}F_x}{15\ kT} \qquad (48)$$

and

$$\frac{\Delta I_z^{+}}{I_z^{+}} = -\frac{\Delta I_z^{-}}{I_z^{-}} = \frac{2c^{-1}<\mu_z(t)\ m_x(t)>}{<\mu_z^{\ 2}(t)>} = \frac{G'_{12}\mu^{(0)}F_x}{c(\alpha_{||}-\alpha_{\perp})kT} \qquad (49)$$

Actually G'_{12} depends on the origin, and one should include the
scattering from the electric quadrupole as well as from $\vec{m}$, and then
G'_{12} is replaced by $G'_{12} + \frac{1}{3}\omega A_{113}$ which is independent of
origin.[18]

The expected magnitude of the differential scattering is of
the order of a few parts in a million, $G'_{12}\ c^{-1}(\alpha_{||} - \alpha_{\perp})^{-1}$ being
$\sim10^{-3}$ for CH_3Cl. Like the Faraday effect, this linear electro-
optic effect should be exhibited by all molecules, but dipolar
molecules should show larger effects than non-polar ones. The
Faraday effect comes from the linear effect of $\vec{B}$ on $\alpha'_{\alpha\beta}$ and linear
electro-optical scattering from the change in $G'_{\alpha\beta}$ induced by $\vec{F}$:

$$\alpha'_{\alpha\beta}(\vec{B}) = \alpha'_{\alpha\beta} + \alpha'^{(B)}_{\alpha\beta\gamma}B_\gamma + \cdots \qquad (50)$$

$$G'_{\alpha\beta\gamma}(\vec{F}) = G'_{\alpha\beta} + G'^{(F)}_{\alpha\beta\gamma}F_\gamma + \ldots \qquad (51)$$

Both $\alpha'^{(B)}_{\alpha\beta\gamma}$ and $G'^{(F)}_{\alpha\beta\gamma}$ are symmetric under parity and time-reversal.

From figure 1 it is clear that the role of F_x is to distinguish up from down, and this may be achieved in other ways, as in
a ferroelectric crystal. Light reflected from the surface of a
ferroelectric should show a phase change proportional to G'_{12}; this
effect is analogous to the magneto-optic effect discovered
by Kerr in 1877 (a rotation of the plane of polarization on reflection of light from a magnetized surface).[19]

SOME TYPICAL MAGNITUDES

Typical magnitudes for the molecular properties described in
this Chapter are tabulated below for molecules of small size.

Property	Gaussian system (esu or emu)		SI
$\mu^{(o)}$	$1D = 10^{-18}$	$=$	3.3356×10^{-30} C m
$\theta^{(o)}$	$1b = 10^{-26}$	$=$	3.3356×10^{-40} C m^2
$m^{(o)}*$	$1\mu_B = 9.274 \times 10^{-21}$	$=$	9.274×10^{-24} A m^2
α	$1\overset{\circ}{A}^3 = 10^{-24}$	$=$	$1.1126 \times 10^{-40} C^2 m^2 J^{-1}$
β	$\pm 10^{-30}$	$=$	$\pm 0.3711 \times 10^{-50} C^3 m^3 J^{-2}$
γ	10^{-36}	$=$	$0.1238 \times 10^{-60} C^4 m^4 J^{-3}$
$G'*$	$\pm 10^{-26}$	$=$	$\pm 3.3356 \times 10^{-34} C^2 m^3 J^{-1} s^{-1}$
A	$\pm 10^{-32}$	$=$	$\pm 1.1126 \times 10^{-50} C^2 m^3 J^{-1}$

*In the Gaussian system of units, electrical quantities are in e.s.u. and magnetic quantities are in e.m.u. The Bohr magneton μ_B is $e\hbar/2m_e c$ in this system and $e\hbar/2m_e$ in SI, so that $G'c^{-1}$ in equations (39), (40) and (49) becomes G' in the Gaussian system.

References

1. A. D. Buckingham, Adv. Chem. Phys. 12, 107 (1967).
2. M. Born and K. Huang, Dynamical Theory of Crystal Lattices, (Oxford: Clarendon Press 1954), p. 18.
3. A. D. Buckingham and J. A. Pople, Proc. Phys. Soc. A, 68, 905 (1955).
4. G. Placzek, Handbuch der Radiologie (ed. E. Marx), vol. 6, part 2 (Leipzig: Akademische Verlagsgesellschaft 1934), p. 205.
5. D. A. Long, Raman Spectroscopy (New York: McGraw-Hill, 1977).
6. A D. Buckingham, Phil. Trans. Roy. Soc. London, A293, 239 (1979).
7. A. C. Tam and C. K. Au, Optics Commun. 19, 265 (1976); H. Hamaguchi, A. D. Buckingham, and M. Kakimoto, Optics Letters 5, 114 (1980).
8. A. D. Buckingham and D. A. Dunmur, Trans. Faraday Soc. 64, 1776 (1968).
9. M. P. Bogaard, B. J. Orr, A. D. Buckingham and G. L. D. Ritchie, J. Chem. Soc. Faraday Trans. II, 74, 1573 (1978).
10. A. D. Buckingham and B. J. Orr, Proc. Roy. Soc. London, A 305, 259 (1968).
11. G. Hauchecorne, F. Kerhervé and G. Mayer, J. Phys. (Paris) 32, 47 (1971).
12. C. K. Miller and J. F. Ward, Phys. Rev. A 16, 1179 (1977).
13. L. D. Barron and A. D. Buckingham, Ann. Rev. Phys. Chem. 26, 381 (1975).
14. E. U. Condon, Revs. Mod. Phys. 9, 432 (1937).

15. A. D. Buckingham and P. J. Stephens, Ann. Rev. Phys. Chem.
     17, 399 (1966).
16. L. D. Barron, Nature 257, 372 (1975).
17. A. D. Buckingham and R. A. Shatwell, Phys. Rev. Letters 45,
     21 (1980).
18. A. D. Buckingham and R. E. Raab, Proc. Roy. Soc. London A
     345, 365 (1975).
19. J. Kerr, Philos. Mag. 3, 321 (1877); 5, 161 (1878).

ROTATIONAL DIFFUSION COEFFICIENTS

José García de la Torre

Departamento de Química Física
Facultad de Ciencias
Universidad de Extremadura
Badajoz, Spain

I.   INTRODUCTION

The time dependence (buildup and decay) of electro-optic properties of macromolecules in solution is determined by the rotational diffusion coefficients (RDC's) of the macromolecule. In this paper we will be concerned only with rigid macromolecules (for flexible polymers see Jernigan's paper in this book) so that other relevant aspects such as chain statistics and internal motions are left out of our presentation. The RDC's enter in the diffusion equation governing the temporal evolution of the orientational distribution function when the electric field is switched on or off. The solution of this equation for the time dependent distribution function is not an easy task, but can be achieved under some physically meaningful simplifying assumptions, and from it one can calculate the expectation values of the electro-optical properties as a function of time provided that the rotational diffusion constants are known.

For instance, the decay of electric birefringence, Δn, of rigid macromolecules of arbitrary shape has been shown by Wegener et al. |1| to contain up to five exponential terms,

$$\Delta n \ (t) = \sum_{i=1}^{5} Q_i \ \exp \ (-E_i t) \tag{1}$$

where the Q_i's depend on the components of the electro-optical tensors and vectors and the E_i's are simple combinations of the three fundamental RDC's, D_r^1, D_r^2 and D_r^3, which will be identified later in this paper with the principal values of the rotational diffusion tensor, $\underset{\sim}{D}_r$. According to Wegener et al.,

$$E_1 = 6D - 2\Delta$$

$$E_2 = 3D + D_r^{\,2}$$

$$E_3 = 3D + D_r^{\,1} \qquad\qquad\qquad\qquad (2.a\text{-}e)$$

$$E_4 = 3D + D_r^{\,3}$$

$$E_5 = 6D + 2\Delta$$

where

$$D = (1/3) (D_r^{\,1} + D_r^{\,2} + D_r^{\,3}) \qquad\qquad\qquad (3)$$

and

$$\Delta = [(D_r^{\,1})^2 + (D_r^{\,2})^2 + (D_r^{\,3})^2 - D_r^{\,1}D_r^{\,2} - D_r^{\,1}D_r^{\,3} - D_r^{\,2}D_r^{\,3}]^{1/2} \quad (4)$$

Expressions similar to (1)-(4) hold for other properties such as linear dichroism |1|, optical rotation |1|, fluorescence depolarization |2| and depolarized light scattering |3|.

Thus, measurements of time-dependent optical properties, in addition to being a primary source of information on dipole moments, polarizabilities and anisotropies can also yield results for hydrodynamic quantities, the RDC's. The hydrodynamic behavior of rigid macromolecules is determined by their shape and size so that a theoretical formalism is needed to ascertain them from the observed properties. Although simple expressions for ellipsoids (exact) and cylinders (approximate) were available a long time ago, a general theory valid for macromolecules of arbitrary shape and able to incorporate structural details in the calculations was required for the accurate calculation of hydrodynamic properties of biopolymers and macromolecular complexes. Such a theory has been developed by several workers over the past five years |4-15|. Here I will present its basic aspects and applications to the evaluation of the RDC's.

To end with these introductory comments, some words about the position occupied by the RDC's in the realm of the hydrodynamic properties are pertinent. While sedimentation coefficients and translational diffusion coefficients are easy to measure, they are little sensitive to structural variations. The intrinsic viscosity depends strongly on the longest dimension in the case of elongated macromolecules, but is almost independent of particle size for compact, globular geometries. By contrast, the RDC's are exceedingly sensitive to changes in both size and shape; indeed at least one of them varies roughly with the third power of the length of the particle in a direction perpendicular to the corresponding rotation axis. Therefore the RDC's are the best choice when one wants to detect subtle changes in the macromolecular conformation induced by alterations in the physicochemical properties of the solvent, binding of small molecules, and so on.

II.   RIGID PARTICLE HYDRODYNAMICS

Although the rotational dynamics of particles with high symmetry can be studied independently of the translational dynamics, in the general case they are coupled to each other. Also, the determination of the hydrodynamic center of the particle makes necessary the previous evaluation of translational quantities. Therefore, we will present here a combined translational-rotational theory put forward first by Happel and Brenner |16,17| and further developed by Harvey and García de la Torre |14|. While the observed properties are the rotational diffusion coefficients, the ones that can be directly calculated are the frictional tensors. Hence, we will treat them separately and the relationship between them will be given later on.

II.1.   Friction

Let us consider a rigid macromolecule which is translating with velocity U_0 measured at an arbitrary origin O and rotating with angular velocity ω. The frictional froce F and frictional torque T_0 at O are given by

$$
\begin{pmatrix} F \\ \\ T_0 \end{pmatrix} = \underbrace{\begin{pmatrix} \Xi_t & \Xi_{0,c}^T \\ \\ \Xi_{0,c} & \Xi_{0,r} \end{pmatrix}}_{R_0} \cdot \begin{pmatrix} U_0 \\ \\ \omega \end{pmatrix}
\tag{5}
$$

R_0 is the so-called 6x6 resistance matrix |16| and can be partitioned in four 3x3 blocks,

$\Xi_t \quad \equiv \quad$ translational friction tensor

$\Xi_{0,r} \equiv$ rotational friction tensor

$\Xi_{0,c}$ ≡ frictional coupling tensor

$\Xi^T_{0,c}$ ≡ transpose of $\Xi_{0,c}$

Particular forms of eq. (5) can be written for pure translation ($\omega = 0$) and pure rotation ($U_0 = 0$). Note that the rotational friction tensor and the coupling tensor are origin dependent, while Ξ_t does not depend on the choice of origin. Since Ξ_t and $\Xi_{0,r}$ at any 0 are symmetric tensors, R_0 is symmetric too. However, $\Xi_{0,c}$ is only symmetric at a specific point R, the center of reaction |16|. It can be shown that an unconstrained particle will rotate around R because the energy dissipation is minimal in that way, and therefore $\Xi_{R,r}$ and $\Xi_{R,c}$ are only meaningul when referred to R. If one knows Ξ_t and $\Xi_{0,c}$ at an arbitrary origin 0, the distance from 0 to R, r_{OR}, can be calculated as |13|

$$
\begin{pmatrix} x^1_{OR} \\ x^2_{OR} \\ x^3_{OR} \end{pmatrix} = \begin{pmatrix} \Xi^{22}_t + \Xi^{33}_t & - \Xi^{12}_t & - \Xi^{13}_t \\ - \Xi^{12}_t & \Xi^{11}_t + & - \Xi^{23}_t \\ \Xi^{13}_t & - \Xi^{23}_t & \Xi^{11}_t + \Xi^{22}_t \end{pmatrix}^{-1}
$$

$$
\begin{pmatrix} \Xi^{32}_{0,c} - \Xi^{23}_{0,c} \\ \Xi^{13}_{0,c} - \Xi^{31}_{0,c} \\ \Xi^{21}_{0,c} - \Xi^{12}_{0,c} \end{pmatrix}
\qquad (6)
$$

When the macromolecule has an axis and/or a center of symmetry, R is then called center of hydrodynamic stress |16| and has the property that $\Xi_{R,c}$ = 0. In the latter case R coincides with the symmetry center. If there is just a symmetry axis, R must lie on it, but its location is unknown at the onset and eq. (6) has to be used. Thus, only when R is a center of hydrodynamic stress are the translational and rotational motions uncoupled and, at any rate, if R can not be found by simple inspection, the evaluation of the rotational friction tensor at R requires the previous calculation of the translational and coupling tensors. Happel and Brenner have given expressions for the origin dependence of $\Xi_{0,c}$ and $\Xi_{0,r}$ in their monograph |16|.

II.2. Diffusion

Brenner |17| has shown that the translational-rotational Brownian diffusion can be characterized by means of a 6x6 diffusion matrix, $\underset{\sim}{D}_0$, that links the generalized fluxes and gradients in the six-dimensional coordinate space. As in the preceding case, $\underset{\sim}{D}_0$ is partitioned in four 3x3 blocks,

$$\underset{\sim}{D}_0 = \begin{matrix} \underset{\sim}{D}_{0,t} & \underset{\sim}{D}^T_{0,c} \\ \underset{\sim}{D}_{0,c} & \underset{\sim}{D}_r \end{matrix} \tag{7}$$

where

$\underset{\sim}{D}_{0,t} \equiv$ translational diffusion tensor

$\underset{\sim}{D}_r \equiv$ rotational diffusion tensor

$\underset{\sim}{D}_{0,c} \equiv$ diffusion coupling tensor

A generalized form of the Einstein relationship hold between $\underset{\sim}{R}_0$ and $\underset{\sim}{D}_0$ |14,17|,

$$\underset{\sim}{D}_0 = k_B T \underset{\sim}{R}_0^{-1} \tag{8}$$

which is valid at any origin O. k_B is Boltzmann's constant and T is the absolute temperature.

Eq. (8) can be written more explicitly as

$$\underset{\sim}{D}_{0,t} = k_B T (\underset{\sim}{\Xi}_t - \underset{\sim}{\Xi}_{0,c}^T \cdot \underset{\sim}{\Xi}_{0,r}^{-1} \cdot \underset{\sim}{\Xi}_{0,c})^{-1} \tag{9}$$

$$\underset{\sim}{D}_r = k_B T (\underset{\sim}{\Xi}_r - \underset{\sim}{\Xi}_{0,c} \cdot \underset{\sim}{\Xi}_t^{-1} \cdot \underset{\sim}{\Xi}_{0,c}^T)^{-1} \tag{10}$$

$$\underset{\sim}{D}_{0,c} = k_B T \underset{\sim}{\Xi}_{0,r}^{-1} \cdot \underset{\sim}{\Xi}_{0,c} \cdot (\underset{\sim}{\Xi}_t - \underset{\sim}{\Xi}_{0,c}^T \cdot \underset{\sim}{\Xi}_{0,r} \cdot \underset{\sim}{\Xi}_{0,c})^{-1} \tag{11}$$

Although $\underset{\sim}{D}_{0,t}$, $\underset{\sim}{D}_r$ and therefore $\underset{\sim}{D}_0$ are always symmetric, there is an unique point at which the diffusion coupling tensor is symmetric, namely, the center of diffusion, D.

When R is a center of hydrodynamic stress, $\underset{\sim}{\Xi}_{R,c} = 0$ and R = D since it follows from eq. (11) that $\underset{\sim}{D}_{R,c}^T = \underset{\sim}{D}_{R,c} = \underset{\sim}{0}$. In addition, $\underset{\sim}{D}_{D,t} = k_B T \underset{\sim}{\Xi}_t^{-1}$ and $\underset{\sim}{D}_r = k_B T \underset{\sim}{\Xi}_{R,c}$, which means that translation and rotation are independent of one another. Brenner |17| claimed

that R = D always, but recently Harvey and García de la Torre |14|
have shown that when R is just a center of resistence, R and D do
not coincide. In general cases, the center of diffusion is given by
|14|

$$
\begin{pmatrix} x_{OD}^1 \\ x_{OD}^2 \\ x_{OD}^3 \end{pmatrix} = \begin{pmatrix} D_r^{22} + D_r^{33} & -D_r^{12} & -D_r^{13} \\ -D_r^{12} & D_r^{11} + D_r^{33} & -D_r^{23} \\ -D_r^{13} & -D_r^{23} & D_r^{11} + D_r^{22} \end{pmatrix}^{-1}
$$

$$
\begin{pmatrix} D_{0,c}^{23} - D_{0,c}^{32} \\ D_{0,c}^{31} - D_{0,c}^{13} \\ D_{0,c}^{12} - D_{0,c}^{21} \end{pmatrix} \tag{12}
$$

Expressions for the origin dependence of $D_{0,c}$ and $D_{0,t}$ have been
presented elsewhere |14,17|.

Once $D_{\sim r}$ has been calculated (details of the computational
procedures will be given in the following sections of this paper),
it can be diagonalized to get its eigenvectors and eigenvalues,

$$
D_{\sim r} = M_{\sim} \cdot \begin{pmatrix} D_r^1 & 0 & 0 \\ 0 & D_r^2 & 0 \\ 0 & 0 & D_r^3 \end{pmatrix} \cdot M_{\sim}^T \tag{13}
$$

where $(M^{1k}, M^{2k}, M^{3k})^T$ is the k-th eigenvector (k=1,2,3), defining
one of the principal axis of the rotational friction tensor, and
D_r^k the corresponding eigenvalue. The D_r^k's are to be substituted
in eqs (2)-(4).

III.   CALCULATION OF FRICTION AND DIFFUSION COEFFICIENTS

III.1.   Generalities

The first step in the calculation of hydrodynamic properties
is to build a suitable model of the rigid macromolecule. A very

simple approach is to use building elements, or subunits, with shapes like spheres, ellipsoids, cylinders or disks for which the frictional tensors are given by means of exact or quite approximate equations. The calculation then reduce to just a sum of the frictional tensors of each subunit expressed in a common frame of reference. This method has been followed by several workers |18-22| since it offers versatility in modeling and simplicity in the calculations. It has, however, a major drawback: the hydrodynamic interaction between elements is neglected, which results in unphysical values of the hydrodynamic properties, especially when the model consists of a moderately large number of elements. Therefore, although this method has predictive capability in some cases, we believe that it is less rigorous and less applicable than the one we are going to present next.

The modeling and computational method with full inclusion of hydrodynamic interactions is based on the theory of irreversible transport processes in solutions of macromolecules proposed by Kirkwood and Riseman |23,24| that was primarily intended for its application to flexible polymers. The development of algorithms for handling the more general version of the Kirkwood-Riseman theory and the proposal of accurate ways of accounting for hydrodynamic interactions has prompted its particularization to rigid macromolecules over the past few years. It should be noted that RDC's of rigid particles had been obtained earlier using approximate versions of the Kirkwood-Riseman theory |25-26| but the results were rather unsatisfactory.

Now the macromolecule is modeled as an assembly of spherical elements, or beads, which in turn act as friction sources. There is no restriction on the radius of these elements; they can be equal or different. Obviously, the size and shape of the model must be as close as possible to those assumed for the macromolecule. A good reproduction of the macromolecule's structural details usually requires a large number of elements in the model. In this sense, the ideal way is the so-called shell model |27|, in which a large number of small elements is placed on a shell that reproduces perfectly the surface of the macromolecule. Swanson et al. |7| have shown that shell-model calculations reproduce the exact value of the translational friction coefficient of particles such as ellipsoids, dumbbells and hemispherical caps. Unfortunately, the number of elements, N, is in practice limited by computer time and memory requirements, and usually must be less than one hundred.

The basic equation in the Kirkwood-Riseman theory is

$$F_i + \zeta_i \sum_{j=1}^{N,} T_{ij} \cdot F_j = \zeta_i (u_i - v_i^{\,0}) , \ i = 1, \ldots, N \qquad (14)$$

where, for the i-th element, $\zeta_i = 6\pi\eta_o\sigma_i$ is the Stokes law friction-al coefficient (σ_i being the element's radius), u_i and v_{io} are respectively, the velocity of the element and that of the fluid at its center, and F_i is the frictionsl force exerted by the element on the solvent. Note that if the second term in the left-hand side of eq. (14) were missing, we would have just $F_i = \zeta_i(u_i - v_i^o)$, which is the Stokes law for the spherical element. Therefore, that term accounts for the hydrodynamic perturbation produced by all the other elements (Σ' denotes omission of the term with j=i) on the i-th one. This perturbation is determined by the hydrodynamic interaction tensors, T_{ij}, given by |9|

$$T_{ij} = \frac{1}{8\pi\eta_o R_{ij}} \left\{ I + \frac{R_{ij} R_{ij}}{R_{ij}^2} + \frac{(\sigma_i^2 + \sigma_j^2)}{R_{ij}^2} \left(\frac{1}{3} I - \frac{R_{ij} R_{ij}}{R_{ij}^2} \right) \right\} \quad (15)$$

I in eq. (15) is the 3x3 unit tensor.

Eq. (14) represents a set of linear equations with 3N unknowns, that are the three components of each of the N frictional forces, and can be rewritten as

$$\sum_{j=1}^{N} Q_{ij} \cdot F_j = \zeta_i (u_i - v_i^o) \quad (16)$$

where

$$Q_{ij} = \delta_{ij} I + (1-\delta_{ij}) \zeta_i T_{ij} \quad (17)$$

Here δ_{ij} is Kronecker's delta. If we define a supermatrix Q of dimension 3Nx3N in such a way that its NxN blocks of dimension 3x3 are the Q_{ij}'s, the solution of eqs. (15) or (17) for the forces is

$$F_i = \sum_{j=i}^{N} \zeta_j \cdot S_{ij} \cdot (u_j - v_j^o) \quad (18)$$

where the S_{ij}'s are the blocks of dimension 3x3 of the 3Nx3N supermatrix S given by

$$S = Q^{-1} \quad (19)$$

Thus, from the geometry of the model (σ_i's and R_{ij}'s) one would obtain the T_{ij}'s, Q and the latter would be finally inverted to get S. These calculations constitute the first part of the computational procedures, and once they have been performed, eq.

(18) allows the evaluation of the frictional force at each element for arbitrary velocity fields. Now, we have to distinguish between translational and rotational motions.

III.2. Translation

For translational motion of the macromolecule with velocity $\underset{\sim}{u}$ and $\underset{\sim}{\omega}=0$ in a fluid that would be otherwise at rest, $u_i^o=0$ for all the elements. According to eq. (5), the translational friction tensor is given by

$$\underset{\sim}{F} = \underset{\sim t}{\Xi} \cdot \underset{\sim}{u} \tag{20}$$

with $\underset{\sim}{F} = \sum_i \underset{\sim i}{F}$. From eqs. (18) and (20) we have $|4,5,9|$

$$\underset{\sim t}{\Xi} = \sum_{i=1}^{N} \sum_{j=1}^{N} \zeta_i \underset{\sim ij}{S} \tag{21}$$

III.3. Rotation and Translation-rotation Coupling

Let us place the origin of the coordinate system at an arbitrary point 0. If the particle is undergoing pure rotation [$\underset{\sim}{u}=0$ in eq. (5)] with angular velocity $\underset{\sim}{\omega}$ in a quiescent fluid, then $v_i^o=0$ and $\underset{\sim i}{u} = \underset{\sim}{\omega} \times \underset{\sim i}{R}$, where $\underset{\sim i}{R}$ is the vector joining 0 and the center of element i. in the preceeding case, the $\underset{\sim i}{F}$'s can be obtained from $\underset{\approx}{S}$ via eq. (18). The frictional torques, measured at 0 are given by $\underset{\sim i}{T} = \underset{\sim i}{r} \times \underset{\sim}{F}$. Now, eq. (5) reduces to

$$\underset{\sim}{F} = \sum_{i=1}^{N} \underset{\sim i}{F} = \underset{\sim 0,c}{\Xi} \cdot \underset{\sim}{\omega} \tag{22}$$

and

$$\underset{\sim}{T} = \sum_{i=1}^{N} \underset{\sim i}{T} = \underset{\sim 0,r}{\Xi} \cdot \underset{\sim}{\omega} \tag{23}$$

The final results for the rotational and coupling tensors are $|5,13|$

$$\underset{\sim 0,c}{\Xi} = \sum_{i=1}^{N} \sum_{j=1}^{N} \zeta_j \underset{\sim i}{R} \times \underset{\sim ij}{S} \tag{24}$$

$$\underset{\sim 0,r}{\Xi} = - \sum_{i=1}^{N} \sum_{j=1}^{N} \zeta_j \ \underset{\sim i}{R_i} \times \underset{\sim ij}{S_{ij}} \times \underset{\sim j}{R_j} \tag{25}$$

The crosses in eqs. (24) and (25) denote dyadic products. More explicit expressions for the k, 1-components (k,1=1,2,3) of these tensors are

$$\underset{0,c}{\Xi}^{k,1} = \sum_{i=1}^{N} \sum_{j=1}^{N} \zeta_j \ (S_{ij}^{k-1\,l}\,x_i^{k+1} - S_{ij}^{k+1\,l}\,x_i^{k-1}) \tag{26}$$

and

$$\underset{0,r}{\Xi}^{kl} = \sum_{i=1}^{N} \sum_{j=1}^{N} \zeta_j \ (x_i^{k+1} S_{ij}^{k-1\,l-1} x_j^{l+1} + x_i^{k-1} S_{ij}^{k+1\,l+1} .$$
$$\cdot\, x_j^{l-1} - x_i^{k+1} S_{ij}^{k-1\,l+1} x_j^{l+1} - x_i^{k-1} S_{ij}^{k+1\,l-1} x_j^{l+1}) \tag{27}$$

where x_i^1, x_i^2, x_i^3, are the components of $\underset{\sim i}{R_i}$. In equations (26) and (27) $k-1 = 3$ if $k = 1$, $k+1 = 1$ if $k=3$, and the same convention holds for l.

III.4. Diffusion Coefficients

After calculating the three friction tensors at the arbitrary origin 0, the rotational diffusion tensor, $\underset{r}{D}$, can be obtained from eq. (10). We recall that $\underset{\sim}{D}$ is origin-independent, so that the evaluation of the RDC's does not require us to obtain the coupling and translational tensors at the center of diffusion. Diagonalization of $\underset{\sim r}{D}$ according to eq. (13) yields the RDC's and the principal axes of rotational diffusion directly.

That is all one needs to interpret the time dependence of the electro-optical properties. With little additional effort it is also possible to know the translational diffusion coefficient and the position of the center of diffusion. The former offers the possibility of analyzing other hydrodynamic properties and the latter is necessary, for instance, in the study of rotational effects in quasielastic light scattering |28,29|. From eqs. (9) and (11) $\underset{\sim 0,t}{D}$ and $\underset{\sim 0,c}{D}$ can be obtained. Next, the position vector of the center of diffusion, $\underset{\sim OD}{r}$ id calculated by the use of eq. (12). Then, the coupling and translational diffusion tensors are obtained at D using transformation laws given elsewhere |13,14,17|. The translational diffusion coefficient is given by

$$D_t = \frac{1}{3} \ Tr \ (\underset{\sim D,t}{D}) \tag{28}$$

and, if the molecular weight and the partial specific volume are
known, the sedimentation coefficient can be calculated by means of
Svedberg's equation.

III.5. Computational Methods

The most cumbersome step in the calculation of the hydrodynamic
properties is the inversion of the supermatrix Q|Eq. (19)|. The
time required grows as N^a, $\underline{a}$ being close to three, and the $9N^2$
elements of Q have to be stored in the central memory of the computer
for the procedure to be efficient. Computer time can be reduced by
using fast matrix-inversion routines, but the need of storing Q
limits the computational possibilities of the theory to models with
less than 100 frictional elements. (At the cost of much longer
computer times, this range of N can be expanded using auxiliary
memory devices, as done by Teller and coworkers |7,30|).

In some instances, a substantial saving of computer time (albeit
with the same memory requirements) can be achieved using iterative
algorithms. The Gauss-Seidel method has been found to give a reasonably fast convergence, although this is not always guaranteed.
Computer time for that method is proportional to nN^2, where n is the
number of iterations needed to meet the convergence criterion. If
N is high, and a moderate precision is acceptable, $n < N^{a-2}$ and
Gauss-Seidel iterations are less time-consuming than matrix inversion. The problem with iterative methods is that one has to repeat
the procedure for each set of values of the term in the right-hand
side of eq. (16), while the matrix-inversion method give directly
the solution [after the straightforward product in eq. (18)] for any
set of velocities. In fact, three Gauss-Seidel runs have to be
carried out to get $\Xi_{0,r}$ and three more for Ξ_t. Therefore, a factor
of six favorable to the inversion method has to be included in the
above comparison. More details on the Gauss-Seidel solution of
eq. (16) for the obtention of $\Xi_{0,r}$ and Ξ_t have been presented in a
separate article |15|.

Both computer time and memory requirements can be reduced by
taking advantage of the model's symmetry. Such is the case for
symmetric top macromolecules, which will be studied with some detail
in the next subsection of this paper. An even greater simplifi-
cation takes place when the frictional force is proportional to the
unperturbed velocity. This is true for any linear array of beads
rotating around a perpendicular axis and also for rotation of a
ring around either a diameter or an axis perpendicular to it |31|.
If ρ_i is the distance from the i-th element to the rotation axis,
which for instance is assumed to be x^1, then the modulus of the
unperturbed velocity is $u_i = \rho_i \omega$, ω, being the angular velocity around
x^1. We define the coefficients H_i by means of

$$F_i = \zeta_i \, H_i \, u_i \tag{29}$$

The rotational friction tensor in these cases is diagonal, with $\Xi_{0,r}^{12} = \Xi_{0,r}^{13} = 0$ and

$$\Xi_{0,r}^{11} = \sum_{i=1}^{N} \zeta_i \, \rho_i^{2} \, H_i \tag{30}$$

Also, the translation-rotation coupling tensor is zero at R, so that the rotational diffusion coefficient for rotation around x_1 is just given by

$$D_r^{1} = k_B \, T / \, \Xi_{R,r}^{11} \tag{31}$$

The H_i's are the solutions of the system of linear equations

$$\sum_{j=1}^{N} A_{ij} \, H_j = \rho_i \tag{32}$$

where

$$A_{ij} = \delta_{ij} + (1-\delta_{ij}) \, \frac{3 \, \sigma_j}{4 \, R_{ij}} \left\{ 1 + \frac{\sigma_i^{2} + \sigma_j^{2}}{3 \, R_{ij}^{2}} \cos (\theta_i - \theta_j) + \tag{33} \right.$$

$$\left. + \, 1 - \frac{\sigma_1^{2} + \sigma_j^{2}}{R_{ij}^{2}} \, \frac{r_i \, r_j}{R_{ij}^{2}} \, \sin^2 (\theta_i - \theta_j) \right\}$$

where δ_{ij} is the Kronecker delta and θ_i is the polar angle of the i-th element with respect to the rotation axis.

In previous works |10,12,32,33|, eqs. (29)-(33) were presented without any caution about their limited validity. In fact, they were used not only for linear particles (ellipsoids, lollipops and dumbells) but also for models of T-even bacteriophage |10| and polyhedral arrays |10,12| whose frictional elements are not colinear. In the case of some polyhedral structures, the results from eq. (29)-(33) have been empirically found to be exact |34| by comparison with the rigorous ones [eqs. (10), (21), (24) and (25)], while those obtained for T-even phage deviate only a few percent |35|.

At any rate, eqs. (29)-(33) can be regarded as an approximate method for the calculation of the RDC's of macromolecules with arbitrary shape. The procedure has to be performed three times, one for each RDC. In each run, the number of unknowns is N instead of 3N, so that memory requirements are smaller than in the general method by a factor of nine, and computer time is about one sixth. The author is currently investigating the error introduced by eqs. (29)-(33) for a number of irregularly shaped macromolecules.

Finally, we will comment on a simple formula for the evaluation of RDC's that results from an approximate version of the Kirkwood-Riseman theory. If $x_i'^1$, $x_i'^2$, $x_i'^3$ are the coordinates of bead i referred to the center of resistence R (at the level of this approximation one can assume that R≡D), the RDC's can be estimated as

$$\frac{6 \pi \eta_o}{k_B} \frac{D_r^k}{T} = A^{-1} \left[1 + \frac{3}{4} A^{-1} \sum_i \sum_j \sigma_i \sigma_j \cdot \right.$$

$$\left. \cdot \left(\frac{x_i'^{k+1} x_j'^{k+1} + x_i'^{k-1} x_j'^{k-1}}{R_{ij}} + \frac{x_i'^{k+1} x_j'^{k-1} + x_j'^{k+1} x_i'^{k-1}}{R_{ij}^3} \right) \right] \quad (34)$$

where

$$A = \sum_i \sigma_i \left[(x_i^{k+1})^2 + (x_i^{k-1})^2 \right] \quad (35)$$

Eqs. (34)-(35) were first obtained by Hearst |37| for models composed of identical beads and generalized later by Bloomfield and coworkers |10,25| to elements of different size. If eqs. (34) and (36) are to be used for a rough estimation of the RDC's, one certainly would wish to avoid the calculation of the hydrodynamic center via the translational and coupling tensors. To do so, D_r^k is calculated at several points, and from the resulting values the maximum, which corresponds to the actual RDC, is located. Efficient procedures for this maximization have been presented elsewhere |10,25|. Alternatively, one can use the following approximate formula |13| for the hydrodynamic (resistance or diffusion) center of the macromolecule

$$\mathbf{r}_{OR} \sim \mathbf{r}_{OD} \sim (\sum_i \sigma_i \mathbf{R}_i) / (\sum_i \sigma_i) \quad (36)$$

Eq. (34) should be used only for very rough calculation, and keeping in mind that it can yield quite disparate results. This is in contrast with the good performance of an equivalent expression for the translational diffusion coefficient, the so-called Kirkwood formula |23,27|, and shows that the RDC's are very sensitive not only to the details of the macromolecular shape, but also to the approximations embodied in their calculation.

III.6. Symmetric Top Macromolecules

Following the usual notation for small molecules, we use the qualifier "symmetric top" for rigid macromolecules having a s-fold symmetry axis plus a symmetry plane containing the axis. The N elements composing the hydrodynamic model can be rearranged in N_r rings, each having s identical elements that are equivalent from the point of view of rotation around the symmetry axis and should therefore give the same contribution, expressed in some convenient form, to the hydrodynamic property. A formalism that makes use of symmetry simplifications in the calculation of translational diffusion coefficients and RDC's of symmetric top macromolecules has been recently proposed by M. M. Tirado and the present author |35,36|, and only the most relevant results will be described here.

The equations in section II take simpler form for symmetric tops. If $x^3 \equiv z$ is the symmetry axis, then Ξ_t and $\Xi_{0,r}$ are diagonal with $\Xi_t^{11} = \Xi_t^{22}$ and $\Xi_{0,r}^{11} = \Xi_{0,r}^{22}$, and the only non-zero components of $\Xi_{0,c}$ are

$$\Xi_{0,c}^{12} = - \Xi_{0,c}^{21} \tag{37}$$

Then, equation (12) reduces to

$$x_{OR}^3 = x_{OD}^3 = - \Xi_{0,c}^{12} / \Xi_t^{11} \tag{38}$$

which gives the position of the center of hydrodynamic stress.

In the calculation of the components of the friction tensors, symmetry considerations can be introduced easily if they are expressed in cylindrical coordinates (θ,r,z). An important advance in this direction was the derivation of the cylindrical components of T_{ij} |36,38|. Eqs. (14)-(19) can be rewritten in cylindrical coordinates and then, symmetry relationships are found between the several hydrodynamic vectors and tensors. For instance, we have shown that the shielding tensors for translational motion, defined as

$$F_i = \zeta_i G_i U_i \tag{39}$$

adopt the following form when expressed in cylindrical coordinates

$$\underset{\sim}{G}_i^c = \begin{pmatrix} G_k^{\theta\theta} & 0 & 0 \\ 0 & G_k^{rr} & G_k^{rz} \\ 0 & G_k^{zr} & G_k^{zz} \end{pmatrix} \tag{40}$$

and $\underset{\sim}{G}_i^c$ is the same for all the beads belonging to the same ring, k. The components of $\underset{\sim}{\Xi}_t$ and $\underset{\sim}{\Xi}_{0,r}$ are given, in terms of the shielding tensors, by

$$\Xi_t^{11} = \Xi_t^{22} = \sum_{k=1}^{N_r} \zeta_k n_k (G_k^{\theta\theta} + G_k^{rr}) /2 \tag{41}$$

$$\Xi_t^{33} = \sum_{k=1}^{N_r} \zeta_k n_k G_k^{zz} \tag{42}$$

and

$$\Xi_{0,c}^{12} = - \Xi_{0,c}^{21} = \frac{1}{2} \sum_{k=1}^{N_r} \zeta_k n_k (r_k G_k^{zr} - z_k G_k^{\theta\theta} - z_k G_k^{rr}) \tag{43}$$

where the sums are extended over the N_r rings. n_k is the number of elements in ring k, and is euqal to s except for a single sphere on the z axis, which is regarded as a ring with $n_k=1$. $G_k^{\theta\theta}$, G_k^{rr} and G_k^{rz} are the solutions of a set of $3N_r$ linear equations, and similarly, G_k^{zr} and G_k^{zz} are those of another set of $2N_r$ equations |36|.

The rotation around z-axis can be characterized in terms of coefficients C_k that are the same for all the beads in ring k. They are obtained as the solution of a set of N_r equations |35| and Ξ_r^{33} is just a weighted sum of them:

$$\Xi_r^{33} = \sum_{k=1}^{N_r} n_k \zeta_k r_k^2 C_k \tag{44}$$

The subindex corresponding to the origin has been dropped to indicate that this term does not depend on the choice of origin.

Finally, for rotation around any perpendicular axis (x^1 or x^2) we have found that cylindrical components of the frictional force are harmonic functions of the polar angle,

$$F_i^\theta = E_k^\theta \sin \theta_i \tag{45.a}$$

$$F_i^r = E_k^r \cos \theta_i \tag{45.b}$$

$$F_i^z = E_k^z \cos \theta_i \tag{45.c}$$

where k is the ring were the i-th element is placed. The amplitudes E_k^θ, E_k^r and E_k^z in eqs. (45.a-c) are the unknowns in a set
of $3N_r$ linear equations $|36|$ and give directly the remaining
components of the rotational friction tensor

$$\Xi_{R,r}^{11} = \Xi_{R,r}^{22} = \frac{1}{2} \sum_{k=1}^{N_r} n_k \ (-E_k^\theta z_k + E_k^r z_k - E_k^z r_k) \tag{46}$$

Since for symmetric tops $\underset{\sim}{\Xi}_{R,c} = \underset{\sim}{0}$, we have for the RDC's

$$D_r^1 = D_r^2 = k_B T / \Xi_{R,r}^{11} \tag{47}$$

and

$$D_r^3 = k_B T / \Xi_r^{33} \tag{48}$$

Similar equations hold for the translational friction and diffusion
coefficients.

        If the macromolecule has a center of symmetry, it must
coincide with R, so that only the C_1's and E_k's have to be evaluated for the computation of the RDC's. When R can not be found
by simple inspection, it is necessary to calculate first Ξ^{11} and
$\Xi_{0,c}^{12}$ at any arbitrary origin 0. Next, x_{OR}^3 is obtained using eq.
(38) and the x^3 coordinates are translated to R, to which the fina
calculation of $\Xi_{R,r}^{11}$ is referred.

        It should be pointed out that the symmetry simplifications
introduced in this treatment of symmetric top macromolecules are
essentially exact, so that the values obtained from the above
equations are identical with those from the general method. The
great advantage of this treatment is that the computer require-
ents are reduced by a factor that is proportional to s^2.

III.7. Improvements in the Calculation of Rotational Diffusion
Coefficients

When applied to a single sphere of radius σ (N=1) centered at
the origin 0, eq. (25) [or eq. (30)] yields of anomalous result
$\Xi_{0,r} = 0$, while the correct value is, for any of the diagonal
components, $8\pi\eta_o\sigma^3$. If the sphere is rotating around axis, say,
x^1 and the distance from its center to the rotation axis is r,
eqs. (25) or (30) predict

$$\Xi_{0,r}^{11} = 6\pi\eta_o\sigma\ r^2 \tag{49}$$

instead of the correct result $|16|$,

$$\Xi_{0,r}^{11} = 6\pi\eta_o\ \sigma\ r^2 + 8\pi\eta_o\sigma^3 \tag{50}$$

Similar anomalies may occur for multielement models. If
element i happens to be centered at 0, then $R_i = 0$ in eq. (26) $[\rho_i = 0$
in eq. (30)$]$ and regardless of the non-zero values of the S_{ij}'s
the contribution of that element to the rotational friction coefficients is zero.   This is particularly important when the model is
composed of a few elements, or when the central element at 0 is
appreciably larger than the others so that it dominates the frictional behavior of the model.   The reason for these discrepancies
is that, although the frictional force is spread over the spherical
surface of the element, the evaluation of the frictional torque
assumes a total force, F_i acting on the element's center, so that
the corresponding torque, $T_i = R_i \times F_i$ is zero because $R_i = 0$.

The way to overcome this problem is to move the frictional
forces from the center of the elements to the surface. This could
be done by substitution of the original spheres by shells of very
small spherical elements, as described in subsection II.1.   The
obvious inconvenience is that N grows usually beyond the practical
computational limit. A more feasible remedy put forward by
Bloomfield and coworkers $|12,32|$ is to use polyhedral arrays made
up of a few identical spheres rather than shells.   They can be
put in place of each elements or, at least, of the one closest to
the origin. Bloomfield et al. found that a cubic arrangement of
eight spheres of radius 0.465 σ_i is a good compromise between the
requirement of moderately low N and a good reproduction of the
rotational coefficients of a sphere.   Examples of the application
of this strategy will be presented in the next section.

IV. APPLICATIONS

In recent years, the theory outlined in sections II and III
has been applied to the study of hydrodynamic properties of a

variety of macromolecular structures. ·I will describe here just
those related to the RDC's.

IV.1. Ellipsoids

The RDC's of prolate ellipsoids of revolution are given by the
exact formula of Perrin |39|. The purpose of model calculations
|10| was to check the performance of the theory when applied to
models made up of spherical elements, by comparison of the resulting
values with the exact ones. The ellipsoidal models are built by
placing a sphere at the center of the ellipsoid and smaller spheres
on both sides of the former, in such a way that any sphere is
tangent to its neighbors and to the surface of the ellipsoid. The
centers of the elements are aligned, and eq. (30)-(33) are valid.
The RDC's for rotation around a perpendicular axis agrees well with
Perrin values when the axial ratio q, is high (the deviation is
smaller than 5% when p>5). However, when q approaches unity and the
ellipsoid becames nearly spherical, the RDC diverges to infinity
|10|. This is due to the fact that in those cases the model consists
of a large central sphere flanked by two smaller spheres, one on
each side, so that the anomalies described in subsection III.7 take
place.

If just the central sphere of radius σ is substituted by a
cubic array of spheres with radius 0.465 σ |12| the calculations
reproduce the sphere limit for $q \to 1$ and yield the correct asymptotic
limit, as the earlier calculations |10| did, for $q \to \infty$. In addition,
the deviations in the range 1<q<5, caused by the obvious imperfec-
tions of the model, are smaller than 15% |12|.

These successful comparisons seem to indicate that the theory,
coupled to models that reproduce fairly well the macromolecular
shape and size, and with use of the substitution strategy when
necessary, is able to yield accurate results forthe RDC's. This
gives confidence in calculations for more complex structures, such
as those presented hereafter.

IV.2. Lollipops and Dumbbells

By "lollipop" we denote a structure composed by a large sphere
of radius σ_1 at which a rod of length L and radius σ_2 is attached.
The rod itself is modeled as a string of $L/2\sigma_2$ smaller spheres.
Lollipops are useful to represent some protein and nucleoprotein
complexes as well as bacteriophage. In a preliminary study |10|
the RDC for rotation around a perpendicular axis was calculated as
a function of L, σ_2 and σ_1, and the results turned out to be quite
sensitive to L and σ_1 but weakly dependent on σ_2. It was realized
that when the tail is short relative to the size of the sphere,
the hydrodynamic properties are dominated by the former. Besides,
the center ofhydrodynamic stress is in those instances close to the

center of the large sphere and therefore the failure described above can vitiate the results in this region of small L/σ_1.

The RDC's of lollipops were later recalculated |32| using the substitution strategy. For large values of L/σ_1 the results were the same as those of the first study |10|, while when $L/\sigma_1 \rightarrow 0$ the correct limit, i.e. the RDC of a single sphere (no tail) was obtained. Numerical values of the RDC's of lollipops covering reasonable ranges of L, σ_1 and σ_2 have been reported in graphical |10,32| and analytical |10| form, so that it is easy to interpolate at the desired dimensions.

Completely parallel studies have been carried out for dumb-bells. They consist of two large spheres of radius σ_1 connected by a thinner rod of radius σ_2 and length L, and can serve as structural models for nucleosomes and ribosomes. The first calcu-lations |10| are little changed when the two bulky spheres are replaced by cubic arrays |32| since the center of hydrodynamic stress is now at the middle of the rod, at a distance from either of the two spheres larger than their radii. As in the preceeding case, the RDC for a given set of L, σ_1 and σ_2 can be obtained by interpolation using graphs or analytical expressions presented in the original papers |10,32|.

IV.3. T-even Bacteriophage

Electron microscopy shows that bacteriophage T2, T4 and their relatives consist of a prolate icosahedral head, a cylindrical tail built on a hexagonal plate and six long fibers attached to the base-plate. Depending on the physicochemical properties of the solvent, T-even phage adopt either a fast form, with high sedimentation and diffusion coefficients, or a slow one, having the opposite charac-teristics. The RDC's for rotation around a perpendicular axis are 110–150 and 280–320 s^{-1} for the slow and fast forms, respectively. Phage lacking tailfibers can also be prepared and have a RDC of about 300 s^{-1}.

It is now accepted that the differences in hydrodynamic behavior is caused by a change in the disposition of the tailfibers. These are retracted toward the phage's head in the fast form while in the slow form they are extended away from the body, thereby producing an increase in the frictional drag that is experimentally observed as a diminution in the diffusion coefficients. For the purpose of hydrodynamic calculation, the head is replaced by a sphere with a radius of 55 nm. The tail is modeled with five colinear spheres of 8.0 nm and the baseplate is represented by one 20 nm sphere. Each of the six fibers is modeled as a straight or kinked array of 69 spheres with a radius of 1 nm. These dimen-sions correspond to those observed in electron microscope images.

The RDC's of the two forms and those for fiberless phage calculated by different methods are listed in Table I. Column A corresponds to results form eq. (35), which was used in the early calcu-
lations of Filson and Bloomfield |25|. It is seen how eq. (34)
predicts an important change in the RDC's in the slow-fast transition, but the actual values deviate remarkably from the experimental ones. Data in column B were obtained in Ref. 33 using eqs.
(30)-(33). The concordance with the experimental results is clearly
improved in the case of the slow forms but the values for the fast
and tailfiberless forms are yet unsatisfactory. As pointed out in
subsection III.5, these equations still embody some approximations
since they treat rotation in a monodimensional scheme. A fully
rigorous calculation has been made by Tirado and García de la
Torre |35| considering T-even phage as symmetric tops. The results
are summarized in column C, and it is evident that the deficiencies
of eqs. (30)-(33) are not the reason for the disagreement. In fact,
these equations are exact for the fiberless model, since the spherical elements representing the head, tail and baseplate are colinear.

Table I.   Rotational Diffusion Coefficients,
in s^{-1}, of T-even Bacteriophage.

Calculated[a]

Form	A	B	C	D	Experimental[b]
Slow	190–230	100–120	90–120	–	100–150
Fast	690–970	340–430	380–420	–	280–320
Fiberless	586	470	470	327	300

[a]Range of values obtained for four slightly different geometries of
the fibers in the slow form and four others for the fast form. See
Fig. 1 in Ref. 33.

[b]Range of values observed for T2, T4 and T6.

It was realized by Wilson and Bloomfield |12| that while in the
slow forms the center of hydrodynamic stress is by the middle of the
tail, quite distant from the head, in the fast and fiberless forms
is inside the head itself. As the single sphere representing the
head is by far the largest in the model, the failures of the formalism described in subsection III.7 will affect the calculated RDC's
in those cases. Therefore, they proposed to use a cubic array for
the head instead of a single sphere. As seen in column D the new

value for fiberless phage shows a substantial improvement over the previous ones. Although no concrete results have been reported yet for the fast form, similar improvements must take place, for the reason of the disagreement is the same.

IV.4. Oligomeric Structures

Many proteins in their native form present an oligomeric structure in which a few individual subunits are arranged in a polygonal or polyhedral pattern. The number of subunits in the oligomer can be obtained by straightforward measurements of the molecular weight of the monomer and oligomer, but their spatial disposition is more difficult to ascertain. In this sense, hydrodynamic properties and particularly the RDC's can be useful.

Table II lists values of the RDC's of some polygonal or polyhedral arrays composed of n identical spherical subunits |34|. As the oligomer is not much larger than the individual subunits, these are rather close to the center of hydrodynamic stress. Then, to avoid the problems described above, each spherical subunit of radius σ was in turn modeled as a cubic array of spheres with radius 0.467 σ. In Table II, D_r^3 is the diffusing coefficient for rotation around the axis of highest symmetry, x^3, and $D_r^1 = D_r^2$ correspond to any axis perpendicular to it. For the purpose of comparison we have included in Table II the translational diffusion coefficients, D_t. All the values are normalized to those for a single spherical subunit (monomer). It is evident from Table II that the RDC's are exceedingly more sensitive to differences in n and, for fixed n, in geometry than D_t.

IV.5. Cylinders

Many biopolymers and macromolecular complexes in solution exhibit a rodlike shape, and therefore can be hydrodynamically considered as cylinders of length L and diameter d. The RDC for rotation around a perpendicular axis, D_r^1, can be formulated in terms of the dimension of the cylinder as

$$\frac{\pi \eta_o L^3 D_r^1}{3 k_B T} = \ln p + \delta_1 \tag{51}$$

where $p = L/d$ and δ_1 is a function of p that reaches an asymptotic value when $p \to \infty$ he obtained $\delta_1 = -0.447$, and this value has been confirmed later |45|. In the range of short to moderately long cylinders, he combined theoretical results for large p and experimental values of macroscopic cylinders at low p to get the following

Table II.   Translational and Rotational Diffusion
Coefficients of Oligomeric Subunit Structures

Number of subunits, n.	Geometry	D_r^{1}/D_r^{sph}	D_r^{3}/D_r^{sph}	D_t/D_t^{sph}
1	Sphere	1.000	1.000	1.000
2	Dimer	0.264	0.564	0.725
3	Triangle	0.236	0.175	0.621
3	Colinear	0.108	0.397	0.586
4	Square	0.159	0.119	0.550
4	Tetrahedron	0.165	0.165.	0.564
4	Colinear	0.056	0.306	0.500
5	Pentagon	0.111	0.084	0.491
5	Bipyramid	0.118	0.156	0.520
6	Hexagon	0.081	0.061	0.444
6	Octahedron	0.115	0.115	0.494
6	Trigonal Prism	0.100	0.109	0.482
8	Cube	0.075	0.075	0.433

semiempirical expression:

$$\delta_{\perp} = -0.88 + 7 \left[(\ln 2p)^{-1} -0.28 \right]^{2} \tag{52}$$

Using the formalism for symmetric top macromolecules, we |35| have
recently calculated D_r^{1} for cylindrical models consisting of a stack
of identical rings, each composed of s touching beads of radius σ.
For fixed values of L and p, results corresponding to decreasing σ
were extrapolated to the $\sigma = 0$ limit, where the model reduces to a
perfect cylinder. We were unable to obtain the exact result for
$p \rightarrow \infty$ because for very long lengths the computer calculation is
prohibitive. Our results for moderate p turned out to be in
disagreement with Broersma's, a conclusion similar to that from a
previous study of the translational friction coefficients. Fig. 1
displays our results along with those from other theoretical works

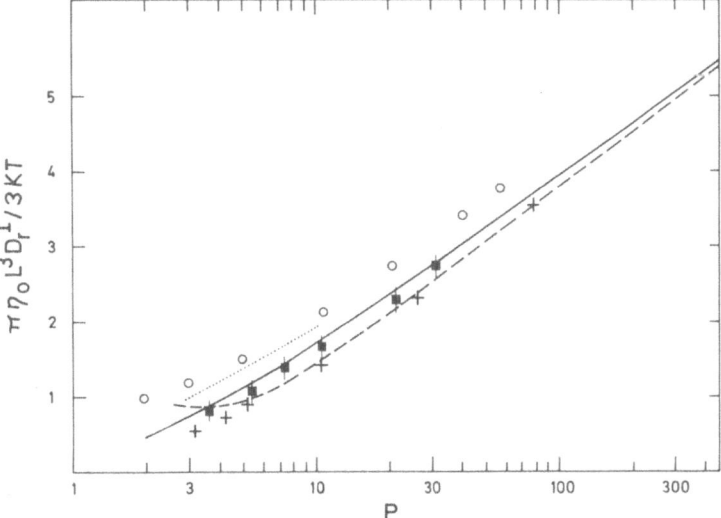

Figure 1. Semilogarithmic Plot of the Rotational Diffusion
Coefficient, Normalized as Indicated of Cylinders Versus the
Length-to-diameter Ratio, p. (■), experimental results by
Broersma |44| with indication of the estimated errors as
vertical bars; (- -) Broersma's semiempirical equation (52);
(.......) results of Yoshizaki and Yamakawa |46| ; (○) results
for a rod of touching spheres obtained by Nakajima and Wada
| 5 | ; (——) results by Tirado and García de la Torre |35| .

and Broersma's experimental data for macroscopic cylinders. For
p < 50 our values are closer to the experimental points than all
the other theoretical predictions, including eq. 52 in which those
points were taken into account. The discrepancies between the
several theories are less serious in the p < 50 region since as p
increases, ln p in the right-hand side of eq. (51) becomes much
greater than $\delta_{\perp}$.

We have also considered |35| the rotation around the cylin-
drical axis. The corresponding RDC, D_r^3, can be related to the
dimensions of the cylinder by means of the following equation:

$$k_B T/A_o \pi \eta_o LR^2 D_r^3 = 1 + \delta_{||} \tag{53}$$

where A_o = 3.841 and $\delta_{||}$ is a function of p that takes values
between 0 for p = ∞ and 0.294 for p = 2.

In a recent paper, Yamakawa and Yoshizaki |46| have calculated
the RDC's of cylinders with rounded ends. Actually, their model
consists of a cylinder capped with hemispheroids at both ends. If
L is the total length, including the caps, they found that the end
effects, i.e., the influence of the shape of the caps, is rather
small even for values of L/d as low as 5.

IV.6. Bent Rods

When a incipient extent of denaturation takes place in a
helical biopolymer, this can no longer be considered as a rod or
cylinder. Instead, it must be regarded as a bent or broken rod.
Broken-rod polypeptides can be syntesized from bifunctional initi-
ators, and also occur naturally such as in the bacteriophage
tailfibers and the myosin rod. The particle may be hinged at the
joint of the two rodlike portions, or arms, and in such a case
flexibility effects would have to be included in the theory. This
has been done recently |20,21| but neglecting hydrodynamic interact-
ions between the two arms.

Other workers, including the present author |11,13,48|, have
studied bent rods as rigid particles with a fixed angle χ between
the two arms, and flexibly hinged rods are treated in a first-order
approach by averaging over χ. The two arms, of diameter d and
lengths L_1 and L_2, are represented by strings of colinear, touching
spheres of diameter d. In Fig. 2 we give as examples of results
the RDC's of bent rods with L_1 = L_2, $(L_1+L_2)/d$ = 72 and 24, and
varying χ. The results are in both cases normalized to the
rotational diffusion coefficinet of a straight rod (χ=180°) of the
same length denoted as D_r^{str}. In this form the results are little
sensitive to the length-to-diameter ratio, as seen in Fig. 2. Axis

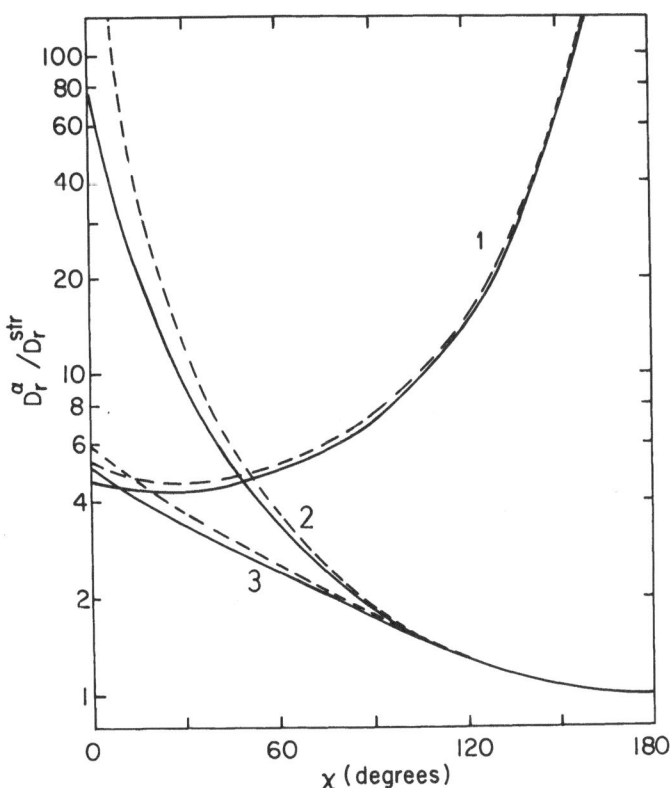

Figure 2. Rotational Diffusion Coefficients D_r^1, D_r^2 and D_r^3 of
Bent Rods with Arms of Equal Length and Varying χ,
Relative to the Rotational Diffusion Coefficient of a
Straight Rod, D_r^{str}.
(————), $L/d = 24$; (– – – –), $L/d = 72$.

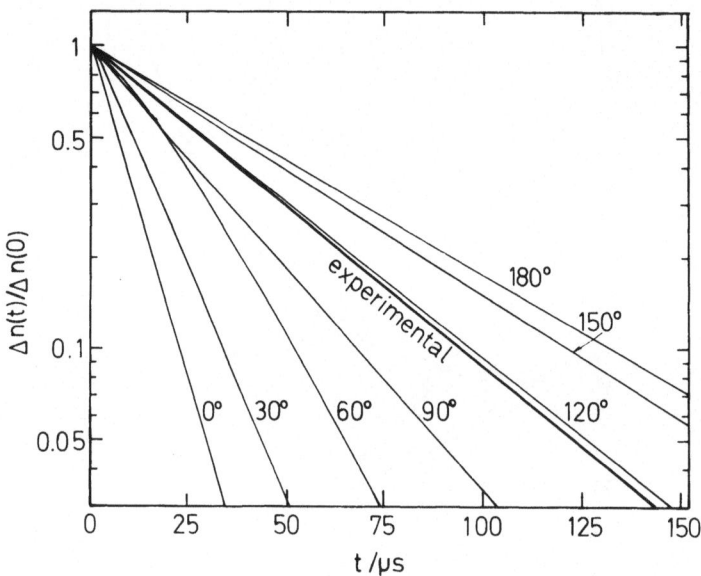

Figure 3. Birefringence Decay of a Bent Rod with L = 1440 Å,
$L_1 = L_2 = 720$ Å at 3°C. Numbers attached to the thin theoretical
curves are the values of the angle between the two arms. The thick
curve, labelled as experimental, corresponds to experimental
results of Highsmith et al. | 50 | for the myosin rod.

x^2 is the bisector of χ, x^1 is perpendicular to x^2 in the particle's plane and finally x^3 is perpendicular to that plane. The values of $D_r^{\,3}$ when $\chi \simeq 0$ and $D_r^{\,1}$ when $\chi \simeq 180°$ should be considered only as rough estimates because the distance from the spherical elements to the rotation axis is not larger than their radius.

Using the expressions of Wegener et al. |1|, the RDC's of bent rods can be combined with electro-optic properties to get the birefrigence decay curve. Figure 3 shows decay curves for a rod 144 nm long, 2 nm thick and kinked at its middle point with a variable angle χ between the two arms. In the calculation a pure permanent-dipole orientation mechanism was assumed and the temperature was taken as 3°C. |49|.

These results are to be compared with experimental data for the myosin rod. Highsmith et al. |50| have found that the electric birefringence of myosin rods decays as a single exponential with a relaxation time of 41.2 μs, which is shorter than expected for a straight rod of that length, and this evidences that the myosin rod is kinked and probably hinged. Recently García de la Torre and Bloomfield |51| have proposed that flexibly hinged rods can be treated in a first approximation as bent rods in which χ corresponds to some conformational average. This average $<\chi>$, can be obtained from hydrodynamic properties, and its departure from 180° is a measure of flexibility. Comparison of the experimental decay with the theoretical ones in Fig. 3 shows that for myosin rod $<\chi> \simeq 110°$, in good agreement withthe distribution of angles observed by electron microscopy |52|. It is also interesting to note that in spite of the high hydrodynamic and electro-optic anisotropy of bent rods, the decay is clearly single-exponential for $\chi > 90°$ and $\chi < 60°$.

ACKNOWLEDGEMENTS

The author is deeply indebted to Professor Victor A. Bloomfield for his incessant encouragement and advice over the past five years. Research of the author at the University of Minnesota (U.S.A.) was supported by grants from NIH (GM 17855) and NSF (PCM-7522728 and PCM-7806777), and fellowships to J.G.T. from the U.S.A.- Spain Joint Committee for Scientific and Technological Cooperation. The collaboration of graduate students M. M. Tirado, J. M. García Bernal, P. Mellado and A. Jiménez at the University of Extremadura (Spain) is also acknowledged.

REFERENCES

1. W. A. Wegener, R. W. Dowben and V. J. Koester, J. Chem. Phys.
    70:622 (1979).
2. G. G. Belford, R. L. Bedford and G. Weber, Proc. Natl. Acad.
    Sci. U.S.A. 69:1392 (1972).
3. R. Pecora, J. Chem. Phys. 49:1036 (1968).
4. J. A. McCammon and J. M. Deutch, Biopolymers 15:1397 (1976).
5. H. Nakajima and Y. Wada, Biopolymers 16:875 (1977).
6. H. Nakajima and Y. Wada, Biopolymers 17:2291 (1978).
7. E. Swanson, D. C. Teller and C. de Haën, J. Chem. Phys. 68:5097
    (1978).
8. D. C. Teller, E. Swanson and C. de Haën, Adv. Enzymol. 61:103
    (1979).
9. J. García de la Torre and V. A. Bloomfield, Biopolymers 16:1747
    (1977).
10. J. García de la Torre and V. A. Bloomfield, Biopolymers 16:1765
    (1977).
11. J. Garcia de la Torre and V. A. Bloomfield, Biopolymers 17:1605
    (1978).
12. R. W. Wilson and V. A. Bloomfield, Biopolymers 18:1205 (1979).
13. J. M. García Bernal and J. Garcia de la Torre, Biopolymers
    19:751 (1980).
14. S. C. Harvey and J. García de la Torre, Macromolecules 13:in
    press (1980).
15. J. García de la Torre and V. A. Bloomfield, Quart. Rev.
    Biophys., to be published.
16. J. Happel and H. Brenner, Low Reynolds Number Hydrodynamics
    (Nordhoff, Leyden, 1973) chapter 5.
17. H. Brenner, J. Colloid Interface Sci. 23:407 (1966).
18. O. Hassager, J. Chem. Phys. 60:2111 (1974).
19. S. I. Abdel-Khalik and R. B. Bird, Biopolymers 14:1915 (1975).
20. S. C. Harvey, Biopolymers 18:1081 (1979).
21. W. A. Wegener, Biopolymers, to be published.
22. M. Tirrell and J. Torkelson, J. Rheol. 23:751 (1979).
23. J. Riseman and J. G. Kirkwood, in Rheology, Vol I., F. Eirich
    Ed., (Academic New York, 1956(, ch. 13.
24. J. G. Kirkwood, J. Polym. Sci. 12:1 (1954).
25. D. P. Filson and V. A. Bloomfield, Biochemistry 6:1650 (1967).
26. V. A. Bloomfield and D. P. Filson, J. Polym. Sci. C25:73 (1968).
27. V. A. Bloomfield, W. O. Dalton and K. E. Van Holde, Biopolymers
    5:135 (1967).
28. G. Koopmans, B. J. Van der Meer, P. C. Hopman and J. Greve,
    Biopolymers 18:1533 (1979).
29. R. W. Wilson and V. A. Bloomfield, Biopolymers 18:1543 (1979).
30. E. Swanson, D. C. Teller and C. de Haen, J. Chem. Phys. 72:1623
    (1980).
31. H. Yamakawa and J. I. Yamaki, J. Chem. Phys. 58:2049 (1973).

32. V. A. Bloomfield, J. García de la Torre and R. W. Wilson, in Electro-optics and Dielectrics of Macromolecules and Colloids, B. R. Jennings, Ed. (Plenum, N.Y. 1979), 183.

33. J. García de la Torre and V. A. Bloomfield, Biopolymers 16:1779 (1977).

34. J. M. García Bernal and J. Garcia de la Torre, Biopolymers to be published.

35. M. M. Tirado and J. Garcia de la Torre, J. Chem. Phys. 73:in press (1980).

36. M. M. Tirado and J. García de la Torre, J. Chem. Phys. 71:2581 (1979).

37. J. E. Hearst, J. Chem. Phys. 38:1062 (1963).

38. J. García de la Torre and V. A. Bloomfield, Biophys. J. 20:49 (1977).

39. F. Perrin, J. Phys. Rad. 5:497 (1934).

40. W. Boontje, J. Greve and J. Blok, Biopolymers 16:551 (1977).

41. W. Boontje, J. Greve and J. Blok, Biopolymers 17:2689 (1978).

42. P. C. Hopman, G. Koopmans and J. Greve, in Electro-optics and Dielectrics of Macromolecules and Colloids, B. R. Jennings Ed. (Plenum, N.Y. 1979), 197.

43. V. A. Bloomfield, K. E. Van Holde and W. O. Dalton, Biopolymers 5:149 (1967).

44. S. Broersma, J. Chem. Phys. 32:1632 (1960).

45. H. Yamakawa, Macromolecules 8:339 (1975).

46. T. Yoshizaki and H. Yamakawa, J. Chem. Phys. 72:57 (1980).

47. A. Teramoto, T. Yamashita and H. Fujita, J. Chem. Phys. 46:1919 (1967).

48. G. Wilemski, Macromolecules 10:28 (1977).

49. P. Mellado and J. García de la Torre, unpublished.

50. S. Highsmith, K. M. Kretzchmar, C. T. O'Konski and M. F. Morales, Proc. Natl. Acad. Sci. USA 74:4986 (1977).

51. J. García de la Torre and V. A. Bloomfield, Biochemistry, in press (1980).

52. A. Elliot and B. Offer, J. Mol. Biol. 123:505 (1978).

THE ROTATIONAL DIFFUSION FUNCTION IN AN ELECTRIC FIELD

S. P. Stoylov

Institute of Physical Chemistry
Bulgarian Academy of Sciences
1040 Sofia, Bulgaria

Modern electro-optic theory is built exclusively on the assumption that the variation in the optical properties of disperse systems (both molecular and colloidal) is caused by the orientation of the molecules or the particles. Practically, there are very few works where orientation not due to the interaction of the electric field with electric moments of the particles or molecules is considered. Other possible causes for orientation like electro-osmosis, electrophoresis, sedimenation, convection, electric field inhomogeneities and interfacial polarizability were thoroughly discussed in the extensive review of Wilfried Heller[1] as early as 1942. Of these, interfacial polarizability (interfacial induced moment), convectional, electrophoretic, and sedimentational orientation deserve special attention. It is not excluded that in many cases of reported peculiar electro-optics effects an explanation could be found in these "non-conventional" orientation effects which, with the rare exception of the interfacial polarizability orientation, have not attracted the necessary attention and are often very easily forgotten.

The orientational effect due to the interaction of the applied electric field with the interfacial electric polarizability or with the interfacial induced dipole moment could lead not only to changes in the magnitude of the electro-optic effect but also to changes in its sign. Along with the dependence of the electro-optic effect on the zeta-potential and the anisotropy of the interfacial charge and potential resulting from the non sphericity of the particles[1] there are corrections to be introduced similar to the relaxation correction in electrophoresis[2]. Systematic theoretical studies on this problem have been done by Dukhin, Shilov and collaborators summarized in ref. 2.

105

When an electric field is applied to a disperse system it is possible to have optical effects which are not connected with molecular or particle orientation. While conformational (structural)[3,4], chemical[5], deformational[4] and permeational[6] electro-optical effects have gotten some attention mainly in connection with their important impact on the functions of biological macromolecules and membranes, the aggregational electro-optic effects, both reversible[7] and irreversible[8], have only been demonstrated. Different electro-optic methods provide several ways of checking the presence of one or other effect in the experiment. When interpreting experimental results from electro-optic studies, especially when peculiar effects are encountered, the possible presence of one or a combination of several non-orientational effects should not be ignored. This very important problem for electro-optic methods will not be treated here since efforts will be mainly concentrated in this paper on the discussion of the rotational diffusion function in an electric field and its impact on electro-optic theory in the different optic and "energetic" (ratio of energy of orientation to the energy of thermal motion) approximations.

The rotational diffusion function in an electric field is important for more than the theory of electro-optic phenomena. The measurement of conductivity[9], dielectric permittivity[10], viscosity[11] and other parameters of disperse systems in an electric field and the respective theories[9,10] are closely connected, like the electro-optic theories, with the rotational diffusion of the molecules or the particles in the electric field. In addition, some theoretical development like the expressions for the transient processes following the switching on and off of the electric field by energies of orientation much higher than the energy of thermal motion (saturating electric fields) were worked out first for the case of conductivity in an electric field[9]. Furthermore, when speaking about the rotational diffusion in an electric field connected with the theory of the electro-optic effects, it should be kept in mind that the majority of the conclusions drawn are valid for the other electro-orientational electro-hydrodynamic and other effects and vice versa.

THE ROTATIONAL DIFFUSION FUNCTION

The knowledge of the distribution of the orientation of the axes of non-spherical molecules or particles as a function of the parameters of the applied electric field and of the time is the principal problem in the calculation of the different electro-optic effects. Very often this distribution is expressed by the so-called orientational distribution function $f(\Theta,t)$, where Θ is the angle between the symmetry axis of the molecule or particle (here the discussion is restricted to the case of rigid, cylindrically symmetric molecules or particles, which as it will be seen

further can lead to qualitative differences in the expressions derived for the electro-optical functions, and t is the time. The
rotational diffusion function in the general case can be derived by
solution of the partial differential equation for the rotational
diffusion process in an electric field:

$$(1/D)\left|\partial f(\Theta,t)/\partial t\right| = (1/\sin\Theta)(\partial/\partial\Theta)\left|\sin\Theta\ (\partial f(\Theta,t)/\partial\Theta)-\right.$$
$$\left.-M\ f(\Theta,t)/kT\right|, \tag{1}$$

where D is the rotational diffusion constant, characterizing the
reorientation of the symmetry axis of the molecule of particle
around its transverse axes, M is the torque produced by the application of the electric field and k, T are respectively the Boltzmann
constant and the absolute temperature.

Equation (1) has been solved for the case of no electric field
applied on the solution i.e. when M=0 even for the case of axial
particles or molecules whose cross section perpendicular to the sym-
merty axis is not a circle[12]. The reader who is interested in the
more extended physical and mathematical treatment of the rotational
diffusion process is advised to consult the excellent review of
Ridgeway[12] or the monograph of Pokrovskii[13]. O'Konski and Shepard[30]
have shown that inertial effects are unimportant for the particles
or molecules most frequently studied under the usual experimental
conditions. In the stationary case, the rotational distribution
function coincides with the Maxwell-Boltzmann distribution function
when $M=M(\Theta)$.

Here and in the entire discussion it is assumed that the concentration of particles is so low that there are no interactions
among the particles or molecules. The interesting case of
rotational diffusion by interacting particles has been treated by
Mazo[14] and there exist some preliminary electro-optical experimental
data[15] of considerable interest.

In the more general non-stationary case, when an electric
field is applied to the system, the molecules or particles orient
to an extent determined by the energy of interaction of the electric
parameters (moments) with the electric field. The orientation takes
place about the center of hydrodynamic drag, which in the general
case may not coincide with the center of gravity. In the case
when some molecules of the solvent (or the disperse medium, respectively), which are assumed to be much smaller than the orienting particles or molecules, are bound to the latter the whole complex is generally regarded as a rigid hydrodynamic unit at ordinary
conditions.

The most uncertain part in the resolution of the diffusion
problem is how to specify the torque $M(\Theta,t)$ in the case of particles

or macromolecules with permanent dipole moments, or with predominantly interfacial polarizability in the case of a conducting
solvent and charged particles or macromolecules; this is by far
the most frequently met experimental situation. It could be supposed that the permanent dipole moment under such conditions will
be screened by ions in the medium, which will decrease the value of
the calculated permanent dipole moment and influence the orientation
dynamics[2] similarly to the internal field function[4]. Particles or
macromolecules with interfacial electric polarizability will not
only change the dynamics of orientation, depending on the relaxation times of the different types of polarizabilities, but also
will require detailed consideration of the dynamics of the interfacial electric charge[2], as is done in the theory of the electrophoresis. A promising attempt of Shilov[2] to overcome this difficulty, using the concepts of irreversible thermodynamics is of
considerable interest. Instead of studying the orientation of the
particles due to their interaction with the applied electric field
he follows the inverse problems of calculating the electric field due
to the rotation of the charged particle[5]; these calculations go beyond the limits of the double layer around the particles. In the
general case of particles or molecules having both permanent dipole
moments and interfacial electric polarizability, it seems that their
consideration as independent parameters, as this is done in the
great majority of theoretical electro-optic work is a rather un-
realistic approximation. It is scarcely necessary to emphasize
here that the interfacial induced dipole moment could not only
directly complicate the analysis of the rotational diffusion process by the orienting torque, but also indirectly by the enhanced
interactions that may become essential at quite low concentration.
The complications connected with the interfacial electric polarizability and the conductance of the medium will not be further considered in this paper. The interested reader is advised to consult
the papers of Dukhin and Shilov or their review in ref. 2.

ELECTRO-OPTIC EFFECTS AND THE ROTATIONAL DIFFUSION FUNCTION

Once the rotational diffusion function (or which is practically the same, the orientational distribution function) $f(\Theta,t)$ is
calculated, the required electro-optic effect α could be derived
by averaging the respective optical function $A(\Theta,t)$ by the orientational distribution function:

$$\alpha = \int_W A(\Theta,t) \ f(\Theta,t) \ dw \ , \qquad\qquad (2)$$

where $\int_W \ .....dw$ indicates integration over all directions.

How complicated the expressions for the electro-optic effects
will be depends essentially on the type of optic function used or,
more correctly stated, on the optical approximation used. On the
other hand the correctness of interpretation of the experimental
data will depend on the extent to which the real molecules or particles measured correspond to the optical approximation used in the
derivation of the respective theoretical expressions. In this
paper some new results will be presented and a review of the current state of the electro-optic theory in this field attempted.
The more complicated problem of the rotational diffusion function
in an electric field for particles or molecules with interfacial
electric polarizability (in conducting medium) will not be treated.

FIRST ATTEMPTS FOR GENERAL ELECTRO-OPTIC THEORY

The vast majority of all the electro-optical theoretical work
has been done for the case of the electric birefringence. Therefore it was quite logical to try to clarify under what conditions
the theory developed for the case of the electric birefringence
could be used for the interpretation of the other electro-optic
effects especially for electric dichroism and electric light scattering. These last two electro-optic effects have no rivals at
the moment among the other electro-optic effects with respect to
the extent of application.

On the basis of general considerations, noting the way expressions for electro-optical effects are obtained by averaging
the respective optical function with the orientation distribution
function, Holcomb and Tinoco[16] for the first time recognized the
possibility of writing equations which are one and the same for
electric birefringence and electric dichroism. It is very surprising to see that this important generalization has scarcely had
any effect on the electro-optic development that followed. In our
opinion, this is explained to some extent by their failure to present their result in sufficient detail. Later Yoshioka and
Watanabe[17] revived this idea and wrote the general expressions in
a much more detailed manner. But even after this work, whose
impact on electro-optical studies proved to be comparatively much
larger, one could encounter reviews (and even monographs) in which
electro-optic theory of the separate electro-optic effects was
presented independently.

The first attempt to include electric field light scattering
in the general theory was made by Stoylov[18]. This was done on the
basis of the observed similarity between existing expressions for
electric birefringence on the one hand and the expressions for
electric light scattering when particles are sufficiently short
(wavelength of light sufficiently long respectively) or the angle
of observation is sufficiently small on the other hand, that is,

when

$$K = (2\Pi/\lambda) \; 1 \; \sin(\Theta'/2) \ll 1 \; , \tag{3}$$

where λ is the wavelength of light in the particle or the molecule, 1 is the length of the particle, assumed to be a very elongated cylinder and Θ' is the angle of observation (see figure 1) i.e. the angle between the incident and scattered beams. In other words, these are weakly scattering particles in the Rayleigh-Debye-Gans[19] approximation. It seems that although it was made on quite formal ground's as it will be seen further, this generalization has played a stimulating role both in the development of the theory and the experiment. It would be of interest in the future to consider the connection of this theory with the original approach of Herbert[20] to the discussion of the low angle approximation of alternating electric field light scattering.

In this approximation for the electric light scattering, all the expressions known for the electric birefringence contained in the majority of the electro-optic literature, are directly applicable. The integration indicated in (2) leads to an expression for the generalized electro-optic effect representing a product of two functions – one function only of the optical and geometrical parameters of the particles or molecules and the other function only of the parameters important for the orientation.

$$\alpha = \alpha_\infty \;\; \bullet \; F \; (\; \frac{U}{kT} \; , \; \frac{t}{\tau}, \; \omega\tau, \; \ldots)$$

$$\begin{cases} (\Delta I/I_o)_\infty = (K^2/9) \; , \\[2mm] \Delta \; n_\infty = 2\pi C_v \; (g_1 - g_2)/n \; , \\[2mm] (\Delta A/A_o)_\infty = (2/3)(\varepsilon_{ap}/\bar\varepsilon) \; \text{or} \; 3 \cos \Psi - I \; , \end{cases}$$

where U is the energy of orientation, τ is the relaxation time for the disorientation of the particles, ω is the angular frequency, ΔI is the variation in the intensity of the scattered light with electric field, I_o is the intensity of scattered light without field, Δn is the electric birefringence, C_v is the volume concentration, g_1 and g_2 are respectively the optical polarizabilities per unit volume along the symmetry axis and perpendicular to it, n is the refraction index of the system, ΔA and A_o are respectively the variation of the absorbance in an electric field and that without the field, ε_{ap} is absorptivity anisotropy, $\bar\varepsilon$ is the average absorptivity, Ψ is angle between the symmetry axis and the transition moment and the subscript ∞ stands for full orientation. For the calculation of the second function F, frequently called

degree of orientation function, the calculation of the rotational
diffusion function is an essential step. Practically, this second
function is the unifying link of the general theory and it is one
and the same for the three electro-optical effects discussed and is
also applicable to other electro-optic effects when an appropriate
definition of the respective optical function is used. For details, the interested reader could consult references 18, 2 and 4.

ELECTRO-OPTIC THEORY IN THE RAYLEIGH APPROXIMATION

The so-called general electro-optic theory discussed above has
two very unpleasant defects. One is that the optical approximations for the different electro-optic effects are different.
Therefore, while electric light scattering is developed in the
Rayleigh-Debye-Gans approximation, electric birefringence and
electric dichroism are in the Rayleigh approximation. The second
and possibly more important drawback is that there is no logical
derivation where more general optical functions are incorporated
in the calculation.

The first problem has practically been solved by Kielich[21]
and Farinato[22] who considered the electric light scattering effect
in the Rayleigh approximation and provide the expressions for the
optical function. It has to be stated here that in the Rayleigh
approximation quite similarly to the case of the "formal" general
theory, the electro-optic effect is expressed as a product of two
functions, each of which is a function only of one type of par-
ameter:optical (in the case of the RDG approximation also geometrical) and orientational. The first function at appropriate
normalization corresponds to the respective optic effect by fully
oriented particles or molecules. The optical function for the
electric birefringence and for the electric dichroism is given by
(4) and for the electric light scattering there are several expressions, depending on the state of polarization of the incident
and the scattered beams. So, for example, for vertically polarized incident and scattered beams at an angle of observation 90°
the following expression[18,21], exists:

$$(\Delta I/I_o)_\infty = (20 \ \delta + 16 \ \delta^2)/(5 + 4 \ \delta) , \qquad (5)$$

where δ is the optical anisotropy which is assumed positive
and coincides with the electrical anisotropy in direction. This
development is restricted to the steady state electric light
scattering effect; some results are contained in ref. 18. As one
can see below the case of the transient effect is not so simple
and attempts have been made[18] to present it earlier. There is a
practical question which arises. What approximation is better to
use when interpreting the electric light scattering measurements:
the RDG (Rayleigh-Debye-Gans) or the Rayleigh? The question is

still more complicated if one takes into consideration the fact that
there are two ways the RDG approximations is used: one, as in the
"formal" general theory following requirement (3), and the other,
the full RDG approximation as used in the electric light scattering
theory which cannot be incorporated in the "formal" general electro-
optic theory[18].

In the case of biopolymers Kielich[21] estimates that for a
longest dimension no longer than 1000 Å the electro-optic theory
in the Rayleigh approximation is quite reasonable, while for dimensions as long as 3000 Å there could be an error of 50%. In the
RDG approximation, there is the possibility to make the misfit less
by decreasing the angle of observation θ'. At an angle of 90° for
biopolymers the "full" RDG theory will certainly lead to smaller
errors for longest dimensions not exceeding 3000 Å while the theory
in the RDG approximation from the "formal" general theory will probably lead to larger discrepancies. In general, one has to prefer the
Rayleigh approximation when working with small, optically anisotropic
particles, for which the knowledge of the geomerty is not very important and the RDG approximation for optical isotropic particles,
when knowledge of their geometry is important. For the future it
could be of some interest to compare data obtained using the different approximations with data obtained from independent methods.
This could be done with existing experimental data.

The second more complicated problem has practically been solved
by Stoimenova[23], who introduced as the "basic" optical function the
light scattering matrix as defined by van de Hulst[24].

$$\begin{pmatrix} E_1 \\ E_r \end{pmatrix} = \begin{pmatrix} S_2 & S_3 \\ S_4 & S_1 \end{pmatrix} \frac{e^{-ikR + ikz}}{ikR} \begin{pmatrix} E_{1_0} \\ E_{r_0} \end{pmatrix}, \tag{6}$$

where E_1, E_r, E_{1_0} and E_{r_0} are, respectively, the electric field of
the light beam perpendicular (r) and parallel (1) to the plane of
observation (plane defined by the incident and the scattered beams)
of the incident and the scattered light. S_1 and S_4 are the four
components of the light scattering matrix. R is the distance to
the observer, z is the direction of propagation of the incident
light and k is the wave number $k = (2\pi/\lambda)$ – see figure 1. An
essential result of this study is that outside the Kerr region
(i.e. beyond low degrees of orientation) even for monodisperse
particles whose cross sections perpendicular to the symmetry axis
is a circle, the decay curve (following the switching off of the
electric field) could be biexponential–equal to $(\exp(-6DT)$ plus
$\exp(20\ Dt))$, the magnitude of the second exponent growing with
the increase of the particle optical anisotropy. Details of

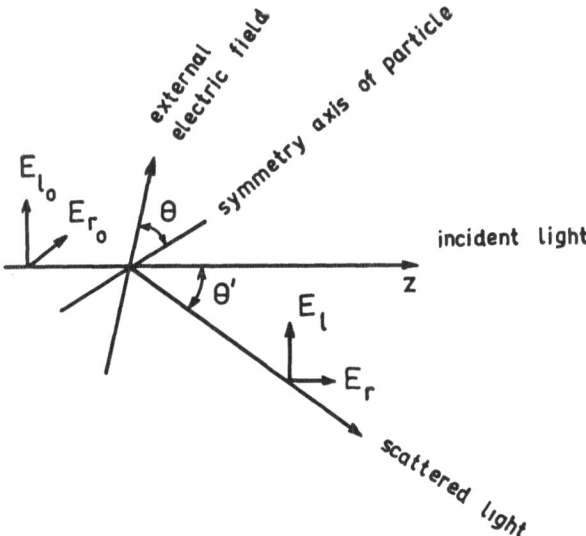

Fig. 1:   Definitions of Quantities Used in the Discussion of
          Electric Field Light Scattering.

this analysis can be found in ref. 25. Of interest is the fact that similar biexponential decays have to be expected also for the electric birefringence and the electric dichroism in so far as light intensities are measured in these effects as well.

ELECTRO-OPTIC THEORY FOR PARTICLES OR MOLECULES LARGER THAN THE RAYLEIGH RESTRICTION

The widely used current theory for the electric birefringence and for the electric dichroism is a theory in the Rayleigh approximation. It is of considerable interest to know how large a particle could be well described by this approximation and what are the deviations from the existing expressions. An analysis of this problem has been made by Stoimenova[23] both in Stevenson (developing the opticalfunction in series by wave number) and Ikeda (developing the optical function in series by dielectric permitivity of the particle) approximations. Analytical expressions have been derived for the three electro-optical effects both in the stationary and in the transient regimes. In this way an essential generalization of the electro-optic theory has been achieved, this time not a "formal" but a real one.

The most important result of this analysis is the conclusion that for particles comparable in dimensions to the wave length of light, the decay of birefringence or other electro-optic effect after the electric field is turned off for monodisperse systems with high degrees of orientation is not monoexponential[23]. This contradicts the conclusion of Ravey[26], who on the basis of similar calculations concluded that both for low and high degree of orientations where will not be significant differences from the interpretation in the RDG approximation.

These new results make necessary the reconsideration of methods suggested for polydispersity characterization from the decay curves of electro-optical effects of fully oriented particles. These new results are also of importance for the other electro-optical and electrical jump techniques, where nonmonoexponential decays are a source of chemical or conformational information. In other words, the application of a high electric field leading to high degrees of orientation 'DEFORM' the decay curves. More information on this question can be found in references 27 and 28. A practical question is when this' 'deformation' is probable how it can be detected. In general, with an increase in the values of geometrical and optical anisotropies, the 'deformation' should grow. In addition, when there are inexplicable fast transients or when the ratio of relaxation times is 3.2 in an experiment, one has to be aware that this could be connected with the 'deformation'.

ELECTRO-OPTIC THEORY AT LOW DEGREES OF ORIENTATION

All complications in the electro-optic theory are very noticebly
diminished in the case of low degrees of orientation i.e. when the
energy of interaction of the electric parameters of the particles
or molecules with the electric field (i.e. plus orientation energy)
is much smaller than the energy of the thermal motion. In this
case, developing the expression under the integral in (2) in series
using the electric field strength as parameter, it is possible in
the most general case to have the expressions for the electro-optic.
effect as a product of two functions: one function only of the
geometrical and optical parameters and the other-function only of
the orientational parameters.

In this very important, "energetic", approximation the decay
curves are monoexponential for monodisperse systems of spheroidal
particles for all optical approximations. It is possible to derive
even more general relations which show that already calculated
orientation functions for the dispersion and for the transient processes[4] could be applicable for particles[29] arbitrary in dimensions
and with axial symmetric optical properties. It is also logical
to expect that in this approximation, the complications due to the
interfacial nature of the orienting torque, could become less
critical. It seems that following ref. 2, a really simplifying approximation in this respect is to work at low surface electric potential (or low surface charge) of the particles or molecules which
could be achieved by appropriate choice of the experimental conditions.

CONCLUSION

The main problem in the calculation of the rotational diffusion function in an electric field is to find the correct form
of the orientational torque for the interfacial induced dipole moment
of conducting particles or molecules; such molecules are the usual
ones in the majority of the experimental electro-optic studies.
Here, one way for overcoming the complications is to use electrophoretic theory. However one must remember that electrophoretic
theory does not take correctly into account the collective dynamics
of the interfacial electric charge. So a corrected electrophoretic
approach has to be followed. Supplementary complications could
come from the permanent dipole moment, whose screening in a conducting medium makes the manifestation of the interfacial electric
polarizability and permanent dipole moment mutually interrelated.
The experimental testing of the theories with well defined samples
at low degrees of orientation and low surface electric potentials
is of great importance for progress in this field of the electro-
optics.

Another theoretical and experimental problem of great interest is the study of rotational diffusion of interacting particles. Some preliminary results of ours seem quite encouraging.

Serious complications arise in the theory when high degrees of orientations and large particles are encountered. These complications deserve further attention since in a number of cases the higher degrees of orientations and larger particles are unavoidable in order to obtain sufficiently high signal to noise ratios in experiments. Numerical calculations and tabulation of the different approximation, appropriately experimentally verified could serve as a firmer basis for broader applications of electro-optics and for elucidation of the relative specificity of each electro-optic method.

REFERENCES

1. W. Heller, Rev. Mod. Physics 14:390 (1942).
2. S. P. Stoylov, V. N. Shilov, S. S. Dukhin, S. Sokerov, I. Petkanchin, (S. S. Dukhin, ed.), Electro-optics of Colloids Naukova Dumka, Kiev (1977).
3. H. Watanable, K. Yoshioka, in "Electro-optics and Dielectrics of Macromolecules and Colloids", (B. R. Jennings, ed.), Plenum Press, N.Y. (1979) p. 345.
4. C. T. O'Konski, S. Krause, in "Molecular Electro-optics", part I, (C. T. O'Konski, ed.), Marcel Dekker, N. Y. (1976) chapter 3.
5. E. Neumann, in "Elektro-optiks and Dielectrics of Macromolecules and Colloids", (B. R. Jennings, ed.), Plenum Press, N. Y. (1979) p. 233.
6. P. Lindner, E. Neumann, K. Rosenhek, J. Membrone, Biol. 32:231 (1977).
7. S. Sokerov, T. Vorobeva, S. Stoylov, J. Polym. Sci. 44:147 (1974).
8. S. Sokerov, T. Vorobeva, Proc. Intern. Conference on Colloid and Interface Science, Budapest 1:391 (1975).
9. G. Schwarz, Z. Physik 145:563 (1956).
10. S. Kielich, Chem. Phys. Letters 6:72 (1970).
11. B. Jacobson, J. Am. Chem. Soc. 77:2919 (1955).
12. D. Ridgeway, in "Molecular Electro-optics", part I, (C. T. O'Konski, ed.), Marcel Dekker, N. Y. (1976) Chapter 4.
13. V. N. Pokrovski, Statistical Mechanics of Diluted Suspensions, Nauka, Moscow (1978).
14. R. M. Mazo, J. Stat. Phys. 1:89 (1969).
15. S. P. Stoylov, B. Tenchov, in preparation.
16. D. N. Holkomb, I. Tinoko, J. Phys. Chem. 67:2691 (1963)
17. K. Yoshioka, H. Watanabe, in "Physical Principles and Techniques of Protein Chemistry", Part A, (S. J. Leach, ed.), Academic Press, N.Y. (1969).
18. S. P.Stoylov, Adv. Colloid. Interface Sci. 3:41 (1971).

19. M. Kerker, The Scattering of Light and other Elektromagnetic Radiation, Academic Press, N.Y., (1969).

20. T. H. Hebert, in "Elektro-optics and Dielektrics of Macromolecules and Colloids", (B. R. Jennings, ed.) Plenum Press, N.Y., (1979) p. 29.

21. S. Kielich, Acta Phys. Polon. A37:447 (1970); J. Coll Interface Sci. 34:228 (1970).

22. R. S. Tarinato, J. Coll. Interface Sci. 53:402 (1975).

23. M. Stoimenova, Ph.D. Thesis, Sofia (1979).

24. H. C. van de Hulst, Light Scattering by Small Particles, John Wiley, N.Y., (1957).

25. M. Stoimenova, S. Stoylov, J. Coll. Interface Sci. 73:566 (1980).

26. J. C. Ravey, J. Coll. Interface Sci. 56:540 (1976).

27. M. Stoimenova, S. P. Stoylov, J. Coll. Interface Sci. 76:502 (1980).

28. M. Stoimenova, L. Labaki, S. Stoylov, in press in J. Coll. Interface Sci.

29. S. Sokerov, Ph.D. Thesis, Sofia (1980).

30. C. T. O'Konski and L. S. Shepard, in "Molecular Electro-Optics", Part II, (C. T. O'Konski, ed.) Marcel Dekker, N. Y. (1976) Appendix B.

THEORY OF THE KERR CONSTANT

Chester T. O'Konski

Department of Chemistry
University of California
Berkeley, California 94720

INTRODUCTION

Kerr effect data on gases and liquids provided important results in studies of molecular structure as mentioned in the Chapter
concerning the history of molecular electro-optics (the first Chapter of this book). The development of theories of the Kerr constant
of gases, liquids, solutions and suspensions was very closely involved with that of the theories of the dielectric constant. Here,
however, we describe highlights of theoretical developments of only
the Kerr constant. In this, as in other areas of physical chemistry,
there has been a strong interplay between experimental studies
guided by theoretical concepts and the selection of physical models
with their mathematical treatments. The theory of the Kerr constant
is relatively complete and rigorous for dilute gases, where the
complications of molecular interactions are absent (1). It is also
in quite a good state for most liquids and dilute solutions of
solutes in nonpolar solvents, but not so complete for polar molecules -- especially hydrogen bonding ones, which interact strongly
in liquids.

Studies of macromolecules in solution by the Kerr effect began
about three decades ago. In macromolecular solutions, the orientational relaxation times can readily be measured and they give valuable information about the size and shape of the macromolecules, as
discussed in the chapter on electric birefringence dynamics.

In this chapter we shall deal with the magnitude of the Kerr
constant, defined on page 3, in gases and liquids and in solutions of rigid macromolecules. This furnishes a basis for treating
other electro-optic effects. In the chapter by Jernigan (2), the

statistical theory recently developed for flexible macromolecules
is presented separately.

THE KERR CONSTANT OF A GAS

The conventionally defined Kerr constants are B which is
given by $\Delta n/\lambda E^2$, where $\Delta n = n_{||} - n_{\perp}$, the birefringence of light
of wavelength λ in the medium produced by a field of intensity
E, and K which is given by $\Delta n/nE^2$. The Kerr constant of the gas
phase is the easiest to calculate. When the molecules are subjected
to an electric field, they are to some extent oriented by the interaction of their dipole moments and their anisotropic polarizabilities, opposed by random thermal motion, and the macroscopic sample
then refracts polarized light differently when its electric vector
is vibrating along the direction of the electric field ($n_{||}$) than
when it vibrates perpendicularly ($n_{\perp}$). The early theory was due to
Cotton, Mouton, Born and Gans and was further developed by Debye
(3, 4), leading to the following equation for dilute gases (5):

$$K = 3\pi\nu(k_1 + k_2) = K_1 + K_2$$

where ν is the number of molecules per cm^3, and k_1, and k_2 are
given by

$$k_1 = (1/45kT) \ [(\alpha_a - \alpha_b) \ (b_a - b_b) + (\alpha_b - \alpha_c) \ (b_b - b_c)$$

$$+ (\alpha_c - \alpha_a) \ (b_c - b_a)] \ ,$$

$$k_2 = (1/45k^2T^2) \ [(\mu_a^2 - \mu_b^2) \ (b_a - b_b) + (\mu_b^2 - \mu_c^2) \ (b_b - b_c)$$

$$+ (\mu_c^2 - \mu_a^2) \ (b_c - b_a)] \ .$$

Here the components of the electric molecular polarizability along
molecular axes a, b, c, are given by α_a, α_b, α_c, whereas those of
the optical molecular polarizability are b_a, b_b, b_c, and those of
the permanent electric dipole moment are μ_a, μ_b, μ_c, respectively.
T and k have their usual meanings.

The Kerr constant thus has two parts. It can be seen that the
anisotropy term, K_1, due to the induced polarization orientation
effect, is generally positive and has an inverse T dependence --
positive because any electric polarizability component in most molecules, α_i, is about 1.05 b_i; i.e., the "atomic" polarizability due

to "vibrational" polarization is only about 5% of the electronic part responsible for the optical b_i. In contrast, the dipole term K_2 may be positive or negative depending upon the direction of the dipole moment vector relative to the direction of the maximum optical polarizability, and it has an inverse T^2 dependence. In a more exact treatment (6), there is a temperature independent contribution too.

It is instructive to examine a representative set of numbers for different molecular geometries and polarities. In Table 1 we have a compilation of B values by Buckingham (6) of a set of small molecules which illustrate different symmetry and polarity types. Kerr effect data, light scattering and refractive index measurements are needed to calculate all the molecular parameters. The molecular dipole moment and optical polarizability components can be calculated from ab initio or semi-empirical molecular quantum theory. The quantum theory has been reviewed by Ha (7). The molecular electric polarizability components involve the optical or electronic polarizability or molecular deformation. Little attention has been given to theoretical computations of the latter.

A comparison between the theory of the Kerr constant and experiments was made by Dunmur, et al. (7a) with measurements as a function of temperature and pressure for gaseous Ar, Kr, Xe, CH_4, CF_4, Me_4C, and SF_6. The data were analyzed in terms of a virial expansion giving 3 coefficients, relating to molecular electric dipole hyperpolarizabilities and the polarizabilities of interacting pairs and triplets of molecules. Chue, et al. (7b) have made a theoretical study of the optical polarizability tensors of $n - C_mH_{2m+2}$ (m = 4 to 20), calculated by a fragment method and found results in good agreement with experiment.

THE KERR CONSTANTS OF LIQUIDS

Liquids and solutions are more difficult to treat because of interactions between the molecules, and between the induced electric moments produced by the oscillating optical fields. Equations have been summarized by Buckingham (6) for various types of molecules; hyperpolarizabilities and local field effects have been considered. Buckingham and Raab (8) derived an equation for the Kerr constant of a polar liquid using Onsager's Theory and model and apply it to water. They have also reviewed the approximations of earlier theories. Buckingham, Stiles and Ritchie (8a) have developed a simplified treatment of the effects of interaction in pure liquids and in solution on the Kerr constant, where dipolar interactions are incorporated in correlation tensors related to anisotropy in

Table 1. Kerr Constants of Gases at 1 atm (1.01325×10^5 J m^{-3}) for Light of Wavelength λ_0 = 633 nm.[a]

Gas	T(K)	$B(10^{-19}$ V^{-2} m$)$	$B(10^{-12}$ esu$)$
He	296	0.024	0.022
Ne	296	0.047	0.042
Ar	296	0.53	0.48
Kr	296	1.3	1.1
Xe	296	3.5	3.2
CH_4	273.6	1.43	1.29
CF_4	288.8	0.70	0.63
SF_6	296	1.1	0.97
H_2	245.4	0.74	0.67
	332.4	0.48	0.43
N_2	248	3.63	3.26
	334	2.19	1.97
CO_2	252	30.8	27.8
	337	19.0	17.1
C_2H_6	255	5.13	4.61
	318	3.68	3.31
C_2H_4	262	16.7	15.0
	334	10.3	9.2
C_2H_2	254	26.1	23.5
	334	17.1	15.4
Cyclopropane	265	6.76	6.08
	318	5.30	4.77
CH_3F	250.8	107.0	96.0
	318.9	50.6	45.5
CH_2F_2	244.0	−17.5	−15.7
	303.2	−9.2	−8.3
CHF_3	245.5	−89.1	−80.2
	308.9	−40.8	−36.7

[a]Taken from Buckingham (6).

the distribution of neighbor molecules. Large solvent effects are
possible due to dipole-dipole, hydrogen bonding and ion-dipole
interactions.

The Kerr constants of a few representative liquids are given
in Table 2. It can be seen that the magnitude of B depends
strongly upon molecular symmetry and dipole moment. The negative
value for chloroform shows that the two equal polarizabilities in
the trigonal plane of the Cl atoms are greater than the component
along the C-H axis which is in the direction of the dipole moment.

Table 2. Kerr constants of fourteen simple liquids at 25°C,
λ = 589 nm (Values from Ref. 6).

Liquid	$B(10^{-16} \ V^{-2} \ m)$
n-hexane	5.2 ± 0.1
n-octane	9.6
n-decane	11.3
cyclohexane	6.0
benzene	46.0 ± 1.0
toluene	86.0
chlorobenzene	1,380.
o-dichlorobenzene	3,800.
p-dichlorobenzene	290.
carbon disulfide	339.
pyridine	2,270.
nitrobenzene	44,000.
carbon tetrachloride	9.2 ± 0.1
chloroform	-358.0

The theory of the Kerr constant of solutions of Buckingham and
Raab (8) proceeds along the general lines of Onsager's dielectric
theory (8b); several simplifying assumptions are necessary for its
use, however - for example, a spherical cavity for the solute is
often assumed. Recently, Patz and Raetzsch (8c) calculated the
Kerr constants of various binary organic liquid mixtures using the
Scholte continuum model (8d) of dielectric theory which allows for
nonspherical shapes of molecules, and found good agreement between
theoretical and experimental results. Battaglia (8a) reviewed some

of the complications in interpreting condensed phase birefringence data, including the effects of dispersion and hyperpolarizability and reanalyzed data for CS_2, C_6H_6 and C_6F_6. It was concluded that by carefully combining additional experimental data and theory, state-independent molecular properties and information on liquid structure and dynamics can be obtained. Three sets of studies have been made of the Kerr constants of various solutes in aqueous solutions (8f, 8g, 8h). Specific interaction effects between molecules limit the applicability of the continuum dielectric models.

THE KERR CONSTANTS OF MACROMOLECULES

Flexible Polymers

The Kerr effect theory of polymer molecules in solution has been treated using several models for flexible polymers (9a-9h). Jernigan and Thompson have provided a comprehensive review of the theory of the static and the dynamic electric birefringence of flexible polymers (9i). Recent developments are reviewed in the chapter of this volume by Jernigan (2).

Non-Conducting Rigid Macromolecules

Consider a rigid macromolecule of arbitrary structure in a solvent of uniform optical and electrical properties. The effects on the electric and optic properties of the solution which will be produced by this macromolecule may be described in terms of its electric dipole moment and a polarizability tensor, and its response to the applied electric field will be seen as an orientation effect. In this section we treat the low-frequency or static Kerr constant only. The polarizability tensor is a well-known 3 x 3 matrix, and it can be diagonalized by proper choice of the Cartesian axis system, x, y, z. The dipole moment vector may always be resolved into three components, one along each of the respective axes, and we shall designate those μ_x, μ_y, μ_z. The hydrodynamic and the electric polarizability principal axis systems do not necessarily coincide.

In a dilute solution of macromolecules to which no field is applied, all possible orientations of any macromolecular axis system will be equally probable because of the random Brownian motion, and the solution will be optically isotropic. After an electric field, E , is applied and the system has come to a steady-state orientation distribution, we may calculate from the Boltzmann distribution function the relative populations of molecules at various orientations, if we know how the energy varies with the orientation.

Knowing how any one molecule in a given orientation will affect the optical properties of the medium, we then calculate the optical properties of the solution as an integral over the entire population of macromolecules in the ordered system. This follows the Born-Langevin orientation theory (1, 10).

It is instructive to write out the expression for the birefringence of a dilute solution of rigid axially symmetric macromolecules in terms of the orientation distribution function, $f(\theta)$, before presenting more complicated expressions for macromolecules of lower symmetry. Peterlin and Stuart's equation for the birefringence Δn may be written

$$\Delta n = \frac{\pi C_v (g_a - g_b)}{n} \int_0^\pi f(\theta) (3 \cos^2 \theta - 1) 2\pi \sin \theta \, d\theta \tag{1}$$

Here $f(\theta)$ is the probability per unit solid angle of finding a molecule with its symmetry axis at angle θ, with respect to the direction of the applied electric field. The distribution function is

$$f(\theta) = e^{-U/kT} \left[\int_0^\pi e^{-U/kT} 2\pi \sin \theta \, d\theta \right]^{-1} \tag{2}$$

where U is the energy of the macromolecule as a function of angle, C_v is the volume fraction of the particles, and the term $(g_a - g_b)$ is the optical anisotropy factor which can be calculated by application of the equation for ellipsoids due to Maxwell and given by Peterlin and Stuart (11,12). This equation is

$$g_j = \frac{1}{4\pi} \frac{n_j^2 - n^2}{1 + (n_j^2/n^2 - 1) A_j} \tag{3}$$

where A_j is the depolarization factor, n is the refractive index of the solvent, and n_j ($j = a, b$) are refractive indexes of the particle for light with electric vector along the axes a and b, respectively. The depolarization factor of an ellipsoid is, in general

$$A_j = \int_0^\infty \frac{abc}{2(j^2 + s) [(a^2 + s) (b^2 + s) (c^2 + s)]^{1/2}} \, ds \tag{4}$$

Here, a, b, and c are the lengths of the semiaxes of the particle,

and s is a variable of integration. The integrals have been tabulated for ellipsoids of revolution (13). Graphs (14) and special
formulas also are available (15) for the A_j.

To compute $f(\theta)$, we require an expression for the energy U
of the macromolecule as a function of angle. In insulating media,
this may be written

$$U = -\mu EB_a \cos\theta - \frac{1}{2}(\alpha_a - \alpha_b) E^2 \cos^2\theta \tag{5}$$

Here μ is the dipole moment, which is assumed to lie along the
symmetry axis of the macromolecule, designated by the subscript a,
$\mu_b = \mu_c = 0$, and B_a is the internal field function which expresses
the ratio of the local field acting upon the macromolecule along the
direction a to the component of the external field along that axis.
The polarizabilities of the macromolecule in excess of the solvent
it replaces are α_a and α_b along the symmetry axis a , and the
transverse axes, b = c , respectively. If the particle can be
represented as an ellipsoid in an insulating homogeneous dielectric
continuum, the internal field functions B_j may be calculated from
the relation

$$B_j = \left[1 + \left(\frac{\varepsilon_j}{\varepsilon}\right) - 1 \ A_j\right]^{-1} \tag{6}$$

in which ε and ε_j are the dielectric constants of the solvent
and of the ellipsoid along the axis j , respectively.

From Eqs. (1), (2), and (5) one can readily obtain expressions
for the birefringence as a function of the applied electric field
and for the Kerr constant in weak fields, that is, U << kT .
Except for the explicit inclusion of the internal field factor B_a
by O'Konski (16), the results were previously given by Peterlin
and Stuart (12) and Benoit (17). The result for the specific Kerr
constant, defined in Chap. 1, is

$$K_{sp} = \frac{2\pi(g_a - g_b)}{15\ n^2}\left[\frac{\alpha_a - \alpha_b}{kT} + \frac{B_a^2\ \mu^2}{k^2\ T^2}\right] \tag{7}$$

Here μ is not the vacuum dipole moment, but rather the dipole
moment of the solvated macromolecule in the solution, which is in
general different because of solvent orientation and polarization.

For the general ellipsoid with dipole moment components along
all three axes, the Kerr constant was obtained by Holcomb and

Tinoco (18) and it may be expressed as follows, after inclusion of the internal field functions B_a, B_b, and B_c, for the various axes:

$$K_{sp} = \frac{\pi}{15 \, n^2 \, kT} \, [(g_a - g_b) \, (\alpha_a - \alpha_b) + (g_b - g_c) \, (\alpha_b - \alpha_c)$$

$$+ (g_a - g_c) \, (\alpha_a - \alpha_c)] + \frac{\pi}{15 \, n^2 \, k^2 \, T^2} \, [(g_a - g_b)$$

$$(B_a^2 \, \mu_a^2 - B_b^2 \, \mu_b^2) + (g_b - g_c) \, (B_b^2 \, \mu_b^2 - B_c^2 \, \mu_c^2)$$

$$+ (g_a - g_c) \, (B_a^2 \, \mu_a^2 - B_c^2 \, \mu_c^2)] \qquad (8)$$

The intermediate case of an ellipsoid of revolution (axes a and b = c) with dipole moment components μ_a and μ_b was calculated earlier (19).

The above equations apply for a homogeneous population of macro-molecules. If we are dealing with a heterogeneous population, or a distribution of conformers, and are interested in properties mea-sured by techniques which are fast compared to the rate of inter-change among the conformations, we may, in the dilute solution case, express the Kerr constant simply as a sum over various species, that is,

$$K_{sp} = \sum_i C_{v,i} \, K_{sp,i} \qquad (9)$$

where $C_{v,i}$ is the volume fraction of the species i and $K_{sp,i}$ is the specific Kerr constant of that species.

Electrically Conducting Systems

In this section we shall consider the Kerr effect of metallic suspensions, emulsions, and polyelectrolytes in solution. Many aspects of modern macromolecular science grew out of earlier stud-ies of colloid systems, and there is no clear line of demarcation between a colloidal suspension and a macromolecular solution.

Metallic Suspensions. Several studies were made of the Kerr effect in metallic sols near the beginning of the century (12,13). Interest in the Kerr effect in metallic suspensions has been stim-ulated by the development of devices for controlling the phase and

polarization of microwave radiation (22). The theory for suspensions
of metallic particles (e.g., gold sols), which are small compared to
the wavelength of light, is essentially the same as that for microm-
eter-size particles in centimeter wavelength radiation fields, except
for the computation of the optical anisotropy factors, which apparently have not been treated for metallic particles large compared
to the wavelength.

Metallic suspensions may be considered as highly conducting
particles in an insulating medium. So far as the orientation effect
is concerned, this simplifies the treatment which becomes similar to
that for an insulating particle of infinite dielectric constant
immersed in a medium of finite dielectric constant.

The excess electric polarizabilities of an anisotropic dielectric ellipsoid are given by the relation

$$\alpha_j = \frac{v}{4\pi} \frac{\varepsilon_j - \varepsilon}{(\varepsilon_j/\varepsilon - 1) A_j} \tag{10}$$

in which ε_j and ε are defined above, and v is the volume of
the ellipsoid. Taking $\varepsilon_j \gg \varepsilon$, this leads to the following electric anisotropy term:

$$\alpha_a - \alpha_b = \frac{v\varepsilon}{4\pi} \left[\frac{1}{A_a} - \frac{1}{A_b} \right] \tag{11}$$

An exact treatment of the optical part of the problem is quite
complex for metals, as in the Mie theory for the scattering of electromagnetic radiation from metallic particles at optical frequencies
(23). Here we shall assume that we are dealing with radiation of a
frequency such that the particles may be regarded as perfect conductors. This approximation is expected to be useful for metallic
colloidal particles at infrared and microwave frequencies.

The optical anisotropy factor is obtained from Eq. (3) by insertion of $n_j^2 \gg n^2$ with the result

$$g_a - g_b = \frac{n^2}{4\pi} \left[\frac{1}{A_a} - \frac{1}{A_b} \right] \tag{12}$$

Inserting Eqs. (11) and (12) into Eq. (8), and noting from Eq. (6)
that $B_a = B_b = B_c = 0$, the expression for the Kerr constant of
metallic suspensions becomes

$$K_{sp} = \frac{v\varepsilon}{240\pi kT} \left[\left(\frac{1}{A_a} - \frac{1}{A_b} \right)^2 + \left(\frac{1}{A_b} - \frac{1}{A_c} \right)^2 + \left(\frac{1}{A_a} - \frac{1}{A_c} \right)^2 \right]$$

(13)

So far as we are aware, this approximate theory (24) has never been tested experimentally.

Liquid Droplets. A Kerr effect will be produced in an aerosol or emulsion by the deformation of the liquid droplets by an electric field. Interfacial tension tends to keep the droplets spherical, but the electric free energy will, in general, be lowered by deformation from a spherical shape. The deformation of liquid droplets of arbitrary dielectric constant and conductivity in a suspending liquid of arbitrary dielectric constant and conductivity has been calculated (25) on the assumption that the Maxwell-Wagner boundary conditions (26) are applicable, and that the deformations are small, so the deformed particle may be approximated as an ellipsoid of revolution with symmetry axis along the field. The electric free energy of a conducting ellipsoid of revolution in a conducting medium has been calculated (25). Electro-optic effects may be appreciable for aerosol or cloud droplets in atmospheric electric fields. Equations have been derived for the droplet deformation (27) and the birefringence produced by the deformed droplets can be calculated from the eccentricity and the refractive indexes of the particle and the surrounding medium. This is a case of form birefringence and the equations of Wiener (28) can be used.

To calculate the Kerr effect when the wavelength of the radiation is long compared to the diameter of the particle, one may use the expressions above for the optical anisotropy factor and simplify them for the case where the eccentricity is small. This would lead to the expressions for the Kerr constant, e.g., for cloud droplets at microwave frequencies. So far as we are aware, those calculations have not yet been carried to completion, although a microwave refractometer method for aerosols was considered by Thacher (Ref. 9 of Ref. 27).

Because of power dissipation effects and the necessity to apply rather large electric fields, the Kerr effect in emulsions would be of interest mainly for solvents of low electric conductivity (e.g., water in oil suspensions) and droplets which are micrometer size or larger.

Polyelectrolytes. By definition, polyelectrolytes are ionizable macromolecules, and therefore, their solutions are electrically conducting. Biological polyelectrolytes, such as the nucleic acids, proteins, and polysaccharides, are ordinarily dissolved in solutions

with high ionic conductivity. Physiological solutions are around
0.15 M salt. Both synthetic and natural polyelectrolytes may have
local ion densities considerably higher than that, and therefore
the conductivity effect is extremely important. In dilute aqueous
solutions of low salt content, a high-density polyelectrolyte such
as DNA may be regarded as a more or less uniformly charged macromolecule with quite high surface-charge density arising from the
ionizable phosphate groups, surrounded by a swarm of counterions
in a relatively insulating dielectric continuum. It was logical to
expect, therefore, that the counterion atmosphere polarization
effects would play a significant part in the orientation of such
macromolecules by electric fields (29). This is because the transport of ions profoundly influences the distribution of electric
fields near a macromolecule, and equations derived for insulating
systems are generally inapplicable. A comprehensive set of experiments conclusively established that most of the orienting torque
in a model polyelectrolyte system, tobacco mosaic virus (TMV), must
be due to the counterion polarization effect (30). This was concluded by showing that the dipole moment of TMV is not its main
orienting factor, and that the specific Kerr constant is around 50
times larger than calculated from the Peterlin-Stuart theory for
insulating systems; also, it is depressed by the addition of electrolyte and affected by pH changes. Earlier, birefringence studies
on inorganic colloids had led to the suggestion that counterions
were important in the orientation effect (31).

An important polyelectrolyte to consider would be a protein
molecule of nonuniform charge density (various groups) along a
flexible chain. It would require a very elaborate theory to treat
such a system, so we restrict ourselves now to rigid macromolecules
of uniform charge density.

A rigid polyelectrolyte, such as a short section of DNA or a
virus, has a configuration of fixed charges, some of which bind
counterions, and it will in general be surrounded in solution by a
swarm of ions distributed diffusely according to the Poisson-
Boltzmann distribution law (32). Applying an electric field per-
turbs the counterion distribution. When the field is removed, the
perturbation relaxes. The relaxation of the counterion distribution gives rise to frequency dependences of the Kerr effect, the
dielectric constant, and the conductivity, all related to the
counterion atmosphere relaxation time. In 1955 the relaxation time
was calculated for a model system -- a sphere with a uniform surface
conductivity to model the transport of charge by the mobile counterions (33). Subsequently, a dielectric and conductivity theory based
upon extensions of this model was developed by extending the Maxwell-
Wagner theory to include explicitly the internal and external volume
conductivities and the interfacial conductivity (13). It was shown
that a surface conductivity of value λ (ohms^{-1}) at the interface
of a sphere is electrically equivalent to an internal conductivity,

κ' , of value $2\lambda/\underline{a}$ on a sphere of radius $\underline{a}$. In general, for
anisometric particles, the surface conductivity will contribute dif-
ferent amounts to the effective conductivity along the different
axes. Each axial conductivity may be written

$$\kappa_j = \kappa_j^0 + \kappa_j' \qquad\qquad\qquad (14)$$

wnere the term κ_j^0 is the internal volume conductivity along the

axis j and κ_j' is the effective contribution along the axis j

from the surface conductivity associated with the counterions and
any mobile surface charges. The terms κ_j' have been evaluated for

various-shaped particles (13). Prior to this theory the nearest
treatment had been a calculation of the electrical conductivity and
its frequency dependence for a suspension of spherical particles
carrying a concentric shell of different electrical properties (34).
It can be shown through the relationships between the complex dielectric constant and the complex conductivity that this treatment
gives equivalent results in the limit of a negligibly thin shell of
equal surface conductance (35).

Examples of physical situations which are expected to lead to
surface conductivity contributions have been discussed (13). They
include ion atmosphere polarization (29) or the mobility of the
counterions of the ion atmosphere (13, 33, 36), the proton mobility
on the surface of proteins in solution (37), and electronic semi-
conductivity. The external or "atmospheric" contributions from
mobile counterions are important for synthetic polyelectrolytes
and also have been studied by various techniques (38-42) in nucleoproteins (13, 30) and nucleic acids (42-46).

To treat the static or low-frequency Kerr constant, we can
neglect not only the problem of the rate of the orientation of the
macromolecules because it is fast compared to one period at low
frequencies, but we may also ignore the dielectric dispersion and
treat the problem as a situation involving steady-state conduction,
when the charge transport relaxation or ionic polarization effects
often are fast compared with the frequency.

Let us consider a general ellipsoid with axes a, b, and c ,
having respective internal volume conductivities κ_a, κ_b, and κ_c

and dielectric constants ε_a, ε_b, ε_c . This assumes that the prin-

cipal axis systems of the dielectric and conductivity tensors coincide with the geometric axes. Considering that conditions of
steady-state electric conduction apply, and using the Maxwell-Wagner
boundary conditions, it was shown (47) that the equation for the
electric-free energy of this ellipsoid in a solvent of dielectric

constant ε and conductivity κ, subjected to an applied field E is

$$U(\theta,\phi) = -E(\mu_a B_a \cos\theta + \mu_b B_b \sin\theta \cos\phi - \mu_c B_c \sin\theta \sin\phi)$$

$$- \frac{\varepsilon v E^2}{8\pi} \left[\cos^2\theta \left\{ \left(\frac{\kappa_a}{\kappa} - \frac{\varepsilon_a}{\varepsilon}\right) B_a^2 + \frac{(\kappa_a - \kappa) B_a}{\kappa} \right\} \right.$$

$$+ \sin^2\theta \cos^2\phi \left\{ \left(\frac{\kappa_b}{\kappa} - \frac{\varepsilon_b}{\varepsilon}\right) B_b^2 + \frac{(\kappa_b - \kappa) B_b}{\kappa} \right\}$$

$$\left. + \sin^2\theta \sin^2\phi \left\{ \left(\frac{\kappa_c}{\kappa} - \frac{\varepsilon_c}{\varepsilon}\right) B_c^2 + \frac{(\kappa_c - \kappa) B_c}{\kappa} \right\} \right] \quad (15)$$

Here μ_a, μ_b, and μ_c are components of the permanent dipole moment of the solvated macromolecule along the respective ellipsoid axes, and v is the volume of the ellipsoid. The functions B_a, B_b, and B_c are internal field functions expressed in terms of electric conductivities as follows:

$$B_j = \left[1 + \left(\frac{\kappa_j}{\kappa} - 1\right) A_j \right]^{-1} \quad (16)$$

and θ and ϕ are two of the Eulerian angles in the coordinate system used to specify the orientation of the ellipsoid axes. The dielectric constant and conductivity of the solvent are ε and κ. From this expression one obtains a value for the specific Kerr constant which is

$$K_{sp} = \frac{\pi}{15 \, n^2} \left[(g_a - g_b)(P_a - P_b + Q_{ab}) \right. \quad (17)$$

$$\left. + (g_b - g_c)(P_b - P_c + Q_{bc}) + (g_c - g_a)(P_c - P_a + Q_{ca}) \right]$$

where

$$P_i = \frac{B_i^2 \, \mu_i^2}{k^2 \, T^2}, \quad i = a, b, \text{ or } c$$

$$Q_{ij} = \varepsilon v \left\{ (\kappa_i/\kappa - \varepsilon_i/\varepsilon) B_i^2 + (\kappa_i - \kappa) B_i/\kappa \right.$$

$$\left. - (\kappa_j/\kappa - \varepsilon_j/\varepsilon) B_j^2 - (\kappa_j - \kappa) B_j/\kappa \right\} / 4\pi\kappa T$$

and other terms are defined above. In these equations the subscripts
i and j label the quantities referring to the axes a, b, or c .
Coefficients of the form $(\kappa_i/\kappa - \varepsilon_i/\varepsilon)$ in Eq. (17) are proportional
to the surface charges which accumulate, under the influence of an
externally applied field, at the interface of two conductors having
different conductivity and dielectric constant ratios; this is the
essence of the Maxwell-Wagner polarization phenomenon (13,47). The
accumulation of interfacial charges at the ends of an elongated
particle which is more conducting than the solvent produces a contribution to the induced dipole, and it may arise from either of
the terms on the right side of Eq. (14), where κ_j' arises from
mobile counterions (13).

     In the original development of the surface conductivity model
it was pointed out that the Maxwell-Wagner boundary conditions may
not be adequate to represent the details of the interfacial polarization (13). Rectification effects, due to transport barriers to
ions at the interfaces, requires a more general approach to be followed, strictly speaking. That qualification holds even for large
particles. Another limiting factor in the quantitative accuracy
of this theory of the Kerr constant is the fact that with macromolecules, the ion atmosphere thickness may be the order of the
particle dimensions, rather than being negligible. The test of the
theory made by comparing experimental data for tobacco mosaic virus
with the predicted Kerr constant -- made on an absolute basis by
using free ion mobilities to get the K values -- gave agreement
within a factor of two (47). This was about the best that could be
expected for the relative dimensions involved at the low ionic
strengths.

     Another model, namely, a linear polyion with mobile bound
counterions subject to diffusion along a line was introduced by
Schwarz (48) and has been extended to estimate the induced polarization of polyelectrolytes in solution, and to develop equations for
dielectric polarization and dispersion (49). The model has been
developed for a counterion fluctuation theory of dielectric dispersion in polyelectrolytes by Oosawa (50). Hornick and Weill have
discussed it and compared it with other linear model treatments to
interpret data on electro-optic studies of DNA fragments (51). They
concluded that Oosawa's treatment is most satisfactory. It allows
for counterion repulsions but does not give quantitative agreement.

     The electric birefringence of a rodlike polyelectrolyte has
been calculated using Mandel's linear model (49), without counterion
repulsion, but over a wide range of electric field intensity, by
Kikuchi and Yoshioka (52). Shirai and this author have developed a
model which includes a term for counterion repulsion, but numerical
calculations are necessary so its evaluation is yet in progress (53).

Another approach is the theory of Manning (54) for counterion condensation on a highly charged polyelectrolyte cylinder. It is asserted that after the condensation or binding occurs, it is justified to use a solution of the Poisson-Boltzmann equation within the Debye-Hückel approximation. This theory has been employed recently by Manning to obtain an expression for the anisotropy of the polarizability of DNA (55), which has since been employed (56) to discuss the dependence of electric dichroism on the ionic strength of the DNA solution. Charney is engaged in applying this approach further for orientation effects (57). Further tests are needed.

A critical comparison of the Manning theory with a cylindrical Gouy model for a polyelectrolyte has been made by Stigter (58) who has tested the applicability of the Debye-Hückel approximation with reference to cylindrical solutions of the exact Poisson-Boltzmann equation (59). He also utilized experimental Donnan equilibria and electrophoresis data on DNA in various salt solutions. It appears that a Gouy, or better -- a Gouy-Stern -- model is needed for a quantitative fit to the data and reasons were ferreted out by Stigter in the theoretical analysis. It may be anticipated that similar refinements will be necessary in electro-optic theories to adequately predict or explain the effects in DNA and other polyelectrolytes quantitatively.

Expressions are obtained for the induced dipole moment of the bound counterions in the linear polyion theories mentioned above. In the region of low electric field intensities, these may be used to compute corresponding polarizabilities for the polyelectrolyte. Usually the flows of counterions off of the polyelectrolyte from the solvent, and vice versa, are ignored. This can cause major errors for electrolyte solutions. The extended Maxwell-Wagner model (13, 47) puts in these flows approximately. Explicit treatment of ion flows in the Gouy-Stern model (58) and the solvent in an applied field, with numerical integrations of the electric free energy functions for the conducting model and solvent (60) may be required for a quantitative theory for polyelectrolytes.

DISPERSION OF THE KERR CONSTANT IN AN ABSORPTION BAND

Buckingham (61) treated the theory of the optical dispersion of the Kerr constant of gases and liquids in the vicinity of an absorption band. In a dilute solution of macromolecules, the specific Kerr constant, K_{sp} , will vary with wavelength via the optical parameters, g_a, g_b, and g_c [Eqs. (3), (7), and (9)]. A study of the dispersion of the Kerr constant can yield information on the structure of the macromolecule. The transition moments responsible for the absorption band can be assigned a direction with respect to the principal axes of the macromolecules. Unfortunately, no systematic study of review of optical dispersion seems to exist in the literature. However, Orttung (62) studied the Kerr constant of horse oxyhemoglobin and methemoglobin at 546, 436, and 365 nm. He also calculated the expected Kerr constant at these wavelengths using the known spectra and structure of the macromolecule assuming (a) orientation of the macromolecule with its long axis parallel to the electric field, and (b) orientation with the short axis parallel to the electric field. The data fitted case (b), and this allowed a dipole moment parallel to the short axis of the macromolecule to be surmised.

Powers (63) studied the dispersion of the Kerr constant of acridine orange-polyglutamic acid complex from 400 to 550 nm and concluded that the long axis of the dye molecule is approximately parallel to the long axis of the polymer helix in the complex. The Kramers-Kronig relations between birefringence dispersion and dichroism were developed by Kuball and coworkers (64) and were applied to the DNA-proflavine complex by Houssier and Kuball (65).

An early infrared study in the vicinity of absorptions due to weak vibrational overtones was made by Charney and Halford (66). Lutskii, et al. (67) have examined the dependence of K on optical frequency for aniline and nitrobenzene and found measured values in disagreement with those calculated from six empirical formulas.

EXPERIMENTAL KERR CONSTANTS OF MACROMOLECULES

The Kerr constants of some macromolecules of various types are shown in Table 3. It is seen that there is a tremendous range of K_{sp} values. The polyelectrolytes, especially the large rigid ones, have the largest Kerr constants. The relatively rigid polypeptide, poly-λ-benzyl-L-glutamate (PBLG) also has a very large value, whereas the flexible coil polymer, polystyrene, has a very low specific Kerr constant. Additional data are given in "Molecular Electro-Optics," Part 2 (68). The Kerr constant is very sensitive to structure as well as molecular weight.

The synthetic polypeptide, PBLG is of special interest because of extensive measurements of this system as a nonpolyelectrolyte model for protein structure. TMV, a large, rigid, rodlike particle 3000 A long and 150 A in diameter, became a valuable model system in studies of the mechanisms of polarization which produce the electric field orientation effect. Comparisons between the experimental data and earlier theory led to the recognition of the importance of counterion polarization effects and stimulated development of the counterion polarization theory (13). The polyelectrolyte properties of TMV are known from the studies of the amino acid composition of the TMV protein, and from direct titration measurements. Thus, it

Table 3. Specific Kerr Constants of Some Macromolecules

Macromolecule	Mol. Wt.	K_{sp} (esu)	Ref.
Tobacco mosaic virus (in distilled water, 0.049 g/liter, 25.5°C) (TMV)	4.3×10^7	1.36×10^{-3}	30
Collagen (0.11 g/liter and 0.22 g/liter, in 2.9×10^{-3} N acetic acid, 25°C)	2.80×10^5	4.3×10^{-5}	69
Poly-γ-benzyl-L-glutamate (in ethylene dichloride, 0.1 to 4 g/liter) (PBLG)	3.5×10^5	1.3×10^{-5}	70
Sodium polyethylene sulfonate (in water, extrapolated to zero conc.)	2.4×10^4	1.0×10^{-5}	71
Poly-γ-benzyl-L-glutamate (in ethylene dichloride, 0.2 g/liter, 28.5°C)	2.0×10^5	3.3×10^{-6}	71
Poly-γ-benzyl-L-glutamate (in ethylene dichloride, 1 to 15 g/liter)	8.4×10^4	2.3×10^{-6}	70
Fibrinogen (pH 8.1, 2 g/liter, 25°C)	3.3×10^5	9.3×10^{-7}	71
γ-globulin (2.0 g/liter, in 10^{-3} M phosphate buffer at pH 7, 30°C)	1.5×10^5	3.4×10^{-8}	72
Ribonuclease (9.8 g/liter, in 10^{-3} M phosphate buffer at pH 7, 30°C)	1.4×10^4	1.9×10^{-8}	73
Bovine serum albumin (in water, extrapolated to zero conc., 25°C)	6.7×10^4	1.70×10^{-8}	74
Polystyrene (in carbon tetrachloride, extrapolated to zero conc.)	2.5×10^5	8.7×10^{-14}	75

was possible to obtain figures for the surface densities of counterions. Also, the specific optical properties of TMV are known from both flow birefringence and electric birefringence saturation studies [71].

Currently, attention is being given to the electro-optic properties of DNA which can now be prepared homogeneously over a broad
range of macromolecular weights and in linear, closed circular and
superhelical forms. Although apparent Kerr constants have been
obtained for various preparations, this author was reluctant to put
numbers into Table 3 because they depend strongly on many factors:
Macromolecular weight and polydispersity, often unknown; ionic
strength; buffer composition; presence of divalent ions or other
binding agents, and electric field intensity. Systematic studies
clearly have been needed for two decades. As background for understanding the effect of electric field intensity we shall deal in
the next section with the theory of saturation of the orientation
in solutions of rigid macromolecules. In this study institute we
shall be hearing current results on DNA from several laboratories
– results which are certain to play a leading role in any further
development of the theory of electric field orientation effects of
all polyelectrolytes.

HIGH FIELD BIREFRINGENCE THEORY

When the applied electric field is intense enough, all systems
will deviate from the Kerr law because the macromolecules tend to
become completely aligned. Experimental measurements up into this
saturation region are useful because they provide a basis for
separately determining the electrical properties and the optical
properties which are both involved in the Kerr constant. This
method was reported in 1959, and to implement it, the theory of
saturation of the Kerr effect was extended to include mixed permanent and induced dipole mechanisms [71]. Earlier theories were by
Gans [77] who obtained a birefringence saturation expression for
purely permanent moment orientation, and by Peterlin and Stuart
[11], who obtained an equation for a purely induced moment orientation mechanism. The method was tested and illustrated by measurements and analysis on TMV, the synthetic polypeptide, PBLG, and a
synthetic polyelectrolyte, polyethylene sulfonate [71]. We shall
summarize the theory and illustrate its application in the next section. Again we shall begin by considering a solution of axially symmetric macromolecules. Because the optical anisotropy due to orientation is an integral over the product of $f(\theta)$, the distribution
function, with $P_2(\cos\theta) = (3 \cos^2\theta - 1)/2$, where θ is the angle
defined in equations (1) and (2) above, it is convenient to define
a parameter Φ to characterize the extent or degree of orientation
of the macromolecules as follows

$$\Phi = \frac{1}{2} (3 <\cos^2\theta>_{av} - 1) \tag{18}$$

This is often called the "order parameter;" it is zero in a randomly
oriented system and tends toward unity as the axes tend toward

complete alignment along the electric field direction. It can
readily be seen from Eq. (1) that the birefringence may then be
expressed by

$$\Delta n = \frac{2\pi C_v}{n} (g_a - g_b) \Phi \tag{19}$$

By use of Eqs. (2) and (5), Φ was calculated from Eq. (18) with
the result [71]

$$\Phi = \frac{3 \int_{-1}^{1} u^2 \exp(\beta u + \gamma u^2) \, du}{2 \int_{-1}^{1} \exp(\beta u + \gamma u^2) \, du} - \frac{1}{2} \tag{20}$$

where we introduced new symbols

$$u = \cos \theta \tag{21}$$

$$\beta = \mu B_a E/kT \tag{22}$$

and

$$\gamma = \frac{\alpha_a - \alpha_b}{2kT} E^2. \tag{23}$$

At low fields the exponentials of Eq. (20) may be expanded in a
power series of E and then the degree of orientation is

$$\Phi = \frac{\beta^2 + 2\gamma}{15} \tag{24}$$

when terms beyond the first are dropped, in accord with the Kerr
law, and with Eq. (7). Calculations based upon Eq. (20) were
carried out (71) for the case where $\alpha_a - \alpha_b > 0$. The integrals
were expressed in terms of a standard set of functions which has
been tabulated. Later, to explain the reversal of the sign of the
birefringence of montmorillonite (polyelectrolyte clay mineral)
particles in terms of this kind of theory, the case where
$(\alpha_a - \alpha_b) < 0$ was calculated, and similar expressions were obtained
(78). In both treatments, tables of $\overline{\Phi}$ were calculated and graphs
were shown.

Figure 1 shows the behavior of Φ as a function of $\beta^2 + 2\gamma$, which is proportional to E^2, for a wide variety of cases (79). Matsumoto, et al. (80) have published extensive tables of Φ and $15\Phi/(\beta^2 + 2\gamma)$ for various values of $\beta^2/2\gamma$. The latter function is very useful (71) for analyzing experimental results, and it is shown for several cases in Fig. 2. The earlier saturation theories (77,11) treated permanent and induced moments as separate cases.

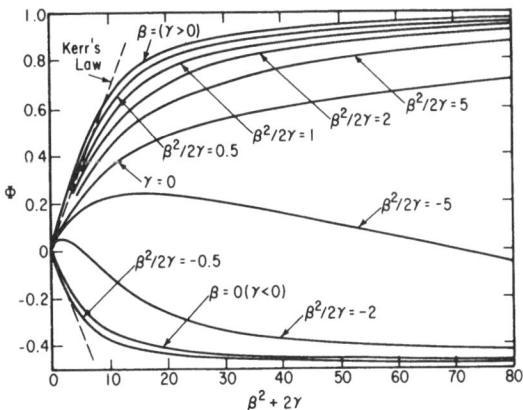

Fig. 1.   Orientation factor Φ vs. $|\beta^2 + 2\gamma|$ for various values of $\beta^2/2\gamma$ ($= b^2/2c$). Deviations from Kerr's law occur at high field strengths. (Reproduced from results given in Ref. 79.)

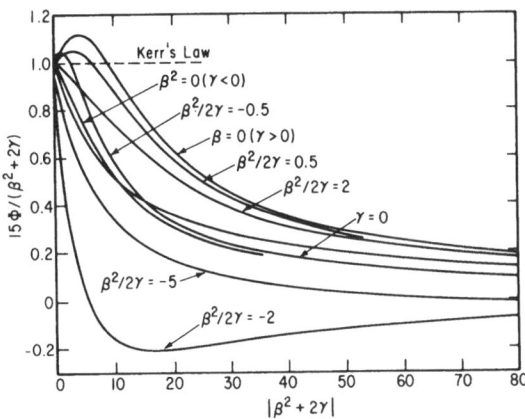

Fig. 2. $15\Phi/(\beta^2 + 2\gamma)$ vs. $|\beta^2 + 2\gamma|$ for various values of $\beta^2/2\gamma$ ($= b^2/2c$). These curves are useful for analyzing experimental data. (Reproduced from results given in Ref. 79.)

Calculations have been made (81) for a model in which both the electric and optical polarizabilities need not have any symmetry and the permanent dipole may have any orientation. The internal field functions defined above may be inserted into the dipole terms as before (16,47), if they were not included.

EXPERIMENTAL STUDIES OF KERR EFFECT SATURATION

The onset of electric birefringence saturation can be seen in the early experiments on TMV (17,29) and also in the results for DNA and vanadium pentoxide sols (17). Benoit gave a first-order saturation term in the development of his equations, but his experimental saturation data were not quantitatively analyzed.

Experiments which extended into the region of strong electric orientation were performed on TMV, PBLG in dichloroethane solution, and sodium polyethylene sulfonate (NaPES) in water (71). The saturation curve for TMV was consistent with the shape expected for an induced polarization orientation mechanism, in accord with earlier observations by other methods (30). An improved fit of the data was obtained on the assumption of two macromolecular components. The optical anisotropy factor, $g_a - g_b$, was found to be 5.9×10^{-3} and this was essentially unchanged upon extrapolating the data to an infinite field condition. The electric anisotropy term $\alpha_a - \alpha_b$, was found to be 3.3×10^{-14} cm^3 at pH 7 in 1.5×10^{-4} M phosphate buffer. This is in reasonable accord with other values discussed in the original paper, in view of the variation of the electric polarizability with conditions. The data on PBLG were consistent with a permanent dipole moment orientation mechanism, and led to a value for the dipole moment of 2700 D for the sample of molecular weight 195,000. For NaPES, the optical anisotropy factor was found to be 8×10^{-3}, and the orientation behavior was that of a permanent dipole system with $\mu = 3.1 \times 10^3$ D for a sample of MW 24,000. The dipole moment predominance was surprising in view of the polyelectrolyte nature of the macromolecule and the expectation of a large counterion polarization and it led to the suggestion that a non-symmetrical polyelectrolyte chain should have an intrinsic configurational dipole moment (16,46,71).

The first measurements of a macromolecule which permitted explicit determination of a dipole moment from the stauration behavior of a mixed type, that is, with appreciable components from both permanent and induced orientation effects, was collagen (82). Previous reports of permanent moments of polyelectrolytes are generally unreliable. This macromolecule has a MW of 280,000, a length of approximately 2800 Å, and is a triple helix. The optical anisotropy factor was found to be 1.7×10^{-3}, roughly in agreement with the value of 2.4×10^{-3} observed in flow birefringence measurements (83). The anisotropy of the electric polarizability was found to be 2.7×10^{-15} cm^3, a large value as expected from an ion atmosphere polarization in a rodlike polyion, and the permanent dipole moment was estimated to be 1.5×10^{4} D. For additional data and interpretations in comparison with other literature the original reference should be consulted. This case was of special interest because earlier workers had no way of separating induced and permanent contributions to the Kerr constant, which made reported dipole moments uncertain.

Shah and coworkers (84) observed that fractionated bentonite suspensions (montmorillonite clay mineral) showed saturation behavior at modest electric fields (ca. 1 kV/cm) and exhibited a reversal of the sign of electric birefringence at low electric field intensities. I proposed that saturation effects in this system might involve both permanent and induced polarization mechanisms to account for the birefringence reversal on the basis that the two orientation torques oppose each other. Since the permanent dipole moment torque varies linearly with the field, whereas the induced polarization torque varies as the square of the electric field, as in Eq. (5), a dipole moment perpendicular to the plate-like particles would oppose the high counterion polarizability along the surface of the plate. Shah and coworkers (84) found evidence to support this idea by square-wave experiments, and subsequently by extension of the saturation theory mentioned above (78). The agreement between experiment and theory was quantitatively limited by the polydispersity of the preparations, but was qualitatively satisfactory.

Saturation effects have also been observed in the electric birefringence of a variety of systems by Yoshioka and coworkers. These include potassium polystyrene sulfonate in water, and dioxane-water mixtures (85), PBLG in mixed solvents (86), and potassium polystyrene sulfonate in various solvents (87,83). Strong saturation effects also were observed in solutions of poly-(L-glutamic acid) (PLGA) in methanol-water mixtures (88,89) and also in a variety of other organic solvent systems. Electric birefringence saturation has also been studied in DNA and polynucleotides by Stellwagen (90). In her experiments, and also in those of Yoshioka and coworkers on polyelectrolytes, large deviations from the Kerr law were observed at low electric field strengths. This may be an induced moment saturation phenomenon (52,53).

The development of the electric birefringence saturation method has made the Kerr effect a much more efficient tool for studies of the electric and optical properties of macromolecules. As

experimental methods are improved, many more biological systems
currently of interest can be studied by this technique. Saturation
measurements are also important for the interpretation of dichroism
experiments, which require high degrees of orientation for good
results. In studies of flexible polymers, one needs the rigid
macromolecule theory for comparison with data, in order to identify
the aspects due to flexibility, and similarly it is needed in cases
where the electric field is sufficiently influential to alter the
structure of a macromolecule. The last topic entered with birefringence studies of DNA (91) in high electric fields and has been
discussed elsewhere (92,93,94). We are likely to hear more of it
in relation to studies of membrane biomolecules such as receptors
and ion channels (95,96).

ACKNOWLEDGEMENT

This chapter includes adapted excerpts of Chapter 3 by C. T.
O'Konski and S. Krause, Electric Birefringence and Relaxation in
Solutions of Rigid Macromolecules, In: "Molecular Electro-Optics,
Part 1, Theory and Methods," C. T. O'Konski (ed.), by courtesy of
Marcel Dekker, Inc. (1976).

REFERENCES

1. M. Born, Ann. Physik., 55:177-240 (1918).
2. R. Jernigan, chapter in this volume.
3. P. Debye, and E. Marx (eds.), "Handbuch der Radiologie," Vol.
    6, Leipzig (1925), pp. 597-790.
4. P. Debye, and H. Sack, in: E. Marx (ed.), "Handbuch der
    Radiologie," Vol. 6, Leipzig (1934).
5. J. R. Partington, "An Advanced Treatise on Physical Chemistry,"
    Vol. 4, Physico-Chemical Optics, Longmans, Green and Co. Ltd.,
    London and Toronto (1953), p. 280.
6. A. D. Buckingham, Electric Birefringence in Gases and Liquids,
    Chap. 2, in: "Molecular Electro-Optics," Part 1, Theory and
    Methods, C. T. O'Konski (ed.), Marcel Dekker, Inc., New York
    (1976), pp. 27-62.
7. Tae-Kyu Ha, Quantum Theory and Calculation of Electric Polar-
    izability, Chapter 14, in: "Molecular Electro-Optics," Part 1,
    Theory and Methods, C. T. O'Konski (ed.), Marcel Dekker, Inc.,
    New York (1976), pp. 471-513.
7a. D. A. Dunmur, D. C. Hunt, and M. E. Jessup, Mol. Phys., 37(3):
    713-724 (1979).
7b. Q. D. Chue, and M. M. Raikhshtat, Dokl. Akad. Nauk SSSR (Phys.
    Chem.), 217(6):1347-1350 (1974).
8. A. D. Buckingham, and R. E. Raab, J. Chem. Soc., 2341 (1957).

8a. A. D. Buckingham, P. J. Stiles, and G. L. D. Ritchie, Trans. Far. Soc., 67:577 (1971).

8b. L. Onsager, J. Am. Chem. Soc., 58:1486 (1936).

8c. R. Patz, and M. T. Raetzsch, Z. Phys. Chem. (Leipzig), 260(4): 769-787; 788-794 (1979).

8d. Th. G. Scholte, Physica, 15:437 (1949).

8e. M. R. Battaglia, Proc. Int. Meet. Soc. Chim. Phys., 31st (1978), pp. 237-250.

8f. W. H. Orttung, and J. A. Meyers, J. Phys. Chem., 67:1911 (1963).

8g. M. S. Beevers, J. Chem. Soc. Far. Trans. II, 75:679 (1979).

8h. G. Khanarian, Thesis, Ph.D., Univ. of Sydney (1980).

9a. A. Peterlin, Acad. Znanosti Umetnosti, 3:111 (1947).

9b. H. A. Stuart, and A. Peterlin, J. Polymer Sci., 5:551 (1950).

9c. Y. Y. Gotlib, Zh Tekhn. Fiz., 27:707 (1957).

9d. M. V. Vol'kenshtein, and Y. Y. Gotlib, in: "Configurational Statistics of Polymer Chains," M. V. Vol'kenshtein (ed.), Interscience Publishers, a Division of John Wiley & Sons, Inc., New York, Chap. 7 (1963).

9e. W. H. Stockmayer, and M. E. Baur, J. Am. Chem. Soc., 86:3485 (1964).

9f. D. A. Dows, J. Chem. Phys., 41:2656 (1964).

9g. K. Nagai, and T. Ishikawa, J. Chem. Phys., 43:4508-4515 (1965); K. Nagai, J. Chem. Phys., 51:1091-1101 (1969); T. Ishikawa, and K. Nagai, Polymer J., 2:263-273 (1971).

9h. R. L. Jernigan, and P. J. Flory, J. Chem. Phys., 50:4178 (1969).

9i. R. L. Jernigan, and D. S. Thompson, Flexible Polymers, Chap. 5, in: "Molecular Electro-Optics," Part 1, Theory and Methods, C. T. O'Konski (ed.), Marcel Dekker, Inc., New York, pp. 159-206 (1976).

10. P. Langevin, Oeuvres Scientifiques.de Paul Langevin, "Service des publications du CNRS," Paris, pp. 369-391 (1950); Radium 7:249 (1910).

11. A. Peterlin, and H. A. Stuart, Z. Physik, 112:129-147 (1939).

12. A. Peterlin, and H. A. Stuart, "Hand- und Jahrbuch der Chemischen Physik," Vol. 8, Part 1B, A. Eucken and K. L. Wolf (eds.), Akademische Verlagsges., Leipzig, pp. 1-115 (1943).

13. C. T. O'Konski, J. Phys. Chem., 64:605-619 (1960).

14. J. A. Osborne, Phys. Rev., 67:351 (1945).

15. H. C. van de Hulst, "Light Scattering by Small Particles," Wiley, New York (1957).

16. C. T. O'Konski, "Encyclopedia of Polymer Science and Technology," Vol. 9, Wiley-Interscience, New York, p. 551 (1968).

17. H. Benoit, Ann Phys., 6:561-609 (1951).

18. D. N. Holcomb, and I. Tinoco, Jr., Biopolymers, 3:121-133 (1965).

19. I. Tinoco, Jr., J. Am. Chem. Soc., 77:4486-4489 (1955).

20. C. Bergholm, and Y. Björnstahl, Physik. Z., 21:137-141 (1920).

21. Y. Björnstahl, Phil. Mag., 2:701-732 (1926).

22. H. T. Buscher, R. M. McIntyre, and S. Mikuteit, IEEE Trans. Microwave Theory Techniques, 19:950 (1971).

23. G. Mie, Ann. Physik, 25(4):377-445 (1908).
24. C. T. O'Konski, note in preparation.
25. C. T. O'Konski, and F. E. Harris, J. Phys. Chem., 61:1172 (1957).
26. A. R. von Hippel, "Dielectrics and Waves," Wiley, London, pp. 228-235 (1954).
27. C. T. O'Konski, and H. C. Thacher, Jr., J. Phys. Chem., 57: 955 (1953).
28. O. Wiener, Ber. Sächs. Ges. Wiss. (Math. Phys. Klg.), 62:256 (1910).
29. C. T. O'Konski, and B. H. Zimm, Science, 3:113-116 (1950).
30. C. T. O'Konski, and A. J. Haltner, J. Am. Chem. Soc., 79:5634 (1957).
31. J. Errera, J. Th. G. Overbeek, and H. Sack, J. Chem. Phys., 32:681-704 (1935).
32. S. A. Rice, and M. Nagasawa, "Polyelectrolyte Solutions: A Theoretical Introduction," Academic, New York (1961).
33. C. T. O'Konski, J. Chem. Phys., 23:195 (1955).
34. J. B. Miles, and H. P. Robertson, Phys. Rev., 40:583 (1932).
35. H. P. Schwan, Adv. Biol. Med. Phys., 5:147 (1957).
36. J. P. McTague, and J. H. Gibbs, J. Chem. Phys., 44:4295-4301 (1966).
37. J. G. Kirkwood, and J. B. Shumaker, Proc. Nat. Acad. Sci., U.S., 38:855 (1952).
38. U. Schindewolf, Naturwissenschaften, 40:435 (1953).
39. U. Schindewolf, Z. Elektrochem., 58:697 (1954).
40. M. Eigen, and G. Schwarz, J. Colloid Sci., 12:181-194 (1957).
41. A. Katchalsky, S. B. Sachs, A. Raziel, and H. Eisenberg, Trans. Faraday Soc., 65:77 (1969).
42. G. A. Johnson, and S. M. Neale, J. Polymer Sci., 54:241 (1961).
43. J. M. Neale, and D. A. Weyl, Proc. Roy. Soc. (London), A291: 368 (1966).
44. S. Takashima, Adv. Chem. Ser., 63:232-252 (1967).
45. S. Takashima, Biopolymers, 5:899 (1967).
46. C. T. O'Konski, N. C. Stellwagen, and M. Shirai, manuscript in preparation.
47. C. T. O'Konski, and S. Krause, J. Phys. Chem., 74:3243 (1970).
48. G. Schwarz, Z. Physik. Chemie N.F., 19:5/6 (1959).
49. M. Mandel, Mol. Phys., 4:489 (1961): F. van der Touw, and M. Mandel, Biophys. Chem., 2:218 (1974).
50. F. Oosawa, Biopolymers, 9:677-688 (1970).
51. C. Hornick, and G. Weill, Biopolymers, 10:2345-2358 (1971).
52. K. Kikuchi, and K. Yoshioka, Biopolymers, 15:583-587 (1976).
53. C. T. O'Konski, and M. Shirai, in preparation.
54. G. S. Manning, J. Chem. Phys., 51:924, 3249 (1969).
55. G. S. Manning, Biophys. Chem., 9:65 (1978).
56. E. Charney, K. Yamaoha, and G. S. Manning, Biophys. Chem., 11:167-172 (1980).
57. E. Charney, Biophys. Chem., 11:157-166 (1980); E. Charney and C. H. Lee, Macromolecules, 13:66-88 (1980).

58. D. Stigter, J. Phys. Chem., 82:1603-1606 (1978).

59. A. D. MacGillivray, J. Chem. Phys., 56:80, 83 (1972); 57:4071, 4075 (1972).

60. C. T. O'Konski, and F. E. Harris, J. Phys. Chem., 61:1172 (1957).

61. A. D. Buckingham, Proc. Roy. Soc. (London), A267:27 (1962).

62. W. H. Orttung, J. Am. Chem. Soc., 87:924 (1965); J. Phys. Chem., 73:2908 (1968).

63. J. C. Powers, Jr., J. Am. Chem. Soc., 88:3679 (1966); 89:1780 (1967).

64. H. G. Kuball, Z. Naturforsch., 22a:1407 (1967); H. G. Kuball and R. Göb, Z. Physik. Chem. (Frankfurt), 62:237; 63:251 (1968); H. G. Kuball, and D. Singer, Ber. Bunsenges. Physik. Chem., 73:403 (1969); H. G. Kuball, W. Galler, R. Göb, and D. Singer, Z. Naturforsch., 24a:1391 (1969).

65. C. Houssier, and H-G. Kuball, Biopolymers, 10:2421 (1971).

66. E. Charney, and R. S. Halford, J. Chem. Phys., 29:221 (1958).

67. A. E. Lutskii, B. A. Veretenchke, and I. S. Romadanov, Isv. Vyssh. Uchebn. Zavd., Fiz. 17(9):156-157 (1974).

68. C. T. O'Konski (ed.), "Molecular Electro-Optics," Part 2, Applications to Biopolymers, Marcel Dekker, New York (1978); (References to proteins and polypeptides in Ch. 17 by K. Yoshioka; to polynucleotides and nucleic acids in Ch. 18 by N. C. Stellwagen.)

69. K. Yoshioka, and C. T. O'Konski, Biopolymers, 4:499-507 (1966).

70. I. Tinoco, Jr., J. Am. Chem. Soc., 79:4336 (1957).

71. C. T. O'Konski, K. Yoshioka, and W. H. Orttung, J. Phys. Chem., 63:1558-1565 (1959).

72. A. Haschemeyer, and I. Tinoco, Jr., Biochemistry, 1:503 (1962).

73. S. Krause, and C. T. O'Konski, Biopolymers, 1:503 (1963).

74. P. Moser, P. G. Squire, and C. T. O'Konski, J. Phys. Chem., 70:744 (1966).

75. C. G. LeFevre, R. J. W. LeFevre, and C. M. Parkins, J. Chem. Soc., 1958:1468 (1958).

76. N. C. Stellwagen, Ch. 18, Electro-Optics of Polynucleotides and Nucleic Acids, in: "Molecular, Electro-Optics," Part 2, Applications to Biopolymers, C. T. O'Konski (ed.), Marcel Dekker, Inc., New York, pp. 645-683 (1976).

77. R. Gans, Ann. Physik, 64:481 (1921).

78. M. J. Shah, J. Phys. Chem., 67:2215 (1963).

79. K. Yoshioka, and H. Watanabe, "Physical Principles and Techniques of Protein Chemistry," Part A, S. J. Leach (ed.), Academic, New York, pp. 335-67 (1969).

80. M. Matsumoto, H. Watanabe, and K. Yoshioka, Sci. Papers Coll. Gen. Educ. Univ. Tokyo, 17:173-202 (1967).

81. D. N. Holcomb, and I. Tinoco, Jr., J. Phys. Chem., 67:2691-2698 (1963).

82. K. Yoshioka, and C. T. O'Konski, J. Polymer Sci., A-2(6):421 (1968).

83. K. Kikuchi, and K. Yoshioka, Rep. Prog. Polymer Phys. Japan, 10:19-22 (1967).

84. M. J. Shah, D. C. Thompson, and C. M. Hart, J. Phys. Chem., 67:1170 (1963).

85. H. Nakayama, and K. Yoshioka, Nippon Kagaku Zasshi (J. Chem. Soc. Japan), 85:177-182 (1964).

86. H. Watanabe, K. Yoshioka, and A. Wada, Biopolymers, 2:91-101 (1964).

87. H. Nakayama, and K. Yoshioka, J. Polymer Sci., A-3:813-825 (1965).

88. M. Matsumoto, H. Watanabe, and K. Yoshioka, J. Phys. Chem., 74:2182-2188 (1970).

89. M. Matsumoto, H. Watanabe, and K. Yoshioka, Biopolymers, 6:905-915 (1968).

90. N. C. Stellwagen, Configurations of Sodium Deoxyribonucleate and Sodium Phosphate in Solution, Thesis, University of California, Berkeley (1967).

91. C. T. O'Konski, and N. C. Stellwagen, Biophys. J., 5:607-613 (1965).

92. C. T. O'Konski, and S. Krause, Ch. 3, Electric Birefringence and Relaxation, in: "Molecular Electro-Optics," Part 1, Theory and Methods, C. T. O'Konski (ed.), Marcel Dekker, Inc., New York, pp. 63-120 (1976).

93. K. Kikuchi, and K. Yoshioka, Biopolymers, 15:1669-1676 (1976).

94. G. Schwarz, Ann. N. Y. Acad. Sci., 303:190-197 (1977).

95. R. D. O'Brien, Problems and Approaches in Noncatalytic Biochemistry, in: "The Receptors," R. D. O'Brien (ed.), pp. 311-335 (1979).

96. G. Ehrenstein, and H. Lecar, Electrically Gated Ionic Channels in Lipid Bilayers, Q. Rev. Biophys., 10:1-34 (1977).

ELECTRIC BIREFRINGENCE DYNAMICS

Sonja Krause

Department of Chemistry
Rensselaer Polytechnic Institute
Troy, New York

and

Chester T. O'Konski

Department of Chemistry
University of California
Berkeley, California

The rate at which macromolecules in a solution orient under the influence of an applied electric field determines the time de-.pendence of the optical birefringence. It is well known from studies of dielectric dispersion and nuclear magnetic resonance in ordinary liquids that molecular reorientation processes of small molecules normally occur with lifetimes of 10^{-11} to 10^{-9} sec. Longer relaxation times are encountered with macromolecules in so-lution and the direct measurement of these relaxation times through electro-optic effects is now an important method for studying the sizes and shapes of macromolecules.

When an electric field is applied or changed across a solution, there will be an accompanying change in the orientation distribu-tion function of the macromolecules, $f(\theta, t)$. Because of fric-tional and inertial effects, the orientation will not change instan-taneously, so that transient effects in the birefringence will be observed. The orientation takes place about some center within the macromolecule, the center of hydrodynamic drag, and any motion about this center may be represented as the sum of rotations about three perpendicular axes. In general, some solvent molecules will be bound to the macromolecule, and for our purposes they may be regarded as part of the rigid hydrodynamic unit. The hydrodynamic properties may be described in terms of frictional coefficients for rotation of the macromolecule about various axes through the center of drag, i.e., three orthogonal axes for the principal frictional coefficients, or more conventionally, in terms of the principal components of the rotational diffusion tensor. These are related to the frictional coefficients by the Einstein equation,

$$\Theta_{ii} = \frac{kT}{\zeta_{ii}} \tag{1}$$

where ζ_{ii} is a frictional coefficient, Θ_{ii} is the corresponding rotational diffusion coefficient, or "constant," at temperature T, k is Boltzmann's constant, and i labels one of the principal axes.

We restrict the initial discussion to cylindrically symmetric molecules. To calculate the time dependence of the birefringence, $\Delta n(t)$, we need to find the time dependence of the orientation distribution function, $f(\theta, t)$, since

$$\Delta n(t) = \left[\frac{\pi C_v (g_a - g_b)}{n}\right] \int_0^\pi f(\theta, t)\, (3 \cos^2 \theta - 1)\, 2\pi \sin \theta\, d\theta \tag{2}$$

Here $f(\theta, t)$ is the probability per unit solid angle of finding a molecule with its symmetry axis at angle θ with respect to the direction of the applied electric field. The distribution function is

$$f(\theta, t) = e^{-U/kT} \left[\int_0^\pi e^{-U/kT}\, 2\pi \sin \theta\, d\theta\right]^{-1} \tag{3}$$

where U is the energy of the macromolecule as a function of angle, C_v is the volume fraction of the particles, and the term $(g_a - g_b)$ is the optical anisotropy factor which can be calculated by application of an equation for ellipsoids due to Maxwell and given by Peterlin and Stuart [1,2]. This equation is

$$g_j = \frac{1}{4\pi} \frac{n_j^2 - n^2}{1 + (n_j^2/n^2 - 1) A_j} \tag{4}$$

where A_j is the depolarization factor, n is the refractive index of the solvent, and n_j ($j = a, b$) are refractive indexes

of the particle for light with electric vector along the axes a
and b , respectively. The depolarization factor may be calculated
from the expression

$$A_j = \int_0^\infty \frac{abc}{2(j^2 + s) \ [(a^2 + s) \ (b^2 + s) \ (c^2 + s)]^{1/2}} \ ds \tag{5}$$

Here, a, b, and c are the lengths of the semiaxes of the particle, and s is a variable of integration. The integrals have been
tabulated for ellipsoids of revolution [3]. Graphs [4] and special
formulas also are available [5].

In order to calculate $f(\theta, t)$ as a function of time, one must,
in general, consider the orienting torque due to the electric field,
the random Brownian motion which causes the particles to undergo a
random walk in the orientation sense, the friction between the
particle and the surrounding medium, and inertial effects. A
straightforward calculation reveals that for ordinary sized macromolecules in common liquids, the inertial effects are quite negligible; that is, a rotational motion is very quickly dissipated as
the molecule expends its inertial energy on viscous drag in the
surrounding medium [6]. Hence, we will be concerned here with the
formulation of the problem as a Brownian diffusion process biased
by the application of an electric field which produces a torque,
M , acting on the macromolecule. The partial differential equation
for the diffusion process may be written [7-9]

$$\frac{1}{\Theta} \frac{\partial f(\theta, t)}{\partial t} = \frac{1}{\sin \theta} \frac{\partial}{\partial \theta} \left[\sin \theta \left\{ \frac{\partial f(\theta, t)}{\partial \theta} - \frac{Mf(\theta, t)}{kT} \right\} \right] \tag{6}$$

where θ is the angle between the symmetry axis of the macromolecule under consideration and the applied electric field, t
is the time and k , T , and $f(\theta, t)$ have the meanings stated
above. Θ is the diffusion constant for the reorientation of the
symmetry axes of the molecules by a rotation about their transverse
axes. This equation has been solved for situations in which
M = 0 , corresponding to field-free relaxation, M = M(θ) , corresponding to a steady electric field, and also for M = M(θ, t) as
in time dependent electric fields.

FIELD-FREE RELAXATION THEORY AND RESULTS

It was shown [7,8] that in a static electric field, the solution of Eq. (6) in the steady state may be expressed

$$f(\theta) = a_0 + a_1 P_1 (\cos \theta) + a_2 P_2 (\cos \theta) + a_3 P_3 (\cos \theta) + \ldots \tag{7}$$

where the $P_n(\cos \theta)$ are the Lengendre polynomials of order n, and a_0, a_1, and a_2 are constants. The term of this series that contributes to the polarization tensor characterizing the birefringence is the one involving $P_2(\cos \theta) = (3 \cos^2 \theta - 1)/2$. Orthogonality of all other terms with respect to this term in the integral of Eq. (2) leads to zero contributions from them. It also was shown that the field-free decay of the birefringence follows a first-order decay law which may be written

$$\Delta n(t) = \Delta n^o \, e^{-t/\tau} \tag{8}$$

where the birefringence relaxation time, τ, is given by [9,20]

$$\tau = \frac{1}{6\Theta} \tag{9}$$

THE ELECTRIC BIREFRINGENCE RELAXATION METHOD AND APPARATUS

A convenient and versatile way of studying the electric, optical, and hydrodynamic properties of macromolecules is the transient electric birefringence or the electric birefringence relaxation method. It was introduced at the University of California at Berkeley in 1948, and about the same time in the Soviet Union [10], and in France [7,11,12]. In the Berkeley work, the inspiration for the method was a desire to study the sizes and shapes of macromolecules in solutions by optical observations, and a recognition of the need to better understand the puzzling electro-optic behavior of colloidal systems observed in sinusoidal fields of various intensities and frequencies [13-21]. Since the systems to be investigated were aqueous solutions displaying electric conductivity and electrophoresis, it was apparent that it would be an advantage to apply short rectangular pulses, thus minimizing these effects, and to observe the relaxation effects directly. Experiments were carried out using an electronically generated antisymmetric square wave of the kind illustrated in Fig. 1a [8].

Since pulses of equal and opposite polarity were applied, the net electrophoretic motion of the particles was zero. The pulse duration, amplitude, and the period between pulses was adjustable, so that the duty cycle could be reduced to minimize the effects of Joule heating of the solution. The early Strasbourg work was done with a motor-driven rotating switch [7,11,12].

The first pulsed-field observations at the University of California at Berkeley were made on TMV , which was a fortunate choice because it was a rigid macromolecule which could be obtained

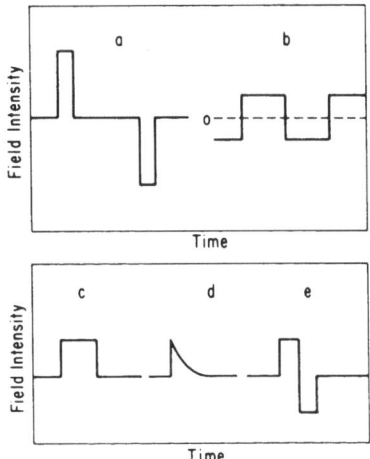

Fig. 1. Electric field waveforms for electro-optic experiments.
(a) Continuously repeated pulses of opposite polarity
(see Ref. 3); (b) continuous square wave; (c) single pulse
(d) exponential pulse; and (e) reversing field pulse.

in homogeneous preparations, and it was large enough to permit convenient measurements on an oscilloscope. It also turned out that
with TMV the relaxation effects associated with rotational diffusion were well separated in the time domain from those produced by
the electrical mechanisms involved in orienting the particles [22].
It was found [8] that the relaxation time for the field-free decay
of the electric birefringence of TMV was approximately 0.6 msec,
and that dilute solutions exhibit simple exponential build-up and
decay curves. Additional experiments were conducted in which the
field consisted of regular square waves, shown in Fig. 1b (note
that the field never is zero). It was found that the birefringence
was substantially unchanged as the frequency of the field was increased from values below to values above the reciprocal of the
observed birefringence time constant. Also, there was no transient
upon field reversal. These observations led to the proposal that
the mechanism of orientation was primarily not a permanent dipole
moment coupling effect, as previously supposed [15], but rather an
induced polarization effect in the dilute solutions [8]. More concentrated solutions of TMV showed a complex build-up curve and a
steady-state birefringence of negative sign. Values of steady-state
optical retardation as a function of applied field strength at
various concentrations are shown in Fig. 2. These results led us
to suggest that the rods may orient across the direction of the
electric field at higher concentrations. This could be caused by
a hydrodynamic orientation effect, or by a small transverse dipole
moment, which probably becomes relatively more important when the
ion atmospheres overlap. This should tend to reduce induced polarization orientation.

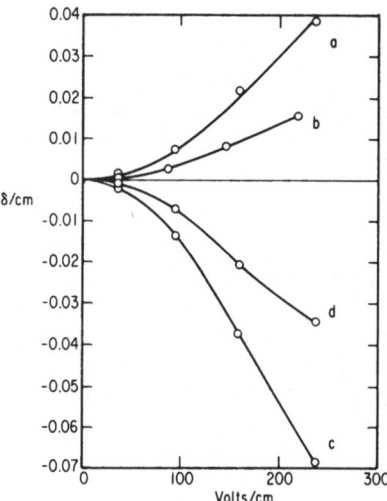

Fig. 2.   Peak values of optical retardation in radians per centimeter of solution versus applied field strength: (a)
0.073% TMV, 1.5×10^{-4} M buffer; (b) 0.073% TMV, 1.0 x
10^{-3} M buffer; (c) 1.4% TMV, 1.5×10^{-4} M buffer; (d)
2.75% TMV, 1.5×10^{-4} M buffer.   (Reproduced from results
given in Ref. 8.)

        Subsequently, to minimize effects occurring as a result of
Joule heating, the single-shot square pulse technique was introduced; also, exponential pulses and reversing pulses were employed
(see Figs. 1c, d and e) [25,26,43).   The square pulse allows observations of the build-up of the birefringence under controlled conditions, as well as the decay, whereas the exponential pulse (Fig.
1d) is particularly simple to obtain by the discharge of a condenser through the solution or through a resistor switched across
it.   With the exponential pulse, the field is decaying even as the
birefringence is building up.   A first-order kinetic analysis has
been given [22].   Additional information can be obtained by applying a pulse and suddenly reversing its polarity, as shown in Fig.
1e.   Besides the possibility of following the build-up and decay
kinetics, one can observe the transient which may occur when the
electric field is rapidly reversed [22].   If the orientation is due
to permanent electric dipoles, then reversing the field will cause
the molecules, which are preferentially oriented in a given direction, to diffuse until there is a preferential orientation in the
reverse direction.   This will produce a transient in the electric
birefringence.   On the other hand, if the orientation is due to an
induced polarization which is fast compared to the rate of reorientation of the macromolecules, then reversing the field will at the

same time reverse the polarization and leave the orienting torque
unchanged so that no transient will result [22]. Detailed calculations of reversing field transients were carried out by Tinoco
and Yamaoka [23] and by Matsumoto, et al. [24].

THEORY OF THE BUILDUP AND FIELD REVERSAL TRANSIENTS IN WEAK ELECTRIC FIELDS

Equations for the time dependence of the birefringence in a
suddenly applied electric field were given by Benoit for the case
of cylindrically symmetric macromolecules with a dipole moment μ_a
and an electric anisotropy $(\alpha_a - \alpha_b)$ along the symmetry axis. For
electric fields sufficiently weak so that the orientation energy is
small compared to kT , Benoit [7] obtained

$$\frac{\Delta n(t)}{\Delta n(\infty)} = 1 - \frac{3P}{2(P + Q)} e^{-2\Theta t} + \frac{P - 2Q}{2(P + Q)} e^{-6\Theta t} \qquad (10)$$

where time is measured from the instant of application of the field.
Here, $P = P_a = B_a^2 \mu_a^2 / k^2 T^2$, and $Q = (\alpha_a - \alpha_b)/2kT$ (see Chapter
on Kerr Theory for a discussion of the P and Q terms). Equa-
tions for the less symmetric models were later derived by Holcomb
and Tinoco [24].

For particles of the same hydrodynamic and polarizability
symmetry, but with transverse as well as axial dipole moment components, Tinoco and Yamaoka [23] derived an expression for the birefringence transient produced by rapid reversal of the applied field.
It may be written

$$\frac{\Delta n(t)}{\Delta n(\infty)} = 1 - \frac{3P_a}{P_a - P_b/2 + Q} e^{-2\Theta_b t} \qquad (11)$$

$$+ \frac{6\Theta_b}{5\Theta_b - \Theta_a} \frac{P_b}{P_a - P_b/2 + Q} e^{-(\Theta_a + \Theta_b)t}$$

$$+ \frac{3P_a - 6\Theta_b P_b/(5\Theta_b - \Theta_a)}{P_a - P_b/2 + Q} e^{-6\Theta_a t}$$

Here Q and P_a are the same as Benoit's and P_b is a dipole term
similar to P_a but perpendicular to the symmetry axis of the molecule. Time is measured from the instant of field reversal. Tinoco

and Yamaoka also derived equations for the case where the induced
dipole polarization is not fast, for example, when the transverse
and longitudinal counterion polarization relaxation times are of the
same order of magnitude as $1/\Theta$. The expressions are complex and
are not reproduced here. They presented useful curves based upon
numerical calculations over a wide range of parameters. Further
examples have been presented elsewhere [24]. See the Kerr Theory
chapter for later inclusion of the internal field functions.

REPRESENTATIVE DATA AND INTERPRETATION

     Accurate measurements on homogeneous solutions of TMV mono-
mer gave a 0.50 msec relaxation [25,26]. There was a discrepancy
between this relaxation time, measured in dilute solutions, and the
value calculated from the known dimensions of the rodlike macromolecule using the Burgers equation for the hydrodynamic friction coefficient [26]. The constancy of the relaxation time with varying
ionic strength showed that there is no great effect of the ion atmosphere, so far as rotational diffusion is concerned [22]. After
examination of alternative possibilities, it was suggested that the
hydrodynamic equation might not be sufficiently accurate. Later,
Broersma made a study of the problem and decided that indeed a more
extensive calculation than that of Burgers, based upon the Oseen
hydrodynamic theory, was required for cylindrical macromolecules.
He derived a more accurate expression for the frictional coefficients [27]. With it, a value of 305 nm was obtained for the
length of the macromolecule from the birefringence data, compared
with 298 nm for the dry length measured in an electron microscope
[28]; whereas the older Burgers equation had given about 342 nm.
The excellent agreement between the improved theory and experiment
was an early victory of the electric birefringence method [26,27,
29]. More recent electric dichroism measurements [30] gave
τ = 0.49 msec, verifying the earlier electric birefringence value
to 2%, which corresponds to 0.7% in macromolecule length.

     The Kerr constant of TMV turned out to be extraordinarily
large compared with the value one would predict from electrostatic
theory, and it was observed that the concentration of electrolyte
had an appreciable effect on it [22]. These observations helped
to stimulate the development of a new model for electrically conducting systems such as polyelectrolytes [3]. In rapidly reversing
pulsed fields, no field reversal transient was observed in the
common strain of TMV , confirming that the orientation was not due
to a permanent dipole. This confirmed that the counterion polarization was predominant. Later, a quantitative theory of the low
frequency Kerr effect was developed [31] using the model just mentioned [3], and a comparison between experiment and theory became
possible [31]. Thus the homogeneous preparations of the crystal-
lizable TMV played a key role in the development of electric

birefringence relaxation techniques and theory. The Holmes Rib
Grass strain of TMV was observed to give a large field reversal
transient [32], indicating a permanent dipole moment for that virus.

DISPERSION OF THE BIREFRINGENCE IN SINE WAVE FIELDS

    The anomalous dispersion of the Kerr constant (a decrease of
K with increasing frequency of the applied field) in sine wave
fields was first observed by Raman and Sirkar [33]. It has been
discussed for nonconducting dilute solutions of rigid molecules at
low fields by a number of authors [2,22,34,35]. The molecules
were assumed to have an axis of symmetry (rodlike or ellipsoidal)
along which a dipole moment may or may not be present. Also, the
principal axes of the dielectric tensor were assumed to lie along
the molecular axes. Peterlin and Stuart [2], Benoit [34], and
O'Konski and Haltner [22] considered monodisperse systems only,
while Thurston and Bowling [35] also discussed polydisperse systems.

    To calculate the birefringence in sine-wave fields, there is
no change in the birefringence expression, Eq. (2), but the distribution function for the macromolecular orientation, Eq. (3), becomes time- and frequency-dependent. An additional factor appears
in this expression, as discussed by Plummer and Jennings in connection with the light-scattering equations for rodline particles in
alternating electric fields [36], and in connection with the distribution function discussed by Stoylov [37].

    Let the applied field be $E_o \sin \omega t$, where E_o is the maxi-
mum amplitude of the alternating electric field with angular frequency, ω . Then the calculated value of birefringence of the
solution, Δn , at low field and at a time after turning on the
field, t (which is long with respect to molecular rotation times)
can be separated into a birefringence, $(\Delta n)_i$, due to the induced
dipole moment alone, and a birefringence, $(\Delta n)_p$, due to the permanent dipole moment only. The induced component is

$$(\Delta n)_i = (\Delta n)_{i,av} \left[1 \pm \frac{\cos(2 \omega t - \phi_i)}{(1 + 4 \omega^2 \tau^2)^{1/2}} \right] \tag{12}$$

where $(\Delta n)_{i,av}$ is the average birefringence observed. The birefringence consists of two components; one of these is constant with
time and the other alternates with twice the frequency of the
applied field. Here again τ is the birefringence relaxation time
of the molecule, and ϕ_i is the phase angle between the birefringence and the applied field, given by

$$\tan \phi_i \;=\; 2\omega\tau \tag{13}$$

The value of $(\Delta n)_{i,av}$ is the same as that which would be found if a steady field equal to $E_o/\sqrt{2}$ or E_{rms} , were applied to the same system. When the frequency of the applied field becomes very high, only $(\Delta n)_{i,av}$ will be observed. At extremely high frequencies, such as the optical frequencies obtainable using high-power lasers, the induced dipole can probably no longer be assumed to vary instantaneously when the applied field changes, and the observed birefringence, $(\Delta n)_i$, may drop below $(\Delta n)_{i,av}$. At lower frequencies, $(\Delta n)_i$ fluctuates between maximum and minimum values given by Eq. (12) with the numerator of the last term equal to ± 1 .

Experimental curves of such maximum and minimum values of $(\Delta n)_i$ for a solution of TMV (a molecule with a negligible permanent dipole moment) are shown in Fig. 3. The above theory explains the lower frequency region and does not treat the counterion relaxation responsible for the "anomalous" dispersion around 0.1 – 1 MHz, which is explained in the last paragraph of this section.

The reason for the phase angle, ϕ , for the doubled frequency component of the birefringence, and for the general variation of the birefringence with frequency can be visualized as follows:

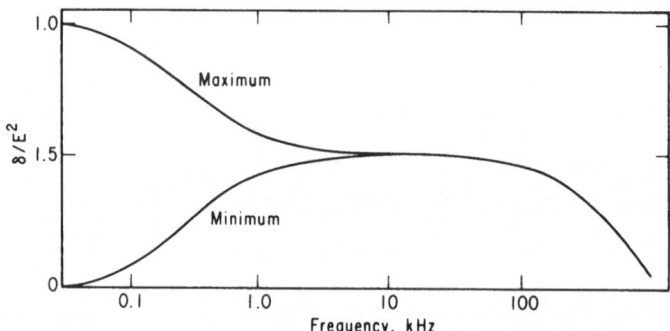

Fig. 3. Birefringence dispersion for a solution of TMV . At low frequencies the macromolecules orient in phase with the field. At 100–1000 Hz, they begin to lag, and above 10^4 Hz, they maintain a steady-state orientation, which decreases because of ion-atmosphere relaxation as frequency is increased. (Reproduced from Ref. [22], courtesy of American Chemical Society Publications.)

At normal frequencies, the induced dipole can be assumed to vary instantaneously with the field intensity, fluctuating between zero and a maximum magnitude which is the same for $\pm E_o$; it therefore fluctuates between zero and its maximum absolute value twice during each cycle of the applied field. At low frequencies, the molecular orientation follows the applied field intensity, but lags behind it because rotational diffusion takes a finite time. At high frequency, the molecular orientation time is so slow with respect to changes in electric field strength, that the molecular orientation no longer changes with time and the observed birefringence equals that which would be observed in a steady field, as discussed above.

The birefringence due to a dipole moment along the molecular symmetry axis is

$$(14)$$

$$(\Delta n)_p = (\Delta n)_{p,0} \; \frac{1}{1 + 9\,\omega^2\tau^2} + \frac{\cos(2\,\omega t - \phi_p)}{(1 + 9\,\omega^2\tau^2)^{1/2}\,(1 + 4\,\omega^2\tau^2)^{1/2}}$$

Equation (14) takes no account of the birefringence due to the induced dipole moment; it is strictly a dipole term, and the dipole phase angle, ϕ_p , is given by

$$\tan \phi_p = \frac{5\omega\tau}{1 - 6\,\omega^2\tau^2} \tag{15}$$

At any frequency, ω , the permanent dipole birefringence, Δn_p , consists of two components, one of which is constant with time, and the other of which alternates with twice the frequency of the applied field, just as in the case of an induced dipole. In the permanent dipole case, however, both the time-independent and the time-dependent components decrease to zero at high frequency. At lower frequencies, $(\Delta n)_p$ varies between maximum and minimum values given by Eq. (14) with the numerator of the last term equal to ± 1 . The value of $(\Delta n)_{p,o}$ is the birefringence that would be obtained in a steady field equal to E_{rms} for pure dipole orientation.

The behavior of the pure dipole $(\Delta n)_p$, as the frequency of the applied field changes, can be explained as follows: At normal frequencies, the torque on the molecules is in one direction when the field is positive, and in the opposite direction when the field is negative. The molecules are induced to turn first in one direction, and then in the opposite direction during a single cycle of the applied field. The finite rotational diffusion time of the molecules again causes a time lag, and this produces a phase angle,

ϕ_p , between the birefringence and the applied field. As the frequency rises, the molecules can no longer follow the field, either
in magnitude or in direction, and the total birefringence, Δn_p ,
falls to zero. At low frequencies, while the molecules are turning,
there comes a time, twice in each cycle of the applied field, when
the molecules have oriented mainly perpendicularly to the applied
field, and the birefringence becomes negative. Also, twice in each
cycle of the applied field, namely, when $(2 \omega t - \phi_p)$ =
$\pi/2$, $3\pi/2$, ... , the birefringence will change its sign. This can
also be seen from Eq. (14).

Since a molecule with a dipole moment also has polarizability,
the total birefringence for such a molecule will be:

$$\Delta n = (\Delta n)_i + (\Delta n)_p \tag{16}$$

At low frequencies, $(\Delta n)_p$ predominates, while at very high frequencies, only $(\Delta n)_i$ remains, in the form of $(\Delta n)_{i,av}$.

Thurston and Bowling [35] have analyzed the behavior of Δn
and the observed phase angle as a function of frequency for different ratios of induced to permanent dipole moments. They also showed
how to determine τ from experimental data. When no dipole moment
is present, τ can be determined much more simply [22] from the
slope of a $\tan \phi_i$ vs. ω curve.

Thurston and Bowling [35], in effect, combined Eqs. (12) and
(14) to give

$$\Delta n = \Delta n_{st} + \Delta n_{alt} \cos(2 \omega t - \phi_{alt}) \tag{17}$$

where Δn_{st} is the steady component and Δn_{alt} is the magnitude
of the alternating component of the birefringence having phase angle
ϕ_{alt} . Equation (17) is an alternate version of Eq. (16). If the
permanent dipole is along the axis of symmetry of the molecule,
then

$$\Delta n_{st} = \frac{\Delta n_o}{r + 1} \left\{ r + \frac{1}{1 + 9 \omega^2 \tau^2} \right\} \tag{18}$$

where Δn_o is the low-frequency limiting value of Δn_{st}

$$\Delta n_o = (\Delta n)_{i,av} + (\Delta n)_{p,o} \tag{19}$$

and

$$r = kT \frac{(\alpha_a - \alpha_b)}{\mu^2} \qquad (20)$$

where α_a and α_b are the excess polarizabilities of the macro-molecule parallel and perpendicular to the symmetry axes, respectively, and μ is the dipole moment of the solvated macromolecule assumed to be directed along its symmetry axis. The factor r (called R by Thurston and Bowling [35]) characterizes the relative magnitude of induced and permanent dipole moment effects in determining the orientation of the molecules.

Figures 4 and 5 show the frequency dependence of Δn_{st} for various values of r. Figure 4 shows the curves for values of r ("P" of Thurston and Bowling) from 0 to ∞ and from -1 to $-\infty$, for which the sign of Δn_{st} does not change with frequency. Figure 5 shows the curves for values of r from 0 to -1, for which there is a reversal of sign with frequency for the steady component. Further discussion is given by Thurston and Bowling [35].

The above theory does not account for effects due to ionic relaxation, as no account of them was taken in the model.

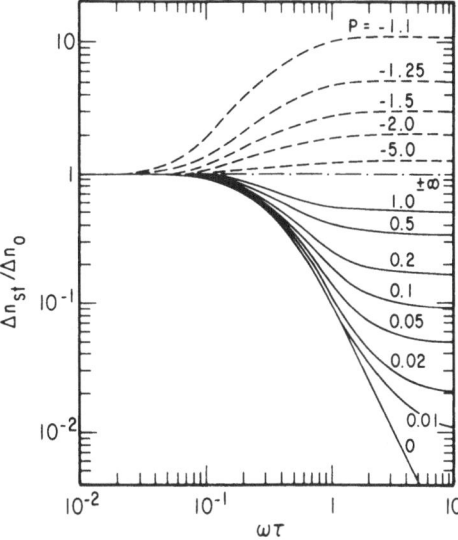

Fig. 4. Frequency dependence of the steady component of the bire-fringence for $r \equiv P = 0$ to $+\infty$ and from -1 to $-\infty$. (Reproduced from results given in Ref. 35, by courtesy of Academic Press, Inc.)

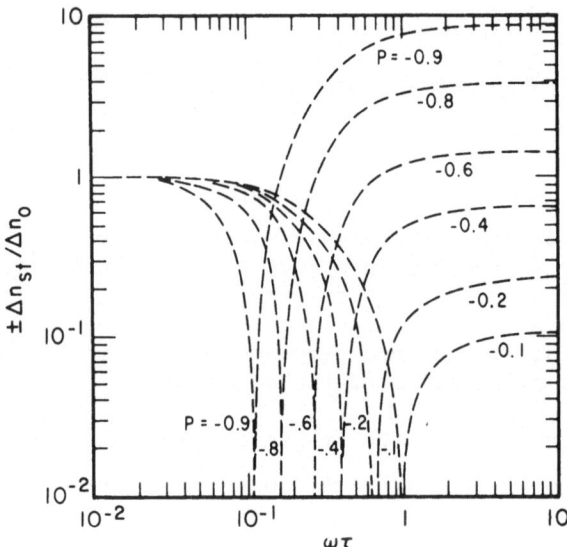

Fig. 5. Frequency dependence of the steady component of the bire-
         fringence for r ≡ P = 0 to -1. The birefringence ratio is
         positive at the lower frequencies and negative at the
         higher frequencies. (Reproduced from results given in
         Ref. 35.)

     The experimental setup for the study of Kerr effect in sine
wave fields is very similar to that using square pulses. In studies of electrolyte solutions, it is often an advantage to use pulsed
rather than continuous sine-wave fields [22,38] to minimize joule
heating of the solutions. The birefringence and the phase angle,
ϕ , have been measured using Lissajou's figures [34,39] or a lock-
in amplifier [35,40].

     Some of the polyelectrolytes studied, using alternating elec-
tric fields, have been TMV [22,34,35], bentonite [35,41], poly-
riboadenylic acid, and polyribouridylic acid [38]. Also the polypeptide poly-λ-benzyl-L-glutamate [42] was studied.

     O'Konski and Haltner [22] and Thurston and Bowling [35] found
that the birefringence of some materials, noticeably TMV solutions, decreased at high frequency. O'Konski and Haltner [22] attributed this anomalous dispersion to the relaxation of the ion atmosphere around the molecules. The theory of the "static" or low
frequency orientation effect for charged macromolecules has been
treated elsewhere [31], but high frequency dispersion equations
have not been developed, so far as we are aware.

ACKNOWLEDGEMENTS

This Chapter has been adapted from a portion of Chapter 3 by the same authors, Electric Birefringence and Relaxation in Solutions of Rigid Macromolecules, in: "Molecular Electro-Optics, Part 1, Theory and Methods," C. T. O'Konski (ed.), by courtesy of Marcel Dekker, Inc. (1976). We thank R. S. Farinato for comments on the proof.

REFERENCES

1. A. Peterlin and H. A. Stuart, Z. Physik, 112:129-147 (1939).
2. A. Peterlin, and H. A. Stuart, Hand- und Jahrbuch der Chemischen Physik, Vol. 8, Part 1B, in: "Akademische Verlagsges.," A. Eucken and K. L. Wolf (eds.), Leipzig (1943), pp. 1-115.
3. C. T. O'Konski, J. Phys. Chem., 64:605-619 (1960).
4. J. A. Osborne, Phys. Rev., 67:351 (1945).
5. H. C. van de Hulst, "Light Scattering by Small Particles," Wiley, New York (1957).
6. C. T. O'Konski, and L. S. Shepard, Appendix B in: "Molecular Electro-Optics, Part 1, Theory and Methods," C. T. O'Konski (ed.), Marcel Dekker, Inc., New York (1976).
7. H. Benoit, Ann. Phys., 6:561-609 (1951).
8. C. T. O'Konski, and B. H. Zimm, Science, 3:113-116 (1950).
9. F. Perrin, J. Phys. Radium, 5:497 (1934).
10. N. A. Tolstoi, and P. P. Feofilov, Dokl. Akad. Nauk SSSR, 66:617-620 (1949).
11. H. Benoit, Compt. Rend., 228:1716-1718 (1949).
12. H. Benoit, Compt. Rend., 229:30-32 (1949).
13. J. Errera, J. Th. G. Overbeek, and H. Sack, J. Chem. Phys., 32:681-704 (1935).
14. C. E. Marshall, Trans. Faraday Soc., 26:173-189 (1930).
15. M. A. Lauffer, J. Am. Chem. Soc., 61:2412-2416 (1939).
16. H. Mueller, Phys. Rev., 55:508 (1939).
17. H. Mueller, Phys. Rev., 55:792 (1939).
18. H. Mueller, and B. W. Sakman, Phys. Rev., 56:615-616 (1939).
19. H. Mueller, and B. W. Sakman, J. Opt. Soc. Am., 32:309-317 (1942).
20. F. J. Norton, Phys. Rev., 55:668-669 (1939).
21. W. Heller, Rev. Mod. Phys., 14:390-409 (1942).
22. C. T. O'Konski, and A. J. Haltner, J. Am. Chem. Soc., 79:5634 (1957).
23. I. Tinoco, Jr., and K. Yamaoka, J. Phys. Chem., 63:423-427 (1959).
24. M. Matsumoto, H. Watanabe, and K. Yoshioka, J. Phys. Chem., 74:2182-2188 (1970).
25. A. J. Haltner, Ph.D. Thesis, Univ. of California, Berkeley (1955), 108 pp.

26. C. T. O'Konski, and A. J. Haltner, J. Am. Chem. Soc., 78:3604-3610 (1956).

27. S. Broersma, J. Chem. Phys., 32:1626-1631 (1960).

28. R. C. Williams, and R. L. Steere, J. Am. Chem. Soc., 73:2075 (1951).

29. A. J. Haltner, and B. H. Zimm, Nature, 184:265 (1959).

30. F. S. Allen, and K. E. Van Holde, Biopolymers, 10:865 (1971).

31. C. T. O'Konski, and S. Krause, J. Phys. Chem., 74:3243 (1970).

32. C. T. O'Konski, and R. M. Pytkowicz, J. Am. Chem. Soc., 79:4815 (1957).

33. C. V. Raman, and S. C. Sirkar, Nature, 121:794 (1928).

34. H. Benoit, J. Chim. Phys., 49:517-521 (1952).

35. G. B. Thurston, and D. I. Bowling, J. Colloid Interface Sci., 30:34 (1969).

36. H. Plummer, and B. R. Jennings, J. Chem. Phys., 50:1033-1034 (1969).

37. S. P. Stoylov, Adv. Colloid Interface Sci., 3:45-110 (1971).

38. S. J. Jakabhazy, and S. W. Fleming, Biopolymers, 4:793-813 (1966).

39. N. A. Tolstoi, A. A. Spartakov, and A. A. Trusov, Kolloid Zh., 28:735-741 (1966).

40. B. R. Jennings, and B. L. Brown, Eur. Polymer J., 7:805 (1971).

41. M. J. Shah, D. C. Thompson, and C. M. Hart, J. Phys. Chem., 67:1170 (1963).

42. V. N. Tsvetkov, Yu. V. Mitin, V. R. Glushenkora, A. Ye. Grischenko, N. N. Boitsova, and S. Ya. Lyubina, Vysokomolekul. Soedin., 5:453 (1963).

43. C. T. O'Konski, and J. B. Applequist, Nature, 178:1464-1465 (1956).

KERR EFFECTS OF FLEXIBLE MACROMOLECULES

Robert L. Jernigan and Sanzo Miyazawa

Laboratory of Theoretical Biology
National Cancer Institute
National Institutes of Health
Bethesda, Maryland 20205

Detailed calculations of the effects of different conformations on the Kerr constant have become available only recently. These calculations can be utilized to evaluate the usefulness of Kerr constant measurements in characterizing configurations of macromolecules. Experimental scientists would like to know, in particular, if Kerr measurements are more sensitive to changes in conformation than other methods and for what types of molecules. These sensitivities must be judged against the results for more common characterization methods.

Kerr effect measurements are more complicated properties than most more common methods. In particular, as we will see, they depend upon both the dipole moments and the optical and static polarizability anisotropies of the molecules. This dependence occurs in two terms, one which dominates for polar molecules, involving the optical anisotropy and the square of the dipole moment and another which is normally important only for nonpolar molecules. The Kerr constant usually is most sensitive to conformations in molecules with large dipole moments; sensitivity is greatest if there is a large difference in the dipole moments of the configurations being compared. There is a greater sensitivity to differences in dipole moments, because of the square dependence, than to differences in optical anisotropies.

The presentation here consists of the following parts: 1) The formalism for the detailed calculation of the configurational average of the Kerr constant is briefly outlined. 2) The question of the tensor additivity of bond polarizabilities and the range of their interactions are taken up in some detail for the n-alkanes.

3) Results are reviewed which indicate that the Kerr constants for polar vinyl chain molecules depend strongly upon their stereoregularity. 4) Polypeptide Kerr constants and their dependences upon composition are treated, as well as their changes through the helix-coil transition.

CONFIGURATIONAL AVERAGE OF THE KERR CONSTANT

The usual approach is to develop molecular properties in a series expansion in the external electric field and retain only the lowest order terms. Nagai and Ishikawa (1965) originally developed the expressions appropriate for treating Kerr constants. Their expression for the molecular anisotropy at a temperature T is

$$\Delta\alpha = (1/10)\ [<\underline{\mu}^T\hat{\underline{\alpha}}\ \underline{\mu}>\ (E/kT)^2 + <\text{Trace}(\hat{\underline{\alpha}}\ \hat{\underline{\alpha}}')>\ (E^2/kT)] \quad (1)$$

μ is the molecular dipole moment, α is the molecular polarizability, the $\wedge$ indicates the traceless form, E is the magnitude of the electric field, the prime on the last α designates that it is the static polarizability, T indicates the row vector, and the brackets $<>$ indicate the average, in the absence of the electric field, of the quantities enclosed. The molar Kerr constant is obtained from eq (1) with a proportionality constant (Le Fevre and Le Fevre, 1955; Le Fevre, 1965)

$$_m K = (2\pi N/135kT)\ [\ <\underline{\mu}^T\hat{\underline{\alpha}}\ \underline{\mu}>/kT + <\text{Trace}(\hat{\underline{\alpha}}\ \hat{\underline{\alpha}}')>\ ] \quad (2)$$

where N is Avogadro's number.

The molecular dipole moments and optical anisotropies are themselves composed of contributions from each of the molecular subunits. There is an alternative approach of combining atomic polarizabilities (Applequist et al., 1972); results for alkanes have been presented by Ladanyi and Keyes (1979). Here, however, we take sums of bond or group contributions.

$$\begin{aligned} \underline{\mu} &= \Sigma\ \underline{\mu}_i \\ \text{and}\ i & \\ \hat{\underline{\alpha}} &= \Sigma\ \hat{\underline{\alpha}} \\ & \quad i \end{aligned} \quad (3)$$

The details of the matrix formalism for calculating these averages have been presented in several places, most usefully in the monograph by Flory (1969) and in the review by Jernigan and Thompson (1976). We will not repeat the details here. The independence implicit in eq (3) has been a matter of some controversy. For the bond polarizabilities, this additivity is often termed the valence optical method. The validity of the additivity of bond

dipoles is much better established than is the additivity of bond
polarizabilities. Next it is our purpose to attempt to indicate
the effects of interactions between bond polarizabilities and to
explore the range of the interactions which may affect this additivity.

INTERACTIONS BETWEEN POLARIZABILITY TENSORS

Here the form of the interactions between polarizabilities is
to be the induced dipole-induced dipole interactions of Silberstein (1917) and presented in a formalism similar to that of
Mortensen (1968). The total electric field at bond j will be the
intrinsic field E^o plus that arising from interactions with all
other bonds i.

$$\underline{E}_j = \underline{E}^o + \sum_{i \neq j} (-\underline{\mu}_i/r_{ij}^3 + 3\underline{r}_{ij}\underline{r}_{ij}^T\underline{\mu}_i/r_{ij}^5) \qquad (4)$$

μ_i is the dipole moment of bond i and r_{ij} is the distance between
bond centers. The bonds to be included in the sum over index i
are arbitrary to a significant extent. Rowell and Stein (1967)
treated ethane and included interactions between all bonds; their
result showed large enhancements of the polarizabilities. We have
chosen here to exclude interactions between bond pairs whose distances are fixed. Since the only conformational variables here
will be backbone bond rotations, interactions between bonds on the
same atom will not be explicitly included. This is equivalent to
taking the intrinsic bond polarizability to be an effective one
that already includes the effects of configurationally invariant
interactions. Practical considerations dictate that an upper bound
be imposed upon the interaction range. A large body of evidence
supports the premise that macromolecules at theta conditions can
be well represented by considering only short range interactions
(Flory, 1969). For the linear alkanes, this range corresponds to
interactions among three and four neighboring carbon-carbon bonds
and their pendant carbon-hydrogen bonds. The conformations of
these fragments are specified by one or two backbone rotation
angles. Those considerations to limit the range of the interactions are appropriate for van der Waals interactions; here we
are assuming this same range of interaction to be appropriate for
induced dipole-induced dipole interactions. We hope that the
range of interactions treated here is adequate. Our calculations
should correspond most closely to either pure liquids or theta
conditions.

If the intrinsic bond polarizability of bond j is $\underline{\alpha}_j^o$, then the
moment induced in bond j is given by

$$\underline{\mu}_j = \underline{\alpha}_j^o \underline{E}_j = \underline{\alpha}_j^o [\underline{E}^o + \sum_{i \neq j} (-\underline{I}/r_{ij}^3 + 3\underline{r}_{ij}\underline{r}_{ij}^T/r_{ij}^5) \underline{\mu}_i] \qquad (5)$$

with I as the identity matrix of order 3. By defining a dipole-
dipole interaction tensor, this becomes

$$\underline{E}^o = (\underline{\alpha}_j^o)^{-1} \underline{\mu}_j + \sum_{i \neq j} \underline{J}_{ji} \underline{\mu}_i \tag{6}$$

which can be reexpressed in matrix form as

$$\underline{M} = \underline{A}^{-1} \underline{E}^o \tag{7}$$

with

$$\underline{A} = \begin{bmatrix} (\underline{\alpha}_1^o)^{-1} & \underline{J}_{12} & \cdots & \underline{J}_{1n} \\ \underline{J}_{21} & (\underline{\alpha}_2^o)^{-1} & \cdots & \underline{J}_{2n} \\ \vdots & \vdots & & \vdots \\ \underline{J}_{n1} & \underline{J}_{n2} & \cdots & (\underline{\alpha}_n^o)^{-1} \end{bmatrix} \tag{8}$$

The entries in this matrix are the inverse of the intrinsic polarizability tensors, on the diagonal, and the dipole-dipole interaction tensors with the interacting bonds indicated by the pair of
subscripts on J.

$$\underline{M} = \begin{bmatrix} \underline{\mu}_1 \\ \underline{\mu}_2 \\ \vdots \\ \vdots \\ \underline{\mu}_n \end{bmatrix} \tag{9}$$

The effective bond polarizabilities are derived by identifying them
with the dipole moment induced by unit electric field; this means
that the effective or net polarizability is identical to the inverse of the matrix A.

We will apply this formalism to treat the interactions within
alkane chains. For the ten bond fragment $C_f-(CH_2)_{i+1}-(CH_2)_{i+2}-$
$(CH_2)_{i+3}C_{i+4}$ we have determined with the method above the polarizability tensor for the group $(CH_2)_{i+2}-C_{i+3}$, with interactions
between the bond polarizabilities permitted. Terminal groups are
treated with a similar 10 bond fragment, in which one of the C-C
bonds has been replaced by a C-H bond. The terminal C-C bond on
each end is taken to be in a fixed configuration; therefore for
the terminal group, there is only one rotation angle and the number of interactions is reduced. Results are given in Table I.
The mean polarizabilities, $\bar{\alpha} = (1/3)$ Trace(α), have been increased
by the interactions quite uniformly by approximately 20%; whereas the
square anisotropies, γ^2, display a greater than fourfold increase over the independent case. Furthermore, the value of γ^2 for

Table I. Nearest Neighbor Conformational Dependence of
the Polarizability Tensors of (CH_2-C)

Conformation	$\overline{\alpha}$	$\overline{\gamma^2}$
trans, trans	2.10	2.99
trans, gauche+	2.13	3.36
gauche+, trans	2.11	3.07
gauche+, gauche+	2.15	3.54
gauche+, gauche-	2.15	3.60

for Le Fevre Bond Polarizabilities,

$$\alpha_{C-C} = (0.99, 0.27, 0.27)$$
$$\alpha_{C-H} = (0.64, 0.64, 0.64)$$

the (CH_2-C) group is almost determined by the rotation angle included
within this group, with the preceding bond rotation having less
effect. That is, 2.99 is similar to 3.07; whereas 3.36, 3.54 and 3.60
are alike. Of course, these invariants do not give a complete indication of how the individual tensor components of the polarizability tensor vary. Also we should report that more anisotropic bond
polarizabilities result in larger interactions and consequently in
much stronger dependences on configuration. Now we wish to average
the configuration dependent polarizability tensors over all available configurations. The alkanes have the advantage of having well
studied configuration statistics (Flory, 1969). These are briefly
outlined below.

The n-Alkane Configurational Statistics

The configurational variations in the flexible n-alkanes are
adequately described in terms of two simple parameters which can
be ascribed to three bond and four bond interactions (Flory, 1969).
The statistical weights for three bond interactions are 1 for the
planar trans form and σ for each of the two gauche states. Four
bond interactions are accounted for with an additional statistical
weight of ω for adjacent pairs of gauche states of opposite
sense. These statistical weights can be organized into a matrix to
permit generation of a partition function as follows:

$$\underline{U} = \begin{bmatrix} 1 & \sigma & \sigma \\ 1 & \sigma & \sigma\omega \\ 1 & \sigma\omega & \sigma \end{bmatrix} \tag{10}$$

The order of both the row and column matrix indices is trans,
gauche+, and gauche-. The partition function is generated by an
appropriate number of U's in a product.

$$Z = (1\ 0\ 0)\ U^{n-2} \begin{bmatrix} 1 \\ 1 \\ 1 \end{bmatrix} \tag{11}$$

where n is the number of C-C bonds in the backbone. In the calculations to be presented here we have taken $\sigma = 0.424$ and $\omega = 0.0323$ for a temperature of 20 .

A partition function of this type has been utilized in conjunction with the geometry and methods set forth in the monograph by Flory (1969) to calculate averages of the requisite molecular properties. These are, specifically, the mean square polarizability anisotropy (Jernigan and Flory, 1967) and the Kerr constant (Flory and Jernigan, 1968). For the nonpolar alkanes we need only the nonpolar term in eq (2).

In the usual formalism (Jernigan and Thompson, 1976; Flory, 1969) for calculating the averages, each term in the above statistical weight matrix U is multiplied by the polarizability tensor corresponding to the bond type indicated by the index on U and formed into a larger generator matrix. Here we have obtained polarizability tensors that depend upon a pair of backbone bond rotations; each of these is placed into the larger matrix after being multiplied by the corresponding statistical weight from U for that pair of rotational angles. This results in more complicated matrices, but they are no larger in size than those required for previous calculations with configuration independent bond polarizability tensors. In a manner of speaking, this calculation removes some redundancy in the usual equations and takes more complete advantage of the matrix formalism.

Results for n-Alkane Polarizability Interactions

The choice of bond polarizability tensors, even for such simple bonds as carbon-carbon and carbon-hydrogen is a difficult matter (Jernigan and Thompson, 1976). We have chosen for the results here to use the bond polarizabilities of Le Fevre et al. (1966) and Bunn and Daubeny (1954). The Le Fevre C-H bond is taken to be isotropic with a bond polarizability of $0.64\ A^3$; the C-C bond has a parallel component of $0.99\ A^3$ and a perpendicular one of $0.27\ A^3$. These yield 0.72 for the value of $\Gamma = \Delta\alpha_{C-C} - 2\ \Delta\alpha_{C-H}$ which is similar to the somewhat lower value determined experimentally by Patterson and Flory (1972). The Bunn and Daubeny values are: for a C-C bond $\alpha_\| = 0.968$, $\alpha_\perp = 0.263$ and for a C-H bond $\alpha_\| = 0.7815$ and $\alpha_\perp = 0.5731$. For all cases we have arbitrarily taken $\alpha' = 1.1\ \alpha$. The units of all results presented in this paper are as follows, unless otherwise specified: molar Kerr constants in $10^{-10}\ cm^5\ statvolt^{-2}\ mol^{-1}$, polarizabilities in A^3, and

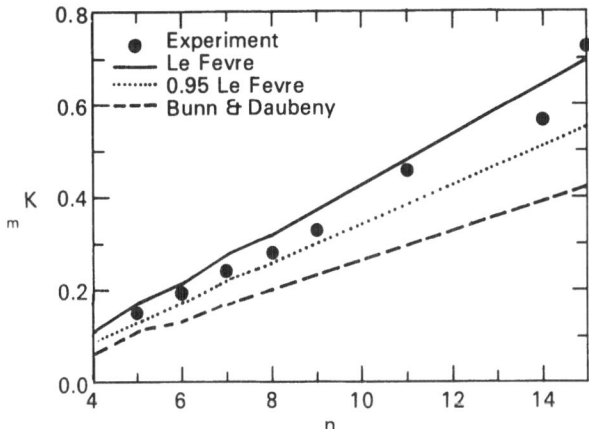

Fig. 1.   Chain length dependence of the Kerr constants for the n-
          alkanes, with short range interactions between bond polar-
          izability tensors. The curves were calculated for the
          indicated sets of polarizability tensors.

mean square anisotropies in A^6. In Fig. 1 we present results for
these bond polarizabilities, with interactions limited in range as
described above, for both the Le Fevre and the Bunn and Daubeny
bond polarizabilities. In addition we have also calculated results
with an arbitrary reduction in each of the Le Fevre bond polarizability values of 5%. The solid curve, for Le Fevre polarizabilities, is slightly higher than most of the experimental values
(Stuart, 1962); whereas a reduction of only 5% in each of the bond
polarizabilities yields results slightly below the experiments.
The strong sensitivity of the results to the exact values of the
bond polarizabilities appears because of the interactions. The
sensitivity of the individual group polarizability tensors to the
conformation is not a strong one. Mostly the interactions serve
simply to increase the overall polarizability anisotropy. The
exact conformational dependences of two invariants of the group
polarizability tensor were given in Table I.

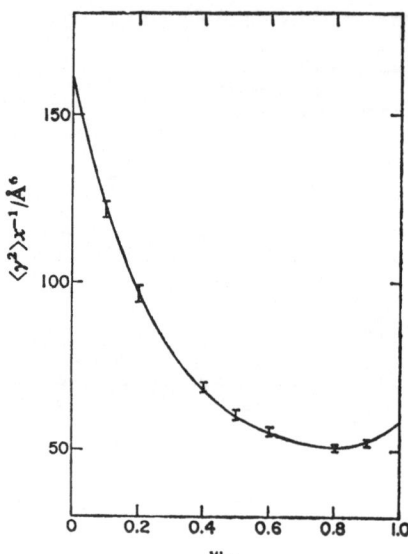

Fig. 2.   Dependence on stereoregularity of the mean square optical
          anisotropy per unit for poly(p-chlorostyrene). w_m is
          the probability of a meso dyad, a dd or ll pair.   The
          value $w_m = 1$ corresponds to a completely isotactic chain.
          Results are averages of 20 Monte Carlo chains, each of
          100 units. (Reproduced by permission, Saiz, Suter and
          Flory, 1977).

VINYL POLYMERS

     The two vinyl polymers of most interest are the polar ones,
poly(vinyl chloride), $+CHClCH_2+$, and poly(p-chlorostyrene), $+CH_2$
$(p-C_6H_4Cl)CH +$. Each unit of these polymers possesses an asymmetric carbon atom. Extensive calculations of dimensions and
dipole moments and their dependences upon composition and stereoregularity have been presented by Mark (1974). Variations in
tacticity can change mean square dipole moments by a factor of 4
or 5. Unfortunately the dipole moments are often nearly constant
over a substantial range of tacticity.

     Tonelli (1977) has calculated Kerr constants for a wide
variety of vinyl copolymers and considered variations of both
stereoregularity and composition. For the nonpolar polypropylene,
stereoregularity in the chain produces a maximum variation in K_m
of 25%. By contrast the highly polar poly(vinyl chloride) ranges
from 0.306 for an isotactic chain and 0.71 for an atactic chain to
−134 for a syndiotactic chain. By comparison with dipole moment
results this is an enormous range. Random 50:50 copolymers of these
two units give values ranging from −6.46 for a syndiotactic molecule to 1.1 for the isotactic form. In general the Kerr constants

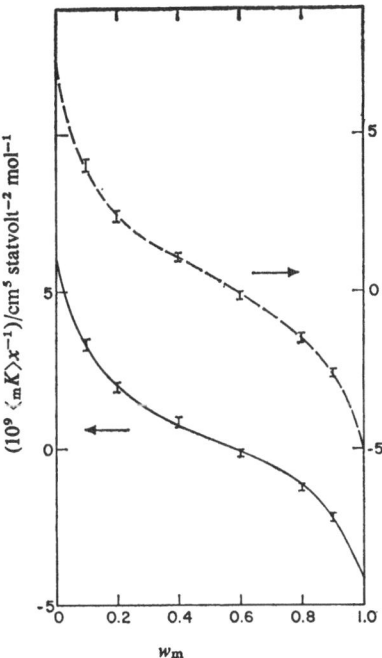

Fig. 3. Molar Kerr constants per unit for para-halogenated poly-
         styrenes. The dashed curve is for poly(p-bromostyrene),
         and the solid curve is for poly(p-chlorostyrene).
         Samples were generated as indicated in the Legend for Fig.
         2. (Reproduced by permission, Saiz, Suter and Flory,
         1977).

appear to show greater sensitivity to both sequence and stereoregularity than do the calculated dipole moments.

In Figs. 2 and 3 are displayed results (Saiz et al., 1977) of
the effect of tacticity on the optical properties of the para-halogenated polystyrenes. The mean square anisotropy evidences behavior
similar to that calculated for other properties of vinyl polymers.
Unfortunately these curves are nearly constant over a broad range
in tacticity, indicating that determination of tacticity by those
methods can be uncertain. From Fig. 2, values of $\langle \gamma^2 \rangle /x$ are very
similar for all values of $w_m > 0.5$. The most interesting feature
evidenced by the Kerr constants in Fig. 3 is their markedly different dependence on tacticity. In particular, the slopes of the
curves indicate that the Kerr constants of these polymers can be
utilized to distinguish between samples of high isotacticity.
Those authors obtained good agreement between experimental Kerr
constants for samples, whose tacticities had been estimated by
depolarized light scattering and other methods, and the calculated

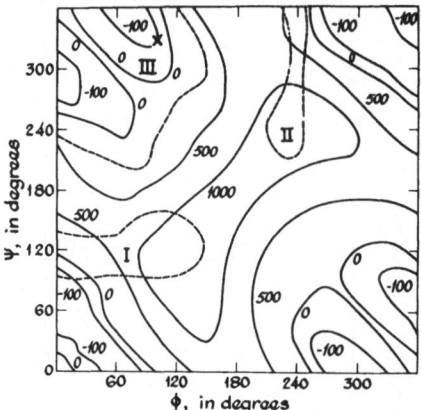

Fig. 4. Kerr constant contour map of the peptide N-acetyl-L-
         alanine-N'-methylamide in units of cm^5 statvolt^{-2} mol^{-1} x
         10^{-12}. ϕ and ψ are the peptide backbone rotation angles
         about bonds N-C$^\alpha$ and C$^\alpha$-C', respectively. The dashed
         lines enclose those energetically favored conformations
         with an energy of 5 kcal mol^{-1} or less. (Reproduced by
         permission, Ingwall, Czurylo and Flory, 1973).

Kerr constants for the corresponding tacticity.

     Other molecules for which calculations have been performed are
chain molecules with oxygen as well as carbon in their backbones
(Ishikawa and Nagai, 1971; Kelly et al., 1977).

POLYPEPTIDES

     A polarizability tensor for the peptide bond was derived by
Ingwall and Flory (1972); recently an alternative, somewhat more
anisotropic tensor has been reported by Khanarian (1980). The calculations reported here were performed with the former values.
Reproduced in Fig. 4 is the Kerr constant contour map for N-acetyl-
L-alanine-N'-methylamide (Ingwall et al., 1973). The areas inside
the dotted lines correspond to the usual favored conformations, I)
right handed alpha helix, II) left handed alpha helix and III)
beta strand. It is noteworthy that most of the helix conformations
have Kerr constants greater than +10; whereas a substantial portion
of the beta strand domain has negative Kerr constants. This strong
sensitivity to conformation makes the method a particularly attractive one for studying polypeptides and proteins.

     We have utilized the Ingwall and Flory polarizability tensors
set forth for glycine and alanine to calculate the effect of com-

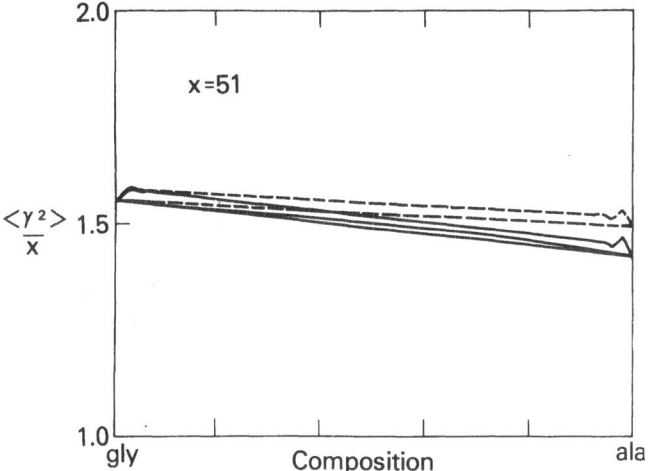

Fig. 5.   Mean square polarizability anisotropy per peptide for var-
          ious copolymers of glycine and alanine. The dotted curves
          correspond to an energy increment of 0.7 kcal mol^{-1} for
          the beta strand region for alanine. The top curve is the
          sum of the scalar anisotropies for the two blocks. The
          other dotted curve is for both the random copolymer and
          the properly calculated block copolymer. The top solid
          curve is for the scalar sum of the anisotropies of the
          two blocks. The middle solid curve is for the properly
          calculated block copolymer, and the lowest curve is for
          the random copolymer.

position on the Kerr constant for an alanine-glycine copolymer, for
both random sequences and for block copolymers. Much larger
effects would be expected for amino acids with polar side chains.
The chain treated is $CH_3-(CO-NH-CHR)_{x-1} - CO-NH-CH_3$. For alanine
it is necessary to specify the value of Γ for the CH_2-C unit; in
these calculations we have taken a value of 0.55. Ingwall et al.
(1973) found that it was necessary, in order to obtain agreement
with the Kerr constants for N-acetyl-L-alanine-N'-methylamide and
N-acetyl-L-leucine-N'-methylamide, to increase the energy of the
region III by 0.7 kcal mol^{-1}; this may correspond to a solvent
effect. We have calculated the energies with the same method of
Brant et al. (1967). In Fig. 5 are shown results for the mean

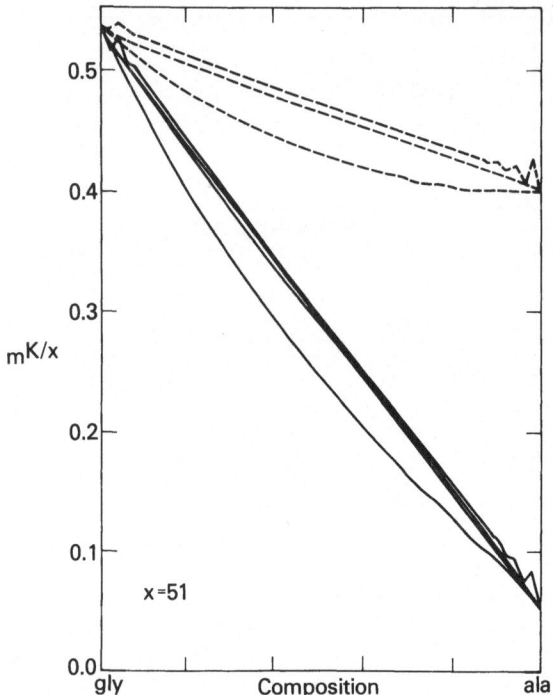

Fig. 6.   Calculated Kerr constants per peptide for copolymers of
          glycine and alanine. As in Fig. 5, the dotted curves
          correspond to calculations in which an energy of 0.7 kcal
          mol^{-1} has been added to the beta strand region for ala-
          nine. The top and bottom solid curves are for the same
          cases as those in Fig. 5. For the solid curves the two
          nearly indistinguishable curves in the middle are for
          properly calculated block copolymers, the lower one for
          $(gly)_i(ala)_{x-i-1}$ and the upper for $(ala)_{x-i-1}(gly)_i$. The
          top dotted curve is for the sum of scalar Kerr constants
          of the two blocks; the middle dotted curve is for properly
          calculated block copolymers; and the lowest dotted curve
          is for the random copolymer.

square anisotropy appropriate to depolarized light scattering
measurements. The results are quite linear and nearly independent of composition, especially for the case with region III at a
higher energy. In Fig. 6 are displayed the corresponding Kerr
constant results. These show substantially more variation in
passing from poly(glycine) to poly(alanine).   In addition, results
for the random sequence copolymer, the lowest curves in each set,
evidence substantial curvature. The mean square polarizability
anisotropies in the previous figure did not indicate significant

similar effects.

Helix-Coil Transitions

We have chosen to treat the helix-coil transition in the simplest possible manner (Poland and Scheraga, 1970). The partition function formulated by Zimm and Bragg(1959) is

$$Z = (1 \ 0) \begin{bmatrix} 1 & \sigma s \\ 1 & s \end{bmatrix}^{x-1} \begin{bmatrix} 1 \\ 1 \end{bmatrix} \tag{12}$$

This generates all occurrences of the two conformations, helix and coil along the molecule. Because of the substantial independence of each virtual bond in a polypeptide, it is possible to treat the random coil by independently averaging each residue over an energy map such as that indicated in abridged form in Fig. 4. We have

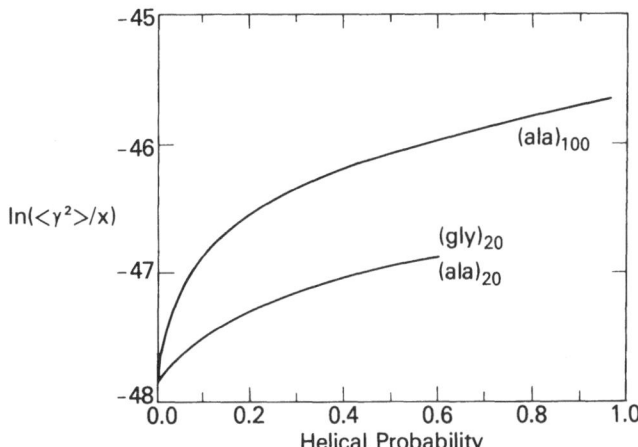

Fig. 7. Values of the mean square polarizability anisotropy per peptide through the helix-coil transition. $\langle\gamma^2\rangle$ is expressed in units of cm^6 . Both glycine and alanine curves are coincident. The lower curve is for 21 peptide bonds and the upper one for .101. The abscissa is the fraction of residues in helical form.

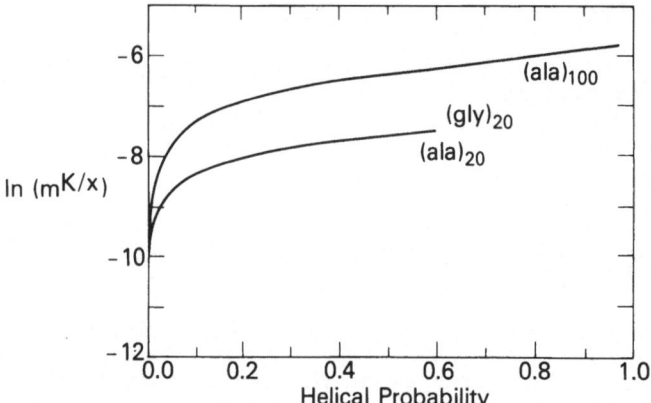

Fig. 8. The Kerr constants for finite polypeptide chains through
the helix-coil transition. $_m$K is expressed in units of
cm^5 statvolts^{-2} mol^{-1}. Designations are the same as those
in Fig. 7.

chosen a Kerr constant value from the contour map in Fig. 4 for
the helix conformation at $\phi =122$, $\psi =133$. Calculations are for
a temperature of 25 and $\sigma = 10^{-4}$. The averages in eq (2) are
easily obtained with the partition function in eq (12). Results
for both the mean square polarizability anisotropy and the molar
Kerr constant are presented in Figs. 7 and 8. Both figures are
for the same set of parameters. The most noteworthy feature is
that the Kerr constant curve increases more sharply at significantly smaller helix fractions than does the mean square polarizability
anisotropy. This occurs because the latter quantity, for a rigid
rod, depends on the square of the molecular weight, but the Kerr
constant, for rigid polar molecules, depends on the cube of the
molecular weight. Hence Kerr constant measurements are more sensitive to initial stages of the onset of rigidity than are most
other methods.

DISCUSSION AND SUMMARY

    From the evidence collected here, it is apparent that the Kerr
constant is most sensitive to conformation in those polymers with
large dipole moments. This effect is especially sizable in the results presented for polar vinyl polymers.

For nonpolar molecules, interactions of the bond polarizabilities, through induced dipole-induced dipole interactions, may be important. The determination of the effective range of such interactions remains as an experimental problem. The results calculated here, for the n-alkanes with short range interactions between bond polarizabilities, fit the experimental data well; however, similar results can be obtained with many combinations of the parameters. Only further careful experimental determinations of bond polarizabilities will be able to resolve this issue.

In polar molecules, the potential for significant interactions is even greater. In addition to induced dipole-induced dipole interactions, there is the possibility of a hyperpolarizability contribution to the induced interactions.

If interactions of these types are in fact important, they may possibly account for the broad range of values of bond polarizabilities reported for nearly identical chemical bonds. A possible alternative to the present approach would be a quantum mechanics evaluation of the dependences of group polarizabilities on conformations.

Also it should be pointed out that the physical state and details of solution conditions can play an important role in determining the relative importance of interactions. Two reports make evident this possibility. Buckingham and Sutter (1976) concluded, from gas phase measurements on n-alkanes, that bond polarizabilities are not additive. In the gas state, the range of interactions for the largest molecules may be longer than at theta conditions because of the possibility that the molecules are more collapsed. By contrast, Meeten (1968) concluded that, for solutions of n-alkanes, the mean second hyperpolarizabilities were additive.

The calculations presented on the helix-coil transition in polypeptides make it clear that different experimental methods may monitor different stages in that transition. Because the molecular weight dependence of the Kerr constant for helix is different than for mean square moments, the Kerr effect is expected to be more sensitive to the initial onset of rigidity.

The polar peptides (Khanarian, 1980) present an excellent opportunity for study of their conformations by Kerr measurements. The abundance of small polypeptides isolated recently from biological sources makes this method particularly important.

REFERENCES

Applequist, J., Carl, J. R. and Fung, K.-K., 1972, An Atom Dipole
        Interaction Model for Molecular Polarizability. Application

to Polyatomic Molecules and Determination of Atom Polarizabilities, J. Am. Chem. Soc., 94:2952.

Brant, D. A., Miller, W. G. and Flory, P. J., 1967, Conformational
Energy Estimates for Statistically Coiling Polypeptide
Chains, J. Mol. Biol., 23: 47.

Buckingham, A. D. and Sutter, H., 1976, Gas Phase Measurements of
the Kerr Effect in Some n-Alkanes and Cyclohexane, J. Chem.
Phys., 64:364.

Bunn, C. W. and Daubeny, R. D. P., 1954, The Polarizabilities of
Carbon-Carbon Bonds, Trans. Faraday Soc., 50:1173.

Flory, P. J. and Jernigan, R. L., 1968, Kerr Effect in Polymer
Chains, J. Chem. Phys., 48:3823.

Flory, P. J., 1969, "Statistical Mechanics of Chain Molecules",
Interscience, New York.

Ingwall, R. T., Czurylo, E. A. and Flory, P. J., 1973, Kerr Constants of Amides and Peptides, Biopolymers, 12:1137.

Ingwall, R. T. and Flory, P. J., 1972, Optical Anisotropy of Polypeptide Chains, Biopolymers, 11:1527.

Ishikawa, T. and Nagai, K., 1971, Theoretical Interpretation of
Kerr Constants of Polymer Chains. n-Alkanes, Poly-
(oxyethylene glycol)s, and Poly(oxyethylene dimethyl
Ether)s, Polymer J., 2:263.

Jernigan, R. L. and Flory, P. J., 1967, Optical Anisotropy of
Chain Molecules. Theory of Depolarization of Scattered
Light with Application to n-Alkanes, J. Chem. Phys., 47:
1999.

Jernigan, R. L. and Thompson, D. S., 1976, Flexible Polymers, in:
"Molecular Electro-Optics", Part 1, C. T. O'Konski, ed.,
Marcel Dekker, New York.

Kelly, K. M., Patterson, G. D. and Tonelli, A. E., 1977, Kerr
Effect Studies of the Poly(oxyethylenes), Macromolecules,
10:859.

Khanarian, G., 1980, The Kerr Effect of Aqueous Solutions, Thesis,
Univ. of Sydney.

Khanarian, G., Mack, P. and Moore, W. J., 1980, Optical Anisotropies of Amides and Peptides in Dioxane, in press.

Ladanyi, B. M. and Keyes, T., 1979, Effect of Internal Fields on
Depolarized Light Scattering from n-Alkane Gases, Molec.
Phys., 37:1809.

Le Fevre, C. G. and Le Fevre, R. J. W., 1955, The Kerr Effect.
Its Measurement and Application in Chemistry, Rev. Pure
Appl. Chem., 5:262.

Le Fevre, R. J. W., 1965, Molecular Refractivity and Polarizability, in: "Advances in Physical Organic Chemistry", Vol. 3,
V. Gold, ed., Academic Press, London.

Le Fevre, R. J. W., Orr, B. J. and Ritchie, G. L. D., 1966, Mole-
cular Polarisability. The Anisotropic Electron
Polarisability of Aliphatic C-C and C-H Bonds, J. Chem.
Soc.(B), 1966:273.

Mark, J. E., 1974, The Use of Dipole Moments to Characterize Configurations of Chain Molecules, Acc. Chem. Res., 7:218.

Meeten, G. H., 1968, Hyperpolarizability Contributions to Kerr Effect in Liquids, Trans. Faraday Soc., 64:2267.

Mortensen, E. M., 1968, Polarizability of Molecules Considering Internal-Field Effects Using a Point Dipole Approximation, J. Chem. Phys., 49:3732.

Nagai, K. and Ishikawa, T., 1965, Internal Rotation and Kerr Effect in Polymer Molecules, J. Chem. Phys., 43:4508.

Patterson, G. D. and Flory, P. J., 1972, Depolarized Rayleigh Scattering and Mean-squared Optical Anisotropies of n-Alkanes in Solution, J. Chem. Soc., Faraday Trans. II, 68:1098.

Poland, D. and Scheraga, H. A., 1970, "Theory of Helix-Coil Transitions in Biopolymers", Academic Press, New York.

Rowell, R. L. and Stein, R. S., 1967, The Internal Field and the Additivity of Bond Polarizabilities, J. Chem. Phys., 47:2985.

Saiz, E., Suter, U. W. and Flory, P. J., 1977, Optical Anisotropies of para-Halogenated Polystyrenes and Related Molecules, J. Chem. Soc., Trans. Faraday Soc. II, 73:1538.

Silberstein, 1917, Molecular Refractivity and Atomic Interaction, Phil. Mag., Ser. 6, 33:92; 1917, Molecular Refractivity and Atomic Interaction II, ibid, 33:521.

Stuart, H. A. and Kuster, R., 1962, Kerr-Konstanten von Flüssigkeiten in: "Landolt-Börnstein Zahlenwerte und Funktionen", 6th ed., Vol. II, Pt. 8, p. 849, Springer Verlag, Berlin.

Tonelli, A. E., 1977, Possible Characterization of Homopolymer Configuration and Copolymer Sequence Distribution by Comparison of Measured and Calculated Molar Kerr Constants, Macromolecules, 10:153.

Zimm, B. H. and Bragg, J. K., 1959, Theory of the Phase Transition between Helix and Random Coil in Polypeptide Chains, J. Chem. Phys., 31:526.

LIGHT SCATTERING IN ELECTRIC FIELDS

B. R. Jennings

Electro-Optics Group, Physics Department
Brunel University
Uxbridge, U.K.

ABSTRACT

Measurements of the intensity of light scattered from molecular
solutions (and particulate suspensions) lead to information about
the molecular weight and radius of gyration of the solute. When an
electric field is applied the intensity changes owing to the resulting orientation, deformation and alignment of the solute molecules.
Measurements of these changes enable the permanent and induced dipole
moments and rotary relaxation times of the molecules to be evaluated.
The most useful equations for the effects are reviewed for rods, discs
and flexible polar chains. The experimental harnessing of the theoretical predictions is described. Representative results are given
which illustrate not only the evaluation of the molecular physical
parameters but also show how such a wealth of parameters obtained in
a single experimental procedure can, in favorable circumstances, enable the characteristics of complex structures and aggregates to be
estimated.

I. INTRODUCTION

As long ago as 1919 Gans[1] foresaw that measurement and analysis
of the intensity of light scattered by colloidal particles would lead
to information on the particle structure. Later, Cabannes[2] showed
that optical anisotropy of the particles, which could be directly
related to their structure, was obtained by analysis of the state of
polarization of the scattered light. Since that time measurement of
the scattered intensity, recorded at various angles Θ to the forward
direction of the incident beam, has become a standard procedure for
evaluating the relative molecular mass M and the size or shape of

solute or dispersed particles in dilute solution or suspension. The
method is particularly valuable for molecules or particles whose dimensions are comparable to the wavelength λ of light. In such cases
the randomly oriented and distributed molecules give rise to significant internal interference effects in the scattered light. It
is the polar scattering diagram which is interpreted to give the
molecular parameters. The interference is generally characterized
in terms of a large particle scattering factor $P(\Theta)$ and inherently
contains information about the molecular size and shape. Math-
ematically, the scattering obeys the equation[3]

$$\frac{KC}{R_\Theta} = \frac{1}{M_w} \left\{ 1 + \frac{16\pi^2}{3\lambda^2} \cdot \overline{S^2}_Z \cdot \sin^2 (\Theta/2) +... \right\} + 2Bc \qquad ...(1)$$

Here, the factor in the brace represents the first two terms of an
expression of the reciprocal of $P(\Theta)$, c is the solution concentration in $g.ml^{-1}$, K an optical constant, B the second osmotic
virial coefficient, R_Θ the Rayleigh ratio and $\overline{S^2}$ the mean square
radius of gyration of the particles. For rods of length L, discs
of diameter d and random coils of statistical end separation r,

$$\overline{S^2} = \frac{\overline{L^2}}{12} = \frac{\overline{d^2}}{8} = \frac{\overline{r^2}}{6} \qquad ...(2)$$

Experimental data are analyzed in terms of equation (1) and are
conveniently displayed in the form of a Zimm[3] plot to yield $\overline{M_w}$ and
$\overline{S_Z^2}$. The subscripts denote weight and Z average parameters for polydisperse systems.

The foregoing equations were derived for solutions in which the
molecules adopt a random orientational and positional array. Alignment of the constituent particles in an electric field affects the
scattered intensity in a number of ways. Firstly, the inherent
anisotropy in the scattering properties of the individual particles
is enhanced upon alignment[4]. This is true even if the particles are
small, and results in changes in both the depolarization and the
intensity of the scattered light at any angle. Secondly, for relatively large particles, the internal interference effects embodied
in the parameter $P(\Theta)$ change with particle alignment. Hence the
scattered intensity also changes[5]. This effect is relatively large
and it is for this reason that the majority of electric field light
scattering measurements have been made on macromolecular or colloidal media. It is with these changes that this paper is concerned. Thirdly, whereas the scattered intensity may appear to be
constant over a long period of time, it does in fact exhibit fluctuations due to the chaotic random motions of the constituent solute
molecules. Measurement of these fluctuations gives rise to the so-

called 'photon correlation spectroscopy' method from which translatory diffusion coefficients can be obtained[6]. These fluctuations
are also affected by electric fields and measurements of them lead
to the evaluation of electrophoretic mobility[7]. Such experiments
are described in another contribution to this volume.

II.   THEORETICAL AND EXPERIMENTAL BACKGROUND

Electrically induced intensity changes Δi depend upon three
groups of factors. Firstly, the electrical parameters of the
molecules; that is their inherent permanent dopole moments μ and
any induced dipole moment arising from an anisotropy $\Delta\alpha$ in the
electric polarizability α due to charge separation within the constituent molecules or at the solvent interface. Secondly, the size
and shape of the molecules are important because of the strong dependence of $P(\Theta)$ on these parameters. By considering the interference between wavelets scattered from extremities of the molecule,
Fig. 1 indicates the scattered intensity changes. The collective
effect from all molecules in a solution gives rise to $P(\Theta)$ and must
be obtained by integrating for all molecules present. Changes Δi
in the scattered intensity (i) are expressed through changes $\Delta P(\Theta)$
as

$$\Delta i/i_o \;=\; \Delta P(\Theta) \,/\, P(\Theta) \qquad\qquad \ldots (3)$$

For clarity i_o signifies the intensity with no applied field.
Thirdly, the characteristics of the applied field, namely its intensity and frequency, are important. At low degrees of molecular
alignment, Δi is a function of the square of the strength of the applied electric field. Hence, for an alternating field $E = e\sin \omega t$
with e the peak amplitude and ω the angular frequency,

$$\Delta i \propto E^2 \propto e^2 \sin^2 (\omega t) \propto e^2/2 \,(1 - \cos 2\,\omega t) \qquad \ldots (4)$$

Hence, the scattered intensity consists of two components, one of
which is 'steady' and independent of frequency whilst the other alternates at twice the frequency of the applied field. Both components can be seen in the transient response of Fig. 2. Hereafter we shall assume that the 'alternating' component is eliminated
experimentally and is thus not included in the ensuing theoretical
and experimental descriptions.

All three factors listed above cannot be embraced in a single
orientation function as is often employed in other electro-optical
methods. Independent equations must be derived for molecules of
various shape, flexibility and electrical characteristics. Useful
equations are thus presented below with illustrative experimental

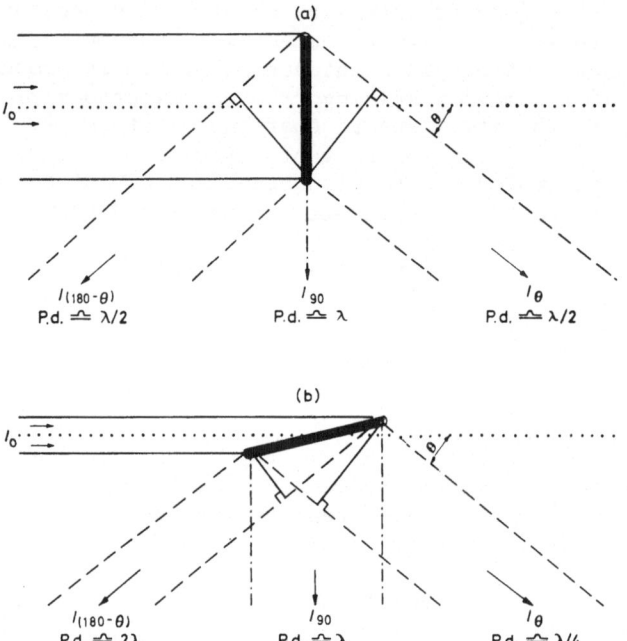

Fig. 1:   Illustration of Scattered Intensity Variations
          with Particle Orientation.
          The upper and lower situations give rise to dif-
          ferent angular interference and hence polar scat-
          tering diagrams; p.d. represents the optical path
          difference between wavelets scattered from the
          rod molecule extremities.

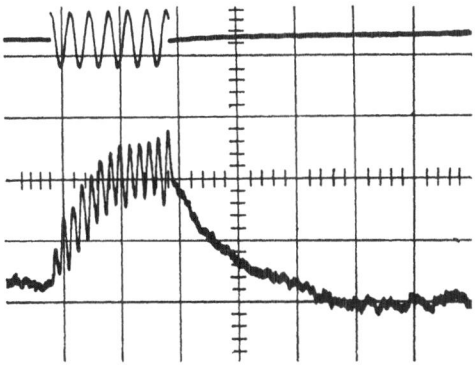

Fig. 2:   Transient Scattering Response Illustrating the Double
          Frequency Component in Addition to the Steady Level
          Change.
          Data for an aqueous suspension of Laponite synthetic
          clay, with $\bar{E}$ = 80 V cm^{-1} and f = 60 H$_z$. (Reproduced
          from ref. 8, courtesy of the Institute of Physics).

data for specific molecular shapes.

An important feature of electric field scattering is that it gives supplementary molecular information to that obtained from scattering data prior to the field application. It is most prudent therefore to have an apparatus that records both pre-field and in-field data at as many angles of observation as possible. A host of molecular physical parameters are then accessible from a single set of experiments. Conventional scattering photometers consist of a light source followed by suitable collimating optics so as to bring a parallel beam of monochromatic, polarized light into a scattering cell which holds the test fluid. An optical detector, capable of observing the sample at pre-chosen angles over a wide angular range, receives the scattered light. The electronic output is passed through various d.c. amplifiers and displayed on a meter or printer.

Additional features for electric field scattering measurements are as follows (Fig. 3). Suitable electrodes need to be incorporated in the cell in such a manner as to give the widest choice of field geometry without any occlusion of the incident or observed light beams. (Typical simple arrangements are shown in Fig. 4). A field generator capable of giving a.c. voltages of up to 10 kV cm^{-1} is required. If continuous a.c. voltages are to be applied it is advantageous to have a light source which runs from a d.c. power supply. Should the field generator be pulsed, and transient changes in the scattered light be recorded, then further considerations are needed. The ripple and noise on the light source need careful attention and suppression. In addition the photodetector and amplifier must be as free from electronic noise as possible and have fast response times if the molecular rotary relaxation times are to be evaluated from the decay rates of the scattering changes following the removal of the field pulse. It is convenient to display the photodetector output directly on to an oscilloscope and to record the display photographically. Alternatively the output can be fed into a transient digitizer and analyzed immediately. As with all scattering experiments cleanliness of solutions is an important factor.

II. RIGID-ROD MOLECULES

(i) Small Degrees of Orientation

A. Theory. The most comprehensive equations have been developed by Plummer and Jennings[9,10] and more recently by Herbert[11]. The intensity changes are expressed by

$$\frac{\Delta i}{i_o} = (1 - 3 \cos^2\Omega) \; C \left\{ S + A^{\frac{1}{2}} \sin 2\omega t \right\} \qquad \ldots (5)$$

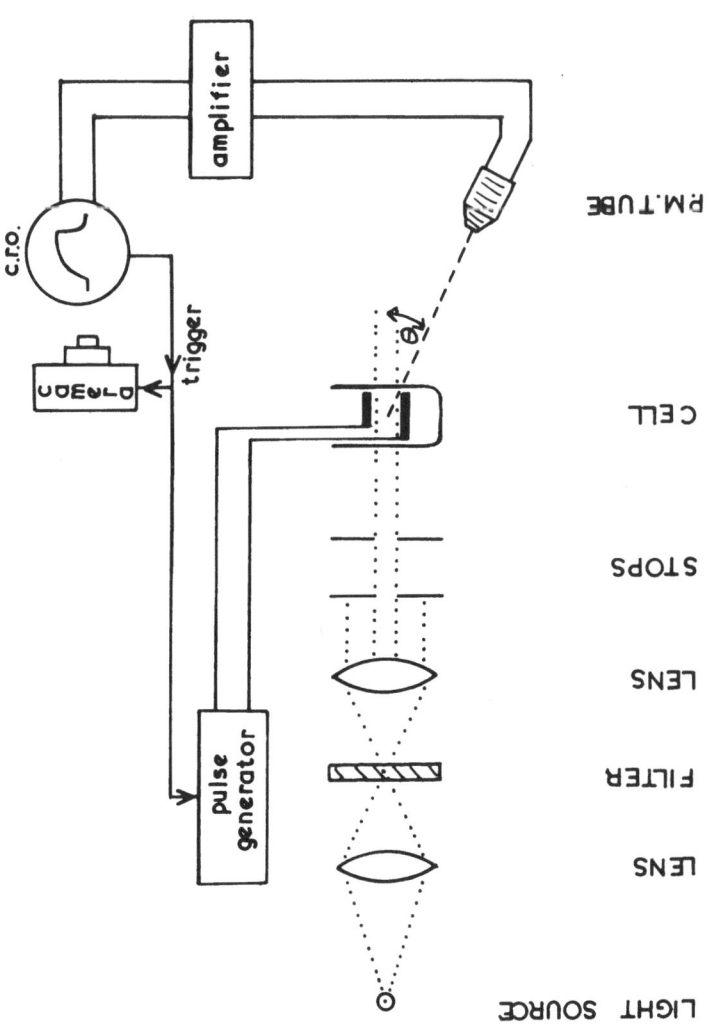

Fig. 3: Diagrammatic Representation of the Apparatus for Pulsed Electric Field Scattering Measurements. The oscilloscope is represented by c.r.o., and p.m. tube indicates the photodetector.

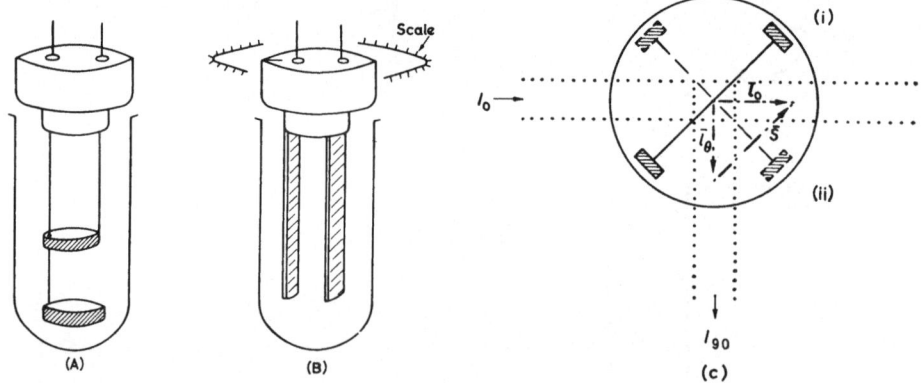

Fig. 4:   Sample Cells with Electrodes.
           (A) Electrodes with $\Omega = 90°$ and Θ variable.
           (B) Vertical electrode assembly allowing
           variable Ω and Θ.   (C) Two positions of
           vertical electrodes allowing measurement
           of ratio R with $\Theta = 90°$. At positions (i)
           and (ii), $\Omega = 0°$ and $90°$ respectively.

Here Ω is the angle between the applied field and the scattering $\bar{s}$ vector which is formed by unit vectors in the directions of the scattered and incident beams. For a stipulated angle of observation, C is a constant dependent on the molecular size and observation angle and is given by

$$C = \frac{1}{12} + \frac{1}{8P_o(\Theta)} \left\{ \frac{\sin 2x}{2x^3} - \frac{1}{x^2} \right\} \qquad \dots (6)$$

where $x = (2\Pi L/\lambda) \sin \Theta/2$.

A represents the amplitude of the double frequency 'alternating' component. It is not discussed here. The parameter S is the 'steady' term given by

$$S = \frac{p^2}{1+(3\omega\tau/2)^2} + q \qquad \dots (7)$$

where

$$p = \frac{(\mu_3^2 - \mu_1^2) <E^2>^{\frac{1}{2}}}{kT} \quad \text{and} \quad q = (\alpha_3 - \alpha_1) \frac{<E^2>}{k^2T^2}$$

Subscripts 3 and 1 represent the major and a transverse minor axis of the rod molecules; k is the Boltzmann constant and T the absolute temperature.

From these equations we note the following. Firstly, Δi is proportional to E^2. Secondly, the factor $(1-3 \cos^2\Omega)$ characterizes the dependence of Δi on the applied field direction. Maximal changes will be encountered as decreases in Δi when $\Omega = 0$. Thirdly, equation (7) indicates that the induced dipole contribution is effectively independent of frequency (for f < 10 GHz), whilst the permanent dipole moment is very frequency dependent. This can be used to isolate the two contributions. Experimental data can be recorded at various frequencies and p and q evaluated from the frequency dependence. Alternatively, Δi can be recorded when a pulse of d.c. field is followed by a pulse of a.c. field of the same r.m.s. amplitude. The magnitudes of the two transient responses yield values of (p^2+q) and q respectively. Fourthly, the rotary relaxation time τ can be evaluated from the frequency dependence. An alternative procedure is to analyze the decay of the transient response of Δi following a pulsed field termination. For small monodisperse molecules, the decay is exponential according to the equation [5,12]

$$\Delta i = \Delta i' \exp - (t/\tau) \qquad \dots (8)$$

Here $\Delta i'$ is the scattered intensity change at the instant (where t=0) of the field termination.

B. Data Handling. Measurements are made on solutions of different concentration at various field strengths and frequencies. The results are extrapolated to infinite dilution. Values of $\Delta i/i_o$ are conveniently plotted as a function of E^2. Linearity should be obtained at low field strengths. From pre-field measurements and via the Zimm plot procedure values of $P_o(\Theta)$ and L are known. Hence x is known. Two methods are available for evaluating the molecular electrical parameters.

The first is to analyze the initial slope of the graph of $\Delta i/i_o$ versus E^2. From equations 5, 6 and 7 the parameter S/E^2 can be determined at the given field frequency. The parameters p and q can be determined explicitly, firstly by measuring the field dependence of Δi, using d.c. fields, and then subsequently using bursts of high frequency a.c. field. For the d.c. case S/E^2 will lead to the sum $(p^2 + q)$. Corresponding analysis under high frequency conditions yields q alone. Hence $(\mu_3{}^2 - \mu_1{}^2)$ and $(\alpha_3 - \alpha_1)$ can be evaluated explicitly. It is important to note that the sign of $(\mu_3{}^2 - \mu_1{}^2)$ and $(\alpha_3 - \alpha_1)$ indicates the direction of the predominant dipole moment and polarizability.

An alternative method of evaluating the parameters is by analysis of the angular dependence of the scattered intensity. Measurements are made at a series of angles Θ. In the absence of an electric field, at low angles of observation,

$$\lim_{c \to o} \left(\frac{Kc}{R_\Theta}\right)_o = \frac{1}{M}\left\{1 + \frac{x^2}{9} + \ldots\right\} \qquad \ldots (9)$$

whilst, for a finite E

$$\lim_{c \to o} \left(\frac{Kc}{R_\Theta}\right)_E = \frac{1}{M}\left\{\frac{1}{P(\Theta)+\Delta P(\Theta)}\right\} \qquad \ldots (10)$$

Combining these two equations, expanding $P(\Theta)$ and C and limiting them for small x, then

$$\lim_{c \to o} \left(\frac{Kc}{R_\Theta}\right)_E = \frac{1}{M}\left[1 + \frac{x^2}{9}\left\{1 + \frac{S}{15}(1 - 3\cos^2\Omega)\right\}\right] \qquad \ldots (11)$$

If m_E and m_o are the initial slopes of the zero concentration curves, then

$$\frac{m_E - m_o}{m_o} = \frac{S}{15} (1 - 3 \cos^2 \Omega) \qquad \qquad \ldots (12)$$

from which S, and hence values of p and q are again found.

C. <u>Illustrative Results</u>. An interesting study has been reported for solutions of polyhexyl isocyanate in tetrahydrofuran.
Viscosity and sedimentation studies[13] had indicated that the polyisocyanates were very still in this solvent and appeared to be rodlike at high molecular weights. Plummer and Jennings[14] studied a
sample and obtained values for $\bar{M}_w$ and S^2. The field dependence of
Δi can be seen in Fig. 5. In such an organic solvent intensity
changes of up to 6 per cent were recorded. In the manners outlined
above the following data were obrained

$$\bar{M}_w = 2.8 \times 10^6 \qquad \qquad \tau = 4.2 \times 10^{-4} s$$

$$<s^2>_Z^{\frac{1}{2}} = 290 \text{ nm} \qquad \qquad \bar{L}_Z = 1.0 \text{ } \mu m$$

$$\mu_3 = 9.7 \times 10^{-26} C \text{ m} \qquad \qquad (\alpha_3 - \alpha_1) = 3 \times 10^{-26} C \text{ m}^2 \text{ V}^{-1}$$

These results correspond to values of 0.4 Å and $1.0 \times 10^{-29} C$ m
respectively for the rod length and dipole moment per monomer unit.
Furthermore the dipole moment was directed predominantly along the
rod backbone. These results are of significance for the following
reasons. Firstly, such a small monomer projected length is less
than the carbon-carbon bond length and indicates that the molecule
must indeed adopt some helical conformation. Secondly, the existence of a dipole moment arising predominantly from the C = 0 group,
and being directed predominantly along the rod backbone indicates
that the polyisocyanate molecules adopt a <u>cis</u> rather than <u>trans</u>
configuration (see Fig. 6). From this study one sees that electric
field light scattering data are useful not only for the evaluation
of molecular parameters but also for the structural information
they provide owing to the large number of parameters determined.

<u>E.coli</u> bacteria are extended prolate structures. They are
multi-component organisms consisting of cell membranes and intra-
cellular fluid and components. They scatter light strongly. Their
major dimension however exceeds λ as their length is of the order of
1.5 μm. They can be studied by raising λ into the near infra-red
spectral region[15]. Alternatively the samples are generally so
polydisperse that errors involved in exceeding the wavelength
conditions are not serious[16]. Their significant surface charge
would appear to be the reason why they give strong induced dipole

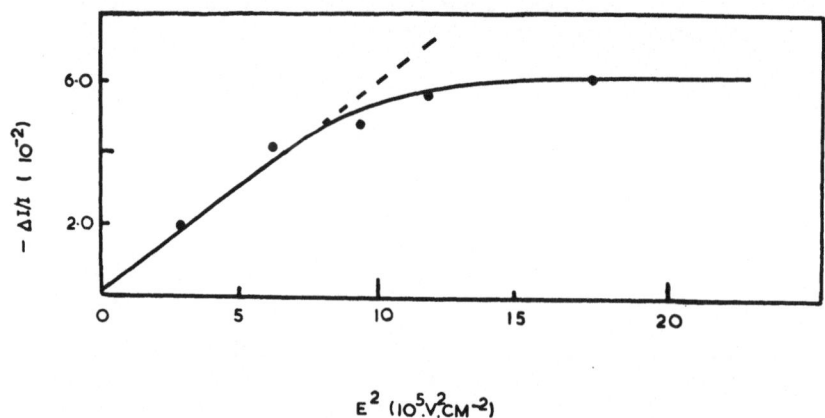

Fig. 5:   Electric Field Strength Dependence of the Scattered
          Intensity Change for a Sample of Polyhexyl Isocyanate
          (M = 2.8 x 10^6) in Tetrahydrofuran.
          Data for Θ = 90°, Ω = 0°, c = 5 x 10^{-3}g ml^{-1} and
          f = 100 H$_z$.   (Reproduced from ref. 14, courtesy of
          Pergamon Press).

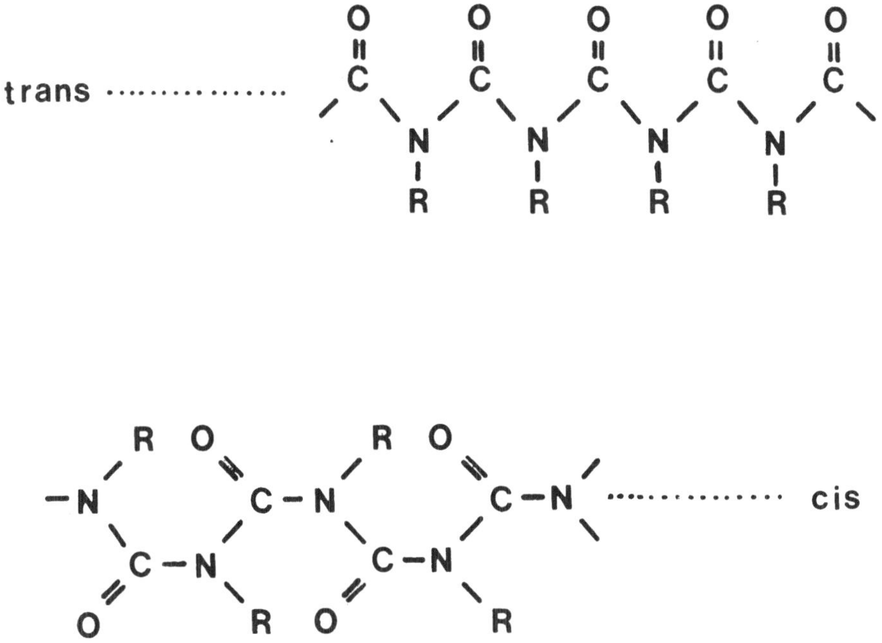

Fig. 6: Alternative Configurations for Polyisocyanates.
Electric field scattering data indicate the cis
form to be appropriate in tetrahydrofuran
solutions.

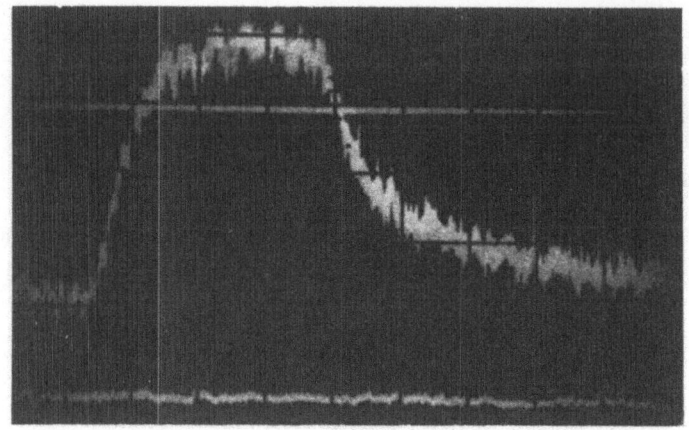

Fig. 7.   Transient Intensity Change for an Aqueous E.coli
          Suspension.
          $E = 60$ V cm^{-1}, $f = 600$ H$_z$, pulse duration 1.6 s,
          $\lambda = 1.06$ μm and $\Theta = 30°$. (Data from ref. 16,
          courtesy of Academic Press Inc.)

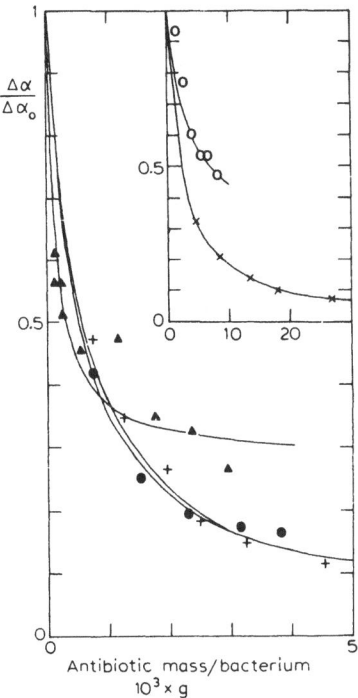

Fig. 8:   Effect of Antibiotics on the Electrical Polarizability
          Anisotropy ($\Delta\alpha$) of E.coli.
          Vertical cross, penicillin; full circle, streptomycin;
          triangle, polymixin B. On the inset: open circle,
          bacitracin; diagonal cross, neomycin. (Reproduced from
          ref. 17, courtesy of Elsevier Biomedical Press).

moments due to interfacial polarization.

Transient scattering studies have been made on dilute aqueous suspensions. Typical responses are shown in Fig. 7. Using relatively high frequency electric fields the anisotropy of polarizability ($\Delta\alpha = \alpha_3 - \alpha_1$) has been determined. Because of the interest in colloid electric double layers and the polarizability of the same, attempts were made to modify this double layer by the addition of various chemotherapeutic agents. Without affecting the geometry of the micro-organisms (verified by the absence of change in τ) measurements of the polarizability indicated the bacterium-additive interaction. A remarkable set of data is shown in Fig. 8 where the effect of various antibiotics on the bacteria can be seen. It would appear that surface polarizability is an extremely sensitive indicator of the interaction of antibiotics with bacteria. The ability of transient electro-optical methods to measure electrical properties simultaneously with particle geometry is an important feature which should be exploited.

(ii) Full Orientation

Scheludko and Stoylov[18] have developed a theory for rod molecules which is applicable to arbitrary degrees of orientation. The important feature of this theory is that one can predict the scattered intensity changes that can be encountered when full orientation (as for high field strengths) can be obtained. In this limiting case[19] the equations reduce to very simple form and are independent of the electrical properties of the molecules. In general

$$\frac{\Delta i}{i_o} = \frac{1}{P_o(\theta)} \left\{ \frac{\sin\,(x\,\cos\,\Omega)}{x\,\cos\,\Omega} \right\}^2 - 1 \qquad \ldots\,(13)$$

for arbitrary values of Ω. In particular, for $\Omega = 90°$,

$$\left(\frac{\Delta i}{i_o}\right)_{\Omega=90°} = \left\{ \frac{1}{P_o(\theta)} - 1 \right\} \qquad \ldots\,(14)$$

This equation is particularly attractive as it affords an independent means of determining $P_o(\theta)$. To date however measurements under high field conditions are few. One reason for this is the fear that under such high fields the molecular system may be influenced. Fig. 9 does display data both for disc-shaped particles and for E.coli bacteria which show a near attainment of complete molecular alignment. It would be interesting to extend these studies in the future, particularly in the case of non-aqueous and non-conducting solvents.

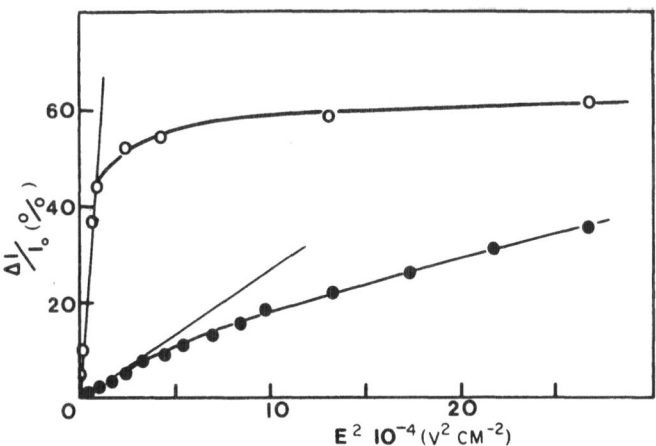

Fig. 9:   Scattered Intensity Changes Approaching Orientation
          Saturation.
          Closed circles for E.coli bacterial filaments and
          open circles for bentonite clay discs. (Reproduced
          from ref. 20, courtesy of the authors).

IV. RIGID DISC MOLECULES

A. _Theory_. Stoylov[21] has developed an equation for d.c.
fields and low degrees of orientation. Using subscript 3 to indicate the unique disc axis and with $y = (2\Pi/\lambda) . d . \sin \Theta/2$, and
thus the analogue of x used in the rod theory, then

$$\frac{\Delta i}{i_o} = \left[\frac{y^2}{180 \, P_o(\Theta)}\right] (1 - 3 \cos^2\Omega) \left\{q - p^2\right\} \quad \ldots (15)$$

From this equation we note that Δi is again characterized by the
factor $(1 - 3 \cos^2\Omega)$. Furthermore the signs of the p^2 and q contributions to Δi are opposite to those for the equivalent equations
for rods. This is a reflection of the fact that subscript 3 designates the minor geometrical axis for discs. The scattered intensity changes are again a function of E^2 and an alternating component can be detected.

The influence of alternating electric fields can easily be
grafted into equation (15) using the generalized theory of Peterlin
and Stuart[22]. However these authors only considered the two specific
cases where the molecules have either a permanent or an induced dipole moment, but not both simultaneously. In the case of dipolar,
isotropic particles (i.e. $p_3 \neq 0$ and $\alpha_3 = \alpha_1$), and ignoring the
double frequency component,

$$\frac{(\Delta i)_\omega}{(\Delta i)_{\omega=o}} = \frac{1}{1 + (3\omega\tau/2)^2} \quad \ldots (16)$$

For non-polar anistropic particles ($p_3 = 0$; $\alpha_3 \neq \alpha_1$), the 'steady'
component does not change with frequency.

These equations are similar to those for rods, and by analogy
with the rod theory, Jennings et al.[23] have expanded equation (15)
to the form

$$\frac{\Delta i}{i_o} = \frac{y^2}{180 \, P_o(\Theta)} \cdot (1 - 3 \cos^2\Omega) \left\{q - \frac{p^2}{(3\omega\tau/2)^2}\right\} \quad \ldots (17)$$

It is emphasized that this equation is valid only for the steady
component. It does however afford a means of isolating q from p
for disc-like particles.

B. _Data handling and illustrative results_. The procedures
are essentially those described earlier for rod molecules. Measurements are recorded on solutions of different concentration

at various field strengths and frequencies. From the pre-field
data, a Zimm plot is drawn and $P_o(\Theta)$ and the disc diameter d evaluated. After verifying that Δi has a linear dependence on E^2, the
factor of proportionality between these entities is evaluated.
Using equation (15) and knowing Ω, the factor $(q-p^2)$ can be evaluated.
Repetition of the experiments for high frequency fields yields q
alone. Alternatively the angular dependence of Δi can be analyzed
for data extrapolated to low Θ. For discs the equivalent expression
to equation (12) is as follows

$$\frac{m_E - m_o}{m_o} = \frac{(1-3\cos^2\Omega)}{30}\left\{ q - \frac{p^2}{\left[1 + (3\omega\tau/2)^2\right]^2} \right\} \qquad \dots (18)$$

Once again the parameters of q and p can be isolated from measurements made at different frequencies. To date, the theory has been
used primarily in the study of flake-like clay minerals. The following provides an example.

Montmorillonite is a smectite clay which readily disperses in
water as thin discs[24]. Prior to the application of electric fields
the angular dependence of the scattered intensity was measured, the
data expressed in the form of a Zimm plot and the average particle
mass and radius of gyration evaluated[25] (Table I). From S an equivalent thin disc diameter of 308 nm was calculated, although the
particles were shown from the scattering properties to be better
represented as thin, flat plates of 625 x 120 x 0.9 nm dimensions.
Electric fields gave rise to significant changes. Furthermore,
Δi was strongly frequency dependent (see Fig. 10). This figure

TABLE I

Experimental Data for Fresh and Aged Montmorillonite

Fresh Sample	Parameter	Aged Sample
1.2	particle mass $(10^{-6}g)$	10.6
184	radius of gyration (nm)	554
0.42	relaxation time (ms)	1.7
12	dipole moment $(10^{-26}Cm)$	7.8
5	polarizability anisotropy $(10^{-30}Fm^2)$	1.8

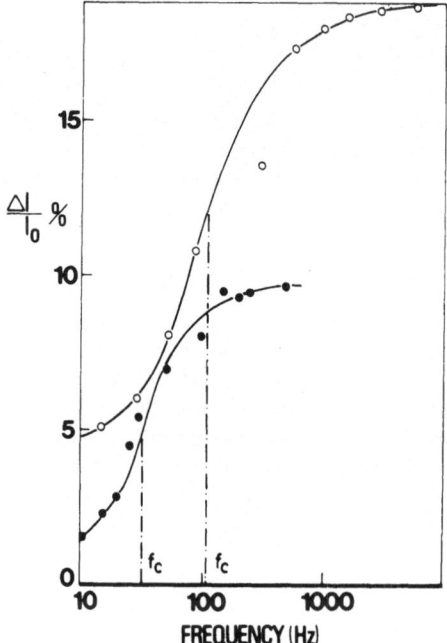

Fig. 10:   Frequency Dependence of Δi for Fresh (open circles)
           and Aged (filled circles) Montmorillonite Aqueous
           sols. (Reproduced from ref. 25, courtesy of Academic
           Press Inc.).

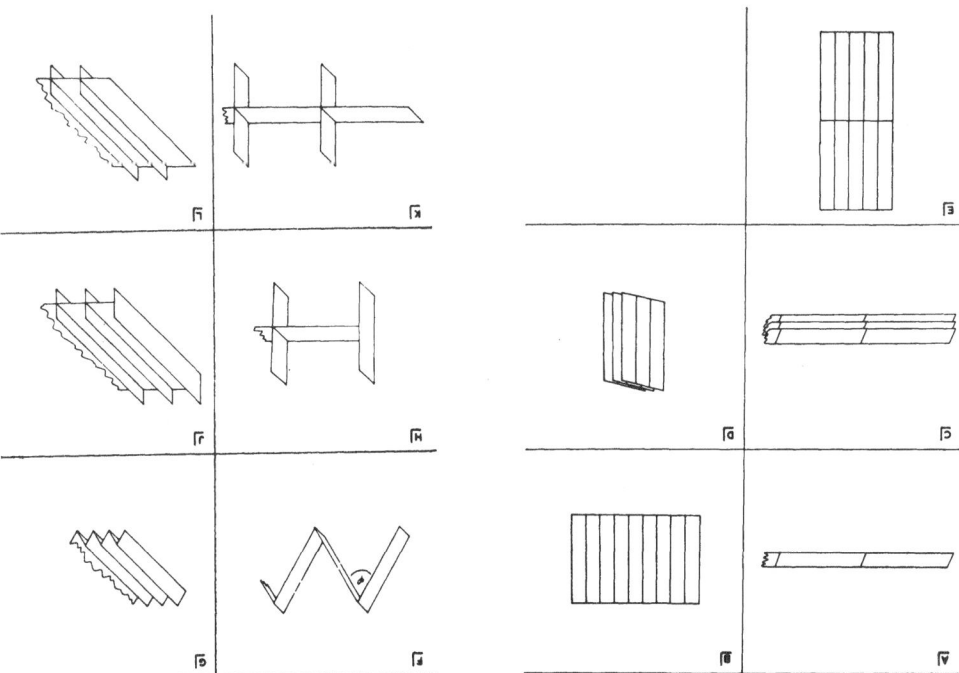

Fig. 11: Alternative Associations of Nine Montmorillonite Flakes.
The experimental values of $\overline{S^2}$ and τ could only be satisfied by the face-to-edge stacks represented in frame J
(or L). (Reproduced from ref. 25, courtesy of Academic
Press Inc.).

is instructive because of the apparent increase of Δi with increasing
frequency. Such an increasing dispersion phenomenon is possible with
light scattering changes and contains information about the direction
of the dipole moments. The high frequency asymptote corresponds to
the contribution to Δi from q alone. At lower frequencies this is
<u>reduced</u> by a permanent dipolar contribution which tends to turn the
plate in the opposite sense. Hence it decreases the contribution
from q. Thus p and q act in quadrature. From the data, both the
amplitudes and directions of μ and Δα were evaluated along with τ.
These are also given in Table I. In aqueous solutions, montmorillonite has the tendency to gel. Even in dilute solution small local
microgel can form upon standing. Experiments were repeated on an
'aged' solution which had been left for a few hours. On the assumption that the data then represented the average microgel particle, all of the molecular, geometrical, optical and electrical
parameters were re-determined. These data are also listed in
Table I. Fig. 10 shows the frequency dispersion for the 'aged'
as well as the 'fresh' material. One notices the following features.
Firstly, the increased size of the aggregated particles gives rise
to a lower frequency dispersion. Secondly, the changes in particle
mass give an estimate of the number of individual clay particles in
an aggregate. This can be seen to be of the order of 9 or 10.
Thirdly, measures of both τ and S for the aggregate and the isolated
platelets were determined. A series of models were constructed
consisting of 9 or 10 individual platelets and the parameters τ
and S calculated for each model in the hope of isolating a unique
structure for the aggregate. Fourthly, any such structure had to
satisfy not only the geometrical considerations, but the compilation
of the vectorial dipole moments and the tensorial polarizabilities
of the individual particles into the aggregate. Of the 11 models
considered, the one that satisfied the experimental data to within
experimental error is that in which the individual montmorillonite
discs associate edge-to-face in what is termed the 'card-house'
structure (see Fig. 11).

This particular study not only illustrates the use of electric
field scattering for the characterization of disc molecules, but
also demonstrates how the method can be used to infer very complex
molecular geometry.

V. POLYMER COILS

A. <u>Theory</u>. The theory for these systems is complicated due
to the fact that Δi results from deformation and ultimate rotation
of the chains. Equations have been developed by Wippler and
Benoit[26] and Isihara et al.[27]. Both of these treatments have been
extended more recently by Wallach and Benoit[28]. Only d.c. field
effects have been considered for low field strengths. The chains
are supposed to consist of N segments of identical length (a) with

$a < \lambda$. The distribution of segments is assumed to be Gaussian and each link is considered to be perfectly flexible. Hence the electric field is able to rotate the links independently due to their possession of individual permanent or induced dipole moments.

It is convenient to define a parameter R which characterizes Δi at various Ω.

$$R = \frac{(\Delta i)_{\Omega=0}}{(\Delta i)_{\Omega=90}} = \frac{(\Delta P)_{\Omega=0}}{(\Delta P)_{\Omega=90}} \qquad \ldots (19)$$

From equations (5) and (15), by substituting values of $\Omega = 0°$ and $\Omega = 90°$ one sees that R has the value -2 for these rigid molecules. Experimentally R can be determined by measuring Δi for two electrode configurations in the cell. These are illustrated by A and B in Fig. 4 for which $\Omega = 0°$ and $90°$ respectively. One measures Δi under these two conditions. As will be seen below, for non-rigid molecules R departs drastically from the value -2. It therefore affords a means of detecting molecular flexibility.

(i)   Flexible Polar Molecules

Various models have been considered[26] since the original equations by Wippler[29]. All cases may be condensed into two equations given below. Theories have only been derived for specific field directions with $\Omega = 0°$ or $90°$. Under these conditions

$$\Delta P(\Theta) = \frac{N^2 Z^2 a^2}{108} \left(\frac{E}{kT}\right)^2 \quad f \quad \text{for } \Omega = 0° \qquad \ldots (20)$$

and

$$\Delta P(\Theta) = \frac{N Z^2 a^2}{270} \left(\frac{E}{kT}\right)^2 \quad F \quad \text{for } \Omega = 90° \qquad \ldots (21)$$

Here $Z = (4\Pi/\lambda) \sin \Theta/2$. The functions f and F are related to the dipole moments of the segments and will be expanded below. From equations (20) and (21) we note the following; Δi is proportional to E^2 in both cases; it is also proportional to Z^2 and hence to $\sin^2(\Theta/2)$. Furthermore Δi is much larger for $\Omega = 0°$ owing to the dependence on N^2. For $\Omega = 90°$ Δi is proportional to N. Hence the changes no longer depend upon Ω according to the factor $(1-3 \cos^2\Omega)$ as in the case of rigid molecules. The parameter R thus affords an indication of flexible molecules as

$$R = -\frac{5N}{2} \cdot \frac{f}{F} \qquad \ldots (22)$$

Four slightly different polar chain models have been considered
theoretically. These are represented in Fig. 12. Type A is for the
molecule whose segments bear dipole moments of equal magnitude directed in a head-to-tail manner along the chain zig-zag. Type B
consists of dipoles directed along the zig-zag but with alternate
segments bearing dipole moments of different amplitude. The
factors f and F are given in Table II. Should p_1 be in opposite
direction to p_2, as in the case of polyvinyl halides and oxides,
then f = R = 0 and Δi is finite only for Ω = 90°. The resultant
dipole moment is then at right angles to the chain backbone and
Δi is proportional to N rather than N^2. The scattering changes are
then small. Type C consists of segments bearing dipole moments of
the same magnitude but directed at an angle β to the segment direction. The parameters f and F are given in Table II and R then
has the value

$$R = -\frac{5N}{2} \cdot \frac{1}{(1-\frac{1}{2}\tan^2\beta)} \qquad \ldots (23)$$

The original formulation in Wallach and Benoit's[28] paper was
erroneous probably due to a printing error. Type D is the most
general model in which alternate dipoles are neither coincident with

TABLE II

The Parameters f and F for polar, Flexible chains

Model	f	f
A	μ^2	μ^2
B	$(\mu_1 + \mu_2)^2 /4$	$(\mu_1^2 + \mu_2^2)/2$
C	$\mu^2 \overline{\cos^2\beta}$	$\mu^2(3\overline{\cos^2\beta}-1)/2$
D	$\frac{1}{4}\left\{(\mu_1 \overline{\cos\beta_1} + \mu_2 \overline{\cos\beta_2})\right\}^2$	$\frac{1}{4}\left\{\mu_1^2(3\overline{\cos^2\beta_1}-1)+\mu_2^2(3\overline{\cos^2\beta_2}-1)\right\}$

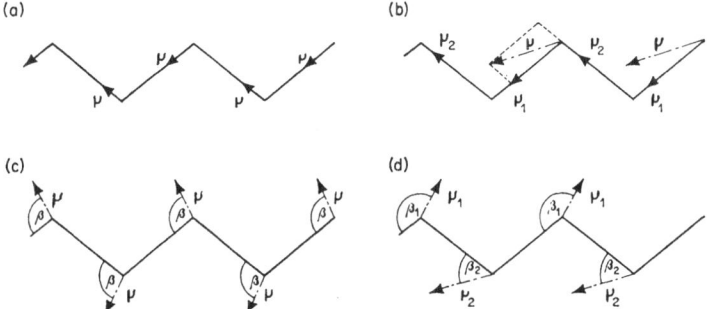

Fig. 12: Models for Flexible, Polar Coils, with Segmental Dipole
Moment μ_i Directed at Angles β_i to the Bond Direction,
with i = 1,2.

the segment axis nor are they of equal amplitude to their neighbors. The parameters f and F are given in Table II and R becomes

$$
R = - \frac{5N}{2} \left\{ \frac{(p_1 \overline{\cos\beta_1} + p_2 \overline{\cos\beta_2})^2}{p_1^2 (3 \overline{\cos^2\beta_1} - 1) + p_2^2 (3 \overline{\cos^2\beta_2} - 1)} \right\} \quad \ldots (24)
$$

From all of these equations it can be seen that the parameter R is essentially a function of $-5N/2$. For macromolecules where $10^3 < N < 10^6$ this is significantly greater than the ratio -2 obtained with rigid molecules.

(ii) Other coil Equations

Although primarily of academic interest alone, an equation exists[29] for flexible non-polar electrically anisotropic molecules. In this case

$$
\Delta P(\theta) = \frac{NZ^2 a^2}{270} (1 - 3 \cos^2\Omega) \frac{E^2 \Delta\alpha'}{kT} \quad \ldots (25)
$$

where $\Delta\alpha'$ is the polarizability of an individual segment.

An equation also exists[29] for completely frozen polar coils. These are supposed to consist of segmental dipoles directed along the segment axis but with such steric hindrance that the configuration is completely frozen. In such cases

$$
\Delta P(\theta) = \frac{N^2 Z^2 a^2}{324} (1 - 3 \cos^2\Omega) \left(\frac{\mu E}{kT} \right)^2 \quad \ldots (26)
$$

B. Typical Results. Here again measurements are recorded on solutions of different concentration at a variety of angles θ and for various values of E and frequency ω. In addition measurements are recorded under conditions of $\Omega = 0°$ and $90°$. This can be performed with cell C in Fig. 4. By considering the ratio R one can immediately tell whether the molecule is flexible or rigid. If this ratio equals -2, it is quite likely that a rigid rod, ellipsoid or disc is indicated. If however (Δi) $\Omega = 90°$ is effectively zero, then one should suspect a very flexible molecular system. A check is afforded through the following consideration. From the pre-field Zimm plot data the molecular weight M can be obtained. From the analysis of the decay of the transient changes one can evaluate τ. Stockmayer and Baur[30] have derived an equation

$$
\tau = \chi M [\eta] \eta_o / RT \quad \ldots (27)
$$

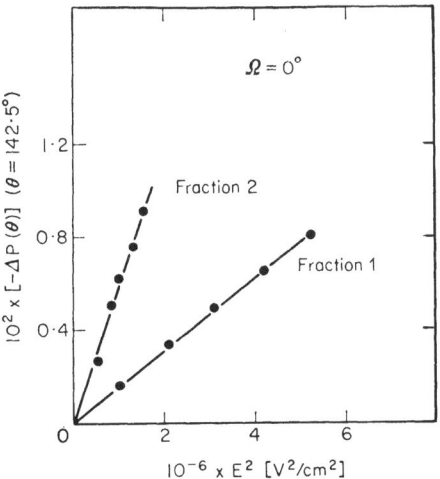

Fig. 13: Relative Scattered Intensity Changes for two Fractions of Nitrocellulose in Ethyl Acetate. (Reproduced from ref. 28, courtesy of Messrs. Wiley Interscience).

in which coefficient χ has the values 0.43 and 0.61 for non-draining and free-draining coils respectively. Here R is the gas constant, [η] the solution intrinsic viscosity and η_o the solvent viscosity.

Few experiments have been made to date on solutions of randomly coiled polymers. Wallach and Benoit[28] studied nitrocellulose solutions. Typical data are given in Fig. 13 of the change in the P(Θ) function as a function of E^2. For these experiments the solvent was ethylacetate and the molecular weights of fractions 1 and 2 were 1.5 and 3.96×10^5 respectively. Values of Δi were extremely small and of the order of 1 per cent. However, by analyzing the initial slopes of Fig. 13, assuming that the polymer was flexible and that the dipoles were directed along the chain contour, these authors obtained values of 4.0 and 4.2 debye respectively for fractions 1 and 2. Alternative analysis in terms of the angular dependence of the scattering yielded 2.9 and 3.5 debye respectively. Considering the uncertainty of the data and of the equations used, these results are most encouraging.

The only use of the parameter R to indicate molecular flexibility has been in a study[31] of an aqueous solution of poly-L-proline form II. For a high molecular weight sample of this polypeptide α-helix, measurements were made of $\Delta i/i_o$ at various field strengths and for $\Omega = 0°$ and $90°$. Only when $\Omega = 0°$ were measurable changes detected, corresponding to an infinite value of the flexibility parameter R. With this polypeptide and for $\Omega = 0°$ scattered intensity changes of some 12 per cent were recorded.

VI.   CONCLUSIONS

The major advantage of electric field scattering experiments is the extensive list of molecular parameters that can be obtained from a single experiment. Combining M with S obtained prior to field application, along with values for the molecular dipole moments, polarizabilities and relaxation times, important conformation and structural information may be inferred. In addition the measurements are relatively easy to make. Commercial scattering photometers exist which can be converted for electric field experiments with relative ease and little expense.

From the examples cited herein it is also clear that complex molecular structures, aggregates and interactions can be studied. In particular, criteria for the determination of molecular flexibility is an important advantage in polymer and biopolymer science. Further studies on flexible molecules would be welcome.

There are three disadvantages of the light scattering method. Firstly, along with all electro-optical techniques, difficulties may be encountered with highly conducting solutions. Secondly, the method is not as sensitive as electric birefringence. The reason for this is that, with birefringence measurements, the apparatus is arranged to indicate the transmission of light about a zero light level condition. In scattering experiments an ambient intensity exists prior to the application of the field, resulting in reduced sensitivity. Thirdly, the degree of sample cleanliness is greater than with most other electro-optical methods. Nevertheless the number of electric field scattering measurements have increased drastically over the last ten years and there is every reason to suspect this will continue.

Finally, a comprehensive list of the materials studies by electric field scattering is given in Table III.

TABLE III

Materials Studies by Electric Field Scattering

POLYMERS

Nitrocellulose	5, 28, 38
Poly alanine	29
Poly benzyl glutamate	39, 40
Poly ethylene oxide	41, 42
Poly ethylene oxide – polystyrene copolymer	41
Poly isocyanate	14
Poly proline	31
Poly vinyl chloride	27

BIOPOLYMERS

Chromaffin granules	43
DNA	44–46
Tobacco Mosaic Virus	5, 47–51
Tobacco Rattle Virus	52

MICROORGANISMS

E.coli	20, 16, 53–55, 17, 56
T. ferroxidans	16

Numerals refer to references

MINERALS

Attapulgite	32
Bentonite	42, 5, 58, 33, 20
Crocidolite	34
Hectorite	5, 35, 36
Halloysite	37
Laponite	23, 8
Montmorillonite	25
Palygorskite	54, 71, 72

COLLOIDS

Benzopurpurine	61, 57, 51
Graphite	57
Stearic Acid	61
Silver Iodide	61
Vanadium pentoxide	57, 51

REFERENCES

1. R. Gans, Annln. Phys. 62:331 (1919).
2. J. Cabannes, 'La Diffusion Moleculaire de la Lumière', Presses
    Universitaires, Paris (1929).
3. B. H. Zimm, J. Chem. Phys. 16:1099 (1948).
4. S. Kielich, Acta. Phys. Polon. A37:447 (1970).
5. C. Wippler, J. Chim. Phys. 53:316 (1956).
6. B. Chu, 'Laser Light Scattering', Academic Press, New York
    (1974).
7. B. R. Ware and W. H. Flygare, Chem. Phys. Lett. 12:81 (1971).
8. J. F. Schweitzer and B. R. Jennings, J. Phys. D. (Appl Phys.)
    5:297 (1972).
9. H. Plummer and B. R. Jennings, J. Chem. Phys. 50:1033 (1969).
    A minor error in the original equation has been pointed out
    on p. 290 of ref. 10.
10. B. R. Jennings, Ch. 8 in 'Molecular Electro-Optics', Vol. 1,
    (C. T.O'Konski-Ed.), Dekker, New York (1976).
11. T. J. Herbert, J. Coll. Int. Sci. 69:122 (1979).
12. H. Benoit, Ann. Phys. (Paris) 6:561 (1951).
13. W. Burchard, Makromol. Chem. 67:182 (1963).
14. H. Plummer and B. R. Jennings, Europ. Polymer J. 6:171 (1970).
15. H. J. Coles, B. R. Jennings and V. J. Morris, Phys. Med. Biol.
    20:225 (1975).
16. B. R. Jennings and V. J. Morris, J. Coll. Int. Sci. 50:352
    (1975).
17. V. J. Morris and B. R. Jennings, Biochim. Biophys. Acta. 497:253
    (1977).
18. A. Scheludko and S. P. Stoylov, Kolloid Z.u.Z. Polymere, 199:36
    (1964).
19. S. P. Stoylov, Coll. Czech. Chem. Comm. 31:2866 (1966).
20. S. P. Stoylov, S. Sokerov, I. Petkanchin, and N. Ibroshev,
    Dokl. Akad. Nauk. SSSR 180:1165 (1968).
21. S. P. Stoylov, Izv. Inst. Fizkhim. 6:79 (1967).
22. A. Peterlin and H. A. Stuart, Handbuch und Jahrbuch der
    Chemischen Physik. 8:Section 1 b (1943).
23. B. R. Jennings, B. L. Brown and H. Plummer, J. Coll. Int. Sci.
    32:606 (1970).
24. R. E. Grimm, 'Clay Mineralogy', McGraw-Hill, New York (1953).
25. J. F. Schweitzer and B. R. Jennings, J. Coll. Int. Sci., 37:443
    (1971).
26. C. Wippler and H. Benoit, Makromol. Chem. 13:7 (1954).
27. A. Isihara, R. Koyama, N. Yamada and A. Nishioka, J. Polymer
    Sci. 17:341 (1955).
28. M. L. Wallach and H. Benoit, J. Polymer Sci. A2,4:491 (1966).
29. C. Wippler, J. Polymer Sci. 23:199 (1957).
30. W. H. Stockmayer and M. Baur, J. Amer. Chem. Soc. 86:3485 (1964).
31. B. R. Jennings, Brit. Polymer J. 1:70 (1969).
32. H. J. Plummer and B. R. Jennings, Brit. J. Appl. Phys. 1:1753
    (1968).

33. B. R. Jennings and H. G. Jerrard, J. Chem. Phys. 42:511 (1965).
34. B. R. Jennings and M. Bhanot, Clay Minerals 12:217 (1977).
35. B. R. Jennings and H. Plummer, J. Coll. Int. Sci. 27:377 (1968).
36. B. L. Brown, B. R. Jennings and H. Plummer, J. Appl. Optics
     8:2019 (1969).
37. M. Bhanot and B. R. Jennings, J. Coll. Int. Sci. 56:92 (1976).
38. B. R. Jennings and J. F. Schweitzer, Europ. Polymer J. 10:459
     (1974).
39. M. L. Wallach and H. Benoit, J. Polymer Sci. 57:41 (1962).
40. B. R. Jennings and H. G. Jerrard, J. Phys. Chem. 69:2817 (1965).
41. C. Picot, C. Hornick, G. Weill and H. Benoit, J. Polymer Sci.
     (C) 30:349 (1970).
42. C. Picot and G. Weill, J. Polymer Sci. (Polymer Physics)
     12:1733 (1974).
43. K. Rosenheck, P. Lindner and I. Pecht, J. Membrane Biol. 20:1
     (1975).
44. A. Scheludko and S. P. Stoylov, Biopolymers 5:723 (1967).
45. B. R. Jennings and H. Plummer, Biopolymers 9:1361 (1970).
46. C. Hornick and G. Weill, Biopolymers 10:2345 (1971).
47. B. R. Jennings and H. G. Jerrard, J. Chem. Phys. 44:1291 (1966).
48. S. P. Stoylov, J. Polymer Sci. (C) 16:2435 (1967).
49. S. P. Stoylov in 'Chemistry, Physics and Application of Surface
     Active Substances;, p. 172, Gordon and Breach, New York
     (1967).
50. S. P. Stoylov and S. Sokerov, J. Coll. Int. Sci. 27:542 (1968).
51. B. R. Jennings and J. F. Schweitzer, Polymer 13:164 (1972).
52. H. Plummer, Ph.D. thesis, London University (1969).
53. V. J. Morris and B. R. Jennings, J. Chem. Soc. (Faraday II),
     71:1948 (1975).
54. V. J. Morris, B. R. Jennings, N. J. Pearson and F. O'Grady,
     Microbios 17:133 (1976).
55. V. J. Morris and B. R. Jennings, J. Coll. Int. Sci. 55:143 (1976)
56. G. J. Brownsey, B. R. Jennings and V. J. Morris, J. Coll. Int.
     Sci. 63:597 (1978).
57. R. Subramanya and M. R. Rao, Proc. Ind. Acad. Sci. 29A:442
     (1949).
58. A. Scheludko and S. P. Stoylov, Izv. Inst. Fiz. Chim. 2:191
     (1962).
59. S. P. Stoylov and I. B. Petkanchin, Comp. Rend. Bulg. 24:487
     (1971).
60. S. P. Stoylov and I. Petkanchin, J. Coll. Int. Sci. 40:159
     (1972).
61. R. S. Subrahmanya, K. S. Doss. and B. S. Rao, Proc. Ind. Acad.
     Sci. 19:405 (1944).

WAVELENGTH DEPENDENT ASPECTS OF ELECTRIC BIREFRINGENCE AND DICHROISM

Elliot Charney

Laboratory of Chemical Physics

National Institute of Arthritis, Metabolism and
Digestive Diseases, National Institutes of Health
Bethesda, Maryland 20205

INTRODUCTION

The optical properties of matter arise from the oscillations
in the charge and current distribution induced by the incident
radiation as Buckingham has pointed out in another paper in this
symposium (1). Both the incident radiation and the response of the
molecular charge distribution have a time dependence that may be
described analytically in terms of their oscillation frequencies
which, though correlated, need not be equal nor necessarily linearly
proportional to the radiation frequency. It is our object to
discuss some aspects of these properties when the molecular system
is perturbed not only by the electromagnetic optical fields, but
also by a static or pulsed external electric field, a condition which
produces anisotropic properties, such as dichroism and birefringence,
in a molecular system otherwise isotropic in the absence of the
field. Similar perturbations may, in fact, be observed in already
anisotropic systems, as in crystals for example, but as this is a
special more complex case to which no new general phenomena may be
ascribed, it will not be discussed here.

Although far less used as an analytical and investigative device
than measurements at a fixed optical frequency, the measurement of
the wavelength dependence of electric field birefringence and
dichroism and its theoretical interpretation have contributed
enormously to our understanding of electric-field induced molecular
orientation and to electric field perturbation of energy states (Stark
splitting of degenerate states and frequency or intensity changes in
optical transition degenerate states). Except for brief allusions,

we will limit ourselves here to a discussion of the optical
dispersion associated with electric field orientation phenomena,
the latter generally known as Kerr effects, although this appellation
is sometimes restricted to the region where the orientation is
quadratic in the applied static field strength, E.

It is instructive to consider the origins of our current knowledge of the phenomena and perhaps a bit amusing to look at this by
means of a dispersion curve, a dispersion in time rather than
frequency. Thus figure 1 is a graphic display of developments in
this subject, limited to those that have a bearing on the wavelength
dependence. The amplitudes of the peaks, of course, represent a very
subjective point of view. The accompanying Table I, however, is
non-committal in this respect. The table contains abbreviated
descriptions of the contributions corresponding to the peaks in the
figure. In the discussion which follows, further reference will be
made to some of them. Optical and electrical tensors, both of which
are termed polarizabilities, describe the respective properties and
specify the way in which the induced moments depend on the spatial
relationship between the applied (time-dependent) optical and
(static) electrical fields and the molecular axes. In the case of a

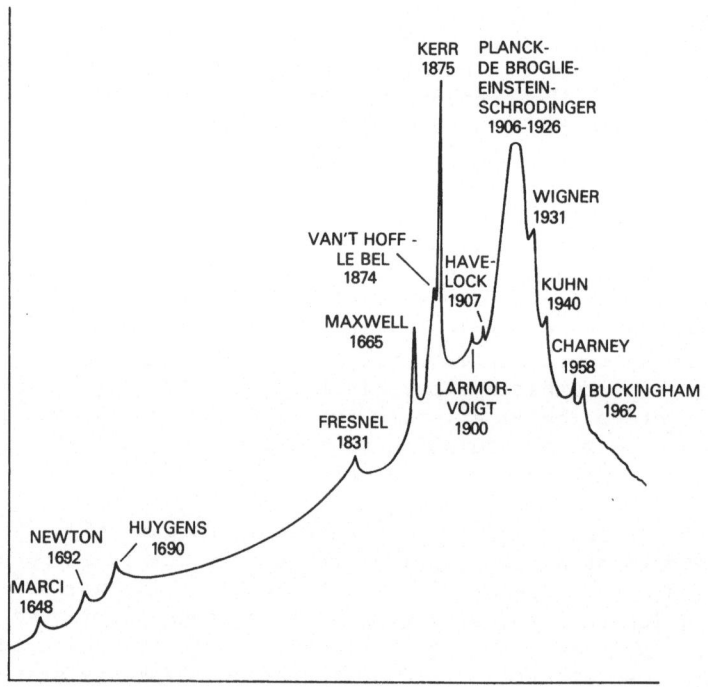

Figure 1. Chronology of developments in the wavelength dependence
of electro-optic orientation phenomena.

Table 1. Chronology of Developments in the Optical Frequency
Dependence of Electric-Field Induced Orientation Phenomena‡

1648	Discovery of the dispersion of light.	Johannes Marci
1672	Recognition that dispersion and refraction are related.	Issac Newton
1690	Experimental polarization of light.	Christiaan Huygens
1831	Recognition of the wave properties of light.	Auguste Fresnel
1865	Recognition that light waves are transverse to the direction of propagation; theoretical description of the electro-magnetic fields of radiation.	James Clerk Maxwell
1875	Discovery of electric birefringence and the recognition that it arises from a difference of the bulk refractive indices in perpendicular directions transverse to the direction of light propagation.	John Kerr
1874-'75	Development of the structural characteristics of chemical bonding.	J. H. van't Hoff and J. A. LeBel
1900	Recognition that the properties of individual molecules determine the directional character-istics of bulk refractive indices.	J. Larmor and W. Voigt
1907	The association of the wavelength dependence of electric birefringence of transparent materials with their refractive index.	T. H. Havelock
1906-'26	The quantum theory of radiation and the recog-nition of the directional character of transi-tional character of transition moments.	Louis De Broglie Max Planck Albert Einstein Erwin Schrodinger
1931	Development of the group theory relations between molecular symmetry and the directional character of transition moments.	Eugene Wigner
1939	Orientational dichroism in electric fields.	Walter Kuhn
1958	Semi-classical theory of the wavelength dependence of electric birefringence in absorbing regions of the spectrum; analysis of the sign dependence on the transition moment directions.	Elliot Charney
1962	Quantum theory for the wavelength dependence of electric birefringence.	A. D. Buckingham

‡Note that developments starting with those of Stark in 1913 on electric-field effects other than predominantly orientational and their frequency dependence are not included here.

unidirectional field such as those generally used in Kerr experiments, the electric tensor accrues some symmetry which makes the analysis of the results less complex than it might otherwise be. We will examine the classical origin of these tensor properties and the frequency dependence of the optical tensor.

We start with the premise that in the presence of an electric field an isotropically distributed material achieves a degree of preferential orientation which may be measured by its birefringence or its dichroism in a plane perpendicular to the direction of propagation of the light. The magnitude of the birefringence depends on the electrical and optical properties of the oriented molecules and on the field strength. Kerr (2) was quite aware right from the beginning that the optical properties, namely, the refractive indices, of the bulk material in directions parallel and perpendicular to the field direction were responsible for the observed birefringence, but it was not, in fact, until 1900, at about the same time that Planck was suggesting the quantization of electromagnetic radiation, that Larmor and Voigt recognized that these bulk properties were, in fact, due to the structural characteristics of the individual molecules (3). It is interesting that this connection was not made until 25 years after van't Hoff and Le Bel developed the concept of the structural characteristics of molecular bonding (4); Nor was there any recognition yet of the relation between the wavelength of the measuring radiation and these characteristics. That came in 1907 when Havelock noticed that in the visible region of the spectrum the electric birefringence of transparent materials was inversely proportional to the wavelength of the incident radiation (5).

Havelock was thus the first to quantitate the dependence of the anisotropic optical properties on the wavelength or frequency of the radiation. Havelock's law is

$$\frac{\delta}{l} \sim \frac{(\eta^2 - 1)^2}{\eta\lambda}$$

(1)

This function is monotonic in λ but goes to infinity as the measuring wavelength approaches zero, i.e., at very high frequencies. Havelock was aware of the fact that the frequency dependence arose from the variation of the refractive index with wavelength and quotes the measurement by Blanchard (6), a year earlier, of the electric birefringence of CS_2 at 56,000 v/cm. Apparently he was unaware of the possibility that the wavelength dependence of the refractive index could be different for different directions of propagation of the radiation with respect to the molecule itself. Thus he arrived at the conclusion that

$$\frac{(\lambda^2 - 1)^2}{\delta\eta\lambda} = C$$

(2)

It is quite true that the frequency dependence is to a good approximation separable from the directional properties if the description is that of the birefringence involving a transition between totally isolated states or if a summation is made over all substates of a dominant state.

We note in passing the hold that Havelock's description has had on measurements in this field. As recently as 1972, for example, LeFevre, who is fully cognizant of the detailed origin of the frequency dependence, nevertheless called attention to the fact that measurements of the wavelength dependence of the electric birefringence of cyclohexane show that the observed values of $(\eta^2-1)^2/\delta\eta\lambda$ are not invariant in the interval between 600 and 375 nm where cyclohexane is optically "transparent" (7). The first demonstration that not only the magnitude but also the frequency dependence of the refractive index could be different in different directions came with Kopferman and Ladenburgh's 1925 measurement of the Kerr effect of Na vapor in the vicinity of the D-line of Na near 589 nm (8), Figure 2.

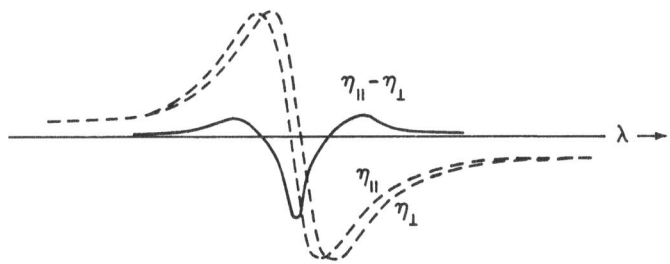

Figure 2. Birefringence of sodium vapor in the vicinity of the D line near 589 nm in a field of 30,000 v/cm. Solid line: birefringence. Dashed lines: the parallel and perpendicular components of the refractive index.

There are several interesting and distinguishing features about this measurement aside from the displacement of the curves on the frequency scale. For one thing, Havelock's law is certainly not obeyed by the measurement near a resonant (or absorbing) frequency. Instead of going to infinity as the wavelength decreases, the parallel and perpendicular refractive indices reverse directions, go through zero, reverse again on the short wavelength side and then behave inversely with λ, but now with a different sign.

Second in the analysis of these observations, Kopferman and Ladenburgh attributed the variation of sign on the two sides of the resonant line entirely to the frequency dependence, their expression

$$\eta_{\shortparallel} - \eta_{\perp} = \frac{\rho}{4\pi\omega_o} \left(\frac{1}{\omega_o^{\shortparallel}-\omega} - \frac{1}{\omega_o^{\perp}-\omega} \right)$$

(3)

where ρ is proportional to the isotropic optical polarizability which is not signed, requires that $\omega_o^{\shortparallel,\perp}-\omega$ be negative when $\omega < \omega_o^{\shortparallel,\perp}$, (see figure 3). As a consequence, the refractive indices $\eta_{\shortparallel}$ and $\eta_{\perp}$ are required to be negative at longer wavelengths and positive at short wavelengths and the difference to be associated solely with a frequency shift of the degenerate components of the absorption line (the D_2 line of Na at 16973.39 cm^{-1}, $^2P_{3/2}\leftarrow 3\,^2\Sigma_{1/2}$).

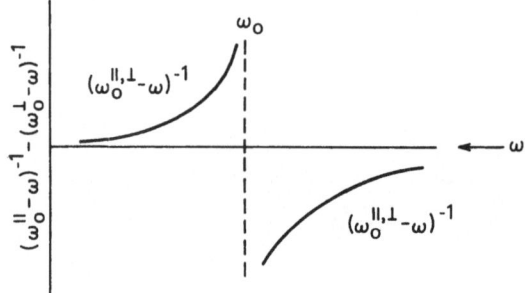

Figure 3. Functional optical frequency dependence of the birefringence from equation (3).

They were in fact nearly correct because this will always be the case for a spherical atom and it was only in 1967 that Bogard, Buckingham, and Orr demonstrated that the origin lies not only in the splitting of the degenerate states, but in different values for the mean hyperpolarizability tensor in the two directions (9).

But what I most wish to call attention to is the fact that the Kopferman and Ladenburgh analysis obscured the possibility that in molecules which are anisotropic in their ground state, the sign difference could arise not only from frequency shifts, but, in fact, more often from orientation effects and depend on the relation between the dielectric and optical tensors of the molecules, the former determining the orientation of the molecules in the Kerr field and the latter, the magnitude and frequency dependence of the component refractive indices and thus the sign of the birefringence.

In 1927, C. V. Raman and K. S. Krishnan published a totally classical treatment of the Kerr effect which examined the relationship between the optical polarizability and the dielectric axes in transparent spectral regions (17). In 1933, Serber did a classical and quantum mechanical analysis of the Kerr effect from which the frequency dependence was specifically excluded because of the desire to give meaning to a frequency independent parameter, the molecular "optical anisotropy" (11).

These theoretical treatments and the measurements of changes in absorption intensity of molecules oriented in a electric field by Kuhn and co-workers in 1939 (12), should have provided the clues to the possibility that orientation in an electric field would give rise to Kerr effects whose sign would depend on the geometric relationship between the polarization of optical transitions and the molecular dielectric tensor. Nevertheless, when we started our investigation in 1951 of the Kerr birefringence in the region of an absorption band, no analysis or measurement of this kind had been made. The formal treatment has since been given a number of times with more or less emphasis on the quantum mechanical details and to different orders of perturbation theory. The 1962 treatment by Buckingham, for example, produced predictions for the dependence of the birefringence on the J and K quantum numbers of the rotational states of diatomic and symmetric top molecules (13). The brief semi-classical treatment I will give here follows our 1958 analysis (14), which is sufficient to demonstrate the frequency response in condensed phases of molecules whose birefringence is caused predominantly by orientation of electrically and optically anisotropic molecules.

Theory

This is a semi-classical treatment. The orientation is treated classically but the quantum mechanical rules for the magnitude and polarization of electromagnetic transitions is substituted for the classical optical polarizability. The frequency dependence is separated into two parts, that appropriate to a frequency range close to an absorption band and the part appropriate to transparent regions of the spectrum. The latter of course is what produces Havelock's dispersion law. We require expressions for the differential refractive indices parallel and perpendicular to the field direction. The classical expression for the isotropic refractive index is:

$$\eta(\omega) \;=\; 1 + \frac{N}{2} \sum_i \frac{e_i^2/m_i\,\varepsilon_o\,(\bar{\omega}_i^2 - \omega^2)}{(\bar{\omega}_i^2 - \omega^2)^2 + \omega^2\lambda_i^2} \tag{4}$$

where e_i is the charge and m_i is the mass of a fictitious particle oscillating with a natural or resonant frequency, ω_i, N is Avogadro's number. ω is the frequency of the measuring radiation; g_i the so-called damping factor, is the width at half-height of the absorption band associated with the natural resonance frequency, ω_i, of the oscillator. If we wish to consider the refractive index in the vicinity of a particular resonant frequency, i.e., a particular electronic or vibrational absorption band, then the second term on the right of this expression can be separated into two terms, one encompassing or running over all other bands, and the other the particular one of interest. In doing so, we can drop the term $\omega_i^2 g_i^2$ from the summation term, since at ω far from the absorption band, ω_i, $(\omega_i^2-\omega^2)^2 >> \omega^2 g_i^2$, g_i generally being small compared even to ω in interesting spectral regions:

$$\eta(\omega) = 1 + \frac{N}{2} \sum_{i \neq j} \frac{(e/m_i \varepsilon_o)}{(\overline{\omega}_i^2-\omega^2)} + \frac{N}{2} \frac{(e_j^2/m_j \varepsilon_o)(\overline{\omega}_j^2-\omega^2)}{(\overline{\omega}_j^2-\omega^2)^2 + \omega^2 g_j^2}$$

(5)

If we now define a dimensionless quantity, the oscillator strength, $f_i = e_i^2 m_i / e^2 m^2$, where e and m are, respectively, the charge and mass of an electron, f_i defines the strength of the optical transitions associated with each of the i resonant frequencies. Then:

$$\eta(\omega) = 1 + \frac{Ne^2}{2\varepsilon_o m} \sum_{i \neq j} \frac{f_i}{(\omega_i^2-\omega^2)} + \frac{Ne^2}{2\varepsilon_o m} \cdot \frac{f_j(\overline{\omega}_j^2-\omega^2)}{(\overline{\omega}_j^2-\omega^2)^2 + \omega^2 g_j^2}$$

(6)

It should be noted that the complex notation, $\hat{\eta} = \eta - i k$, has for simplicity been deliberately avoided here as the discussion is cast in terms of the birefringence and taken only to first order. If the complex notation is included, k, the absorption index, related to the molecular extinction coefficient can be identified as

$$k = \frac{Ne^2\omega}{2\varepsilon_o m} \left(\sum_{i \neq j} \frac{f_i g_i}{(\omega_i^2-\omega^2)} + \frac{f_j g_j(\omega_j^2-\omega^2)}{(\omega_j^2-\omega^2)^2 + \omega^2 g_j} \right)$$

(7)

where the first term in the bracket would be entirely negligible in the vicinity of an isolated absorption band as all the oscillator strengths other than f_k are entirely negligible. The treatment would then proceed in the same manner for dichroism using the differential absorption indices, $k_\parallel - k_\perp$. The classical oscillator strength can only be positive if finite, so that without further analysis the sign of a birefringence or dichroism curve is generated by the difference

between parallel and perpendicular components of the oscillator
strength and by the frequency factors.

Using perturbation theory, it is possible to demonstrate that
the oscillator strengths are proportional to electric dipole (and
much smaller but higher multipoles) transition moment matrix elements,

$$f_k \, \alpha \, |\mu_{mn}|^2 = |\mu_x|_{mn}^2 + |\mu_y|_{mn}^2 + |\mu_z|_{mn}^2 \tag{8}$$

for the electromagnetic transition between the states $|m>$ and $|n>$.

In an isotropic medium, the space fixed components averaged over
all orientations are equal. If, however, the medium is anisotropic
as in a crystal or field oriented system, the space-fixed transition
dipole moment components are not the same in all directions and the
familiar phenomenon of dichroism, and, of course, birefringence
appears when the system is viewed with polarized light.

The important point is that the space-fixed transition moment
components of the frequency and sign of the birefringence and
dichroism can now be separated and, if the orientation axes of the
molecule are known or can be determined, their direction in the
molecule can be specified. The differential refractive index, for
example, from which the birefringence is obtained, is given by
(for a uniaxial oriented system):

$$\eta_\| - \eta_\perp \sim \frac{N}{21\lambda\varepsilon_0} \left\{ \sum_{i \neq j} \left[\Omega_i^\| <\mu_i^\|>^2 - \Omega_i^\perp <\mu_i^\perp>^2 \right] + \left[\Omega_j^\| <\mu_j^\|>^2 - \Omega_j^\perp <\mu_j^\perp>^2 \right] \right\} \tag{9}$$

where the Ω's are the frequency dependent factors. If the j index
refers to a single isolated transition, or to an unresolvable sum
over substates, then the parallel and perpendicular components of Ω
have the same frequency dependence and the second term may be written
as:

$$\Omega(\omega_j) (<\mu_j^\|>^2 - <\mu_j^\perp>^2) \tag{10}$$

In this case, the sign is controlled both by the frequency dependence, which, as we have seen, is negative at low frequencies and
positive at high frequencies, and by the relative magnitudes of the
orientationally averaged projections of the μ_j transition moment on

space-fixed directions parallel and perpendicular to the field
direction.   This is equally true for the first term summed over all
transitions.   As a result, four main types of birefringence dispersion curves are possible in the vicinity of an isolated optical
transition; two in dichroism, as the first term does not measurably
contribute in dichroic spectra when the jth transition is isolated
from the rest (figure 4).

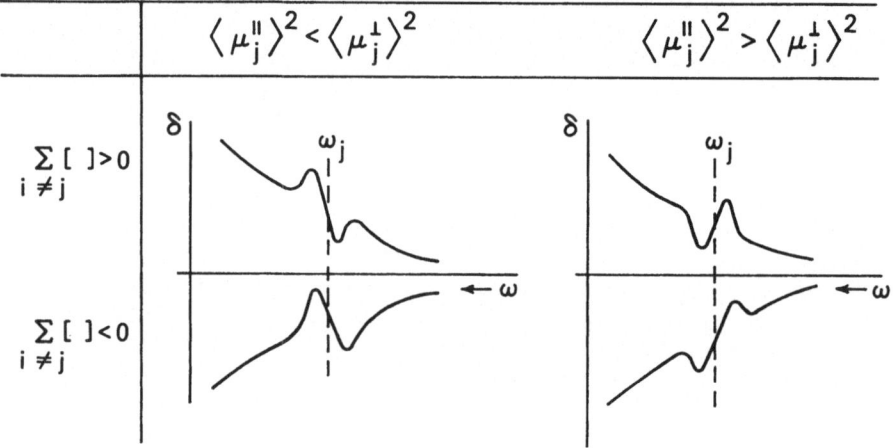

Figure 4. The functional dependence of the birefringence in an
isolated absorption band.   The birefringence is shown superimposed
on the background birefringence from still higher frequency optical
transitions.

When there are elements of planar or rotational symmetry present in
the oriented molecule, it is relatively easy to fix the direction of
the transition moments in the molecule from these curves and optical
selection rules.   The electric field measurements can then be used
in exactly the same way as linear dichroism measurements on crystals
to determine the polarization of optical transitions.

Experimental Examples

In the several examples to be considered below, the optical
transitions are, in fact, not completely isolated. They nevertheless
demonstrate the power of the method which to a large extent results
from the fact that a uniform electric field produces uniaxial
orientation so that two measurements, one parallel and one perpendicular to the field, contain all the optical information. The
dichroic spectra of polybenzyl-L-glutamate (PBLG) produced by its
orientation in an electric field is an example of the potentiality
of the method (15).

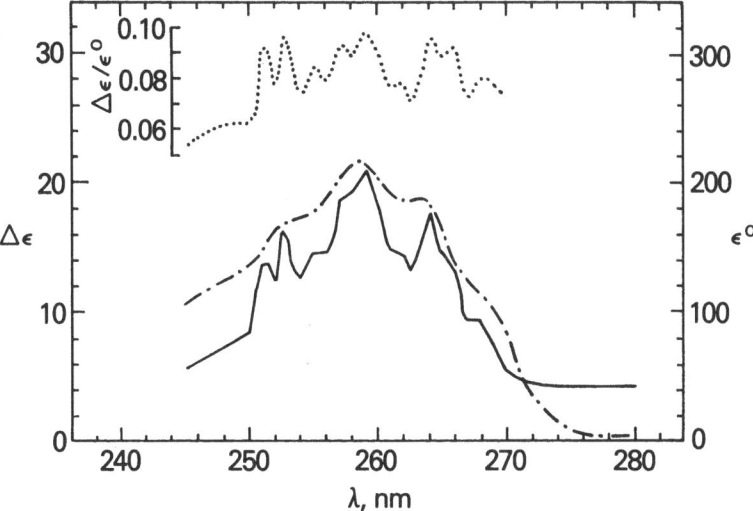

Figure 5. a) Isotropic (-.-) and dichroic (---) spectrum, and
reduced dichroism (....) of PBLG (15).

PBLG is a polypeptide known to be in an ʌ-helical conformation
in solution in simple halocarbons. It's ultraviolet spectrum
exhibits a band of moderate intensity near 260 nm, similar to that
observed in the ordinary spectra of mono substituted benzenes in
that region. The aromatic ring in which this band originates is on
a side-chain separated from the polypeptide backbone by six bonds
about which moderately-free rotation (somewhat restricted by
potential minima) can occur. As a consequence, orientation of the
α-helical polypeptide would not necessarily result in a strong
preferential orientation of the aromatic ring, but, in fact, it does,
indicating that despite the possible rotational freedom about the
intervening bonds, the aromatic ring exhibits a very strong prefer-
ence for a fixed orientation with respect to the α-helical axis (15).

Part of the evidence for this lies in the strongly structured
dichroic spectrum obtained in the electric field (figure 5). Since
the aromatic ring is planar, rather strict selection rules require
that all transitions be polarized in the plane of the ring or
perpendicular to it. As a consequence, the oriented dichroic
spectrum of PBLG in an electric field contains maxima and minima,
resulting from the juxtaposition of the transitions of differing
polarizations in- and out-of-plane of the ring, and each of these may
be associated with a species of the proper symmetry. With the help
of the known spectral assignments of this band in toluene vapor and
of symmetry considerations, the electric-field induced spectrum of
PBLG has been analyzed in detail and used as an aid in defining the
PBLG structure (15).

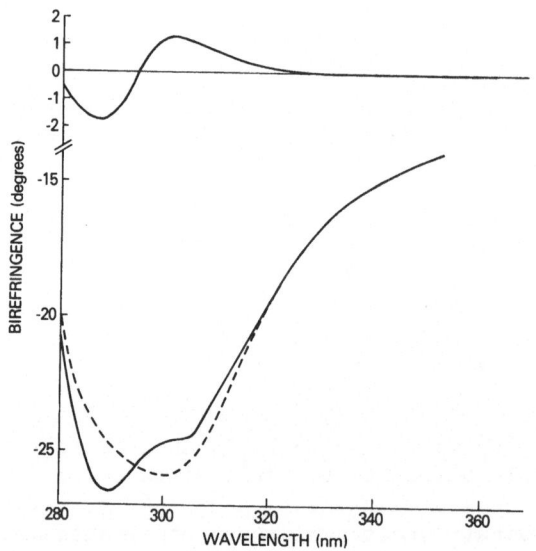

Figure 6. Electro-opticl birefringence of polyribocytidylic acid,
solid-line (——) is the experimental birefringence. Dashed-lines
(---) are the resolution of the experimental data;   Upper: the
parallel out-of-plane component which gives rise to the inflections
in the 290-300 nm region. Lower: the major negative birefringence
component arising predominantly from the strong perpendicular
in-plane transition near 260 nm.

The second example is taken from some recent electric birefringence measurements of polyribocytidylic acid (16). The principal
near ultraviolet transition of this polymer corresponds to the well-
known electronic transition of cytidine near 260 nm. This transition
is required by symmetry to lie virtually in the plane of the hetero-
aromatic ring system, the latter of which is nearly perpendicular to
the helical axis in the double-stranded complex. As the complex is
oriented parallel to the applied electric field, the transition (or
more correctly, its moment) is oriented perpendicular to the applied
electric field and consequently, produces a negative birefringence.
Only the low frequency negative lobe of this transition is observed
and illustrated in figure 6. Note, however, that a clear inflection
is seen in the birefringence spectrum near 295 nm. This inflection
is produced by a weak transition which gives rise to a positive
birefringence as illustrated by the decomposition into components
in figure 6. The existence and location of this transition and its
assignment has long been in question (17,18). It is uniquely proven
by this technique.

Molecular orientation can be achieved in crystallites and
liquid crystals by electric fields with similar results. For
example, Jennings and Forewaker (18, 19) have shown this way that
the smooth isotropic absorption band--near 610 nm of copper
phthalocyanine crystallites--consists of at least two transitions
polarized in different directions and separated by about 15 to 20 nm.

Electrochromism and Other Induced Effects

Electrochromism was first predicted by Platt (20) and the
theoretical analyses developed various degrees of completeness and
sophistication by Labhart (21), Liptay and Czekalla (22), Yamaoka
and Charney (23) and Dows and Buckingham (24). These treatments,
generally second-order in perturbation theory, predicts that at high
fields there can be intensity changes associated with changes in the
excited state dipole moment in the presence of the field as well,
of course, as Stark splitting of degeneracies. These can be
detected using polarized radiation by electric birefringence, or
absorption techniques, and by modulating the field and using phase-
sensitive detection, looking only at the changes due to the exciting
field. Some of these effects have been observed (see for example,
25-27).

While the semi-classical theory discussed earlier in this paper
does predict that changes occur in absorption bands and that they
are responsible for electric-optic phenomena, it is not explicit in
its description of molecular changes in structure and quantum states,
with or without strong orientation effects. Buckingham in his 1962
paper on the frequency dependent Kerr constant (13) and again in the
1964 paper with Dows (24) was able to demonstrate that at least for

small molecules in the gaseous state, molecular distortion of the
molecule by the field is the dominant effect in an absorption band.
For example, it is possible to induce birefringence in molecules
with isotropic polarizability such as CCl_4. This results from field-
induced distortion which breaks down selection rules applicable to
the isotropic structure and makes it possible for normally forbidden
transitions to appear. Using a quantum theoretical analysis, they
showed that for gaseous diatomic and for symmetric top molecules,
there is very strong dependence on the angular momentum quantum
numbers which should appear in the frequency dependence of Kerr
dispersion of the rotation and rotation-vibration lives of gaseous
molecules. The very precise perturbation theory prediction for the
birefringence in the vibration-rotation band of HCl has not yet been
measured, but distortion effects have been observed in electric
field induced spectra by Terhune and Peters (25), Brown, Buckingham
and Ramsey (26), Hochstrasser (27) and others.

REFERENCES

1.   A. D. Buckingham, this volume.
2.   J. Kerr, Phil. Mag. 50:337 (1875).
3.   J. Larmor, Aether and Matter 1:351 (1900); W. Voigt, Ann. Physik
     4:197 (1901).
4.   J. H. van't Hoff, Arch. Neer 9:445 (1874); ibid, Bull. Soc.
     Chim. Fr., 23:295 (1875); J. A. Le Bel, Bull. Soc. Chim. Fr.
     22:337 (1874).
5.   T. H. Havelock, Proc. Roy. Soc. (London) A80:28 (1907).
6.   H. L. Blanchard, Amer. Acad. Proc. 41:647 (1906).
7.   R. J. W. Le Fevre and R. K. Pierens, Aust. J. Chem., 25:413
     (1972).
8.   H. Kopferman and R. Landberg, Ann. Physik, (4)78:659 (1925);
     ibid, 79:96 (1926).
9.   M. P. Bogard, A. D. Buckingham and B. J. Orr, Mol. Phys.,
     13:533 (1967).
10. C. V. Raman and K. S. Krishnan, Phil. Mag. 3:713,724 (1927).
11. R. Serber, Phys. Rev. 43:1003,1011 (1933).
12. W. Kuhn, H. Durkhop and H. Martin, Z. Physik. Chem., B45:121
     (1939).
13. A. D. Buckingham, Proc. Roy. Soc., A267:271 (1962).
14. E. Charney and R. S. Halford, J. Chem. Phys., 29:221 (1958).
15. E. Charney, J. B. Milstien and K. Yamaoka, J. Amer. Chem. Soc.
     92:2657 (1970).
16. H. H. Chen and E. Charney, Biopolymers (in press).
17. A. Rich and M. Kasha, J. Amer. Chem. Soc., 82:6197 (1960).
18. M. Gellert, J. Amer. Chem. Soc., 83:4665 (1961).
19. B. R. Jennings and A. R. Forewaker, Spectrosc. Lett., 7:371
     (1974).
20. J. R. Platt, J. Chem. Phys., 34:862 (1961).
21. H. Labhart, Chimia, 15:20 (1961).

22. W. Liptay and J. Czekalla, Z. Elekrochem, 65:721 (1961).

23. K. Yamaoka and E. Charney, J. Amer. Chem. Soc., 94:8963 (1972).

24. D. W. Dows and A. D. Buckingham, J. Mol. Spect., 12:189 (1964).

25. R. Terhune and C. Peters, J. Mol. Spect., 3:138 (1959).

26. J. M. Brown, A. D. Buckingham and D. A. Ramsey, Can. J. Phys.
     49:914 (1971).

27. R. M. Hochstrasser and L. J. Noe, J. Chem. Phys., 48:514 (1968).

POLARIZED FLUORESCENCE IN AN ELECTRIC FIELD

G. Weill

Centre de Recherches sur les Macromolecules
    and Université L. Pasteur
Strasbourg, France

INTRODUCTION

As for the other electro-optics effects, the change of polarization of fluorescence observed in an electric field results from the orientation of the molecules. The specific advantages which were looked for when we started to develop the technique[1] were, when compared to the more classical electric birefringence and dichroism:

- the possibility of measuring not only the second, but also the fourth moment of the orientation function and therefore to test the separation of the electrical and optical factors and the mechanism of orientation.

- the possibility of obtaining separate information about the conformations of chromophores with essentially overlapping absorption bands but different quantum yields of fluorescence; they will appear as a single class in electric dichroism but can eventually be separated in a polarized fluorescence experiment, as for example, if one of them has zero quantum yield.

Since the literature on the subject contains only a few papers, but since the presently available quantitative formulae deal with a specific model and should not be used without care outside of its range of validity, this paper will mostly deal with the basics of the phenomenon and of the model. Results thus far obtained will be surveyed, referring to the original papers for experimental details.

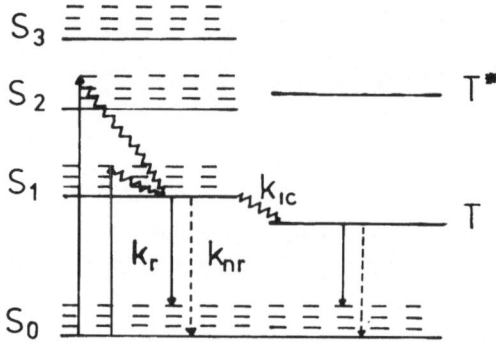

Fig. 1:   Radiative and non-radiative decay from singlet excited
          states

I.   Some Fundamentals of Fluorescence

     The different possible decays of an excited singlet state are re-
presented schematically on the diagram below where it is recalled
that:
     a) Electronic excitation to the first singlet state takes place
between the 0 vibrational level of S_o to higher vibrational levels
of S_1. A rapid relaxation brings the system into the 0 vibrational level of S_1.
     b) Excitation to a higher singlet state $S_2, S_3 \ldots$ is rapidly
followed by a non-radiative relaxation to the same 0 vibrational
level of S_1.
     c) The decay from S_1 to S_o takes place through the following
processes:
          - radiative decay from S_1 to S_o (fluorescence); rate con-
stant k_r;
          - non radiative decay to S_o due to energy dissipation by
interaction with the solvent modes; rate constant k_{nr};
          - non-radiative decay due to collisions with specific
quenchers at a concentration [Q] (quenching), rate constant k_Q [Q];
          - intersystem crossing to the lower lying triplet state T
with a subsequent decay to S_o by a radiative (phosphorescence) or
non-radiative process; rate constant k_{ic}.

     The quantum yield of fluorescence is then:

$$\phi = \frac{k_r}{k_r + k_{nr} + k_Q[Q] + k_{ic}} \qquad (1)$$

and the fluorescence life time τ, which governs the probability of
emission or the decay of the fluorescence intensity $I(\tau)$ of an
assembly of molecules excited by a fast pulse, accordingly, has the
form

$$I(t) = I_o \, e^{-t/\tau} \qquad\qquad (2)$$

$$\tau = k_r^{-1} \qquad\qquad (3)$$

It is related to the natural life time τ_o which would result from
Einstein spontaneous emission in the absence of any other decay
process by

$$\tau = \phi\tau_o \qquad\qquad (4)$$

τ_o can be calculated from the absorption spectrum and lies generally in the $10^{-9} - 10^{-7}$ sec. range.

One big interest of fluorescence is the polarization of the
emitted light. It can be easily understood starting first from the
case of a rigid medium and for emission at right angles to the incident beam as shown in Fig. 2.

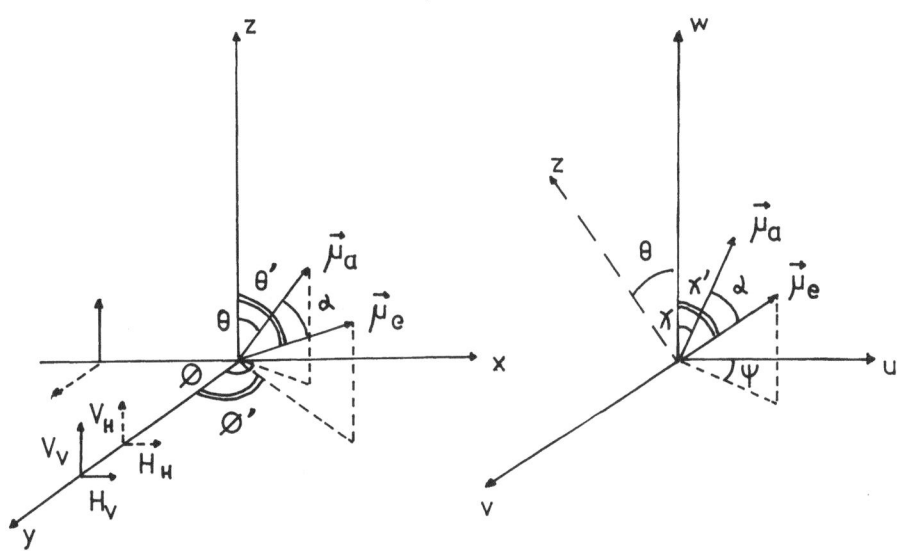

Fig. 2: Polarized components of fluorescence and their relation
with the orientation of absorption and emission transition moments.

We define four components of fluorescence V_V, H_V, V_H and H_H where the index denotes the polarization of the incident beam and the capital letter stands for the position of the analyzer, either perpendicular to (V) or in (H) the plane of the incident and emitted beams. Denoting by $\vec{\mu}_a$ the absorption transition moment and by $\vec{\mu}_e$ the emission transition moment and recalling that the absorption and emission probabilities are proportional to the square of the transition moments, one obtains for the relative values of the polarized components of fluorescence:

$$V_V \propto <\mu^2_{az} \ \mu^2_{ez}>$$

$$H_V \propto <\mu^2_{az} \ \mu^2_{ey}>$$

$$V_H \propto <\mu^2_{ax} \ \mu^2_{ez}> \qquad (5)$$

$$H_H \propto <\mu^2_{ax} \ \mu^2_{ey}>$$

The brackets denote averages over the distribution of orientations of the assembly of molecules.

The angle α between $\vec{\mu}_a$ and $\vec{\mu}_e$ in a rigid medium is due to a difference of orientation of the transition moments associated with either different vibrational levels ($S_0 \to S_1$ excitation) or with different electronic levels ($S_0 \to S_2$ excitation). Using the relation between θ, θ' and α as defined in Fig. 2.

$$\cos \theta' = \cos\theta \cos\alpha + \sin\theta \sin\alpha \cos (\phi - \phi') \qquad (6)$$

It is easy to calculate the different averages for random orientation. One gets in particular the well known Perrin formula for the emission anisotropy r_0

$$r_0 = \frac{V_V - H_V}{V_V + 2H_V} = \frac{2}{5} \left(\frac{3 \cos^2\alpha - 1}{2}\right) \qquad (7)$$

It shows that V_V is different from H_V so that the emitted light is partially polarized.

If the medium is not sufficiently rigid, ie., if the molecule rotates with a rotatory correlation time τ_r, smaller or of the same order of magnitude as τ, the calculation in relation (5) must be modified, using the orientation of $\vec{\mu}_a$ at time zero and the orientation of $\vec{\mu}_e$ at the time t of emission. If the time evolution of $\vec{\mu}_e$ is characterized by an angle $\beta(t)$ the time dependent emission anisotropy following the excitation becomes, for the collection of molecules:

$$r(t) = r_0 <\frac{3\cos^2 \beta(t) - 1}{2}> \qquad (8)$$

Under steady state illumination, the time averaged emission anisotropy becomes

$$\bar{r} = r_0 \int_o^\infty \langle \frac{3\cos^2\beta(t) - 1}{2} \rangle \; e^{-t/\tau}d\frac{t}{\tau} \quad (9)$$

For an isotopic rotatory motion

$$\langle \frac{3\cos^2\beta(t) - 1}{2} \rangle = e^{-6Dt} \quad (10)$$

where $D = (6\tau_r)^{-1}$ is the rotatory diffusion constant and

$$\bar{r}^{-1} = r_0^{-1} (1 + 6D\tau) \quad (11)$$

If D is large, there is essentially no correlation between $\vec{\mu}_a$ at
time zero and $\vec{\mu}_e$ at time t and the fluorescence is unpolarized.

When dealing with macromolecules, which can bear several
chromophores, it is necessary to recall that for close chromophores
($R \leq \sim 30$ Å) energy migration can take place through either excitonic coupling or the weak dipole-dipole Försters mechanism. The
emission can then take place from a molecule distant from the absorbing one and weakly correlated with it in orientation, a source
for fluorescence depolarization.

II. Polarization of Fluorescence in a Non-randomly Oriented and
Non-rigid System

1) General Aspects:

If the molecules are submitted to an external field, both their
orientation and their rotatory motions can be affected. A general
formulation of the components of the emitted light along arbitrary
directions and analyzer orientation can be obtained from a generalization of the preceeding paragraph. It has been mostly developed
as a tool in the study of the orientation and dynamics of deformed
polymer films and the relevant literature can be found in papers
by Frehland[2] and Jarry[3]. Using the formalism of the latter one
defines the direction and polarization of the beams by the
polarization vector $\vec{P}$ and $\vec{A}$ of the analyzer. The orientation
distribution of $\vec{\mu}_a$ is given by $N(\Omega_0) \, d\Omega_0$, the probability of finding $\vec{\mu}_a$ in a small solid angle around Ω_0. The orientation of $\vec{\mu}_e$
is given by the conditional probability $P(\Omega, t/\Omega_0)$ to find $\vec{\mu}_e$
along Ω at time t if $\vec{\mu}_a$ lies along Ω_0 at time zero. The intensity
emitted at time t after a pulse at time zero with direction and
polarization specified by $\vec{P}$ and $\vec{A}$ is then given by:

$$i \, (\vec{P},\vec{A},t) \; \alpha \; \int_\Omega \int_{\Omega_0} N(\Omega_0) \; P(\Omega, t/\Omega_0) \; (\vec{P} \cdot \vec{\mu}_a)^2 (\vec{A} \cdot \vec{\mu}_e)^2 \; d\Omega_0 d\Omega$$

$$(12)$$

Under steady state illumination :

$$i \, (\vec{P},\vec{A},\tau) \, \alpha \int_0^\infty i(\vec{P},\vec{A},\tau) \, e^{-t/\tau} d\frac{t}{\tau} \tag{13}$$

The four terms in the integral (12) can be expanded in spherical harmonics and, using the orthogonality relations, reduced to a
linear combination of 36 angular functions which describe the more
general anisotropy in both orientation and rotatory motion. In the
case of uniaxial orientation, the one of interest for electric field
orientation, it reduces to 6, one giving in fact the total emitted
intensity and the five remaining giving the following space and
time averaged correlation functions:

$$\frac{1}{2} <3\cos^2\theta - 1> \quad ; \qquad \frac{1}{4} <(3\cos^2\theta - 1)\,(3\cos^2\theta' - 1)>$$

$$\frac{1}{2} <3\cos^2\theta' - 1> \quad ; \qquad \frac{9}{16} <\sin\theta \, \cos\theta \, \sin\theta' \, \cos\theta' \, \cos(\theta - \theta')>$$

$$\frac{9}{64} <\sin^2\theta \, \sin^2\theta' \, \cos^2(\theta - \theta')> \tag{14}$$

The determination of these five quantities requires in general measurements not limited to V_V, H_V, V_H, and H_H but performed in out of
plane directions. If the second moment of the orientation function
$\cos^2\theta$ (or $\cos^2\theta'$) is directly available, determination of the fourth
moment requires the introduction of a specific rotatory motion, i.e.
an explicit expression for $P(\Omega, t/\Omega_0)$, compatible with the three remaining correlation functions. Therefore, the results can equally
be used to characterize the orientation distribution and the dynamics of the transition moments. Extra manipulation of the spherical
harmonics is required in order to obtain the orientation function
and the dynamics of the molecules to which the chromophores are
bound, especially if they have some degree of rotatory freedom. If
the molecules are flexible, an extra averaging over the molecular
conformations is needed. For that reason we will, in what follows,
restrict our calculation to a rod-like molecule bearing chromophores
with specified degrees of freedom and recalculate the components of
the fluorescence directly in this restricted case. The general
formulation outlined above should however help to understand the
problems raised by its substitution to the general case in non-
allowed situations such as highly flexible (both statically and
dynamically) molecules.

2)   The rod-like model[4]

     The long axis correlation time is supposed $>>\tau$, so that its
motion during the fluorescence life time is neglected. The rotatory
motion around this axis is unaffected by the electric field,

as will be considered the possible motion of the chromophore with respect to the molecular axis. With these simplifications the details of the Brownian motion will be embraced by an effective angle between $\vec{\mu}_a$ and $\vec{\mu}_e$ measured from the anisotropy in the absence of field and all the information will be contained in the four polarized components V_V, V_H, H_V and H_H. Their calculation can be carried out in terms of the angles θ, ϕ and γ which characterize the rod axis orientation, the angle χ which characterizes the orientation of $\vec{\mu}_a$ with the rod axis, and the angles χ' and ψ which characterize the orientation of $\vec{\mu}_e$ with respect to respectively the rod axis and the plane of the rod axis and $\vec{\mu}_a$ (Fig. 2).

After a cylindrical averaging over the coordinates ϕ and γ, one gets from relation (5) the expression of the polarized components as a weighted sum over five terms A_i depending on χ and χ' with coefficients B_e containing the second and fourth moment of the orientation function in the form of the two mean values $\sin^2\theta$ and $\sin^4\theta$ (Table I).

$\langle\sin^2\theta\rangle$ can be related to the usual orientation factor in electric birefringence and dichroism

$$\langle\sin^2\theta\rangle = \frac{2}{3}(1 - \phi) \tag{15}$$

Another orientation factor Δ related to the fourth moment can also be defined through

$$\langle\sin^4\theta\rangle = \frac{8}{15}(1 - \Delta) \tag{16}$$

Changes in the polarized components of fluorescence result from the change in ϕ and Δ with the electric field. Steady state and transient values of Δ have been carried out[5] for induced dipole moment and permanent dipole moment orientation mechanism to supplement the already known expressions for ϕ. From these calculations three interesting results emerge.[5]

1. In the low field Kerr region $\Delta = \frac{10}{7}\phi$ a result which seems to hold in fact up to values of ϕ of the order of 0.3.
2. At very high fields $\langle\sin^4\theta\rangle = 3\langle\sin^2\theta\rangle^2$ for both orientation mechanisms. This tells us in fact that the relation between Δ and ϕ is not very sensitive to the mechanism of orientation, a useful remark to design a strategy to get χ and χ' from the experimental observations at any field.
3. The presence of the fourth order Legendre polynomial in $\langle\sin^4\theta\rangle$ complicates the high orientation field free decay of

$$\Delta_{E \to \infty} = \frac{10}{7}e^{-6Dt} - \frac{3}{7}e^{-20\,Dt} \tag{17}$$

The calculation of $\langle\sin^2\theta\rangle$, $\langle\sin^4\theta\rangle$, χ, χ' and ψ from any set of experiments make generally use of
- a starting value of $\langle\sin^2\theta\rangle$ deduced from a parallel measurement of electric dichroism;

TABLE I

Calculation of the Polarized Components of Fluorescence as a Function of Orientation

Values for A_J

B_ℓ	$\sin^2\chi \sin^2\chi'$	$\sin^2\chi \cos^2\chi'$	$\cos^2\chi \sin^2\chi'$	$\cos^2\chi \cos^2\chi'$	$\sin\chi\cos\chi\sin\chi'\cos\chi'$
V_V	$\frac{1}{8}\sin^4\theta\,(1+2\cos^2\psi)$	$\frac{1}{2}(\sin^2\theta - \sin^4\theta)$	$\frac{1}{2}(\sin^2\theta - \sin^4\theta)$	$1-2\sin^2\theta+\sin^4\theta$	$2\cos\psi\,(\sin^2\theta-\sin^4\theta)$
H_V	$\frac{1}{4}\sin^2\theta - \frac{1}{16}\sin^4\theta$ $\times(1 + 2\cos^2\psi)$	$\frac{1}{4}\sin^4\theta$	$\frac{1}{2} - \frac{3}{4}\sin^2\theta + \frac{1}{4}\sin^4\theta$	$\frac{1}{2}(\sin^2\theta-\sin^4\theta)$	$-\cos\psi\,(\sin^2\theta-\sin^4\theta)$
V_H	$\frac{1}{4}\sin^2\theta - \frac{1}{16}\sin^4\theta$ $\times(1 + 2\cos^2\psi)$	$\frac{1}{2} - \frac{3}{4}\sin^2\theta + \frac{1}{4}\sin^4\theta$	$\frac{1}{4}\sin^4\theta$	$\frac{1}{2}(\sin^2\theta-\sin^4\theta)$	$-\cos\psi\,(\sin^2\theta-\sin^4\theta)$
H_H	$\frac{1}{64}(24 - 24\sin^2\theta + \sin^4\theta) + \frac{1}{64}\cos^2\psi$ $(-16 + 16\sin^2\theta + 2\sin^4\theta)$	$\frac{1}{64}(16\sin^2\theta-4\sin^4\theta)$	$\frac{1}{64}(16\sin^2\theta-4\sin^4\theta)$	$\frac{8}{64}\sin^4\theta$	$-\frac{16}{64}\cos\psi\,\sin^4\theta$

G.

- the relative changes in intensity of the polarized components of fluorescence ΔV_V, ΔH_V, ΔV_H, ΔH_H with respect to the sum of components $V_V + 2H_V$;

- the fluorescence anisotropy r in the absence of electric field which is a function of $(\chi - \chi')$ and Ψ;

$$r = \frac{1}{5} [3 \cos^2 \Psi (\sin^2\chi \, \sin^2\chi') + 6 \cos \Psi \cos\chi \, \sin\chi \, \cos\chi' \, \sin\chi'$$
$$+ 3 \cos^2 \chi \, \cos^2\chi' - 1] \qquad (18)$$

- a symmetry relation which indicates at low fields:

$$\Delta H_V - \Delta V_H = \Delta H_H + 2\Delta \, H_V = - \, \Delta H_H + 2\Delta V_H \quad \alpha (\cos^2\chi - \cos^2\chi') \qquad (19)$$

which is also useful to check the consistency of some experimental corrections.

An example of such calculations will be found in Ref. (4).

III. Instrumentation:

The principle of the instrumentation is very similar to an E jump or T jump instrument. We have modified such a T jump instrument for sensitive detection of the change in fluorescence, using a conventional stabilized xenon source, with a part of the beam used for the compensation of the signal in the absence of field, taking care of the residual fluctuation of the source[4]. Ridler and Jennings[6] have described an instrument with laser excitation. Their paper contains an analysis of the signal to noise ratio leading to the optimistic view that the electric fluorescence is as sensitive as electric birefringence. This origin of this puzzling result has been traced[7] and one indeed calculates a sensitivity $\sim 10^{-3}$ than electric birefringence. Care must be exercised in the design of the cell and of the electrodes to minimize a spurious effect due to pure electric dichroism in the layers of solution preceeding the small fluorescent volume. Displacing this volume in the cell may help to correct for that effect, with the equalities (19) as a check.

IV. Some Applications and Results

1) Work with small pieces of DNA
    Our own applications meet the motivations expressed in the introduction for the development of the technique, selecting problems where fluorescence should definitely add to the knowledge derived from simultaneously measured electric dichroism.

    a) Intercalating dyes:
    It is well known that intercalating dyes can have different quantum yields according to their binding site. We have shown, from

a comparison of life time and quantum yield changes upon binding
that this was the case for Proflavine (Pro) but not for Acridine
Orange (A.O.).   Therefore,

        i) A.O. has been chosen for a comparison of electric fluores-
cence and electric dichroism[4]. Having demonstrated with the use
of relation (18) that $\chi = \chi'$ we have made the assumption that $\chi =$
$\chi' = \frac{\pi}{2}$. Then Δ can be easily measured from a simplified expression
derived from Table I

$$\frac{\Delta V_V}{V_V} = \Delta$$

and compared to ϕ derived from electric dichroism under the similar
assumption

$$\frac{\Delta \varepsilon_{\parallel}}{\varepsilon} = -\phi$$

Low field results on a DNA segment $1500\overset{\circ}{A}$ in length give indeed a
value of ϕ/Δ close to 0.70 and a polarizability anisotropy $\alpha - \beta =$
$0.6 \ 10^{-15} \ cm^3$.

        ii) Repeating the experiment on Pro leads to a very similar
result, demonstrating that the geometry of the binding in the fluorescent and non-fluorescent sites is similar.

        iii) Going then to higher fields, using A.O. we have measured
the field dependence of ϕ and Δ[8]. They both show the same transition from a low field E^2 dependence to a linear E dependence at
relatively low values of ϕ and Δ. This could be interpreted using
the hypothesis of a saturated induced moment μ_∞. A first rough calculation (which does not take in account the fact that molecules
with an axis nearly perpendicular to the field do not see a high
enough field to be treated as having a μ_∞) predicts[8] in a range of
still low enough orientation $\Delta/\phi = \frac{3}{2}$. Our result is closer to 4/3
but should be repeated on shorter fragments of DNA as a way to
trace the origin of μ_∞ and its possible relation to DNA flexibility.

        b) Non intercalating dyes:

        We have been especially interested with the binding of 2 hy-
droxy 44' diamidinostilbene (OHSA) to DNA, and especially to the
interpretation of its "red" and "blue" fluorescence as a manifestation of binding heterogeneity[6]. In fact, both have been shown to
originate from dyes in a similar geometric arrangement, presumably
in the small groove, as a result of a fast rearrangement. The presence of non-fluorescent bound dyes, with a more perpendicular arrangement is inferred from the difference in angles χ obtained from
electric dichroism ($\cos^2 \chi \sim 0.35$) and electric fluorescence ($\cos^2$
$\chi = 0.45 - 0.50$).

2) Miscellaneous work on macromolecules and colloids.

*The Brunel group has been very active in applying the technique to a large number of systems[9]. One should, however, realize from the first part of this paper that only qualitative results, which cannot reasonably be fitted into the rodlike model, can be obtained with flexible macromolecules such as long DNA molecules[10]. Pigments[9] may present the drawback cf energy migration. Dye tagging of the clay mineral sepiolite[11], which is rodlike and where saturation of the orientation can be reached reveals details of the adsorption site.

3) Fluorescence detected linear dichroism.

Hogan, Dattagupta and Crothers[12] have proposed to use the unpolarized emission of fluorescence ($H_V + V_V$) as a way to record simply the increased absorption resulting from linear dichroism and eventually use the differences obtained in direct or fluorescence detected linear dichroism as a way to study binding heterogeneity. It is clear from inspection of Table I, in line with symmetry arguments that it is $V_V + 2H_V$ which should be used in this respect, due to the redistribution of light along the different directions and polarizations. Since $H_V + V_V$ contains terms in χ' and Ψ it is also dependent upon the initial value of the polarization of fluorescence. This casts a doubt on the conclusions drawn on the geometry of the drug intercalation based on these measurements. It would, however, be a simple matter to put the analyzer at the "magic angle", as currently done with the polarizer to eliminate the orientation effect in field jump experiments, and record then a quantity proportional to $V_V + 2H_V$.

References

1. G. Weill and C. Hornick, Biopol., 10,2029 (1971).
2. E. Frehland, Z. Naturforsch, 30a, 1243 (1975).
3. J. P. Jarry and L. Monnerie, J. Pol. Sc. Physics, 16,443 (1978).
4. G. Weill and J. Sturm, Biopolymers, 14,2537 (1975).
5. S. Sokerov and G. Weill, Biophys. Chem., 10,41 (1979).
6. P. J. Ridler and B. R. Jennings, J. Phys. E., 10,558 (1977).
7. S. Sokerov and G. Weill in "Electro-optics and Dielectrics of Macromolecules and Colloids" Brunel 1978, p. 109, B. R. Jennings editor, Plenum Press, 1979.
8. S. Sokerov and G. Weill, Biophys. Chem. 10,161 (1979).
9. P. J. Ridler and B. R. Jennings, SPIE vol. 164, Fourth European Electro-optics Conference - Utrecht, 1978.
10. B. R. Jennings and P. J. Ridler, Chem. Phys. Letters, 45,550 (1977).
11. P. J. Ridler and B. R. Jennings, Clay Minerals 15,xxx (1980).
12. M. Hogan, N. Dattagupta and D. M. Crothers, Biochemistry, 18,280 (1979).

Interaction of Electric Fields with Membrane-Bound Polyionic
Proteins

E. Neumann and K. Tsuji

Max-Planck-Institut für Biochemie

D-8033 Martinsried/München FRG

Abstract

Recent progress in electro-optic instrumentation has led to
experimental results which give new insight into the dynamic
behavior of membrane-bound polyionic macromolecules, such as
bacteriorhodopsin in purple membranes. Electric impulses of high
field intensity (2×10^5 to 3×10^6 Vm^{-1}, 1 to 20 µs duration) cause
transient changes in the optical absorbance of suspended purple
membranes of <u>Halobacterium</u> <u>halobium</u>. The electric dichroism at
1 mM NaCl pH $\simeq$ 6 and at 293 K is dependent on field strength, pulse
duration and wavelength of the monitoring, plane-polarized light
in the range 400 nm to 650 nm. The optically indicated processes
are, however, independent of bacteriorhodopsin concentration, of
ionic strength and of the intensity of the monitoring light. These
data and the analysis of time course and steady state of the reduced
dichroism suggest electric field sensitive, intramembraneous structural changes which involve restricted orientation changes of the
chromophore. A theoretical analysis of restricted orientation is
developed and applied to the electro-optic data. As a result it is
found that the electric dichroism of purple membranes is associated
with a large induced dipole moment up to 7×10^{-26} Cm (2.1×10^4 Debye)
which develops in a cooperative manner; the electric permanent di-
pole moment which is involved amounts to 4.7×10^{-28} Cm (140 Debye).

1. Introduction

Electro-optic and dielectric methods are traditionally applied
in order to study not only electrical but also structural properties
of dipolar and ionic molecules and molecular organizations.
Furthermore, electric field techniques are gaining increasing

importance for the investigation of chemically interacting, ionic
or dipolar systems; for review see ref. [1] and [2]. The measurement
of electric field effects in biological membranes or biomembrane
fragments may, in addition, reveal functionally relevant molecular
information, because almost all biomembranes are the locus of strong
cross-membrane electric fields and of transient changes in these
electric fields. In this context we note that bioelectric signals
like the nerve impulse involve transient changes in the electric
field of excitable membranes; these electric changes are controlled
by local electro-chemical gating processes of as yet largely
unknown nature; for review see refs. [3] and [4].

In the following account, a short digression on some fundamental principles of electro-chemical coupling and on the analysis
of electrically induced optical changes in chemically interacting
systems is given. Using bacteriorhodopsin of purple membranes as
an example, basic aspects of the analysis of electro-optic data
on membrane-bound proteins are discussed. The results of the
theoretical treatment of the experimental data suggest that the
orientation changes of the chromophore in bacteriorhodopsin are
sterically restricted and are accompanied by a change in the optical
extinction coefficient, resulting in a chemical contribution to the
absorbance change. For the data analysis an electro-optic theory
for restricted orientation is developed. The main result of this
study is that external electric fields induce intramolecular
structural changes in purple membranes which involve restricted
rotation of the chromophore coupled to a very large induced dipole
as well as to an electric permanent dipole moment. The order of
magnitude of the electric moments points to cooperatively stabilized
clusters of bacteriorhodopsin in purple membranes. An electric
impulse is able to cause a cyclic change through at least five
different conformations of the protein[5].

The data on bacteriorhodopsin in purple membranes are suggestive
of a possibly general mechanism for the interaction of electric
fields with membrane-bound proteins, for instance those involved in
the electrical-chemical control of ion flows across excitable biomembranes[3,4].

2.   Primary electric field effects.

The primary effects of electric fields on molecules are fairly
well understood: (1) orientation of dipolar species, deformation
of polarizable systems (and subsequent orientation of induced dipoles in electrically anisotropic particles); (2) movement of ionic
species in the direction of the field vector. Less well explored
is how these polarization and electrophoresis effects are specifically coupled to the various chemical transformations, such as
conformational transitions or dipolar and ionic association-

dissociation equilibria or steady states. Generally we can only
state that polar structures tend to orient in the field direction;
conformations or molecules with large dipole moments increase in
concentration at the expense of those configurations with smaller
electric moments; finally, electric fields increase the dissociation
of weak acidic and basic groups and promote the separation of ion
pairs into the respective dissociated ions or ionic groups (second
Wien effect). For an extended discussion see refs. [1] and [2].

3. Elemental chemical reactions

When molecules carrying permanent or induced dipoles or charged
groups interact with each other and undergo chemical transformations,
three types of electro-chemical coupling may be differentiated:
permanent dipole equilibria, induced dipole equilibria, and ionic
association-dissociation processes. Changes in the concentration
of the reaction partners arise from two types of elemental chemical
reactions: inter- and intra-molecular processes.

Intermolecular steps

If the dipole moments of a ligand L and a binding site B are
larger than that of the couplex LB, then an electric field shifts
the equilibrium

$$L + B \rightleftharpoons LB \tag{1}$$

to the side of the separated reactions partners. A dimerization
equilibrium such as

$$B(\uparrow) + B(\uparrow) \rightleftharpoons B(\uparrow) \cdot B(\downarrow) \tag{2}$$

where the dipole moments of B compensate each other upon complex
formation (as indicated by the arrows) is particularly sensitive
to an electric field because the electric reaction moment is of the
order of the dipole moment of the monomer.

An ionic equilibrium such as

$$LB \rightleftharpoons L^+ \cdot B^- \rightleftharpoons L^+ + B^- \tag{3}$$

where the association step involves ion pair formation, is usually
shifted to the rhs when an electric field is present (dissociation
field effect or second Wien effect).

Intramolecular steps

Suppose the molecules B are able to equilibrate between two
alternate conformations according to

$$B(\uparrow) \rightleftharpoons B'(\uparrow) \qquad\qquad (4)$$

where B' has a larger dipole moment than B, an electric field will shift the equilibrium toward B'. It is recalled that the equilibrium constant for reaction (4) is defined as $K = [B']/[B]$, where the brackets denote concentration.

4. Thermodynamics of chemical field effects

The equilibrium constant (or the steady-state distribution) constant, K, of a chemical equilibrium is dependent on intensive state variables such as temperature T, pressure P, or an external electric field E. For laboratory conditions the state of zero electric field is generally taken as the reference state; thus at $E = 0$, $K = K_o$. When $K(E)$ is the corresponding value in the presence of an electric field, the condition

$$\Delta K = K(E) - K_o \ll K_o \qquad\qquad (5)$$

specifies a linear approximation, generally valid for the small equilibrium shifts which electric fields can cause.

The isothermal-isobaric displacement of dipolar equilibria in electric fields can be described by a general van't Hoff relationship:

$$\left(\frac{\partial \ln K}{\partial E}\right)_{P,T} = \frac{\Delta M}{RT} \qquad\qquad (6)$$

where R is the gas constant and ΔM is the electric reaction moment defined by

$$\Delta M = N_A \left\{ \sum \nu_j \alpha_j (E_i)_j + \sum \nu_j p_j \cdot L(r_j) \right\} \qquad (7)$$

In Eq. (7), N_A is the Avogadro number; ν_i is the stoichiometric coefficient (counting positive for products and negative for reactants), α_j is the electric polarizability, p_i is the permanent dipole moment of species j; E_i is the internal field and $L(r_i)$ is the Langevin function of the ratio $r_j = p_j E_d/kT$, E_d being the directing field.

In the range where Eq. (5) is applicable, integration of Eq. (6) provides an expression for the field-induced shift in K:

$$\frac{\Delta K}{K_o} = \frac{1}{RT} \int \Delta M \, dE \qquad\qquad (8)$$

Ionic equilibria can be treated in an analogous way, see, e.g., refs [1] and [2].

The general relationship in Eq. (8) can be further specified. For instance, if pure induced dipoles interact, Eqs. (7) and (8) yield:

$$\frac{\Delta K}{K_o} = \frac{1}{kT} \int \sum v_j \alpha_j (E_i)_j dE \qquad (9)$$

At field strengths which saturate the induced dipole moments $m_i = \alpha_j (E_i)_j$ to $(m_j)_s$, the system behaves like a permanent dipole equilibrium and

$$\frac{\Delta K}{K_o} \simeq \frac{f(\varepsilon,n)}{6(kT)^2} \sum v_j (m_j)_s^2 \cdot E^2 \qquad (10)$$

where $f(\varepsilon,n)$ is a function of the dielectric constant and the refractive index of the solvent and where we assume that $(E_i)_j = E_i = f(\varepsilon,n) \cdot E$; see refs. [1] and [2].

For the intramolecular step of Eq. (4) we have $v_{B'} = 1$, $v_B = -1$; with the assumption that $(E_i)_{B'} = (E_i)_B = E_i = f(\varepsilon,n) \cdot E$ we obtain $M_{B'} = N_A \cdot \alpha_{B'} \cdot E_i$, $M_B = N_A \cdot \alpha_B \cdot E_i$. Then, $\Delta M = M_{B'} - M_B = N_A \cdot \Delta \alpha \cdot E_i$ where $\Delta \alpha = \alpha_{B'} - \alpha_B$, and

$$\frac{\Delta K}{K_o} = \frac{f(\varepsilon,n)}{kT} \int \Delta \alpha \cdot E \, dE \qquad (11)$$

Provided that $\Delta \alpha$ is constant, i.e. at small field strengths,

$$\frac{\Delta K}{K_o} \simeq \frac{f(\varepsilon,n)}{2kT} \Delta \alpha \cdot E^2 \qquad (12)$$

and at saturating field intensities,

$$\frac{\Delta K}{K_o} \simeq \frac{f(\varepsilon,n)}{6(kT)^2} (\Delta m)_s \cdot E^2 \qquad (13)$$

where $(\Delta m)_s = (m_{B'})_s - (m_B)_s$. As seen in Eqs. (12) and (13) an induced dipole mechanism is characterized by a temperature variation from T^{-2} to T^{-1}. It should be mentioned that there are several experimentally useful criteria to differentiate the various mechanisms and types of electric interactions [1,2].

The concentration changes described by ΔK may be indicated by optical and/or electrical methods; variations with T and E may then be used to identify type and mechanism of interaction.

Experimentally, recent progress in instrumentation has opened the way for measuring field-induced, rotational and chemical relaxations in parallel, both optically and electrically, even in the nanosecond time range[6]. It is obvious that the analysis of the kinetics and thermodynamics of these relaxations is particularly straightforward when rectangular pulses are applied.

5. Rotational and chemical contributions to absorbance changes

The absorbance per cm of a multi-component system is generally given by Lambert-Beer's law:

$$A = \sum_i \varepsilon_i c_i \tag{14}$$

where ε_i is the molar extinction coefficient and c_i is the molar concentration of species i. It is recalled that the extinction coefficients of molecules which are intrinsically optically aniso-tropic, reflect average values $\bar{\varepsilon}_i$ of all chromophore orientations in the system.

In principle both the ε_i and c_i-terms may be field dependent such that in general,

$$dA = \sum_i c_i \left(\frac{\partial \varepsilon_i}{\partial E}\right)_{c_i} dE + \sum_i \varepsilon_i \left(\frac{\partial c_i}{\partial E}\right)_{\varepsilon_i} dE \tag{15}$$

More specifically, when an electric field is applied to a chemical system which exibits both electrical and optical anisotropy, the field induced absorbance change ΔA_σ, measured with light polarized at angle σ relative to the electric vield vector, may not only involve orientational changes ($\Delta \varepsilon_{i,\sigma}$) but also concentration changes (Δc_i).

In the absence of an electric field the absorbance per cm of such a system is expressed by

$$A = \sum_i \bar{\varepsilon}_i c_i^{\,o} \tag{16}$$

Whereas A is independent of σ, the absorbance in the field is σ-dependent:

$$A_\sigma^E = \sum_i \varepsilon_{i,\sigma}^E c_i^E \tag{17}$$

Denoting the field induced changes in ε_i and c_i by

$$\Delta \varepsilon_{i,\sigma} = \varepsilon_{i,\sigma} - \bar{\varepsilon}_i$$

$$\Delta c_i = c_i^E - c_i^{\,o} \tag{18}$$

the field-induced absorbance change $\Delta A_\sigma = A_\sigma^E - A$ is rewritten in terms of Eqs. (16) and (17):

$$\Delta A_\sigma = \sum_i (\varepsilon_{i,\sigma} c_i^E - \bar\varepsilon_i c_i^0) \tag{19}$$

$$= \sum_i \Delta\varepsilon_{i,\sigma}(c_i^0 + \Delta c_i) + \sum_i \bar\varepsilon_i \Delta c_i ,$$

where the separation of terms depending on σ from those independent of σ is apparent. Introducing the definitions

$$\Delta A_\sigma^{(rot)} = \sum_i \Delta\varepsilon_{i,\sigma}(c_i^0 + \Delta c_i) \tag{20}$$

and

$$\Delta A^{(chem)} = \sum_i \bar\varepsilon_i \Delta c_i , \tag{21}$$

Eq. (19) may be generally written in terms of a rotational and a chemical contribution [7]:

$$\Delta A_\sigma = \Delta A_\sigma^{(rot)} + \Delta A^{(chem)} \tag{22}$$

For axially symmetric systems $\Delta A^{(chem)}$ can be experimentally obtained in two independent ways, using the measured absorbance changes ΔA_σ at the polarization modes $\sigma=0$ and $\sigma=\pi/2$, or $\sigma=0.955$ rad (54.7°). In this case the relationship

$$\Delta A_{||}^{(rot)} + 2\Delta A_\perp^{(rot)} = 0 \tag{23}$$

holds, and for $\sigma=0.955$ rad Eq. (22) reduces to

$$\Delta A_{0.955} = \Delta A^{(chem)} \tag{24}$$

On the other hand, Eqs. (22) and (23) can be combined to yield

$$\frac{1}{3}(\Delta A_{||} + 2\,\Delta A_\perp) = \Delta A^{(chem)} \tag{25}$$

Therefore, the equality

$$\Delta A_{0.955} = \frac{1}{3}(\Delta A_{||} + 2\Delta A_\perp) \tag{26}$$

can be used as a criterion for axial symmetry of the electric dipole moment of the orienting unit.

Note that $\Delta A_\sigma^{(rot)}$ specified in Eq. (20) may contain concentration changes Δc_i. These changes ($\Sigma\,\Delta\varepsilon_{i,\sigma} c_i$), are separable from the pure rotational term $\Sigma\,\Delta\varepsilon_{i,\sigma} c_i^0$, only when the kinetics of the rotational part is faster than the concentration changes [1, 2].

6. Conformational changes induced by electric fields in bacteriorhodopsin

Bacteriorhodopsin is the photoactive protein of the purple membrane which is a specialized part of the cell membrane of many halobacteria[8]. The protein functions as a light-driven proton pump producing an electrochemical proton gradient from which finally the free energy of the light induced ATP synthesis is derived[9]. As an integral membrane protein, bacteriorhodopsin is exposed to the internal electric field of the membrane. In order to study the structural and functional effects of the intrinsic electric field and of variations of the membrane field, isolated purple membrane fragments may be subjected to external electric impulses. Provided that electric-field induced structural changes are accompanied by variations in the chromophore and its protein environment, specified conformational transitions should be optically measurable. Light transmission changes have been indeed observed when field pulses were applied to purple membranes[10-13] as well as to apomembrane fragments[11]. However, type and mechanism of the optical signal changes have not been analyzed.

Orientation of whole membrane fragments will contribute to the optical signal if the applied electric fields last longer[10,12,14]. In rectangular electric pulses of short duration the contribution of fragment orientation is negligibly small and does not interfere with electro-optical signals arising from intramembraneous processes.

6.1 Theory of restricted orientation

· The experimental data indicate that bacteriorhodopsin in purple membranes are electro-optically anisotropic. The chromophore is a part of a rather rigid protein-lipid structure[15], hence orientational displacements of the chromophore within the protein are expected to be sterically restricted. Limited orientation is different from rotations of freely mobile particles and requires a different theoretical analysis[5].

It is known that the angle ψ between the chromophore transition moment μ of bacteriorhodopsin and the membrane normal is about 1.2 rad (70°)[16-18]. The electro-optic data of the present study suggest that an electric field E causes an angular displacement $\Delta\psi$ of μ from the position ψ^o in the absence of a field. As exemplified in Figure 1(a), steric restriction may cause, for instance, that $\Delta\psi$ is limited to a maximum value $\Delta\psi_m$. Generally two limiting cases of orientational restriction may be considered for the membrane bound proteins, when the membrane fragments themselves remain distributed randomly.

In Case I, μ may be displaceable in both directions relative

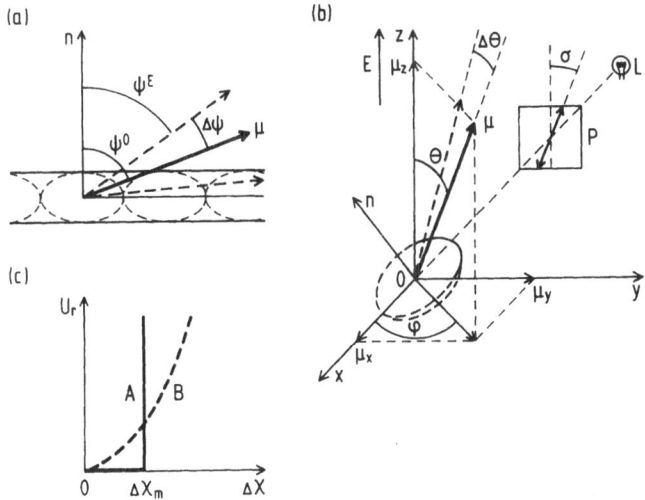

Fig. 1 (a) Cross section of purple membrane; parallel to the mem-
            brane normal n; μ, optical transition moment; Δψ is the
            angular displacement in an electric field E. Case I:
            μ can rotate symmetrically toward and away from n such
            that $\psi^E = \psi^0 \pm \Delta\psi$.   Case II: μ can rotate to only one
            side, e.g., $\psi^E = \psi^0 - \Delta\psi$ leading to positive dichroism.

     (b)   Geometrical relations of electric dichroism in Cartesian
           coordinates. E is in the z-direction, the light beam
           from a source L is in the x-axis and the polarizer, P,
           is in the y, z-plane providing plane-polarized light
           under the angle σ relative to E. The optical transition
           moment μ of a particular chromophore lies under the
           angle θ to E.

     (c)   Dependence of the restriction energy term U_r on the
           angular displacement Δχ for Model A (well-type potential
           energy profile) and Model B (potential energy according
           to Hooke's Law).

to the membrane normal, either increasing or decreasing ψ. Hence, this angular displacement is symmetric and, in the field, we obtain $\psi^E = \psi^O \pm \Delta\psi$.

In Case II, μ can rotate in only one direction. This asymmetric displacement leads to $\psi^E = \psi^O - \Delta\psi$ for positive dichroism and, alternatively, to $\psi^E = \psi^O + \Delta\psi$ for negative dichroism. If Case II applies, only one half of the membrane fragments are affected by the electric field.

Orientation factor

The formalism to calculate the orientation factor for restricted orientation is analogous to that for free, unrestricted orientation. Therefore, the key equations for free orientation[19] are given first.

The absorbance change, ΔA_σ, caused by the electric field, E, in an electro-optically anisotropic system, measured with light that is plane-polarized at the angle σ with respect to E (see Figure 1.(b)), is given by $\Delta A_\sigma = A_\sigma^E - A$ where A_σ^E is the absorbance in the presence of the field and A is the absorbance at E = 0.

The reduced absorbance change is generally expressed in terms of the orientation factor Φ according to

$$\frac{\Delta A_\sigma}{A} = \frac{1}{2}(3\cos^2\omega - 1)(3\cos^2\sigma - 1)\Phi \qquad (27)$$

where ω is the fixed angle between the optical transition moment μ and the electrical symmetry axis of the orienting unit (permanent and induced dipoles)[20].

The reduced dichroism is defined by

$$\frac{\Delta A}{A} = \frac{A_{||}^E - A_\perp^E}{A} = \frac{3}{2}(3\cos^2\omega - 1)\Phi \qquad (28)$$

where $A_{||}^E$ and $A_\perp^E$ are the absorbance values measured in the parallel ($\sigma=0$) and in the perpendicular ($\sigma=\pi/2$) modes of light polarization, respectively.

The orientation factor is given by

$$\Phi = \frac{1}{2}(3\langle\cos^2\chi\rangle - 1) \qquad (29)$$

where χ is the angle between the electric symmetry axis of the orienting unit and the direction of the external electric field.

The average value of $\cos^2\chi$ is defined by

$$\langle\cos^2\chi\rangle = \int_0^\pi \cos^2\chi \, f_\chi \, 2\pi \, \sin\chi \, d\chi \tag{30}$$

and the angular distribution function of the orienting unit for the steady state is given by

$$f_\chi = \frac{\exp(-U/kT)}{\int_0^\pi \exp(-U/kT)2\pi \sin\chi \, d\chi} \tag{31}$$

where k is the Boltzmann constant and T is the absolute temperature.

The potential energy, U, of a freely orienting unit characterized by the permanent dipole moment p and by the excess polarizabilities α_1 (parallel to the electric symmetry axis) and α_2 (perpendicular to it) is given by

$$U = -pE_d\cos\chi - \frac{1}{2}(\alpha_1 - \alpha_2)E_i^2\cos^2\chi \tag{32}$$

where, to a first approximation, we assume that the directing field E_d and the internal field E_i are equal to the external field E; see discussion. Introducing Eqs. (30) to (32) into Eq. (29) yields the familiar relationship

$$\Phi = \frac{3\int_{-1}^1 u^2\exp(\beta u + \gamma u^2) \, du}{2\int_{-1}^1 \exp(\beta u + \gamma u^2) \, du} - \frac{1}{2} \tag{33}$$

where

$$u = \cos\chi$$
$$\beta = pE_d/kT$$
$$\gamma = (\alpha_1 - \alpha_2)E_i^2/(2kT).$$

For restricted orientation the potential energy may be written in two terms:

$$U = U_o + U_r \tag{34}$$

where U_o has the form of Eq. (32) and the restriction term U_r specifies the type of steric restriction. Obviously, for unrestricted, free orientation, $U_r = 0$ and $U = U_o$. In general, with respect to χ, orientation is viewed as an angular displacement $\Delta\chi$ of the electric symmetry axis of the orienting unit from a position χ^o at E = 0 to a position χ^E in the presence of the field; hence

$\Delta\chi = \chi^o - \chi^E \geq 0$. Whereas in unrestricted, free orientation $\Delta\chi$ does not enter into the energy term, restricted orientation may be explicitly expressed as a function of $\Delta\chi$. The further analysis in this study will be confined to the theoretical treatment of two physically simple models.

In <u>Model A</u> the steric restriction is approximated by a well-type energy profile for U_r with infinitely high boundaries. The angular displacement is therefore subjected to the condition $0 \leq \Delta\chi \leq \Delta\chi_m$ where $\Delta\chi_m$ is the maximum displacement which corresponds to a maximum $\Delta\psi_m$ for the displacement of the chromophore μ relative to the membrane normal; see Figure 1(a). The restriction term U_r is then $U_r = 0$ for $0 \leq \Delta\chi \leq \Delta\chi_m$, and $U_r = \infty$ for $\Delta\chi > \Delta\chi_m$, and the total energy is given by:

$$U = - pE\cos\chi^E - \frac{1}{2}(\alpha_1 - \alpha_2)E^2\cos^2\chi^E, \text{ for } (\chi^o - \Delta\chi_m) \leq \chi^E$$

$$\leq (\chi^o + \Delta\chi_m) \qquad (35)$$

$$U = \infty, \quad \text{for } 0 \leq \chi^E < (\chi^o - \Delta\chi_m) \text{ or } \chi^E > (\chi^o + \Delta\chi_m)$$

The potential energy is a function of two variables $\chi^o (0 \leq \chi^o \leq \pi)$ and $\chi^E (0 \leq \chi^E \leq \chi^o)$; note that χ^o is independent of χ^E. Therefore we may introduce a partial orientation factor Φ_{χ^o} for the particular angular position χ^o. Then, the orientation factor Φ can be obtained as the average of Φ_{χ^o} for all angular directions χ^o:

$$\Phi = \frac{1}{2} \int_0^\pi \Phi_{\chi^o}\sin\chi^o \, d\chi^o \qquad (36)$$

Note that the integral is reduced by a factor 1/2 in Case II because for one half of the system there is no orientation.

The partial orientation factor is now obtained from Eqs. (29) to (33) with $\chi = \chi^E$. For the Model A Eq. (35) is introduced into Eq. (31), respectively. The respective partial orientation factor is then:

$$\Phi_{\chi^o} = \frac{3 \int_{u_1}^{u_2} u^2 \exp(\beta u + \gamma u^2)\,du}{2 \int_{u_1}^{u_2} \exp(\beta u + \gamma u^2)\,du} - \frac{1}{2} \qquad (37)$$

with $\mu_1 = \cos(\chi^o + \Delta\chi_m)$ and $\mu_2 = \cos(\chi^o - \Delta\chi_m)$.

For the analytical treatment of model B see ref. [5].

The results of numerical calculations of Eq. (36) for the symmetric
Case I and the asymetric Case II are shown in Figure 2, using Model
A (well-type energy profile) with $\Delta\chi_m$ as a parameter and Model B [5]
(Hooke-type energy profile) with a/kT as a parameter, respectively.
Note that the variable $(\beta^2 + 2\gamma)$ with the permanent (β) and the
induced (γ) dipole terms defined by Eq. (33) is proportional
to E^2. Therefore the field strength dependence of the reduced
dichroism can be curvefitted to the function $\Phi(\beta^2 + 2\gamma)$. Hence,
the mechanism as well as the type of orientation (free or restricted)
may be determined.

   Electric dichroism of restricted orientation. The primary
data of electric dichroism of systems like purple membranes are
related to the angle θ between the optical transition moment μ of
a particular chromophore (or clusters of chromophores) and the
direction of the external electric field; see Figure 1(b). For the
sake of simplicity we shall at first assume that the electric
symmetry axis is parallel to the optical transition moment $(\omega=0)$.
This assumption is in line with the observation of positive dichroism; see Eq. (28). Furthermore, as demonstrated below the
experimental data on membrane-bound bacteriorhodopsin can be con-
sistently analyzed in terms of an axially symmetric overall electric
moment. For $\omega=0$, the boundary conditions for θ are the same as
those for χ discussed in the previous section.

   For Model A the further treatment is particularly simple if
the saturation value Φ_s of the orientation factor is considered.
At sufficiently high external electric fields all chromophore
transition moments which have orientations between $\theta^o = 0$ and $\theta^o =$
$\Delta\theta_m$ are parallel to E, i.e., $\theta^E = 0$. The remaining molecules whose
transition moments are positioned between $\theta^o = \Delta\theta$ and $\theta^o = \pi$
assume the positions $\theta E = \theta - \Delta\theta_m$ for the permanent dipole orien-
tation; while $\theta^E = \theta^o - \Delta\theta_m$ when $\Delta\theta_m \leq \theta^o \leq \pi/2$ and $\theta^E = \theta^o + \Delta\theta_m$
when $\pi/2 \leq \theta^o \leq \pi-\Delta\theta_m$ for the induced dipole orientation.

   The reduced absorbance changes, $\Delta A_\sigma/A$, for maximum angular
displacement of the chromophore are readily obtained. For Case I
the saturation values are:

(i) permanent-dipole orientation

$$\left(\frac{\Delta A_\sigma}{A}\right)_s = \frac{1}{4}(3 - 2\cos\Delta\theta_m - \cos^2\Delta\theta_m)(3\cos^2\sigma - 1) \qquad (38)$$

(ii) induced-dipole orientation

$$\left(\frac{\Delta A_\sigma}{A}\right)_s = (1 - \cos\Delta\theta_m + \sin\Delta\theta_m \cos\Delta\theta_m)(3\cos^2\sigma - 1). \qquad (39)$$

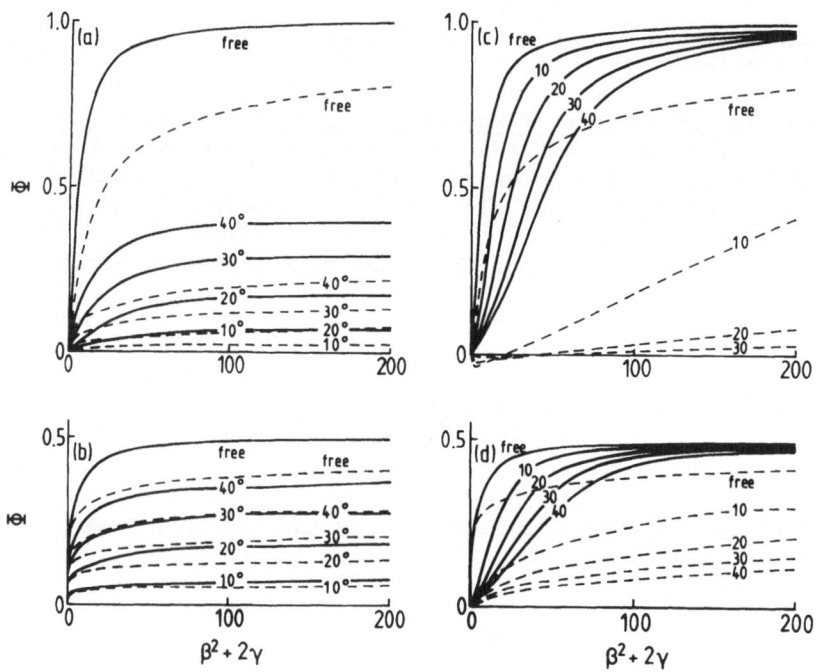

Fig. 2: Orientation factor Φ for restricted angular displacements
$\Delta\chi_m$ of the electric axis as a function of $\beta^2+2\gamma$; for
comparison, unrestricted free orientation is included.
--- refers to Φ due to pure permanent dipoles, i.e.,
$\gamma = 0$; ———— refers to Φ due to induced dipoles, i.e.
$\beta = 0$, $\gamma > 0$.
The parts (a) and (b) refer to Cases I and II, respectively,
assuming the restriction energy Model A for various maximum
angular displacements $\Delta\chi_m$ as a parameter. Parts (c) and
(d) refer to Cases I and II, respectively, assuming the
restriction energy Model B for various values of $\underline{a}/kT$.

These relationships permit an experimental determination of $\Delta\theta_m$. It should be mentioned that the assumption of $\omega=0$, involved in the derivation of Eqs. (38) and (39), provides a minimum estimate for $\Delta\theta_m$. For this case, $\Delta\theta_m = \Delta\chi_m = \Delta\psi_m$, and the maximum angular displacement relative to the membrane normal is obtained.

6.2 Experimental

The concentration of bacteriorhodopsin in purple membranes isolated from the Ml strain of Halobacterium halobium was determined on the basis of the extinction coefficient $\varepsilon = 63000$ $M^{-1}cm^{-1}$ at 570 nm[14].

The optical density of purple membrane suspension for the electric pulse measurements was adjusted to about 0.3 in 1 mM NaCl solution at 293 K, unbuffered (pH $\sim$ 6), unless otherwise stated. The optical density value of 0.3 corresponds to a concentration which yields sufficiently large signals without serious light scattering contributions by the membrane fragments. The electro-optic measurements were performed with an electric-field-jump apparatus recently developed by Schallreuter, Rohner and Neumann[6]. Rectangular electric pulses up to 35 kV and of various durations between 1 μs and 100 μs were used. All measurements were carried out at 293 K. The temperature increase due to Joule heating can be calculated. For the highest voltage V = 35kV and the longest pulse duration of $\Delta t = 100$ μs, the temperature increase $\Delta T = 3.5$ K is negligibly small, because the absorbance change from 293 K to 296.5 K is less than 3%.[5]

The change in the absorbance ΔA_σ, defined above is calculated from the light transmittance at a given wavelength and a fixed light polarization mode σ:

$$\Delta A_\sigma = - \log(1 + \Delta I_\sigma/I)$$

where I is the transmitted light in the absence of the electric field (independent of σ). The field induced transmittance change is given by $\Delta I_\sigma = I_\sigma^E - I$ where I_σ is the light transmitted in the presence of the electric field. It is noted that in this study ΔI_σ is generally not small compared to I. Therefore, the approximation $\Delta A_\sigma = -0.4343\Delta I_\sigma/I$ cannot be applied.

6.3 Results

In Figure 3 typical signal changes induced by an electric impulse in purple membrane suspensions at two light polarization modes are shown. At an electric field strength of $E = 15.4 \times 10^5$ Vm^{-1} a time independent, steady-state level of the light transmittance in the field (subscript ss) is reached only if the pulse duration is equal to or larger than 6 μs. For the parallel ($\sigma = 0$)

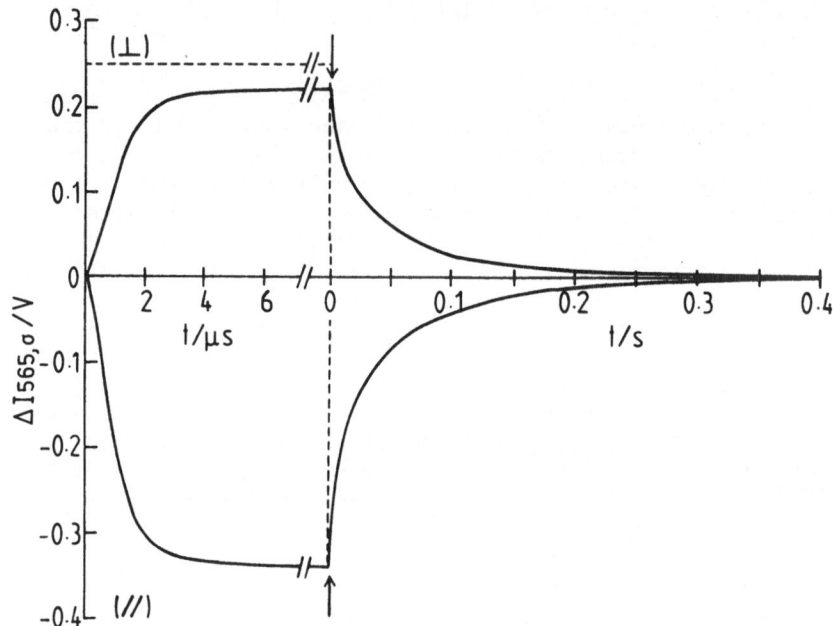

Figure 3:   Electric dichroism of bacteriorhodopsin in purple
            membrane; concentration of bacteriorhodopsin c =
            4.8×10^{-6} M in 1 mM NaCl, unbuffered, pH $\cong$ 6, T = 293 K.
            The light transmittance change ΔI_{σ} at λ = 565 nm is
            caused by a rectangular electric pulse of E = 15.4×10^{5}
            Vm^{-1} applied at t=0 for 20 μs (see arrows). (∥) and
            (⊥) are the signal changes for light plane-polarized
            parallel (σ=0) and perpendicular ($\sigma=\pi/2$) to the electric
            field; the light intensity I at t=0 is 3.0 V.

and the perpendicular ($\sigma=\pi/2$) polarization modes of the monitoring light the reduced absorbance changes calculated according to Eq. (27) are $(\Delta A_{\parallel}/A)_{ss} = 0.32 \pm 0.02$ and $(\Delta A_{\perp}/A)_{ss} = -0.27 \pm 0.02$.

As seen in Figure 3, the signal change, after the electric field is switched off, reflects a continuous relaxation spectrum. At the present accuracy of data recording, the corresponding absorbance curves may be formally described in terms of at least four exponential components according to

$$\frac{\Delta A(t)}{\Delta A_{ss}} = \sum_{i=1}^{n} \delta_i \exp(-t/\tau_i) \quad (n \geqslant 4)$$

where ΔA_{ss} is the steady-state absorbance change (when the pulse is terminated) and δ_i and τ_i are the relative amplitude and relaxation time of component i, respectively. For the curve in Figure 3 it is found that $\tau_1 = 0.3 \pm 0.03$ ms, $\tau_2 = 3.0 \pm 0.3$ ms, $\tau_3 = 26 \pm 4$ ms, $\tau_4 = 100 \pm 10$ ms; $\delta_1 \simeq 0.2$, $\delta_2 \simeq 0.2$, $\delta_3 \simeq 0.3$, $\delta_4 \simeq 0.3$.

Compared to the time range of the rise curve in the presence of the electric field, the field-off response is slower by orders of magnitude. This experimental result already indicates that the optical data cannot reflect simple, free orientation of the chromophore.

Figure 4 shows the field strength dependence of the reduced absorbance change (steady-state) at $\lambda = 565$ nm for three light polarization modes; $\sigma=0$, parallel; $\sigma=\pi/2$, perpendicular; $\sigma=0.955$ rad (54.7°). Note that the absorbance change $\Delta A_{\parallel}$ apparently saturates at $E > 4 \times 10^5$ Vm^{-1} while $-\Delta A_{\perp}$ still increases with increasing field strength.

It is found that the reduced absorbance change (steady-state) depends on the wavelength of the monitoring light, in a different manner for the various polarization modes. In particular, the parallel mode has a pronounced maximum at about 565 nm. The ratio of the absorbance changes, $(\Delta A_{\parallel}/\Delta A_{\perp})$, equals -1.2 ± 0.2 at 565 nm; at 400 nm this ratio is -0.5 ± 0.05. The value of $\Delta A_{0.955}/A$ is larger at 400 nm than at 565 nm. The reduced absorbance changes (steady state) at $\sigma=0$ and $\sigma=\pi/2$ are found to be independent of purple membrane concentration between 1.4×10^{-6} M and 2.0×10^{-5} M and of ionic strength between 1 mM and 3 mM NaCl. The intensity of the monitoring light affects only very slightly the magnitude of the reduced absorbance changes. Thus light induced contributions to ΔA_{σ} are negligibly small.

The relaxation times of the field-off responses are, within

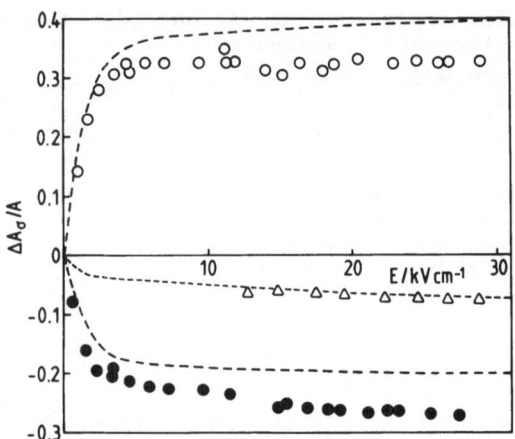

Fig. 4: The electric field strength dependence of the steady-state
value of the reduced absorbance change $\Delta A_\sigma/A$ for the purple
membrane suspension at λ = 565 nm: 0, $\sigma=0$; •, $\sigma =\pi/2$, and
Δ,σ =0.955. The dashed curves were calculated according
to equation (40): — — — refers to $\Delta A_\sigma{}^{(rot)}/A$ at $\sigma=0$ and
$\sigma=\pi/2$; --- refers to $\Delta A^{(chem)}/A$. The other experimental
conditions are the same as in Figure 3.

experimental error, independent of the field strength, pulse
duration, wavelength, light polarization direction, concentration
of purple membrane, ionic strength and the intensity of the monitoring light.

6.4 Discussion

The primary experimental data, ΔA_σ, on purple membranes do not
follow Eq. (23), i.e., $\Delta A_\parallel$ + $2\Delta A_\perp \neq 0$. Therefore, chemical changes
may be involved. If, however, $\Delta A_{0.955}$ is compared with ($\Delta A_\parallel$ +
$2\Delta A_\perp$)/3, it is found that Eq. (26) holds. Hence, the system
exhibits axial symmetry and Eq. (22) in the form

$$\Delta A_\sigma{}^{(rot)} = \Delta A_\sigma - \Delta A_{0.955} \tag{40}$$

can be used to calculate the rotational contributions of ΔA_σ. The
experimental data represented in Figure 4 are evaluated according
to this subtraction procedure and the results are included as
dashed lines.

The wavelength dependence of $\Delta A_\sigma{}^{(rot)}/A$ suggests that the
absorption band between 400 nm and 650 nm arises from at least two
different optical transitions. It is noted, too, that the chemical
contribution ($\Delta A_{0.955}/A$) is larger at 400 nm (−0.14 ± 0.02) than
at 565 nm (−0.055 ± 0.005). It is seen in Figure 4 that the
reduced changes in $\Delta A_\parallel{}^{(rot)}/A$ and $-\Delta A_\perp{}^{(rot)}/A$ increase with

increasing field strength in qualitatively the same way as
$\Delta A^{(chem)}/A$.

The practical coincidence of $\Delta A_\sigma^{(rot)}$ with $\Delta A^{(chem)}$ is not only
reflected in the steady-state but also in the kinetics; the entire
time courses of the relative signal changes in the various polarization modes, i.e., the rotational and the chemical contributions,
are coincident within the margin of experimental error. These
results suggest that $\Delta A^{(chem)}$ and $\Delta A_\sigma^{(rot)}$ reflect two aspects of
one and the same process which is induced by the electric field.

Orienting unit. As is described above, the reduced absorbance
changes are independent of bacteriorhodopsin concentration and of
the ionic strength of the suspension. Therefore, particle-particle
interactions may be considered as negligibly small.

Electric dichroism may be caused by orientation of entire
membrane fragments. Purple membrane fragments are oriented by
long lasting, exponentially decaying electric fields[10,12], or in
relatively low electric field strengths (1×10^4 Vm^{-1}) applied for a
sufficiently long time (1-2 min)[14]. The time constant for fragment
orientation can be estimated using Perrin's equation[21]. Purple
membranes may be considered as oblate discs with diameters between
0.5 μm and 1 μm[22]; the rotational time constants for these dimensions are in the order of 100 ms.

It was recently shown that electric fields of low intensity
and a duration of several seconds orient purple membrane fragments.
The dichroism of fragment orientation is, however, negative and
suggests orientation of the membrane normal parallel to E. [23]

The positive dichroism observed in our study with short-lasting
high field intensity pulses can therefore not come from fragment
orientation. In addition, pure fragment orientation is not expected
to result in a chemical contribution to the absorbance changes.

A further possibility for orientational changes is the bending
of membrane fragments. Because purple membranes are known to be
rather rigid protein lipid lattices[15], it is improbable that the
optical signals are caused by fragment bending.

We are thus driven to conclude that the bacteriorhodopsin
molecules themselves are the origin of the observed electric
dichroism. The rigidity of the hexagonal lattice would, however,
prevent the protein molecule from changing position as a whole
within the membrane[24]. It is more likely that only local parts of
the protein involving the chromophore can move to a limited extent
in the electric field; restricted chromophore motion is also
apparent in the light-induced photocycle[11,25-30].

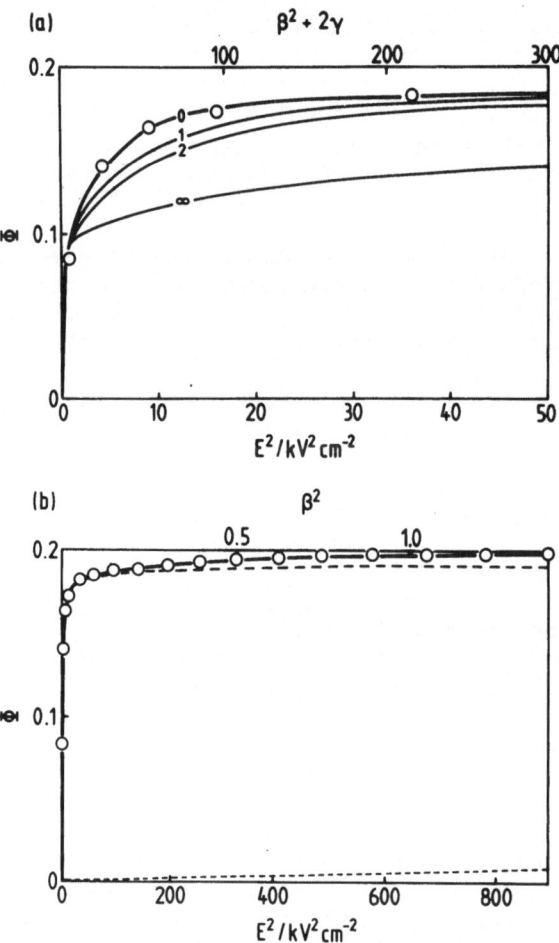

Figure 5: Electric field strength dependence of the orientation
factor Φ in terms of E^2, $\beta^2+2\gamma$ and β^2, for restricted
angular displacement of the electric dipole axis from
χ^0 to $\chi^0-\Delta\chi_m$; $\Delta\chi_m = 0.35$ rad (20°). (a) Low field
strength range for various ratios of $\beta^2/2\gamma$ (————).
(b) Total experimental range of E: — — — refers to the
induced dipole contribution (2γ); ———— refers to the
permanent dipole contributions. The comparison with the
experimental data (0) involves the assumption that $\omega=0$.

In summary, the electro-optic signals described in this study
most probably result from electric-field-induced structural changes
which can be identified as limited orientational changes involving
the chromophore.

Orientation mechanism. Applying the theory for restricted orientation to the field strength dependence of the reduced dichroism
$\Delta A^{(rot)}/A$, we obtain the mechanism for the angular displacement of
the chromophore in bacteriorhodopsin. According to Eqs. (28) and
(33), the orientation factor Φ is proportional to $\Delta A^{(rot)}/A$ and
$|\beta^2 + 2\gamma|$ is proportional to E^2, respectively. Therefore, the
experimental data set $\Delta A^{(rot)}/A$ is fitted to the theoretical curves
of Φ versus $|\beta^2 + 2\gamma|$ (see Figure 2), using proper scale factors[31].
It is found that among the models considered, only Case II of Model
A gives satisfactory results; Case I(A) and Model B as well as free
orientation can be excluded.

According to the observation of positive dichroism we assume
for simplicity that $(\omega=0)$. In detail, Figure 5(a) shows the comparison of Φ from $\Delta A^{(rot)}/A$ (steady-state) with Φ calculated for
Case II(A) with $\Delta\chi_m = 0.35$ rad as a function of $\beta^2 + 2\gamma$ for various
ratios of $\beta^2/2\gamma$; $(\gamma>0)$. The experimental data fit the curve for
$\beta^2/2\gamma = 0$ up to $E^2 = 5\times10^{11}$ V^2m^{-2}, if at $E^2 = 1\times10^{11}$ V^2m^{-2}, $\beta^2 +
2\gamma = 60$ and if $\Phi = 0.19$ for $\Delta A^{(rot)}/A = 0.57$. The low field
strength range is thus dominantly determined by an induced dipole
moment.

An additional slight increase in Φ becomes apparent at higher
field strengths $(E^2 > 5\times10^{11}$ $V^2m^{-2})$ and can be attributed to a
permanent dipole moment.

Figure 5(b) summarizes the results: the entire experimental
field strength range is quantitatively described by the maximum
angular displacement $\Delta\chi_m = 0.35$ rad, the permanent dipole moment
$p = 4.7\times10^{-28}$ Cm (140 Debye) and the excess polarizability difference
$\alpha_1 - \alpha_2 = 2.4\times10^{-30}$ Fm^2 $(2.2\times10^{-14}$ $cm^3)$ corresponding to an induced
dipole moment of 7×10^{-26} Cm $(2.1\times10^4$ Debye). Note that these
numbers represent only estimates for the apparent values of the
electric moments, because no reliable correction factors for the
internal and the directing fields[32] can be given at present. In
any case, the bacteriorhodopsin molecules in purple membranes
represent a highly polarizable system. In conclusion, the field
strength dependence of the electro-optical data suggest that the
orientational change of the chromophore is restricted both in extent
and in direction. The unidirectional angular displacement of the
chromophore is limited to $\Delta\theta_m = 0.35$ rad (20°) for $\omega=0$; the
orientation direction is toward the membrane normal.

Cooperativity. One may note that the electric dipole moments of the
orientation process, in particular the induced dipole moment, are

comparatively large. The permanent dipole moment may originate in the α-helices and in the paired polar groups (salt bridges) of bacteriorhodopsin. Since the seven α-helical segments are almost parallel to the membrane normal[15], the dipole moment of one α-helix is not compensated by the oppositely directed moment of another α-helix. Furthermore, a segment contains on the average 30 ± 3 amino acid residues[33] and about 70% is α-helical[34]. Because the dipole moment of one hydrogen bonded CO···HN group in an α-helix is 1.1×10^{-29} Cm (3.4 Debye)[35], the permanent dipole moment of these groups in one segment is about 2.3×10^{-28} Cm (69 Debye). Since the estimate of 4.7×10^{-28} from the experimental data is larger, either ionic side groups of the residues contribute appreciably to the electric moment, the orienting unit comprises more than one bacteriorhodopsin molecule, or the directing field is larger than the external field.

The primary sequence data[33] and x-ray crystallographic results[15] suggest that the positively charged side groups of lysine and arginine residues and the negatively charged carboxylate side chains of aspartic and glutamic acid residues could form ion pairs of the type $COO^- \cdots HN^+$. Per bacteriorhodopsin, a maximum of 14 such pairs could be present. An applied field will shift the equilibrium $COOH \cdots N \quad COO^- + HN^+$ to the rhs.

An induced dipole moment could now appear when the applied electric field increases the distance, ℓ, between the charged groups of an ion pair. In such a case the dipole moment changes from the zero-field value p_O to a value $p_E = p_O + \Sigma e_i \Delta \ell_i$ where e_i is the charge and $\Delta \ell_i$ is the field-dependent distance change in an ion pair of type i. The increase in ℓ_i then appears as an electric polarizability leading to an induced dipole moment.

If such an ion pair is formed at contact sites of different bacteriorhodopsin molecules in the membrane lattice, electrostatic coupling can be the origin of cooperative behavior. The magnitude of the measured induced dipole moment suggests that the orienting unit is not a single bacteriorhodopsin macromolecule, but rather represents a larger cluster of cooperatively coupled proteins.

Evidence for cooperativity in purple membranes is frequently reported. For instance, cooperativity is indicated in the binding of retinal to apomembranes[36], in the bleaching process[37], as well as in the proton-pump cycle[11]. It has been suggested that cooperative units larger than the trimer are involved[36]. If indeed larger clusters of bacteriorhodopsins orient the chromophores in a cooperative manner, then the electric axis of the orientation process is the vectorial sum of the various local contributions within the macromolecules. The local changes including the orientational displacement of the chromophore toward the membrane normal are the conformational transitions induced by the electric field, leading to a finite value of $A^{(chem)} = \sum_i \varepsilon_i \Delta c_i$. It is evident that the conformational changes must involve at least one structure with a different extinction coefficient.

Reaction cycle. The enormous difference between the time ranges of the rise curve in the presence of the electric field and of the field-off response (see Figure 3) strongly suggests that the sequence of state changes caused by the electric field is different from the structural transitions occurring in the multi-state field off relaxation. The experimental data, therefore, suggest a cyclic reaction scheme which comprises at least five conformations in cooperative clusters of membrane-bound bacteriorhodopsin. The electric field causes a transition from state bR_1 to state bR_2, whereas the field-off relaxation reflects the passage through at least three, probably more, intermediate conformations.

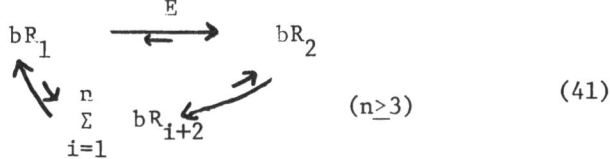

$$\begin{array}{ccc} bR_1 & \underset{E}{\overset{E}{\rightleftarrows}} & bR_2 \\ & & \end{array} \qquad (n \geq 3) \qquad (41)$$

$$\sum_{i=1}^{n} bR_{1+2}$$

In the light of the orientation mechanism discussed above, the step $bR_1 \rightleftarrows bR_2$ may represent a 'synchronized', rapid alignment of ionic groups. These conformational changes appear to occur in the electric field in a concerted way and include a change in the chromophore transition moment from 1.2 rad (state bR_1) to the angle ≤ 0.87 rad (state bR_2).

We may now specify Eq. (19) in terms of these states by

$$\Delta A_\sigma = \{\Delta \varepsilon_{1,\sigma}(1-x) + \Delta \varepsilon_{2,\sigma} x\} c_T + (\bar{\varepsilon}_2 - \bar{\varepsilon}_1) x c_T \qquad (42)$$

where $c_T = [bR_1] + [bR_2]$ is the total concentration of bacteriorhodopsin and $x = [bR_2]/c_T$ is the fraction in state bR_2.

Since the experimental data indicate restricted angular displacement of the chromophore to only one direction, $x \leq 0.5$, at sufficiently high electric field where orientation is saturated, we have $x = 0.5$. For this case the extinction coefficient $\bar{\varepsilon}_2$ is readily calculated from $\Delta A^{(chem)} = (\bar{\varepsilon}_2 - \bar{\varepsilon}_1) x c_T$. For instance, at 565 nm $\bar{\varepsilon}_2 = 54000$ $M^{-1}cm^{-1}$ and at 400 nm $\bar{\varepsilon}_2 \cong 21000$ $M^{-1}cm^{-1}$.

Although at the present accuracy of optical data recording the field-off relaxations appear to reflect at least four processes, the response curve is more probably a continuous spectrum of many local conformational transitions. Whereas the electric field acts simultaneously on all charged groups and dipoles of bacteriorhodopsin to cause a structural transition ($bR_1 \rightleftarrows bR_2$), the rearrangements in the absence of the field are more complex and face higher activation barriers, as is suggested by the larger relaxation times.

At present it is not possible to connect the 'electro-optic cycle' represented in scheme (41) with details of the photocycle. However, some aspects of field and light effects are worth mentioning. It has been recently observed that in dry films of purple membranes there are spectral similarities between the field-induced intermediates and the first intermediate of the photocycle [13].

The conformational changes evident by our electro-optic data clearly involve an orientational displacement of the chromophore toward the membrane normal. Interestingly, on the basis of Raman spectroscopic data it has been suggested that during the photocycle the chromophore rotates out of the membrane plane[38]. A further correlation between the photocycle and the electro-optic data is the appearance of cooperativity.

Although the cyclic nature and the kinetics of the electric field induced structural changes suggest common features with the photocycle, further studies are necessary in order to identify the effect of electric field on the dynamic properties of bacteriorhodopsin.

The electric field effects observed in membrane-bound bacteriorhodopsin are, however, suggestive of a possibly general mechanism for a very effective interaction of electric fields with the ion flow gating proteins in excitable biomembranes [1-4]. The field dependence of the ion permeability changes underlying the nerve impulse and other membrane potential changes would then be molecularly based on a polarization mechanism where the distances between charged groups of the gating proteins depend on the intensity of the membrane electric field.

References

1. E. Neumann, in: <u>Electro-optics and dielectrics of macromole-cules and colloids</u>,(Ed. B. R. Jennings) Plenum Press, New York, p. 233-245 (1979).

2. E. Neumann, in: <u>Topics in Bioelectrochemistry and Bioenergetics</u>, (Ed. G. Milazzo), John Wiley & Sons, London, Vol. 4, p. 113 - 161, (1980).

3. E. Neumann and J. Bernhardt, <u>Ann. Rev. Biochem.</u> <u>46</u>, 117 (1977).

4. E. Neumann, Neurochemistry Intern. <u>1</u>, in press (1980)

5. K. Tsuji and E. Neumann, Intern. J. Biol. Macromol. <u>2</u>, in press, (1980).

6. D. Schallreuter and E. Neumann, in prep. (1980).

7. A. Revzin and E. Neumann, Biophys. Chem. <u>2</u>, 144 (1974).

8. D. Oesterhelt and W. Stoeckenius, <u>Nature New Biol</u>. 233, 149 (1971).

9. D. Oesterhelt and W. Stoeckenius, <u>Proc</u>. <u>Nat</u>. <u>Acad</u>. <u>Sci</u>. <u>USA</u>, 70, 2853 (1973).

10. R. Shinar, S. Druckmann, M. Ottolenghi, and R. Korenstein, <u>Biophys</u>. <u>J</u>. <u>19</u>, 1 (1977).

11. B. Hess, R. Korenstein, and D. Kuschmitz in 'Energetics and Structure of Halophilic Micro-organisms' (Eds. S. R. Caplan and M. Ginzburg), Elsevier, Amsterdam, p. 89 (1978).

12. K. Tsuji and K. Rosenheck in '<u>Electro-Optics and Dielectrics of Macromolecules and Colloids</u>' (Ed. B. R. Jennings), Plenum Press, New York, p. 77, (1979).

13. G. P. Borisevitch, E. P. Lukashev, A. A. Kononenko and A. B. Rubin, <u>Biochim. Biophys. Acta,</u> 546, 171 (1979).

14. M. Eisenbach, C. Weissmann, G. Tanny, and S. R. Caplan, <u>FEBS Lett</u>. <u>81</u>, 77 (1977).

15. R. Henderson, <u>J</u>. <u>Mol</u>. <u>Biol</u>. <u>93</u>, 123 (1975).

16. A. N. Kriebel and A. C. Albrecht, <u>J</u>. <u>Chem</u>. <u>Phys</u>. <u>65</u>, 4575, (1976).

17. T. G. Ebrey, B. Becher, B. Mao, P. Kilbride and B. Honig, J. Mol. Biol. 112, 377 (1977).

18. M. P. Heyn, R. J. Cherry and U. Müller, J. Mol. Biol. 117, 607, (1977).

19. C. T. O'Konski, K. Yoshioka and W. H. Orttung, J. Phys. Chem. 63, 1558 (1959).

20. E. Fredericq and C. Houssier, 'Electric Dichroism and Electric Birefringence', Clarendon Press, Oxford (1973).

21. F. J. Perrin, Phys. Radium, 7, 390 (1926).

22. A. E. Blaurock and W. Stoeckenius, Nature New Biol., 233, 152 (1971).

23. L. Keszthelyi, Biochim. Biophys. Acta, 598, 429 (1980).

24. K. Razi Naqvi, J. Gonzalez-Rodriguez, R. J. Cherry, and D. Chapman, Nature New Biol. 245, 249 (1973).

25. T. Konishi and L. Packer, FEBS Lett. 92, 1 (1978).

26. A. Lewis, M. A. Marcus, B. Ehrenberg, and H. Crespi, Proc. Nat. Acad. Sci. USA 75, 4642 (1978).

27. T. Gillbro, Biochim. Biophys. Acta 504, 175 (1978).

28. R. H. Lozier and W. Niederberger, Fed. Proc. 36, 1805 (1977).

29. K. Schulten and P. Tavan, Nature 272, 85 (1978).

30. J. B. Hurley, B. Becher, and T. G. Ebrey, Nature 272, 87 (1978).

31. K. Yoshioka and H. Watanabe in: 'Physical Principles and Techniques of Protein Chemistry, Part A' (Ed. S. J. Leach) Academic Press Inc., New York, p. 339 (1969).

32. C. J. F. Böttcher, 'Theory of Electric Polarization', Elsevier Sci. Pub. Co., Amsterdam (1973).

33. Yu. A. Ovchinnikov, N. G. Abdulaev, M. Yu Feigina, A. V. Kiselev and N. A. Lobanov, FEBS Lett. 100, 219 (1979)

34. B. Becher and J. Y. Cassim, Biophys. J. 16, 1183 (1976).

35. A. Wada in: 'Poly-α-Amino Acids' (Ed. G. Fasman), Marcel Dekker, New York, p. 369 (1967).

36. M. Rehorek and M. P. Heyn, <u>Biochemistry</u>, <u>18</u>, 4977 (1979).

37. B. Becher and J. Y. Cassim, <u>Biophys</u>. <u>J</u>., <u>19</u>, 285 (1977).

38. A. Lewis, <u>Phil</u>. <u>Trans</u>. <u>R</u>. <u>Soc</u>. <u>Lond</u>., <u>A293</u>, 315 (1979).

THRESHOLD EFFECTS IN FIELD-INDUCED CONFORMATION CHANGES AND

BINDING OF IONS - INCLUDING PEPTIDES - TO POLYNUCLEOTIDES

Dietmar Pörschke
Max-Planck-Institut für
biophysikalische Chemie
Am Faßberg
D-3400 Göttingen/W. Germany

INTRODUCTION

In the past electric fields have been used for the characterisation of macromolecular systems mainly by the analysis of orientation effects. In recent years it has been demonstrated that electric fields can also be very useful for the characterisation of various types of binding reactions and conformation changes in macromolecules. The most important progress for this extended application of electric fields is the development of a reliable technique for the selective indication of reactions in systems with optical anisotropy due to orientation[1-5]. This technique is based upon the so-called "magic angle": absorbance changes detected by plane polarised light oriented at an angle of 54.74° with respect to the electric field vector are not due to physical orientation but only due to "chemical" changes like binding reactions or conformation transitions. By this technique it is possible to characterise the binding reactions and conformation changes of polymers with a large optical anisotropy. Most of the systems of biological interest have such polymers as components. All the systems discussed in the present contribution are characterised by using the magic angle technique.

Two different types of problems will be discussed in the present contribution. In the first part the mechanism of field induced conformation changes will be analysed in a simple model system. The second part is a compilation of different ion binding reactions analysed by the field jump technique. These two subjects appear to be completely unrelated. However, there is a close relation between them, since the mechanism of the reactions in the electric field is very similar.

FIELD INDUCED CONFORMATION CHANGES

General

Conformation changes induced by electric fields are probably an essential step in the processing of electric signals in living organisms. The molecules involved in these processes are very difficult to handle and their reactions are not yet identified completely. At this state of knowledge it will be useful to characterise the mechanisms of conformation changes in electric fields by the investigation of simple model systems. Two different major types of field induced conformation changes may be distinguished:
a) A conformation change A → B may be induced by an electric field, when the conformation B is characterised by a higher dipole moment. It can be calculated that rather high changes of the dipole moment are required to induce an appreciable shift of equilibrium at moderate field strength according to this mechanism (for a discussion of dipole induced reactions, cf. e.g. ref. 6).
b) The second type of field induced conformation changes is found in molecules with ionized groups. It is well known that the equilibrium of ion association can be affected in the presence of electric fields[7,8]. This effect may lead to a change of conformation, when the conformational states are different with respect to their charge density. The amplitude of the reaction induced by the field will increase with an increasing number of ionised groups. Thus the conformation of polyelectrolytes is expected to be particularly sensitive to the presence of electric fields. In fact, most conformation changes induced by electric fields were reported for polyelectrolytes[9-17].

Threshold Effects In Single Stranded Polynucleotides

In the present contribution simple polyelectrolytes are used as model systems to study the amplitude of field induced conformation changes under various conditions. Single stranded polynucleotides are selected as models for several reasons. First the conformation change of these polymers can be followed easily by measurements of the UV-absorbance[18]. Furthermore, the conformation change of single stranded polynucleotides is a rather fast process[19]. This is important for the investigation of field effects, since electric fields cannot be maintained in aqueous salt solutions for long times without problems like decay of field strength and Joule heating of the solution.

When single stranded polynucleotides like poly(A), poly(dA) and poly(C) are subjected to electric field pulses, the absorbance of the polymers increases, indicating a decrease in the stacking interactions Since the stacking reaction is associated only with a minor change of the dipole moment, the large change of the equilibrium observed

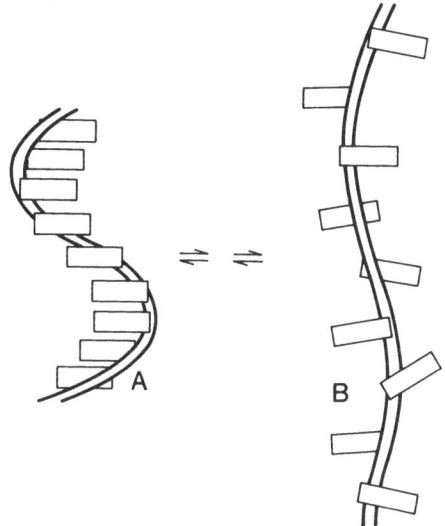

Fig. 1.   Schematic representation: (A) single strand helix;
          (B) extended coil. [from ref. 17].

in the experiments cannot be explained by a dipole-mechanism. Measurements of the polymer absorbance as a function of the ion concentration[4] demonstrate that the conformation of the polymers is dependent upon the degree of ion association: low ion concentrations lead
to unstacking of the bases. This reaction is expected, since the
ordered helical form has a higher charge density than the extended
coil form (cf. Fig. 1). Thus the mechanism of the field induced conformation change will be as follows: counterions are dissociated from
the ion atmosphere of the polymers by the electric field. This reaction results in an increase of the electrostatic repulsion between
the phosphates, which induces a conformation change to the extended
coil form.

The absorbance change induced by electric fields at low ion
concentration is a linear function of the electric field strength E
up to rather high values of E (cf. Fig. 2). When the ion concentration is increased, the absorbance change observed at a fixed field
strength decreases. Moreover, at high ion concentrations it is necessary to exceed a threshold value of the field strength, in order
to induce any appreciable change of the absorbance. At field strengths
above this threshold value again a linear relation is observed between the absorbance change and the electric field strength. The
threshold field strength increases with increasing ion concentration.
This dependence is observed both for monovalent and for bivalent ions
(cf. Fig. 3). The threshold field strength is a linear function of

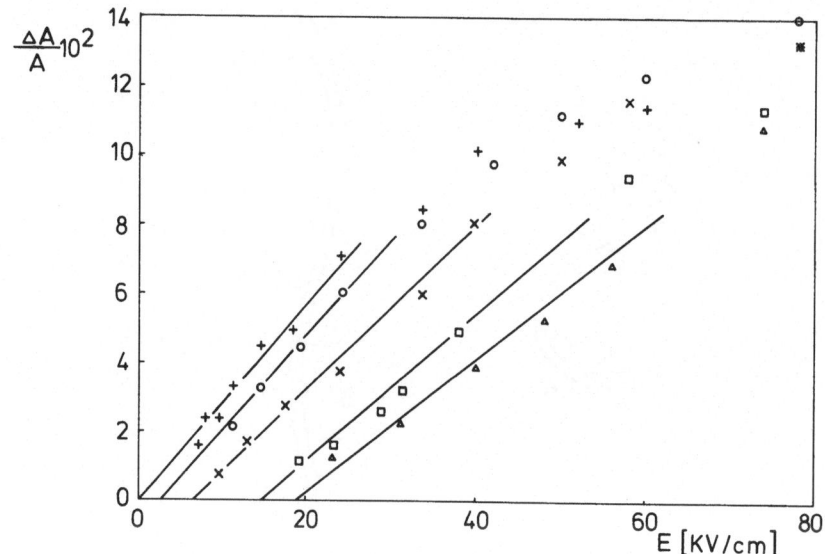

Fig. 2. Field jump amplitudes in poly(A) as a function of the field
         strength at different Tris concentrations: (+) 0.5 mM,
         (o) 1 mM, (x) 3 mM, (□) 6 mM, (Δ) 10 mM (poly(A) concen-
         tration 95 μM, pH 8, 20 °C) [from ref. 4].

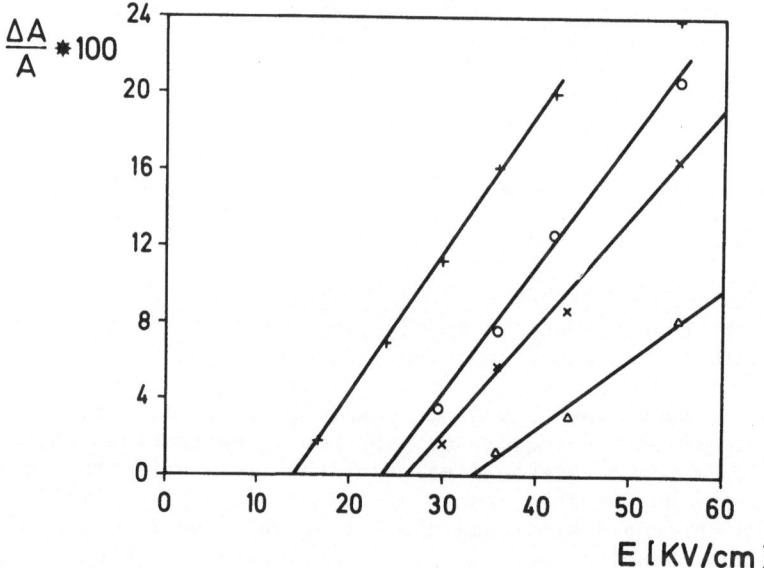

Fig. 3. Field jump amplitudes $\Delta A/A$ at 248 nm in poly(A) as a func-
         tion of the field strength at various Mg^{++} concentrations
         (20 °C, 0.5 mM Tris, pH 8) (+) 5 μM, (o) 10 μM, (x) 25 μM,
         (Δ) 50 μM [from ref. 17].

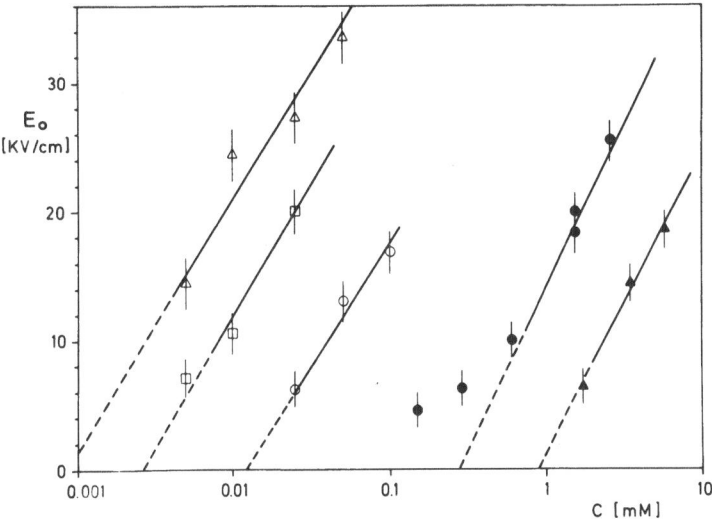

Fig. 4.   Threshold field strength E_0 as a function of the ion con-
          centration c: (Δ) poly(A) + Mg^{++}; (□) poly(A) + Ca^{++};
          (o) poly(dA) + Mg^{++}; (▲) poly(A) + Tris$^+$; (●) poly(dA) +
          Tris$^+$ [from ref. 17].

the logarichm of the ion concentration (cf. Fig. 4) according to

$$E_o = \alpha \ln (1+K_t \cdot c) \tag{1}$$

where c is the ion concentration and K_t as well as α are empirical
coefficients. The observation of a logarithmic depencence seems to
be related to the fact that the electrostatic potential of polyelectrolytes is a function of the logarithm of the ionic strength (cf.
e.g. ref. 20, 21). By extrapolation of the linear dependence between
E_o and ln(c) to $E_o = 0$ values can be obtained for the constant K_t.
K_t-values have been determined for various polynucleotide ion combinations and are compiled in Table 1. For comparison stability constants K determined for the association between the ions and the polynucleotides are also given. Although the agreement between the K_t
and the K-values is not very close in all cases, there is a clear
correlation in the order of magnitude. This correlation demonstrates
that the thresholds are induced by the binding of ions to the polynucleotides. A strong affinity between the ions and the polynucleotides, as in the case of Mg^{++} ions for example, leads the appearance
of thresholds at very low ion concentrations. In the case of low affinity as observed for monovalent ions thresholds are found at high
ion concentrations.

Table 1. Comparison of threshold constants K_t to stability constants
of ion association K determined by titration and relaxation
methods

Polymer	ion	$K_t [M^{-1}]$	$K [M^{-1}]$
poly(A)	$Tris^+$	$1.1 \cdot 10^3$	470
poly(dA)	$Tris^+$	$3.7 \cdot 10^3$	$2.4 \cdot 10^3$
poly(A)	Mg^{++}	$1.2 \cdot 10^6$	$3.8 \cdot 10^6$
poly(A)	Ca^{++}	$3.8 \cdot 10^5$	$3.6 \cdot 10^6$
poly(dA)	Mg^{++}	$8.1 \cdot 10^4$	——

Mechanisms of Field Induced Conformation Changes in Polyelectrolytes

The data described above demonstrate that electric fields induce
a conformation change in single stranded polynucleotides by its influence on the degree of ion association. Almost 50 years ago Onsager
developed a theory for the field induced ion dissociation in the case
of simple electrolytes[8]. According to this theory the change δK of
the dissociation constant K for the reaction

$$AB \rightleftharpoons A^{z+} + B^{z-}$$

increases with the absolute value of the field strength E:

$$\frac{\delta K}{K} = \frac{z_A \cdot u_A + z_B \cdot u_B}{u_A + u_B} \cdot z_A \cdot z_B \cdot \frac{e_o^3}{2\varepsilon k^2 T^2} |E| \tag{2}$$

where z is the valency, u mobility, e_o elementary charge, ε dielectric constant, k Boltzmann constant and T absolute temperature. Although the theory was not developed for polyelectrolytes, a qualitative comparison of the predictions according to equation (2) with the
experimental results is quite instructive. The linear dependence of
δK upon $|E|$ is constant with the experimental data. The change δK_D
is predicted to be proportional to the charge z of the reactants.
The corresponding parameter in polyelectrolytes - the charge density -
usually is rather high, leading to substantial effects in the presence of electric fields. However, the charge density may be reduced
considerably, when ions bind to the polymer. Thus the magnitude of
the dissociation field effect will depend upon the degree of ion association. At high degrees of charge compensation, the equilibrium
change will correspond to that observed for simple electrolytes. The
dissociation field effect will then be very small, but nevertheless
will lead to some increase in the charge density. Further increase of

the field strength will amplify this effect, until a critical value
is attained, where the increased charge density catalyses the dissociation of more ion complexes. This type of "autocatalysis" could be
the reason for the existence of threshold values of the electric
field strength. It is obvious that these thresholds will depend upon
the affinity of the ions to the polymer. Thus the thresholds are expected to be much higher for bivalent ions than for monovalent ions,
if measured at the same ion concentration. Furthermore it is expected that thresholds become aparent, when the ion concentration c exceeds the value of the reciprocal stability constant K for the association of these ions with the polymer. Thus the data obtained in
the present investigation can be readily explained on the basis of
a dissociation effect.

Recently Manning developed a theory for the dissociation field
effect in polyelectrolytes[22]. According to his calculations a given
field pulse will lead to the dissociation of more ions at high than at
low salt concentrations. Nevertheless a field pulse subjected to a
polyelectrolyte at high salt will lead to a smaller change in the
conformation than at low salt, since the stability of the ordered
conformation usually increases with increasing salt concentration.
The theory given by Manning predicts a linear relation between threshold field strength and the logarithm of the ion concentration, in
agreement with the experimental data. Unfortunately, it is not possible to compare the predictions of the theory quantitatively in the
case of single stranded polynucleotides. It seems, however, that the
extent of conformation change observed in the experiments is larger
than that expected according to the theory.

Field induced conformation changes in polyelectrolytes have also
been explained by a different mechanism[10,15]. It is well known that
electric fields induce large dipole moments in polyelectrolytes.
These dipole moments are due to a depletion of counterions at one
end of the polymer and an accumulation at the other end. It has been
argued that this change in ion concentration may lead to conformation
changes. however, a rough estimate of the change in the ion concentration required for the observed dipole moments demonstrates that
these changes are rather small and will hardly be responsable for
the large changes of conformation observed e.g. in single stranded
polynucleotides.

Another mechanism on the basis of the induced dipoles may be
more effective. It is well known that the induction of dipoles leads
to the orientation of the polymer molecules in the direction of the
electric field. The force resulting in orientation will also tend
to elongate the polymer. In the case of single stranded polynucleotides it is apparent that the polymer molecules can be elongated by
a helix-coil transition. However, the magnitude of the "pull effect"
remains for further investigation.

The threshold effects are characterised in the present contribution for the special case of single stranded polynucleotides. However, it is expected that very similar mechanisms will be effective for other polyelectrolytes including membrane structures. These mechanisms may be important for the regulation of bioelectricity. For example the electrical switching potential of certain ion channels may be adjusted by the chemical potential of specific ions. Thus the threshold effects described above could be very useful for the processing of electrical signals in biological systems.

BINDING OF IONS TO POLYNUCLEOTIDES

The conformation and stability of polynucleotides is regulated by the binding of ions[20,21]. Among these ions are simple ones like Na^+ and Mg^{++} or more complex ones like peptides or even proteins. In the following contribution the binding of various ions is analysed by the field jump technique. Compared to other methods used for the analysis of ion binding the field jump technique has the advantage of a high time resolution and thus is able to resolve binding states, that cannot be identified by conventional methods which provide an average picture of the binding states.

Mg^{++} Outer/Inner Sphere Complexes or the Problem of Ion Atmosphere-/Site-Binding

The mode of ion binding to polyelectrolytes is a problem that is under discussion already for a long time. From this discussion it is apparent that it is very difficult to distinguish between ion atmosphere- and site-binding by equilibrium methods. It will be shown below that it is possible to distinguish between these modes of binding by a kinetic approach, at least in the case of Mg^{++} ions. This approach is used for the characterisation of Mg^{++} binding to single stranded polynucleotides.

When single stranded polynucleotides like poly(A) or poly(C) are subjected to electric field pulses, the absorbance is observed to increase (cf. previous section). In the presence of monovalent ions this process is associated with relaxation time constants τ around 1 μs. The τ-values can be represented quantitatively by a two step reaction scheme with a bimolecular process of ion binding followed by coil to helix transition (cf. ref. 5). When Mg^{++} ions are added a separate relaxation process is observed[5]. At low Mg^{++} concentrations the time constant of this process is around 10 μs. From the concentration dependence of the relaxation time the rate constants of Mg^{++} complex formation with poly(A), poly(dA) and poly(C) are found in the range of 1 to 2 * 10^{10} M^{-1} s^{-1} (cf. ref. 5). Thus the reaction is diffusion controlled.

For many reactions rate constants around 10^{10} are nothing particular. In the present case, however, the rate of complex formation is unexpectedly high. The rate of Mg_{23}^{++} binding has been measured for a large number of different ligands. In all cases studied so far the rate of recombination was found to be many orders of magnitude below the limit of diffusion control. The reason for the relatively low binding rate of Mg^{++} is the rather tight binding of the H_2O molecules in their inner hydration sphere. The characteristic rate for the exchange of these tightly bound H_2O molecules against any ligand (including H_2O itself) is 10^5 s^{-1}. Thus the high rate of complex formation in the case of polynucleotides demonstrates that the inner hydration sphere of the Mg^{++} ions remains unaffected upon binding. Since the polynucleotides are the first ligand with a strong binding affinity that does not show inner sphere binding with Mg^{++}, there should be a special reason for this unusual mode of binding.

Most likely the special binding mode is associated with the particular structure of the polynucleotide chain. In the usual conformation of the sugar phosphate backbone the phosphates are separated by 7 to 8 Å. Thus a Mg^{++} ion cannot compensate its two positive charges by site binding to two phosphates simultaneously, since Mg^{++} has an ion radius of 0.7 Å. Upon binding of a Mg^{++} at a single phosphate site, a positive charge of the Mg^{++} would remain uncompensated. Apparently this mode of binding is unfavourable. Thus the Mg^{++} ions do not make a close contact to the polynucleotide. They remain in a rather mobile state ("ion atmosphere binding") as an intact $Mg^{++}(H_2O)_6$ complex, probably with a preferential residence between two adjacent phosphate residues.

The field jump method has also been used to characterise the mode of Mg^{++} binding to oligonucleotides[24]. In the case of oligoriboadenylates a relaxation process in the time range of 50 to 100 μs is detected that clearly demonstrates the presence of inner sphere complexes. A stringent test for the presence of inner sphere complexes is a Mg/Ca exchange experiment. Titration experiments demonstrate that the thermodynamics of the binding process is very similar for both ions. Also the influence upon the conformation of the oligomers is very similar in both cases. Nevertheless the dynamics of binding is quite different: in the presence of Ca^{++} the relaxation time constants are around 1 μs. Ca^{++} is known to have a much higher rate of inner sphere complexation than Mg^{++}. Thus the relatively slow relaxation observed in the case of Mg^{++} cannot be attributed to a slow response of the oligonucleotide structure, but must be associated with inner sphere binding.

The equilibrium constant of inner sphere complexation can be evaluated quantitatively from the experimental data on the concentration dependence of the relaxation time constant. The results are compiled in Table 2. The K_2 values decrease with increasing chain length N, although the overall binding constant $K_1 \cdot (K_2+1)$ increases

Table 2. Mg^{++} binding to riboadenylates
$K_1*(K_2+1)$ overall binding constant $[M^{-1}]$, K_2 inner sphere binding constant, k_{12} rate constant of inner sphere complex formation $[s^{-1}]$, k_{21} dissociation rate constant for inner sphere complexes $[s^{-1}]$.

	$K_1*(1+K_2)$	K_2	k_{12}	k_{21}
A(pA)$_4$	$3.2*10^3$	5.4	$6.5*10^4$	$1.2*10^4$
A(pA)$_5$	$14.3*10^3$	5.0	$4.9*10^4$	$1.0*10^4$
A(pA)$_6$	$19.*10^3$	3.8	$5.0*10^4$	$1.3*10^4$
A(pA)$_7$	$32.*10^3$	0.56	$1.0*10^4$	$1.8*10^4$
A(pA)$_9$	$31.*10^3$	0.27	$0.5*10^4$	$2.0*10^4$
A(pA)$_{17}$	$190.*10^3$	0.04	$0.1*10^4$	$2.7*10^4$
poly(A)	$3800.*10^3$	–	–	–

with increasing N. The rate constants of inner sphere complexation k_{12} decrease with increasing chain lengths. Already at low chain lengths k_{12} is below 10^5 s^{-1}, the usual value for Mg^{++} inner sphere complexation. The decrease in the rate constants seems to indicate some barrier against inner sphere complexation, which increases with increasing chain length. The nature of this barrier is not completely clear yet. Since inner sphere complexation presumably requires a contact of Mg^{++} to two phosphate residues, this reaction will involve binding of the sugar phosphate chain. Apparently there is an increasing hindrance effect against bending with increasing chain length. It remains to be explained, however, how a bending reaction can be correlated to the increase in the CD amplitude induced by Mg binding, which is found to be particularly high for the short oligomers[24].

Field jump experiments with d A(pA)$_5$, I(pI)$_5$, C(pC)$_5$ and U(pU)$_5$ did not provide any evidence for the formation of inner sphere complexes with Mg. Thus both specific base and sugar residues seem to be required for inner sphere complexation. According to these properties Mg^{++} inner sphere complexes may be used as markers for the recognition of specific acid sequences.

Thermodynamic Parameters from Field Jump Amplitudes: Specific Contacts
between   Arginine and Inosine Residues

    The amplitudes observed in field jump experiments contain very
useful information on the thermodynamics of the system under investigation. In the following an example will be given for the evaluation of equilibrium constants from field jump amplitudes[25].

    When the oligonucleotide $A(pA)_5$ is subjected to a field pulse
(in the presence of a monovalent buffer), no change in the transmission is observed. Similarly the peptide $(Lys)_3$ does not show a transmission change in the electric field. However, solutions containing
both nucleotide and peptide exhibit a clear field jump relaxation
(cf. Fig. 5). Obviously this relaxation is due to some interaction
between the nucleotide and the peptide. The thermodynamics of this
interaction can be analysed by measurements of the amplitude at different concentrations. In the present case the transmission change
is measured at a wavelength where only the nucleotide shows absorbance. Thus it is convenient to maintain the nucleotide concentration constant at an optimal absorbance level and to vary the peptide
concentration. An example of experimental data showing the characteristic concentration dependence is given in Fig. 6. The equilibrium
constant can be determined from these data most conveniently by using
a least square fitting procedure.

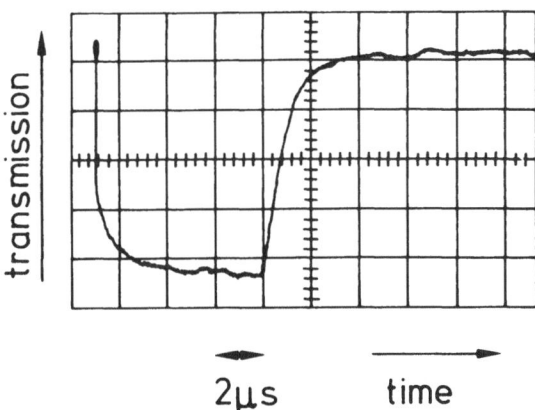

Fig. 5.   Field jump experiment with 18.6 µM $(A)_6$ + 90 µM $(Lys)_3$ in
          1 mM sodium cacodylate pH 5,9, 50 µM EDTA. The field pulse
          of 55.2 KV/cm was on for 7 µs. The slight slope of the 'on-
          field' transmission is due to a small temperature jump
          effect [from ref. 25].

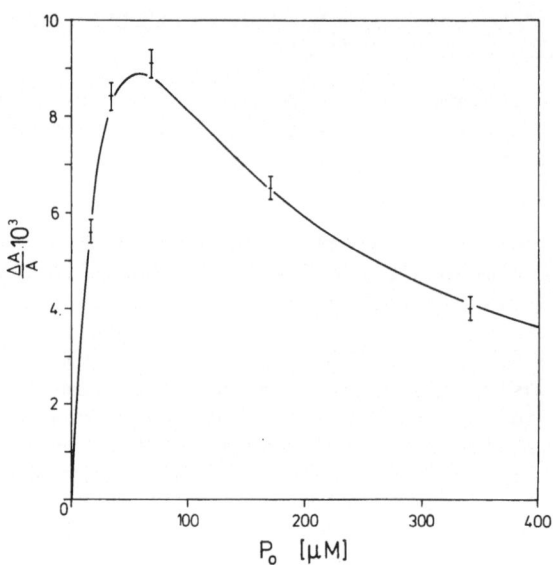

Fig. 6.   Field jump amplitudes $\Delta A/A$ for $A(pA)_5 + (Arg)_3$ as a function
          of the total peptide concentration $[P]_o$ at constant field
          strength ($E = 60$ KV/cm). The total nucleotide concentration
          is $N_o = 15$ µM. The line represents the fit according to the
          theoretical equation [from ref. 25].

        Compared to conventional methods for the determination of equili-
brium constants the amplitude method has several advantages:
1) the absorbance change can be measured with high precision up to
   10^{-4} absorbance units
2) the experiment can be repeated at will without extra solution required and the accuracy can be increased by averaging
3) the absorbance change is measured directly in a single experiment,
   perturbations by instrumental drifts and certain solution arte-
   facts are avoided
4) information on the dynamics is obtained at the same time.

        However, it should also be mentioned here, that the field jump
method cannot be applied in all cases. Its application is restricted
to rather low salt concentration. Furthermore it is difficult to
maintain electric fields at a constant strength for a long time in
aqueous solutions. Long field pulses result in strong heating effects.
Thus the investigation of systems with slow relaxation processes by
the field jump technique is not favourable.

        The amplitude method has been used for the characterisation of
various peptide-nucleotide complexes. Some of the data are compiled

Table 3.   Equilibrium constants $[\text{mM}^{-1}]$ for the association between
various oligonucleotides and the peptide Arg_3 as well as
Lys_3 obtained from field jump amplitudes in 1 mM sodium
cacodylate pH 5.9, 50 μM EDTA

	Arg_3	Lys_3
$C(pC)_5$	9.6	7.4
$U(pU)_5$	15.	8.9
$d\ A(pA)_5$	11.	14.
$A(pA)_5$	38.	17.
$I(pI)_5$	49.	12.

in Table 3, showing the stability constants for the complex formation
of Lys_3 and Arg_3 with various oligonucleotides. Both peptides have
the same number of charges and also a very similar spatial distribution of the charge centers. Thus the electrostatic interaction of
these peptides with oligonucleotides should be very similar. Actually
the stability constants are very similar in most cases. A major exception is observed in the case of $I(pI)_5$, where the binding constant for Arg_3 is larger by a factor of 4 than for Lys_3. This difference is attributed to hydrogen bonding between the guanidino group
of arginine and the hydrogen acceptor positions N_6 and N_7 of inosine.
Specific contacts of this type are expected to be important in the
protein nucleic acid recognition.

Field Jump Experiments with Fluorescence Detection

When one of the components is fluorescent, the system may also
be characterised by field jump experiments with fluorescence detection. In general fluorescence changes can be measured with very high
sensitivity down to much lower concentrations than absorbance changes.
An example of a fluorescence detected field jump relaxation is given
in Fig. 7 for LysTrpLys + poly(A). In this system the fluorescence
of the peptide LysTrpLys is quenched by binding to the polynucleotide poly(A)[26]. As discussed above electric field pulses induce the
dissociation of peptide molecules from the polymer lattice, which
results in an increase of the fluorescence[27].

The binding of LysTrpLys to poly(A) is studied as a model to
characterise the contribution of tryptophane residues in protein
nucleic acid interactions. Measurements of the relaxation time constants over a wide range of concentrations (cf. Fig. 8) demonstrate
that the reaction of LysTrpLys with poly(A) cannot be described by

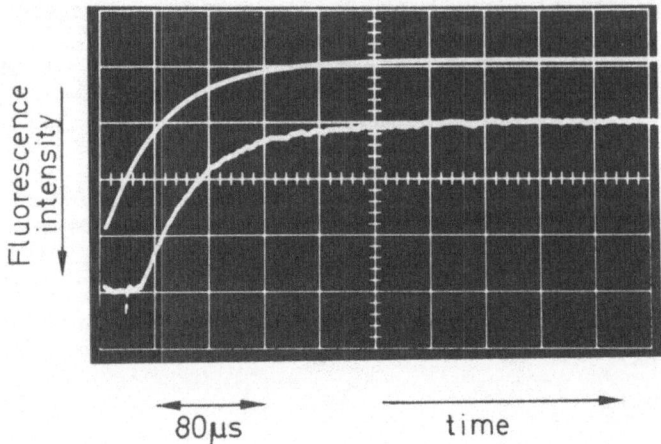

Fig. 7. Field jump relaxation of poly(A) + LysTrpLys detected by
changes in the fluorescence intensity after turning off an
electric field (C_p = 177 μM, C_ℓ = 15.7 μM; 1 mM sodium cacodylate, 1 mM NaCl, 0.2 mM EDTA pH 7; relaxation time
40.8 μs). The lower curve is the measured signal, whereas
the upper curve is simulated electronically to fit the
lower one [from ref. 27].

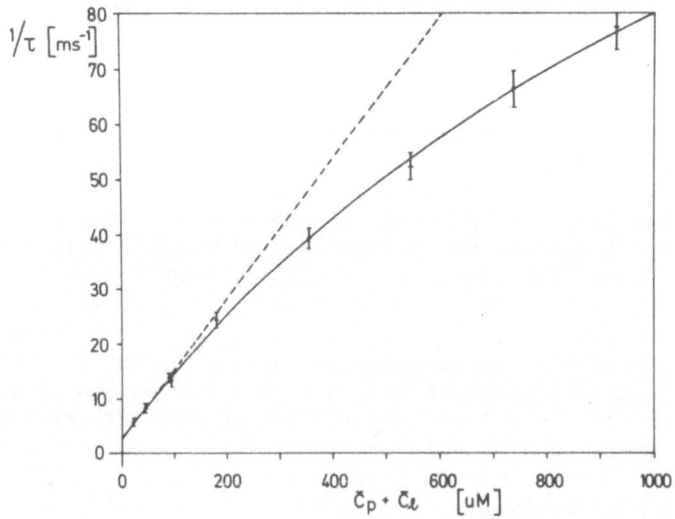

Fig. 8. Reciprocal relaxation time for poly(A) + LysTrpLys as a
function of the free reactant concentration $\bar{C}_p + \bar{C}_\ell$ in 1 mM
sodium cacodylate, 1 mM NaCl, 0.2 mM EDTA pH 7.0. Data
obtained at low degrees of binding Θ < 0.25 [from ref. 27].

a simple $A + B \rightleftharpoons C$ reaction scheme. The experimental data provide evidence for a two step reaction scheme

$$P + L \underset{k_{10}}{\overset{k_{01}}{\rightleftharpoons}} C_I \underset{k_{21}}{\overset{k_{12}}{\rightleftharpoons}} C_{II}.$$

The data show that the first step is a fast pre-equilibrium and the slow process is described by

$$1/\tau = k_{21} + \frac{k_{12} \cdot (C_\ell + C_p)}{(C_\ell + C_p) + k_{10}/k_{01}} . \tag{3}$$

A least squares fit of the experimental data according to equation 3 yields the following parameters

$$k_{12} = 1.5 \cdot 10^5 \ \text{s}^{-1}$$
$$k_{21} = 2.7 \cdot 10^3 \ \text{s}^{-1}$$
$$k_{01}/k_{10} = 660 \ \text{M}^{-1}$$

The forward rate constant of the second step k_{12} is in the range expected for an insertion of an aromatic residue into a stacked structure of adenine bases. However, the stability constant associated with the second step is much larger than expected. Further experiments are required to learn more about the nature of this interaction. Among the various methods available for the characterisation of these complexes, the field jump has the advantage of a high time resolution; i.e. intermediates can be identified which remain unresolved by other methods. Owing to this special advantage the field jump method will be useful in many future investigations.

REFERENCES

1. H. Labhart, Chimica 15:20 (1961).
2. H.H. Grünhagen, Thesis, Universität Braunschweig (1974).
3. M. Dourlent, J.F. Hogrel, and C. Helene, J. Amer. Chem. Soc. 96:3398 (1974).
4. D. Pörschke, Nucleic Acids Research 1:1601 (1974).
5. D. Pörschke, Biophys. Chem. 4:383 (1976).
6. E. Neumann, in: "Ions in Macromolecular and Biological Systems", Proc. 29th Symposium Colston Res. Soc., D.H. Everett and B. Vincent, eds.
7. M. Wien, Phys. Z. 32:545 (1931)
8. L. Onsager, J. Chem. Phys. 2:599 (1934)
9. C.T. O'Konski and N.C. Stellwagen, Biophys. J. 5:607 (1965)
10. E. Neumann and A. Katchalsky, Proc. Nat. Acad. Sci. US 69:993 (1972)
11. A. Revzin and E. Neumann, Biophys. Chem. 2:144 (1974)
12. K. Yoshioka, K. Kikuchi, and M. Fujimori, Biophys. Chem. 11:369 (1980)

13. K. Kikuchi and K. Yoshioka, Biopolymers 12:2667 (1973).
14. T. Yasunaga, T. Sano, K. Takahashi, H. Takenaka, and S. Ito, Chem. Lett. (Jap.) 405 (1973).
15. M. Pollack and H.A. Glick, Biopolymers 16:1007 (1977).
16. M. Fujimori, K. Kikuchi, K. Yoshioka, and S. Kubota, Biopolymers 18:2005 (1979).
17. D. Pörschke, Biopolymers 15:1917 (1976).
18. G. Felsenfeld and H.T. Miles, Ann. Rev. Biochem. 36:407 (1967)
19. D. Pörschke, in: "Mol. Biol. Biochem. Biophys. 24:191", I. Pecht and R. Rigler, eds.
20. G.S. Manning, Quart. Rev. Biophys. 11:179 (1978).
21. M.T. Record, C.F. Anderson,and T.M. Lohman, Quart. Rev. Biophys. 11:103 (1978).
22. G.S. Manning, Biophys. Chem. 9:189 (1977).
23. M. Eigen, Pure Applied Chem. 6:97 (1963).
24. D. Pörschke, Nucl. Ac. Res. 6:883 (1979)
25. D. Pörschke, Eur. J. Biochem. 86:291 (1978).
26. F. Brun, J.J. Toulme, and C. Helene, Biochemistry 14:588 (1975).
27. D. Pörschke, Nucl. Ac. Res. 8:1591 (1980).

POLYELECTROLYTES: A SURVEY

M. Mandel

Department of Physical Chemistry
Gorlaeus Laboratories
University of Leiden
Leiden, The Netherlands

Many electro-optical studies at present deal with a class of
systems which have been named polyelectrolytes, a contraction of
polymers and electrolytes. They are macromolecules which when dissolved in a suitable polar solvent, generally water, bear a large
number of ionized or ionizable groups. Under well defined conditions the macromolecular chain carries a large number of charges
(fixed charges) which are always accompanied in solution by an
equivalent number of small ions of opposite sign (the counterions).
Solutions of polyelectrolytes have for a long time been known to
exhibit certain specific properties which arise from the combination
of the macromolecular and the electrolyte character[1]). The amount
of experimental material accumulated in the course of the last forty
years is quite impressive but its value not always of comparable
level due to the different experimental conditions in which they
have been obtained and the many pitfalls that threaten the experimental physical chemist when dealing with polyelectrolyte systems.
Evidently some qualitative insight in the physical chemical behaviour
of such systems has been gained, sometimes substantiated by more
quantitative theoretical approaches. Nevertheless there still seem
to exist many properties which are far from being well understood
and application of new techniques often only increases the number
of questions which remains unanswered or demonstrates the limited
validity of the theoretical approaches used so far. This is certainly also true for problems arising in connection with the
electro-optical properties of polyelectrolyte systems or related
problems. It is therefore worthwhile to discuss briefly some
aspects of the physical chemistry of polyelectrolyte solutions in
order to realize where we stand. In this survey, which we have
limited to a few topics only (and even for those not in an

exaustive way), we want to emphasize the difficulties arising in connection with the interpretation of polyelectrolyte results and, in as far as possible, their origin.

The specificity of the physical-chemical properties of poly-electrolyte solutions arises from the combination of macromolecular behaviour with the influence of interactions of electrostatic origin as operating in ordinary electrolyte solutions. Let us discuss this briefly and in a qualitative way. For the sake of simplicity we shall consider macromolecules charged negatively only (poly-anions), the negative charge arising from negative elementary charges distributed more or less uniformily along the macromolecular chain. We shall also assume all the small ions in solution (the mobile charges as opposed to the fixed charges) - i.e., the positive counterions and the negative coions in the presence of additional salt - to be monovalent. Generalization to other and more compli-cated situations is, of course, possible. Let us compare a salt-free polyelectrolyte solution with a solution of a low molecular weight electrolyte (mono-mono valent) of the same counterion concentration. In the latter we may use as a simplified model a picture of point charges moving freely through a continuum constituted by the solvent. In the Debye-Hückel picture the average distribution with respect to the coordinate system fixed in a given ion will be determined by the competition between the electrostatic interactions and the free translational motion of all the other ions. This gives rise to a ion atmosphere of spherical symmetry. In contrast in the salt-free polyelectrolyte solution all the negative point charges are now distributed on large, macromolecular chains. Although on the molecular scale the cations will not behave differently as in an ordinary electrolyte solution, the motion of the negative charges is determined by the centre of mass motion of each macromolecule and its internal mobility which depends strongly on the chemical structure. With respect to the coordinate system fixed in a given negative charge a distinction must now be made between the negative charges fixed on the same chain and the others distributed among all other macromolecules. For the former the distribution will be strongly affected by their fixation on the more or less flexible macromolecular backbone. This excludes the possibility of increas-ing their effective distance to the reference charge beyond a certain limiting value determined by the condition of maximal stretch of the chain. The immediate consequence is that, whatever the stoechimetric concentration of the polyelectrolyte solution, on the molecular level there will always be, even on the average, regions of high negative charge concentrations. Interactions with the positive mobile counterions will therefore be much stronger than in the case of a comparable ordinary electrolyte solution. Of course even in such a solution some screening of the electrostatic inter-actions will occur through the distribution of all the counterions and the negatively charged macroions, but necessarily the average distribution of the counterions with respect to the charges on the

macromolecule will deviate with respect to the situation in a low
molecular weight electrolyte solution. The addition of low molecular
weight electrolyte salt to a polyelectrolyte solution will considerably enhance the screening of the coulomb interactions between
macroions and counterions. These effects are borne out by experimental results indeed; e.g., the osmotic coefficient of aqueous
salt-free polyelectrolyte solutions are very low in comparison to
ordinary electrolyte solutions[2] but addition of low molecular
weight electrolyte has as an effect the increase of this thermodynamic parameter characteristic for intermolecular interaction[3].
Also the mean activity coefficient of the added salt is found to be
strongly dependent on the ratio $Y \equiv C_s/C_m$ of salt to polymer concentration (the concentration in monomol per unit volume). If this
ratio is very high $\gamma_{\pm}$ comes close to its value in the electrolyte
solution without polymer but decreases very rapidly to much lower
values indeed when this ratio is decreased[4]. The same is true for
the operationally defined activity coefficient of the counterions
which can be determined even in salt-free polyelectrolyte solutions.
The importance of these effects strongly depends on the average
(linear) density of fixed charges on the macromolecular chain
($|\sigma| \equiv qZ/\ell$, where q is the elementary charge, Z the number of negative
charges per macromolecular and ℓ the contour length of the macromolecule) or the related quantity $A \equiv \ell/Z$ which represents the average
contour distance between two successive fixed charges on the chain.
The higher $|\sigma|$, or the lower A, the more the properties will
deviate from those of an ordinary electrolyte solution.

The accumulation of identical fixed charges on a polyelectrolyte chain will also strongly affect the properties related to the
macromolecule itself. If the backbone is sufficiently flexible
electrostatic repulsions between the charges on a given chain will
tend to increase the average dimensions and may affect the local
flexibility of the polyelectrolyte to turn it into a rigid macromolecule. Again these effects may be reduced by addition of low
molecular weight electrolytes as is borne out by viscosity experiments on polyelectrolyte solutions. At very low Y the viscosity η
at constant polymer concentration increases considerably with
increasing $|\sigma|$ but much less if Y becomes larger. The same is
true for the osmotic second virial coefficient of a polyelectrolyte-salt solution in equilibrium with the salt solution[5].

Before discussing some of these effects more in detail it may
be appropriate to introduce the concept of counterion association
which seems to play an important role in polyelectrolyte literature.
From the qualitative description given so far it will be clear that
in a polyelectrolyte solution the counterions will tend to concentrate in the neighbourhood of the macroions thus apparently
reducing the charge borne by the polyelectrolyte if viewed from a
larger distance. This screening may be modeled by the simple
picture in which a certain fraction of the counterions remains on

the average associated (sometimes one calls it even "bound") with each of the macromolecules thus effectively reducing the number of charges on the chain from Z to Z^+ and taking into account to a large extent the macroion-counterion interaction. This is called "the two phases model" because the counterions are assumed to be distributed over two phases: a) the associated counterions in the macromolecular phases and b) the "free" counterions in the bulk phase[1b,6]. Thus e.g., the low value of the osmotic coefficient of a salt-free polyelectrolyte solution may be thought to be due to the fact that the associated counterions do not contribute to the osmotic pressure[6]. Also transport properties in polyelectrolyte solutions, such as conductivity, sedimentation velocity, etc. can be interpreted by assuming that a fraction of the counterions is dragged along with the macroions[7]. The interest of this rather simple approach lies in the fact that Z^+ or the fraction of associated counterions derived for a given system from different experimental methods has often quite comparable values thus lending some confidence to the concept of counterion association but not necessarily to the simple picture through which it has been introduced. Unfortunately it has also induced, often sterile, discussions about the state of the associated counterions opposing advocates of "site binding" (i.e., binding of the individual counterions to the individual charged groups on the macromolecular chain) and advocates of "domain binding" (i.e., non-localized association within a certain regions around each macroion). Although except in a few case no unambiguous experimental evidence for it is available, site binding certainly cannot be ruled out, but if taken seriously into consideration, would necessitate a much more subtle treatment of the macromolecular domain than is usually done including probably the molecular nature of the solvent.

Let us now turn to a more quantitative approach of the interaction between fixed and mobile charges in a polyelectrolyte solution. It is not surprising that most theoretical attempts have been inspired by the Debye-Hückel treatment of ordinary electrolyte solutions. Many of these start with a Poisson-Boltzmann equation generalized to the case of polyelectrolyte solutions. This implies a slightly different approach and additional assumptions besides those already used for ordinary electrolytes[8]. The most important one is that the average potential ψ (or its reduced equivalent $\phi \equiv q\psi/kT$, where k is the Boltzmann constant and T the temperature in Kelvin) is not defined in a reference system fixed in a given central (point) charge but in a reference system fixed with respect to an assumed average distribution of charged groups on a given macromolecule occupying a certain region of space which may, or may not, be permeable to small ions. Also contributions to ψ around a given macroion due to the fixed charges on other macromolecules are not considered thus imposing certain boundary conditions. The Poisson equation then reads

$$- \Delta\psi = (\rho_p + \rho_m)/\varepsilon\varepsilon_o \tag{1}$$

Here Δ is the Laplace operator, ε_o the permittivity of free space, ε the relative permittivity of the solvent; ρ_p and ρ_m stand for the space charge distribution of the single polyelectrolyte charges and the space charge density due to the average distribution of the mobile charges respectively (the S.I. formalism is used throughout). Here ρ_p is only different from zero within a well defined region of space. In most recent treatments this region of space is chosen as a cylinder impenetrable to mobile charges thus representing the macromolecule as a rigid rod. Although in the absence of low molecular electrolyte and for high dilutions the average conformation of macroion will tend to become rodlike, under other circumstances this is not necessarily the case. Whenever the screening length of the electrostatic forces is much larger than A, the average distance between successive charges along the chain, the macromolecule may however exhibit locally enough stiffness (due to repulsion between the like charges) for the cylindrical model to be justified on that scale. In as far as properties are considered that are not depending on the length of the macromolecular chain (for high enough molar masses at least), such as in generally assumed to be the case for the colligative properties, the rodlike approximation may therefore be reasonable (see however below). The fixed charges may then be assumed to be distributed uniformily along the axis of an impenetrable cylinder of radius a or smeared out uniformily on the surface of the same cylinder and ψ or ρ_m to be a function of the distance r with respect to the cylinder axis only. The Poisson-Boltzmann equation is then obtained from (1) for r>a by replacing $\rho_m(r)$ by a distribution function depending exponentially on the average electrostatic energy of a cation or anion.

$$- \frac{1}{r} \frac{\partial}{\partial r} (r \frac{\partial\phi}{\partial r}) = \frac{q^2}{kT\varepsilon\varepsilon_o} (n_c(R) e^{-\phi} - n_a(R) e^{+\phi}) \quad r>a \tag{2}$$

Here $n_c(R)$ and $n_a(R)$ stand for the concentration at a position r=R of mobile cations and mobile anions respectively where $\psi(R)=0$. Essentially two cases have been studied extensively: a) the polyelectrolyte–salt free solution and b) the polyelectrolyte in the presence of an excess of low molecular weight electrolyte. In the first case no anions are present and in order to exclude the contribution due to the fixed charges on other polyions, each macromolecule is confined to an electroneutral cylindrical subvolume coaxial to the macroion itself and chosen in such a way that it corresponds to the average volume occupied by one macromolecule, i.e., equal to $n_M^o{}^{-1}$ if n_M^o represents the average number of polyelectrolytes per unit volume[9]. The potential ψ as well as the field $\partial\psi/\partial r$ is further assumed to vanish at the surface of this cylindrical subvolume so that its radius corresponds to R as defined above.[¶]

The Poisson-Boltzmann equation reduces for this case to

$$- \frac{1}{r} \frac{\partial}{\partial r} (r \frac{\partial \phi}{\partial r}) = \chi^2 e^{-\phi}$$
(3)

which can be solved analytical. Note that the screening factor $\chi^2 = q^2 n_c(R)/kT\varepsilon\varepsilon_o = 4\pi Q n_c(R)$, where $Q \equiv q^2/4\pi kT\varepsilon\varepsilon_o$ represents the so-called Bjerrum length, is expressed here as a function of the counterion concentration near the boundary of the cylindrical cell where the electrostatic field vanishes. It has been shown that the explicit form of the solution for (3) given the boundary conditions depends on the value of the charge parameter $\lambda \equiv |\sigma| Q/q = Q/A$ (i.e., on the linear charge density on the polyelectrolyte chain expressed in units q/Q), changing from one form for $\lambda < 1$ to another for $\lambda < 1$ in which case the intergration constants β becomes imaginary.

In the second case the counterions due to the dissociation of the polyelectrolyte may be neglected with respect to those provided by the low molecular weight electrolyte. The solution is considered infinitely diluted in macromolecules so that macroion-macroion-interactions can be neglected, corresponding to the boundary condition $\psi(r \to \infty) = 0$. Then the concentrations $n_c(r)$ and $n_a(r)$ both equal the stoichiometric concentration of the electrolyte n_s^o and the Poisson-Boltzmann equation reduces to

$$\frac{1}{r} \frac{\partial}{\partial r} (r \frac{\partial \phi}{\partial r} = \frac{2q^2 n_s^o}{kT\varepsilon\varepsilon_o} \sinh\phi$$
(4)

Here the screening constant $\kappa^2 \equiv 2q^2 n_s^o/kT\varepsilon\varepsilon_o = 8\pi Q n_s^o$ is identical to the Debye-Hückel screening constant for ordinary electrolyte solutions. No analytical solution of (4) exists but this differential

---

[¶]These symmetry conditions follow immediately if the cylindrical macroions are assumed to be arranged in an array of parallel rods. This so-called cell model of the polyelectrolyte solution has been introduced by Katchalsky for the sake of calculation only[10] although it has also been justified by Onsager's treatment of uncharged cylindrical rods with repulsive forces[11]. The use of these boundary conditions however not necessarily implies that in a salt free polyelectrolyte solution such ordering really occurs, even on a local scale. Also the concentration dependence which appears in this P.B. equation and its analytical solution through $R(R^{-2} = \pi \ell n_M^o)$ should be considered with proper care as a prerequisite condition for the treatment considered is that $\partial \psi/\partial r$ should be close to zero in a sufficient wide range[10] thus limiting its applicability to rather low macromolecular concentrations.

equation may be solved numerically$^{12)13)}$. Linearization of the
r.h.s., i.e., the approximation $\kappa^2 \sinh\phi \simeq \kappa^2\phi$, transforms (4) into a
zeroth order modified Bessel equation of known solution but strictly
valid only for $\phi \ll 1$. In both cases the value of the outward radial
electrostatic field at the surface of the cylinder of radius a, as
given by Gauss law, may be used as a second boundary condition, i.e.,
$(\partial\phi/\partial r)_{r=a} = \sigma/2\pi\epsilon\epsilon_0 a$ or $(\partial\phi/\partial r)_{r=a} = -2\lambda/a$. The solution ϕ_L of the
linearized Poisson-Boltzmann equation will be strictly valid only
at distances $r \gg \kappa^{-1}$.

$$\phi_L(r) = 2\lambda \, K_o(\kappa r)/\kappa a \, K_1(\kappa a) \tag{5}$$

Here K_o and K_1 are the zeroth order and first order modified Bessel
functions of the second kind. The exact solution of (4) can be
derived from ϕ_L on replacing λ by an effective charge parameter
$\omega(r,\lambda,\kappa) \equiv \lambda \, g(r,\lambda,\kappa)$ where the correction factor has to be found by
numerical integration of (4) and satisfies the condition $g \to 1$ for
$r > \kappa^{-1}$ $^{13)}$. Note that g is of order unity for a wide range of conditions.

     From the solution of the Poisson-Boltzmann equation the average
distribution of the mobile ions around the macroion can be estimated as well as the electrostatic free energy. The latter can be
used to evaluate the non-ideal contributions to the thermodynamic
properties. The mobile ion distribution functions clearly
demonstrate a considerable accumulation of the counterions in the
neighbourhood of the charged cylinder and a repulsion of the coions
(if present). It also has been shown$^{14)15)}$ that in a region close
to the macroion the high local counterion concentration only very
weakly depends on the salt concentration for macroions of rather
high charge density $|\sigma|$. Nevertheless it remains difficult to
define within the framework of the Poisson-Boltzmann approach a
concept such as counterion association without introducing ad hoc
definitions. Several criteria may be used to define a radius r_b
enclosing a fraction of counterions which may be considered to be
strongly associated to the macroion and although they all lead to
values of comparable order of magnitude their choice is not free
of a certain arbitrariness.

     The usefulness of the Poisson-Boltzmann treatment of poly-
electrolyte solution, the theoretical validity of which has
recently been assessed$^{16)}$, is somewhat reduced by the fact that for
salt containing systems no exact analytical solution is available.
No solution has even been obtained valid for systems where the low
molecular weight electrolyte is not present in excess with respect
to the equivalent concentration of the macroion so that the two
cases a) and b) discussed above cannot be properly connected. This
may explain the success of another approach, proposed by Manning$^{17)}$,
which has the considerable advantage that it leads to simple
expressions and introduces concepts which are easily applicable to

a variety of different situations. Again it is strictly spoken
valid only to the case where macroion-macroion interactions can be
neglected, i.e. to infinitely diluted solutions of polyelectrolyte,
and may be considered as an extension of the Debye-Hückel treatment
of ordinary electrolytes. The rodlike polyion is represented by a
(infinitely long) linear uniform charge density, each elementary
charge of which gives rise to a screened coulomb potential which,
in analogy to the Debye-Hückel potential, is proportional to $e^{-\kappa r}/r$.
Here the screening constant $\kappa^2 \equiv 4\pi Q(n_c^o + n_a^o)$ includes the contribution
not only from the stoichiometric equations of cations and anions
from the salt but also, partially, from counterions provided by the
dissociation of the macromolecule. The latter will depend on the
linear charge density of the polyelectrolyte. For monovalent
counterions when $|\sigma| < q/Q$ or $\lambda < 1$ all the counterions dissociated from
the polyelectrolyte will fully contribute to the Debye-Hückel
screening constant; they are considered "free". For $|\sigma| > q/Q$ or $\lambda > 1$
a certain fraction of these counterions are "condensed" to the
chain[18] reducing its charge density to its limiting value $\lambda_{eff} = 1$
and the concentration of the free ions dissociated from the polyelectrolyte to n_m^o/λ, i.e., $Z^+ = Z/\lambda$. This amounts to using for the
reduced potential around the macroion an expression as given by (5)
in the limit $a \rightarrow 0$ for the case $\lambda < 1$

$$\phi_M = 2\lambda \ K_o(\kappa r) \qquad \lambda < 1 \qquad\qquad\qquad (6)$$

and the same expression with λ replaced by unity for $\lambda > 1$[16]. If
condensation and association are identified then for monovalent
counterions no association will occur for $\lambda < 1$ (not implying however
that the solution behaves ideally) and for $\lambda > 1$ the fraction of
associated counterions is $(1 - \lambda^{-1})$. These considerations lead to
the so-called "limiting laws" of Manning for thermodynamic quantities
which seem to be fairly well satisfied by experimental results
particularly if the ratio Y is high. Considerable deviations e.g.,
in γ_+ have recently been found for values Y^{-1} exceeding unity[4]
We shall not go into the discussions about the validity of this
approach[16)19-21)] nor into recent extensions of the limiting laws
based on a cluster terms expansion for polyelectrolyte-salt solutions combined with the condensation concept[22,23].

It is perhaps too early to make definite statements at the
present but, it seems that the limiting laws and condensation, if
applied with proper care, may remain a useful first approximation
to a very complex problem for which the Poisson-Boltzmann approach
is certainly not the ultimate solution neither. Recent refinements
introduced by i.a. Manning himself[24], such as asigning the condensed counterions to a certain cylindrical volume of radius r_c
and having them contribute to the electrostatic free energy of the
solution beyond the simple fact of reducing the effective charge
density to a limiting value, are based on additional assumptions

and not necessarily constitute an improvement with respect to the
original, more simple, approach.

     As pointed out before the cylindrical model is a good approxi-
mation only for discussing polyelectrolyte properties which in
general are independent of the average dimensions or properties
referring to very specific conditions where the macroion may be
assumed to have an average conformation close to a completely
stretched chain. Obviously such a model cannot be used to predict
the dependence of the average dimensions of a polyelectrolyte on the
chain's charge density or the ionic strength. This problem has
since long been a challenge to physical chemists dealing with polyelectrolytes but the solutions proposed have remained rather dis-
appointing[5,25]. Recent developments seem to be promising however.
Considering only solutions in which interactions between macroions
may be neglected, the main solution to this problem could lie in
the fact that electrostatic repulsions between the fixed charges on
a given chain may give rise to short range and long range effects.
In as far as those effects may be treated separately the former
could lead to an increase of the local stiffness of the chain, the
latter to an enhancement of the so-called excluded volume effect
(repulsion between more distant parts of the chain) with respect
to the uncharged macromolecule). It is clear that here again we
have to deal with a rather complicated problem and simplifying
assumptions are needed to get at a zeroth or first order approximate solution. The dependence of the local stiffness on charge
density and ionic strength of the solution may be approached through
the model of the wormlike chain[26]. In this model the macromolecule
is treated as a sequence of identical linear elements (e.g., the
monomeric units) each making a , generally small, angle θ with its
immediate neighbour along the chain. It has been shown, that the
average dimensions of such a chain under ideal conditions (no
excluded volume effect; unperturbed dimensions) can be expressed
as a function of two parameters only, the contour length ℓ and the
so-called persistence length L. The latter expresses the characteristic length along the chain over which orientational correlation between the segments is maintained and is related to the free
energy of curvature of the chain pictured as a continuous space
curve. A wormlike chain is not necessarily representative for a
macromolecule deviating only slightly from a rigid cylinder, e.g.,
for a bended rod, but may serve to describe a chain which still
has a large number on internal rotational degrees of freedom (large
number of available conformational states) such that in the limit
$\ell/L\to\infty$ it tends to an ideal Gaussian chain but for $\ell/L\to 0$ the rigid
rod limit is approached[27]. In the case of a charged polyelectrolyte the free energy of curvature may be assumed to consist of
two parts, one connected to the intrinsic properties of the
uncharged chain itself, the other resulting from the electrostatic
repulsion between the fixed charges. Consequently the persistence
length will also be composed of two independent parts, the

intrinsic or bare persistence length L_p determined by the former and the electrostatic part L_e determined by the latter and depending on the chain's charge density and the ionic strength of the solution. Skolnick and Fixman[28] and independently Odijk[29] from our laboratory have been able to derive a first order expression for L_e assuming that the fixed charges on the chain interact through a Debye–Hückel potential for $L >> \kappa^{-1}$ and $A << \kappa^{-1}$. The total persistence length L_t is given by [29]

$$L_t = L_p + L_e = L_p + \frac{Q}{4\kappa^2 A^2} f \tag{7}$$

where the factor f takes into account the possibility that the charge on the chain may be effectively reduced through association. If no effective charge compensation occurs $f=1$, otherwise $0<f<1$. If the counterion condensation model is used $f=1$ as long as $\lambda \equiv Q/A$ is smaller than unity; if $\lambda>1$ condensation increases the effective value of A to $A_{eff}=Q(\lambda_{eff}=1)$ and $f=A^2/Q^2$. Thus in general the persistence length will depend on the ionic strength through κ^2. The change in the average dimensions with experimental conditions may be calculated through the variation of the mean square radius of gyration $<S^2>$, a quantity which is directly accessible to experimental determination (in contrast to the mean square end-to-end distance). For unperturbed chains of contour length of the order L_t or smaller $<S^2>_o$ may be evaluated through the formula of Benoit and Doty[30]

$$<S^2>_o = 2 \times \ell^2 (\frac{1}{6} - \frac{x}{2} + x^2 - x^3(1 - e^{-1/x})) \tag{8}$$

where $x \equiv L_t/\ell$. However in many cases, particularly for high molar mass polyelectrolytes, $\ell >> L_t$ and the average dimensions will also be strongly influenced by excluded volume effects. They may be estimated by using an analogous approach as for non charged macromolecules[27] assuming that the polyelectrolyte may be represented as a chain consisting of $\ell/2L_t$ rigid segments of length $2L_t$.[31]

$$<S^2> = \frac{1}{3}\ell L_t \, \alpha_s^2(z). \tag{9}$$

Here α_s is the expansion factor which is a function of the excluded volume parameter z only. The latter can be expressed through ℓ, L_t and the effective volume β excluded from one segment by the presence of others and which may be estimated in case the segments are cylinders interacting only through electrostatic repulsion[32][33]. This yields

$$z \sim \kappa^{-1} L_t^2 \tag{10}$$

and finally for the mean square radius of gyration the following expression[31]

$$<S^2> \simeq \ell^{6/5} \, L_t^{2/5} \, \kappa^{-2/5}. \tag{11}$$

(Note that here and in the following $\sim$ means proportional to and $\simeq$ equal to within a factor of order unity).

As a consequence of these theoretical considerations it follows that the ionic strength dependence of the mean dimensions is complex but may be considerable indeed. Odijk[34] has estimated that e.g, for a DNA molecule (L_B assumed to be 60 nm) of a contour length 7×10^4 nm, at infinite dilution the expansion factor α^2 and total persistence length L_t would increase from 1.145 and 60,2 nm respectively at NaCl 0.2 M to 1.934 and 66 nm at NaCl 0.005 M, in reasonable agreement with experimental viscosity data. For low molecular weight DNA of $\ell=170$ nm only we have recently found from elastic light scattering data that the increase $<S^2>_0$ in the same range of ionic strength and infinite dilution is only a few percent[35]. Here the experimental results seem to obey the theoretical predictions neglecting excluded volume effects but taking condensation into account.

At this stage it is still premature to conclude however that we completely understand the influence of ionic strength on the average dimensions of polyelectrolytes. This difficult problem is of course strongly related to the preceding one but some insight seems to have been gained through the use of the wormlike chain model notwithstanding the introduction of several simplifying assumptions in the theoretical treatment. An important conclusion may be that even at infinite dilution and relative low ionic strength polyelectrolytes are not necessarily fully stretched even if the persistence length is rather large. Over distances of order L_t the chain may roughly be assimilated to rigid cylinders however thus justifying some of the assumptions necessary to estimate colligative properties. But recently it has been suggested that such properties may be influenced by the polyelectolyte chain's flexibility[36]. These considerations may have strong bearing on our understanding of electro-optical properties of polyelectrolytes which may depend on the shape and average dimensions of the macromolecules. In the limit of vanishing salt concentration where the theory outlined above is not strictly valid, the situation is even less clear and here concentration effects may become the dominant factor under usual experimental conditions.

Concentration effects in polyelectrolyte solutions are theoretically difficult to treat. Also from an experimental point of view not too many systematic investigations of the influence of concentration on physical chemical properties have been performed

so that also qualitatively not much information is available. This
is particularly true for the case of polyelectrolyte solutions with
vanishingly small concentrations of added low molecular weight
electrolytes. For such systems it cannot be excluded that some
ordering, eventually on a local scale, may occur due to the strong
electrostatic repulsions between weakly shielded macroions. In
fact ordered, birefringent phases have been observed for the rigid
elongated virus particle TMV[37] and also for more simple polyelectrolytes[38]. No direct experimental evidence is available however
neither for any anisotropy in the lower concentration range nor for
any local ordering occurring on a sufficiently long time scale to
have a considerable influence on the physical chemical properties.
Nevertheless there seems to be some experimental evidence that
several concentration regimes may exist for polyelectrolyte solutions particularly in the absence of added salt[39] as have also
been observed for solutions of uncharged macromolecules. De Gennes
et.al[40] have attempted to describe the concentration dependence of
the average dimensions of polyelectrolytes in the absence of salt
using arguments based on the scaling approach as applied in polymer
solution theory[41]. A critical concentration C_m^* is assumed to
exist above which the polymer solution is isotropic and characterized
by considerable overlap of the macromolecular chains (semi-dilute
regime). Each chain may then be pictured as consisting of sub-
regions (blobs) of an average length ξ which are uncorrelated to
each other by "shielding" through the other macromolecules. The
average dimensions of the complete chain may be calculated assuming
that it may be represented as an ideal Gaussian coil consisting of
segments with a length equal to the <u>correlation length</u> ξ which
should be concentration dependent and has to satisfy two conditions: a) for sufficiently long chains $(\ell \gg \xi)$ ξ must be independent
of ℓ and b) at $C_m \simeq C_m^*$, ξ will be given by the average dimensions
of the chain in the dilute regime $(C_m < C_m^*)$ which are concentration
independent. In the case of polyelectrolyte solutions without
salt, De Gennes et.al.[40] assumed that for chains with a large
charge density rigid cylinders can be expected below C^*. They
distinguish between the situation at extremely high dilutions where
interactions between these cylinders may be neglected (random
orientation) and the one above another critical concentration
$C_m^+ \ll C^*$ where the electrostatic interactions between the chains
lead to a lattice-like ordering not affecting however the dimensions of the polyelectrolytes. Above C_m^* each blob is a linear
segment of length $\xi = \xi(C_m)$ with $\xi(C_m = C^*) = \ell$, which defines the
concentration dependence of the correlation length. Within this
isotropic model of the polyelectrolyte solution the average square
dimensions of the charged macromolecules are expected to decrease
as $C_m^{-\frac{1}{2}}$. As pointed out be these authors C_m^+ is proportional to
ℓ^{-2} and may generally be already near the limit of the sensitivity
of most physical measurements.

Recently Odijk[42] has refined this scaling approach by including changes in the flexibility of the macromolecular chain due to charge interactions within the context of the wormlike chain and condensation model. He considered both the case of polyelectrolyte solutions with and without added salt. For the former in the dilute regime the polyelectrolytes are coils with radii of gyration $<S^2>_d$ satisfying (11) (as long as $\ell >> L_t$, $\kappa^{-1} << L_t$). The transition region will correspond to these monomolecular concentrations where the coils come into contact, i.e.,

$$C_m^* \simeq N_{av}^{-1} (\ell/A) <S^2>_d^{-3/2} \qquad (12)$$

with N_{av} standing for Avogadro's constant. Above C_m^* the chain is represented as an ideal chains of blobs each consisting of a Gaussian coil with g segments of length $2L_t$ with an excluded volume effect such that ξ^2 is given by an equation analogous to (11). (Of course here it must be assumed that $L_t << \xi$ and $Y >> 1$.) It then is found that the average dimensions above C_m^* depend both on ionic strength and polymer concentration according to the relation

$$<S^2>_{s.d.} \sim \ell (Q A N_{Av})^{-\frac{1}{4}} \kappa^{-3/4} C_m^{-\frac{1}{4}} \qquad (13)$$

for the case where $L_e >> L_p$.

In the case of salt-free polyelectrolyte solutions the important additional assumption has to be made that the macromolecule can be described as a wormlike chain with L_t given by (7) where the screening factor κ is defined however in terms of underlined uncondensed counterions only. Considering the case of a polyelectrolyte where each monomeric unit of length A bears an elementary charge, this implies $\kappa^2 = 4\pi Q(Z^\dagger C_m N_{Av}/Z) = 4\pi A N_{Av} C_m$ for $\lambda > 1$. Odijk now distinguishes several concentration regimes in which only the two lowest ones correspond to those discussed by De Gennes et. al.[40]. Below C_m^* the polyelectrolyte is assumed to be an extended rigid chain implying $L_t \simeq L_e >> \ell$. Near C_m^* the macroion lattice postulated by De Gennes et. al. will start to deform as the growing flexibility of the chain (i.e., the decrease of L_e with increasing C_m) will determine the onset of macromolecular overlap. At concentrations above C_m^* the chains are not yet very flexible; the correlation length ξ expressing the average distance between chains is still very small compared to L_e, as in lattice, and the usual blob model cannot be applied. Above another critical concentration C^{**} the persistence length has sufficiently decreased to satisfy the condition $L_e << \xi$ and the polyelectrolyte may again be pictured as an ideal Gaussian chain of blobs as in the case of a polyelectrolyte with salt. The transition from the dilute to the semi-dilute regime at C_m^* can be described by assuming that at C_m^*, ℓ must be of the same order as $L_e \simeq L_t$ (to be evaluated using (7) and the above expression for κ^2) and ξ^t of the same order as the average interrod distance in the ordered lattice d* defined through $(d*)^2 \ell \simeq (\ell/A)$ x $C_m^{*-1} N_{Av}^{-1}$. This leads to the following definition of C_m^*

$$C_m^* \simeq (16\pi QAN_{Av}\ell)^{-1} \tag{14}$$

and the concentration dependence of the square mean radius of gyration.

$$<S^2>_{s.d} \simeq \ell(16\pi QAN_{Av})^{-1} C_m^{-1} \qquad C_m^* << C_m << C_m^{**} \tag{15}$$

This last relation follows from the requirements that $\xi(C_m)$ should be independent of ℓ and that for the semi-rigid chain the square average dimensions should scale like the square of the correlation length, i.e., $<S^2>/\ell^2 \simeq \xi^2/d*^2$. Note that the proportionality of $<S^2>$ to ℓ indicates Gaussian behaviour of the polyelectrolyte above C_m^*; in fact its average dimensions correspond to those of an ideal Gaussian chain of rigid segments with length $L_e (\simeq (16\pi QAN_{Av} C_m)^{-1})$. At much higher concentrations $C_m >> C_m^{**}$ any ordering even on a local scale has disappeared and the chains overlap considerably. From the condition $\xi >> L_e$ above C_m^{**} and the opposite one below C_m^{**}, the transition concentration can be defined as

$$C_m^{**} \simeq 0.04 (4\pi Q^2 AN_{Av})^{-1} \tag{16}$$

which is independent of ℓ. In this higher semi-dilute regime the concentration dependence of the square radius of gyration goes from $C_m^{-5/8}$ as long as $L_e >> L_p$ to $C_m^{-3/8}$ for $L_e << L_p$.

As pointed out before not too many systematic investigations of concentration effects in polyelectrolyte solutions are available, particularly of physical chemical properties which are closely related to the average dimensions of the macromolecules. Therefore it is at present difficult to assess these theoretical predictions. However the different regimes predicted by Odijk[42] for the case of polyelectrolytes without salt seem to correspond to those described by Wolff in connection with viscosity measurements[39]. For poly-electrolytes in the presence of salt we have recently found from quasi-elastic light scattering data of Na-polystyrene sulphonate solutions some evidence for the two regimes predicted by the theory[43]. Clearly it is much too soon to draw definite conclu-sions before more experimental data becomes available. One of the difficulties will be that the dilute region may not always be easily accessible because C_m^* can be rather low. For the case of polyelectrolytes in the presence of salt an increase of the ionic strength may be used to shift C_m^* upwards to some extent (for a vinylic type polyelectrolyte, A=0.25 nm and $L_p \approx 1$ nm, of molar mass 10^6 g mol^{-1} and $\ell \approx 10^3$ nm, (12) predicts $C_m^* \approx 10$ monomol m^{-3} at a 1-1 salt concentration of $C_s = 10$ mol m^{-3} and $C_m^* \approx 60$ monomol m^{-3} at $C_s = 10^2$ mol m^{-3}. For the same polyelectrolyte we find in the absence of salt $C_m^* \approx 2 \times 10^{-1}$ monomol m^{-3} and $C_m^{**} \approx 4 \times 10^1$ monomol m^{-3} according to (14) and (16) respectively). Even if these theoretical considerations will in the future turn out to have been too speculative, they certainly justify a few warnings anyway: a) measurements of physical chemical properties of polyelectrolyte solutions, particularly if they may be expected to depend on average shape and dimensions, should preferentially be performed over a wide range of concentrations. In case the range is limited by experimental difficulties b) the concentration regime should be assessed with proper care and c) extrapolations to zero concentration should be analyzed very cautiously because they not necessarily lead to values corresponding to inifinitely diluted conditions. It goes without saying that these recommendations certainly apply to electro-optical data!

The last topic we wish to discuss is the problem of the electric polarization of polyelectrolyte solutions at vanishing low electric fields (linear dielectric properties) and the directly related one of the dipole moment and electric polarizability of charged macromolecules and their counterions. These dielectric properties have been found to be very specific for polyelectrolyte solutions at very low ionic strength or in the absence of low molecular weight electrolytes, in no way comparable to what is observed for ordinary (uncharged) macromolecules or conventional electrolytes in solution both from a qualitative and quantitative point of view. Whereas for the last two systems the change in the static relative permittivity of the solvent is generally very small, polyelectrolytes at comparable concentrations exhibit often spectacularly high dielectric increments which however disappear rapidly by dielectric dispersion in a frequency range which may extend to very low frequencies indeed. Unfortunately experimental difficulties have so far imposed certain limitations both in macromolecular concentrations and ionic strength as in the frequency range explored.

Relatively high polyelectrolyte concentrations and/or high ionic strengths are generally inaccessible because of the high ohmic conductance of such systems which is often also a limiting factor for the frequency range, particularly downwards. Very low polyelectrolyte concentrations are difficult to investigate due to the lack of sensitivity of the actual measurements which also have often to be corrected for non negligible parasitic effects. Nevertheless some general features have been found which can be summarized as follows for salt-free polyelectrolyte solutions[44]. The relative permittivity of polyelectrolyte solutions exceeds the value of pure water for frequencies below 100 MHz at room temperature. Near 100 MHz in practically all cases investigated the

dielectric increment disappears indicating that at this frequency
no significant contribution of the polyelectrolytes persists to the
polarization of the solution and the polarization of the solvent is
hardly affected by the polyions. Below 100 MHz the increment decreases with increasing frequency exhibiting often two more or less
well separated dispersion regions. Of these the high frequency
one, if present, has been found to be practically insensitive to
the molar mass of the polyelectrolyte in contrast to the low frequency dispersion region. For the latter the amplitude and the
mean relaxation time strongly increase with M at constant monomolar
concentration, as is also the case for the static dielectric increment which often must be determined by an extrapolation procedure
to zero frequency. Both dispersions deviate considerably from a
simple Debye curve but these deviations are not very sensitive
regarding the nature of the polyelectrolyte, its molar mass or its
concentration. The low frequency mean relaxation time has sometimes been found to be of the same order of magnitude as the
rotational relaxation time of the complete macromolecule[44)45)].
The static dielectric increment and the maximum amplitude of the
high frequency dispersion part of the relaxation curve as well as
the two mean relaxation times have been found to be non-linear in
C_m down to values of $C_m \simeq 10^{-1}$ monomol m^{-3}. For DNA in the presence
of 10^{-3} M NaCl an increase of the specific dielectric increment
$\Delta\varepsilon_s/C_m$ with C_m has been found[46)] in contrast to results published
earlier[47)] and for what is observed in the absence of salt (or at
lower C_s) where $\Delta\varepsilon_s/C_m$ decreased with C_m both for DNA[48)49)] and
for other polyelectrolytes. Recently some evidence has been
advanced for the influence of the concentration on the occurrence
of the high frequency dispersion itself[59)51)] which may disappear
below a certain critical concentration close to C_m^+ or C^* in the
case of polyelectrolytes at very low ionic strength or in the
absence of salt. All these findings emphasize the importance of
concentration effects for the linear dielectric properties of polyelectrolyte solutions.

   From this summary it must be clear that we have to deal here
with a very complex problem. For solutions of uncharged macromolecules the contribution of the polymers to the polarization of
the solutions is thought to arise from the permanent dipole moment
and the polarizability of the chains themselves. In view of the
completely different behaviour, both qualitatively and quantitatively, of highly charged macromolecules in aqueous solutions the
importance of these factors in determining their dielectric properties cannot be decisive. In ordinary electrolyte solutions
several effects seem to contribute. The one predicted by Debye and
Falkenhagen[52)], due to the perturbation by the applied electric
field of the ion distribution, also leads to dielectric increments.
It has only been observed at relatively low concentrations (< 20
moles m^{-3} for 1-1 electrolytes) where the increments are very
small[53)]. At higher concentrations it is completely overshadowed

by one or two other effects involving solvent-ion interactions and
leading to dielectric decrements (i.e., lowering of the relative
permittivity of the solution with respect to the solvent): the
kinetic polarization deficiency effect[54] and the dielectric saturation effect[55]. From a theoretical point of view these three
effects must be interrelated but it has not yet been possible to
propose a unified theory from which each of them can be derived.
For polyelectrolyte solutions analogous effects may be thought to
be of importance too but up to now contributions involving the
solvent have to my knowledge not been taken into account. In most
theoretical approaches the dielectric increment of polyelectrolyte
solutions has been related to the mobile ion distribution around
the polyions. Except for a recent attempt by Fixman[56] in which
the total relative motion of the mobile ions with respect to the
polyions was taken into account but under conditions not corresponding to those where most experimental results have been obtained
(low or vanishing salt concentration) they are essentially based on
a two-phases model. Unfortunately all theories presented so far
refer to systems in which macroion-macroion interactions can be
neglected whereas it is far from certain that the experimental data
have been obtained under such conditions. As pointed out before[51]
extrapolation of these data to zero concentration, e.g., by the
empirical procedure proposed previously[57], not necessarily would
yield dielectric data characteristic for the infinitely diluted
situation. This makes the comparison between theory and experiment
rather difficult so that it is not easy to decide whether one
approach is definitely better than another. Therefore we shall
limit this discussion to a few characteristics of the two-phases
model approach only. In practically all cases the polyelectrolyte
is assumed to be stretched and rigid; each macromolecular region
is represented either by a cylinder or by a prolate ellipsoid with
a uniform charge density along the axis or smeared out on the surface of the macromolecule.

Two mechanisms essentially have been proposed to account for
the dielectric increments: a) an interfacial polarization
(Maxwell-Wagner effect)[53] b) an induced dipole moment arising from
the perturbation in the distribution of associated counterions in
the axial direction due to the applied field. In the former
approach each macromolecular region is treated macroscopically and
the associated counterions contribute to the dielectric increment
through a surface conductivity at the interface. This model cannot
in general account for very high increments as observed for large
polyelectrolyte molecules. In the second one several degrees of
sophistication may be introduced but in general internal field
effects are neglected as well as contributions arising from non-
associated counterions, an assumption which cannot be justified in
a simple way. The instantaneous value of the dipole moment of the
macroion and its Z counterions is given by

$$\vec{m}(t) = q \sum_{i=1}^{Z} f_i(t) \, r_i(t) \, \vec{1}_\ell \tag{17}$$

Here r_i gives the position of counterion i along the axis of the polyelectrolyte the orientation of which is defined through the unit vector $\vec{1}_\ell$ and f_i is the association factor which indicates whether i is associated ($f_i=1$) or not ($f_i=0$). To evaluate the contribution of the polyelectrolytes to the polarization of the solution this moment has to be averaged in the presence of the external, vanishingly small, electric field (taking into account that r_i is limited to the, range $-\ell/2 \leq r_i \leq \ell/2$ where ℓ represents the length of the rigid macroion) and summed over the numbers of polyions per unit volume. In the most simple approach interactions between associated counterions are completely neglected and the equilibrium between associated and non-associated counterions is assumed not to be affected by the electric field even not locally[59][60]. This leads to a polarizability and a specific static dielectric increment proportional to ℓ^2 and to $\bar{f}Z$, the average number of associated counterions participating in the effect.

$$\Delta\varepsilon_s/C_M \simeq (q^2 \, \bar{f} \, Z \, \ell^2)/36kT\varepsilon_o \tag{18}$$

In as far as Z is itself proportional to ℓ, the specific increment should show a ℓ^3 depencence. Of course physically speaking the neglect of counterion repulsion along the polyion axis is not very realistic although it may be justified so some extend by cancellation due to other (neglected) effects. Explicit introduction of such interactions has been attempted in various ways[61-67] and generally lowers the absolute value of the polarizability but not necessarily invalidates the ℓ^3 dependence[68] although it must lead to saturation effects. In those approaches often additional parameters are needed and simplifying assumptions about the interaction potential between associated counterions may raise some questions. Quite recently the model has been extended by distinguishing between associated counterions close to the macroion's surface and those further away and considering the possibility of exchange depending on local concentrations in the longitudinal direction[69]. Such an approach seems promising as it brings the two-phases model closer to a more general treatment of dielectric properties analogous to the one proposed by Fixman[36]. In fact a similar attempt by one of my coworkers[70] but using more drastic assumptions leads to a functional dependence of the polarizability on ℓ which strongly resembles the one predicted by Fixman (although this is probably coincidental), changing from a ℓ^3 dependence for relatively low ℓ to a ℓ dependence for realtively large dimensions.

Except for the interfacial polarization model[58] and the model discussed by Meyer and Vaughan[69] all the other have in common that

they can predict only one single dispersion region with a mean
relaxation time determined either by longitudinal counterions distribution relaxation or the rotational relaxation of the rodlike
macroion (or by both). For the first two models mentioned the high
frequency dispersion results from counterion motion around the
cylindrical polyelectrolyte. Another possibility for explaining
that part of the relaxation curve is the assumption that also a
transverse component of the induced electric moment should be taken
into account[66]. None of these mechanisms can however justify the
experimental observations in a consistent way. In particular they
cannot account for the absence of the high frequency dispersion in
the case of really rigid charged particles, such as viruses, and
very low molecular weight polyelectrolytes at very low concentrations[51].

A few years ago with Van der Touw[60] we have proposed a theory
in which a flexible polyelectrolyte is no longer treated as a
rodlike particle but is represented as a sequence of rigid rodlike
subunits. The length b of the subunit corresponds to the average
distance over which an associated counterion could fluctuate more
or less freely in axial direction without being opposed by repulsive
forces arising from important local perturbation in the equipoten-
tial surface on which the counterion moves. In this model both
dispersion regions are predicted and the molecular weight independence of the high frequency dispersion could be understood.
According to this theory the following simple relation should hold

$$\frac{\Delta\varepsilon_s}{\Delta\varepsilon_2} \simeq \frac{<s^2>}{b^2} \tag{19}$$

where $\Delta\varepsilon_2$ is the maximum amplitude of the high frequency dispersion
region. It was found to explain semi-quantitatively a large number
of dielectric results in a consistent way for those systems in
which the polyelectrolyte not necessarily has a rodlike conformation[57][44][51] but has met. some criticism on the grounds that the
model was not very realistic[66][69]. However the recent developments concerning concentration effects, as exposed above, lend some
credit to such a model for polyelectrolytes at not too high dilutions for which b may be assumed to be closely related to the
correlation length ξ[42][51]. Even in the case of very rigid
elongated particles, such as short fragments of DNA at very low
ionic strength, potential barriers opposing the fluctuation of
associated counterions could arise from macroion-macroion interactions above a certain critical concentration where overlap of
the polyelectrolyte particles must occur in an isotropic phase[51].
Of course the model itself remains very crude with too many simplifying assumptions but in view of our present knowledge it may be
worthwhile to attempt further refinements in the future to include

some of the effects neglected so far. This should not exclude other
theoretical approaches to be pursuited. On the contrary! Our
understanding of the dielectric properties of polyelectrolyte solutions has not yet reached a level which would justify theoretical
efforts to be limited to a well defined direction and too many
factors seem to be involved to let one simple model explain all the
effects observed in a satisfactory way.

Extrapolation of these conclusions to problems arising in
connection with electro-optical properties of polyelectrolyte
solutions seems to be justified. Analogous phenomena as in the
case of dielectric properties are involved here with additional
complications arising from the higher field strengths used[71].
Although good arguments have been presented for making the ionic
polarizability of the macroions primarily responsible for the
electro-optical effects observed there are many experimental facts
which are still difficult to understand[72-74] as is the case for
the dielectric properties and probably for the same reasons.

REFERENCES

1. The following books give reviews of the physical chemical
       properties of polyelectrolyte solutions.
       a. S. A. Rice and M. Nagasawa, "Polyelectrolyte Solutions,"
           Academic Press, London, New York, 1961.
       b. F. Oosawa, "Polyelectrolytes," Marcel Dekker, New York,
           1970.
       c. H. Eisenberg, "Biological Macromolecules and Poly-
           electrolytes in Solution," Clarendon Press Oxford, 1976.
       d. E. Sélegny, M. Mandel and U. P. Strauss eds., "Poly-
           electrolytes," D. Reidel, Dordrecht-Boston, 1974.
       e. C. Tanford, "Physical Chemistry of Macromolecules,"
           Wiley, New York, 1961.
2. a. D. Dolar, in ref. 1.d., p. 97.
    b. S. Oman, Makromol. Chem. 178:475 (1977).
3. Z. Alexandrowicz, J. Polym. Sci. 43:337 (1960).
4. For most recent determinations of $\gamma_{\pm}$ see papers by Kwak and
       coworkers
       a. J. C. T. Kwak, J. Phys. Chem. 77:2790 (1973).
       b. J. C. T. Kwak, M. C. O. Brien and D. A. MacLean, J.
           Phys. Chem. 79:2381 (1975).
       c. Y. M. Joshi and J. C. T. Kwak, J. Phys. Chem. 83:1978
           (1979).
5. M. Nagasawa and A. Takahashi in "Light Scattering from Polymer
       Solutions," M. B. Huglin ed., Academic Press, London,
       New York 1972, p. 671.
6. See e.g., A. Katchalsky, Pure and Applied Chemistry, 26:327
       (1971).
7. M. Nagasawa in ref. 1.d., p. 57.

8. M. Mandel in ref. 1.d., p. 39.
9. S. Lifson and A. Katchalsky, J. Polym. Sci. 13:43 (1953).
10. A. Katchalsky, Z. Alexandrowicz and O. Kedem in "Chemical
    Physics of Ionic Solutions," B. E. Conway and R. G. Barradas,
    eds, Wiley, New York, 1966, p. 295.
11. L. Onsager, Ann. N.Y. Acad. Sci. 51:627 (1949).
12. a. L. Kotin and M. Nagasawa, J. Chem. Phys. 36:873 (1962).
    b. L. M. Gross and U. P. Strauss in "Chemical Physics of Ionic
        Solutions," B. E. Conway and R. G. Barradas eds., Wiley,
        New York, 1966, p. 361.
13. D. Stigter, J. Colloid Interf. Sci. 53:296 (1975).
14. D. Stigter, Progr. Colloid and Polymer Sci. 65:45 (1978).
15. M. Gueron and G. Weisbuch, Biopolymers 19:353 (1980).
16. M. Fixman, J. Chem. Phys. 70:4995 (1979).
17. a. G. S. Manning, J. Chem. Phys. 51:924,934,3249 (1969).
    b. G. S. Manning, Ann. Rev. Phys. Chem. 23:117 (1972).
    c. G. S. Manning in ref. 1.d., p.9.
18. The notion of counterion condensation has been introduced by
    Imai.
    a. N. Imai and T. Ohinshi, J. Chem. Phys. 30:1115 (1959).
    b. See ref. 1b.
19. A. D. MacGillivray, J. Chem. Phys. 56:80.83 (1972).
20. a. A. K. Iwasa, J. Chem. Phys. 62:2967 (1975).
    b. A. K. Iwasa, J. Phys. Chem. 81:1829 (1977).
21. K. Iwasa and J. C. T. Kwak, J. Phys. Chem. 80:215 (1976).
22. K. Iwasa and J. C. T. Kwak, J. Phys. Chem. 81:408 (1977).
23. K. Iwasa, D. A. McQuarrie and J. C. T. Kwak, J. Phys. Chem.
    82:1939 (1978).
24. a. G. S. Manning, Biophys. Chem. 7:95 (1977); 9:65 (1978).
    b. G. S. Manning, Quart.-Rev. Biophys. 11:179 (1978).
25. I. Noda, T. Tsuge and M. Nagasawa, J. Phys. Chem. 74:710 (1970).
26. O. Kratky and G. Porod, Rec. Trav. Chim. Pays-Bas 68:1106
    (1949).
27. See e.g., H. Yamakawa, "Modern Theory of Polymer Solutions,"
    Harper and Row, New York, 1971.
28. J. Skolnick and M. Fixman, Macromolecules 10:844 (1977).
29. a. T. Odijk, J. Polym. Sci. (Polymer Phys. Edn.) 15:477 (1977).
    b. T. Odijk, Polymer 19:989 (1978).
30. H. Benoit and P. Doty, J. Phys. Chem. 57:958 (1958).
31. T. Odijk and A. C. Houwaart, J. Polymer Sci. (Polymer Phys.
    Edn) 16:627 (1978).
32. M. Fixman and J. Skolnick, Macromolecules 11:863 (1978).
33. D. Stigter, Biopolymers, 16:1435 (1977).
34. T. Odijk, Biopolymers 18:3111 (1979).
35. M. Mandel and J. Schouten, Macromolecules 13 (1980) to appear.
36. T. Odijk and M. Mandel, Physica 93:298 (1978).
37. J. O. Bernal and I. Fankuchen, J. of Gen. Physiol. 25:111
    (1941).
38. E. Iizuka, Y. Kardo and Y. Ukai, Polym. J. 9:134 (1977).

39. C. Wolff, J. Phys. (Paris) Colloq. 39, C-2:169 (1978).

40. P. G. de Gennes, P. Pincus, R. M. Velasco and F. Brochard,
     J. Phys. (Paris) 37:1461 (1976).

41. a. M. Daoud, J. P. Cotton, B. Farnoux, G. Jannink, G. Sarma,
         H. Benoit, R. Duplessix, C. Picot and P. G. de Gennes,
         Macromolecules 8:804 (1975).

     b. P. G. de Gennes, "Scaling Concepts in Polymer Physics,"
         Cornell University Press, Ithaca, London, 1979.

42. T. Odijk, Macromolecules 12:688 (1979).

43. R. S. Koene, H. E. J. Smit and M. Mandel, Chem. Phys. Letters
     (1980) to appear.

44. See for a summary of some experimental results:
     a. M. Mandel and Van der Touw in ref. 1.d , p. 285.
     b. M. Mandel, Ann. N.Y. Acad. Sci. 303:74 (1977).

45. M. Sakamoto, R. Hayakawa and Y. Wada, Biopolymers 17:1507 (1978).

46. M. Sakamoto, R. Hayakawa and Y. Wada, Biopolymers 18:2769 (1979).

47. S. Takashima, J. Phys. Chem. 70:1372 (1966).

48. M. S. Tung, R. Molinari, R. H. Cole and J. H. Gibbs,
     Biopolymers 16:2653 (1977).

49. T. Vreugdenhil, F. van der Touw and M. Mandel, Biophys. Chem.
     10:67 (1979).

50. S. Zwolle, Thesis, Leiden 1978.

51. M. Mandel in "Dynamic Aspects of Biopolyelectrolytes and
     Biomembranes," (Kyoto Symposium, 1978), F. Oosawa, ed.
     To be published.

52. See i.a. H. Falkenhagen, "Theorie der Elektrolyte," 2nd ed.
     Hirzel, Leipzig, 1971.

53. W. M. van Beek and M. Mandel, J. Chem. Soc. Faraday Trans. I.
     74:2339 (1978).

54. a. J. B. Hubbard, L. Onsager, W. M. van Beek and M. Mandel,
         Proc. Nat. Acad. Sci. 74:401 (1977).

     b. J. B. Hubbard and L. Onsager, J. Chem. Phys. 67:4850 (1977).

55. See e.g., J. B. Hasted, "Aqueous Dielectrics," Chapman and
     Hall, London 1973.

56. M. Fixman, preprints (These papers bear some resemblance to the
     theory presented in S. S. Dukhin and V. N. Shilov,
     "Dielectric Phenomena and the Double Layer in Disperse
     Systems and Polyelectrolytes," Wiley, New York and Keten
     Publishing House, Jerusalem, 1974).

57. F. van der Touw and M. Mandel, Biophys. Chem. 2:231 (1974).

58. C. T. O'Konski, J. Phys. Chem. 64:605 (1960).

59. M. Mandel, Mol. Phys. 4:489 (1961).

60. F. van der Touw and M. Mandel, Biophys. Chem. 2:218 (1974).

61. J. P. McTague and J. H. Gibbs, J. Chem. Phys. 44:4295 (1966).

62. F. Oosawa, Biopolymers 9:677 (1970); see also ref. 1b.

63. J. M. Schurr, Biopolymers 10:1371 (1971).

64. A. Minakata, N. Imai and F. Oosawa, Biopolymers 11:347 (1972).

65. A. Warashina and A. Minakata, J. Chem. Phys. 58:4743 (1973).

66. A. Minakata, Ann. N. Y. Acad. Sci. 303:107 (1977).

67. G. S. Manning, Biophys. Chem. 9:65 (1978).

68. G. Weill and C. Hornick, ref. 1.d., p. 277.

69. P. I. Meyer and W. E. Vaughan, Biophys. Chem. To appear.

70. W. van Dijk, F. van der Touw and M. Mandel. To be published.

71. S. Sokerov and G. Weill, Biophys. Chem. 10:161 (1979).

72. C. Houssier, J. Bontemps, X. Edmonts-Alt and F. Fredericq, Ann. N.Y. Acad. Sci. 303:107 (1977).

73. M. Tricot, C. Houssier, V. Desreux and F. van der Touw, Biophys. Chem. 8:221 (1978).

74. E. Charney, Biophys. Chem. 11:157 (1980).

ELECTRO-OPTICAL INSTRUMENTATION SYSTEMS WITH THEIR DATA ACQUISITION AND TREATMENT

Claude Houssier

Laboratoire de Chimie Physique
Université de Liège (B6)
Sart-Tilman, B-4000 LIEGE (Belgium)

and

Chester T. O'Konski

Department of Chemistry
University of California
Berkeley, CA 94720

INTRODUCTION

The general principles of optical and photoelectronic detection schemes (Fig. 1) are common to all electro-optical instrumentation and have been described in a number of reviews on this subject (1-12). An electric field is applied to the sample, usually in the form of a short duration signal, for example, a single rectangular or bipolar pulse, a sinusoidal pulse train, or a condenser discharge. This results in a transitory perturbation of the sample. The signal, due to the reorientation of the particles, deformation or conformational changes caused by the electric field, is monitored through changes in one or more of several optical properties: absorbance (electric dichroism effect), refractive index (electric birefringence, double refraction, or Kerr effect), fluorescence (fluorescence polarization or intensity), light-scattering, Raman scattering, optical rotation, circular dichroism. Usually, the transient optical signal is detected by a photomultiplier which converts it into an electric signal which is analyzed after suitable amplification through an appropriate detection circuit. The time scales readily covered by these techniques extend, with conventional instruments, down to the microsecond range, and, with fast systems, down to the nanosecond range.

ELECTRO-OPTICAL INSTRUMENT SETTING

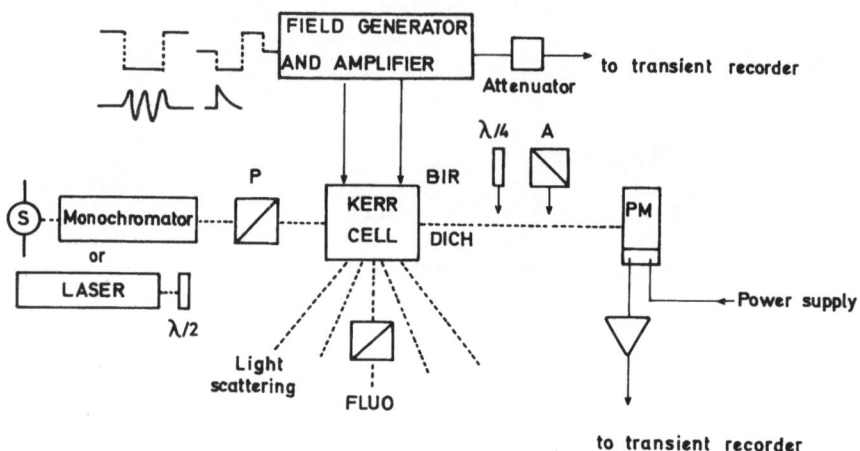

SIGNAL PROCESSING

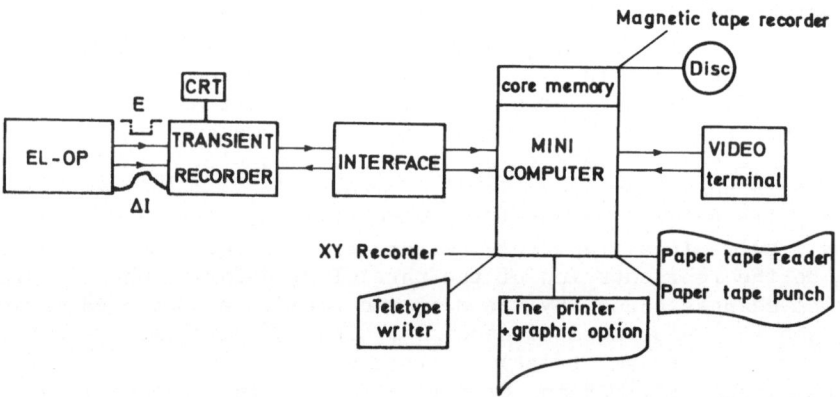

Fig. 1.   General scheme of electro-optical instruments with their
          signal processing system.

Pulsed electro-optics started in the late forties when pulse generators and oscillographic displays became available. Up to the seventies, the transient electro-optical signals were analyzed from photographic recordings of the oscilloscopic traces. With the development of fast transient recorders in the early seventies, the direct recording of the signals in digital form became possible with the concomitant potentiality for their processing through computerized systems (13-15).

This paper describes the present stage of development of electro-optical instrumentation in its different parts (from the light source to the signal acquisition and treatment) for the purpose of helping new experimenters in the design of their own apparatus.

COMPONENTS AND PROCEDURES

A number of factors should be considered when making choices for the most appropriate arrangement of a given electro-optical instrument. A few spectroscopic systems now available commercially are specifically designed for recording fast transient signals: for information consult DiaLog, 4000 Düsseldorf, Deutschland, and Applied Photophysics Limited, London W1X 3HA, U.K. They sometimes offer special accessories which are adaptable for electro-optical measurements. The following comments and recommendations are made in connection with the selection of each component of the system.

Light Source

Whenever a wide wavelength accessibility is desired, as for dichroism and fluorescence measurements, Xe or Xe-Hg light sources associated with a high luminosity monochromator appear appropriate; this in fact corresponds to the present choice made by manufacturers of spectrofluorimeters, circular dichrographs, etc. High stability power supplies are available for these arc lamps (Electro Powerpacs Corp., Cambridge, MA 02139; Schoeffel, Westwood, NJ 07675; Oriel Corp., Stamford, CT 06902). Additional improvement of stability may be attained, if required, by switching after arc ignition, to a very stable D.C. power supply (e.g., Hewlett-Packard models 6267B or 6268B) or to a battery (16). Arc "travelling" may be reduced by properly positioning a magnetic stirring bar in the vicinity of the arc lamp, but this appears to be a tricky adjustment.

In spite of the important progress made during recent years in the technology of lasers, particularly with the availability of tunable lasers, their applicability to electro-optical instruments remains limited mainly to light-scattering and Raman scattering, and only to some limited extent to fluorescence and birefringence (17,18).

In the latter case, the sensitivity achieved by the converging effects of various factors (no requirement for monochromatic light, possibility of increasing pathlength and/or sample concentration, measurement by a nearly null method) makes it usually possible to design a good electric birefringence setting with the aid of a simple projector lamp (tungsten-iodine lamp).

When using a laser as light source, a helium-neon laser (e.g., Spectra-Physics, Model 125) providing the red 632.8 nm line is suitable, since it allows one to study dye-complexes without superposition of dichroism and fluorescence effects to the birefringence or light-scattering signals. Neutral density filters are used to attenuate the incident intensity to the appropriate level.

<u>Optical Arrangement</u>

Electro-optical measurements usually require a polarized incident light beam. Polarizers with high transmittivity in the wavelength range being investigated and with low crossed transmittance, are recommended. Glan-Thompson or Glan prisms with air layers (Bernhard Halle Nachfl., Berlin; Karl Lambrecht, Chicago) are satisfactory. Crossed transmittance should be lower than 10^{-4}. Polacoat sheets may also be used but their specifications are not as good (3M Industrial Optics, St. Paul, MN 55101) allowing as much as 0.5% stray-light to be transmitted in the crossed position. It is advantageous to have precise mounting in graduated holders (adjustable to within 0.02° to 0.1°) to give the potential for various types of measurements and for the necessary controls of the instrument settings (see below). For birefringence measurements, the polarizer is set at 45° to the field direction and a quarter-wave plate precedes the rotatable analyzer to improve the sensitivity of the measurement and to allow the determination of the birefringence sign (7,50). For dichroism, fluorescence polarization and light-scattering measurements, the incident beam is polarized parallel (or perpendicular) to the electric field. In principle, when using a laser light source, the polarizer is not strictly required. However, a $\lambda/2$ plate and a polarizer are desirable to allow an easy adjustment of the incident polarization direction, and to correct instabilities in orientation of polarization.

It was then shown that photomultiplier sensitivity can be optimized in measuring birefringence by using a $\lambda/4$ plate or prism after the Kerr cell, and rotating the analyzer a few degrees off the crossed position; this also has the advantage of introducing a polarity in the intensity change to indicate sign of birefringence, in contrast with the conventional polarizer/cell/crossed analyzer Kerr cell arrangement (50). The derivation for the optimum angle, α_m, of rotation off the crossed position gave $\sin \alpha_m = K_{SL}^{1/4}$, where K_{SL}

is the stray light constant of the optical system with cell and solution in place. That is, $I_{SL} = K_{SL} I_0$, where I_{SL} is the leakage intensity with analyzer in the crossed position and I_0 is the intensity when the analyzer is aligned with the polarizer. The derivation assumes a very stable light source, for example, tungsten operating off a battery-charger combination, or an extremely well regulated supply.

With laser sources, there is the disadvantage of appreciable high frequency instability along with the advantage of higher available light intensity. The more parallel light beam and current improvements in cell windows are allowing lower K_{SL} values, currently around 10^{-6} or less (private communications, J. C. Bernengo and D. Eden). Then the optimization of the signal-to-noise ratio may be based on the practical limitation that the maximum signal intensity has to be consistent with a linear response of the detector. The equation for optimization under this condition has been derived (17) and it shows that for very weak signals ($\delta^2 \ll K_{SL}$) the optimum setting corresponds to $\sin \alpha = K_{SL}^{1/2}$. Concomitantly, some lasers (especially the He/Ne plasma tubes) have been stabilized to around $\pm 0.01\%$ (Bernengo, J. C., private communications). With $K_{SL} \simeq 10^{-6}$, one sets $\sin \alpha \simeq 10^{-3}$. Together with signal averaging (see below) these improvements can lead to a far broader range of applications in future years.

For the dichroism, one may use an arrangement with a Wollaston prism transmitting two beams of orthogonal polarization thus allowing the direct measurement of $A_{||} - A_{\perp}$ from the signals recorded by two photomultipliers (19). This goal may also be reached by the use of photoelastic modulators and a phase sensitive detection circuit (20). These systems, however, yield increased instrument complexity and significant reduction in versatility.

Kerr Cell

Various Kerr cell designs with parallel platinum or gold electrodes have been described for the different types of electro-optical techniques: these include (i) horizontal electrodes for fluorescence and light-scattering measurements (11,21,22); (ii) conducting transparent vertical electrodes (with thin stannous oxide film) for fluorescence and longitudinal dichroism studies (11,23-25); (iii) vertical electrodes for transversal dichroism and birefringence measurements (7,9,11,24,26); and (iv) several special designs for improved sensitivity (11,23-25). A teflon or PTFE holder is most

appropriate for the electrodes which can be conveniently set at a
separation of 1.5 to 3 mm. Distances smaller than 1 mm are difficult to work with for a number of reasons: light beam adjustment,
spark breakdown, bubble holdups, etc.

Special care should be taken in the choice of the quartz cuvette
in which the electrodes are fitted (or of the cell windows) in order
to achieve a very low strain birefringence. Cylindrical windows
seem to give lower strain than rectangular ones.

The Kerr cell may be installed in a thermostatable holder or
directly designed to allow water circulation.

Great care has to be taken in the filling of the cell with the
sample, in order to avoid the presence of air bubbles which favour
sparking at high fields. Solution degazing is highly suitable (if
not indispensable) when using fields above 14-15 kV/cm, especially
when condenser discharges generate the orienting field.

Cells designed for temperature-jump measurements may also be
used but the electrode spearation is usually about 1 cm. This
limits the field strength range covered, unless a hard-pulser or
condenser-discharge circuit is used.

Field Generation

Many wiring diagrams of single rectangular (2,7), bipolar (27),
and sinusoidal pulse generators have been presented in the literature and have been extensively referenced (1,2,7,9,10,12).

It appears to us that, as a general rule, field strengths of
10 to 20 kV/cm should not be exceeded in order to avoid secondary
effects of the field (such as electrolysis, electrophoresis, sparking, heating, perturbations of the particle structures (28), e.g.,
strand separation of polynucleotides (28,29) or nucleic acids (30)).
Studies at ionic strengths up to 0.005-0.01 using relatively long
duration pulses, such as are obtained with commercially available
generators (Cober Electronics, Stamford, CT 06902) may be made,
provided that small electrode surfaces are employed. Fields of up
to 10 to 20 kV/cm can then still be utilized. These generators de-
liver single pulses at 2.5 kV and 12.5 A. With two such generators,
single pulses at 5 kV, 6.2 A, or bipolar pulses at 2.5 kV, 12.5
A, with rise, transition and decay times lower than 1 μs can be
obtained. The manufacturer's specification is ∿30 ns transition
rates but these can only be attained for a very careful Kerr cell
connection via suitable impedance matching using noninductive resistors.

The use of sinusoidal fields is of very appreciable interest for the elucidation of the orientation mechanism, particularly in relation to dielectric studies. Such studies, however, require coverage of a very wide frequency range and have so far been scarcely undertaken. Generally, relatively long bursts of a sinusoidal field of low field strength are applied to the samples (31) so that solutions of very low conductivity must be used. Pulsed sinusoidal fields of short duration are more useful but have been rarely used (11,32,51a).

The use of a condenser discharge to orient the solute particles (33,51a) presents a number of serious inconveniences, even though such generators are readily available on temperature-jump instruments and enable one to reach very high fields (30-40 kV/cm) at relatively high conductivity. Significant heating of the sample solution cannot be avoided; the temperature rise is only a function of the capacitance and voltage used, and of the sample volume; the decay time of the discharge is shorter the higher the conductivity of the sample. The true steady-state cannot be reached. In addition, the relaxation analysis must be performed from the rise of the signals so that its analysis is dependent on the knowledge of the orientation mechanism.

Signal Detection

Many choices are possible for the devices used as light detectors. High sensitivity photomultipliers (e.g., EMI 9558Q) are only required for fluorescence and light-scattering detection. S-5 or S-20 response photomultipliers (e.g., 1P28 RCA or Hamamatsu) are most commonly used on modern spectrophotometers, spectrofluorimeters or dichrographs and are suitable for most electro-optical instruments. In the case of transmission optical arrangements with high intensity sources, improvements of the signal-to-noise have been achieved by operating the detector at a low number of dynodes (34). In birefringence instruments with laser sources, the light intensity reaching the detector is high enough to use photodiodes which may provide higher signal-to-noise ratios and faster responses.

The design of fast response detection circuits requires care to reduce stay capacitances and to use low anode resistance values. For response times down to 0.1 μs, conventional amplifiers may be used; solid state operational amplifiers with D.C. to 100 MHz bandwidths are available (e.g., Analog Device 50J/K). The output has to be of low impedance and proper transmission line terminations must be used to prevent line ringing effects with very short pulses.

The amplification of the small transient signal with respect to the intense D.C. signal in the absence of orienting electric

field requires a annihilation of this D.C. voltage. This can be
done either by inputting a D.C. signal of opposite sign or by using
a reference signal from a second detector on which part of the incident beam is deviated by a beam-splitter. This second procedure
is expected to allow compensation for incident intensity fluctuations. In this case, it is recommended to filter this signal as
much as possible in order to avoid adding noise from the two detector signals which are independent from each other.

Signal Recording

Direct recording of fast transient signals down to the µs range
is achievable with several commercially available transient recorders
(Biomation, Palo Alto, CA 94303; Datalab, Mitcham, CR4 4HR Surrey,
U.K.; Bruker Scientific Inc., Palo Alto, CA 94304; Intercomputer
Electronics Inc.; Tektronix Inc.; Lecroy; Vuko, Elektronische
Geräte GmbH, D-6053 Obertshausen, Germany). With these systems,
1024 to 2048 data points are usually recorded at a maximum amplitude
resolution of 8 bits (0.4% accuracy). Dual channel transient recorders are recommended for the simultaneous recording of both the
attenuated electric pulse and the photomultiplier signal. The
availability of a pretrigger mode is highly advantageous and permits
the recording of the light signal baseline prior to the pulsed signal. For the purpose of reaching the ns range, oscillographic display with the subsequent photographic step still seems to be required (35).

Signal Treatment

Many different approaches are possible depending on the actual
arrangement of the electro-optical instrument.

a) If the signal recording is made only through an oscilloscope display, the processing of the data is generally made from
photographic enlargement of the oscilloscope traces. The calculation of the steady-state amplitudes may then be easily performed
with a simple hand calculator using the relationships given below.
If a more complete analysis of the electro-optical signals (rise
and/or decay) is desired, a more or less automatic digitization of
the traces can be made either directly on the pictures or preferably
after manual smoothing at the enlarger. The computer processing of
the digitized data can then be made as described in c) below. A
visual comparison with electronically simulated curves, on a storage
oscilloscope or a video graphic system, is an alternative, relatively fast and easy way of analyzing decay signals (15,36).

b) A partially automatic processing of transient signals may be realized by storage of the transient recorder data on magnetic or paper tapes for further computer treatment (13,15).

c) Minicomputers now have become readily available for direct access use in the laboratory instrumentation. Properly interfaced and programmed minicomputers, with transient recorder storage of the signals, allow automatic on-line processing of the experiments (14,15) at various levels of sophistication: simple signal amplitude calculations, relaxation time analysis, determination of optical and electrical parameters from a fitting of the field strength dependence of the electro-optical signals, signal averaging and smoothing. The time required to obtain the final result after signal recording ranges from less than one second for simple amplitude determinations, to several minutes or more for more detailed analysis when there is really complex treatment.

Most of the minicomputer systems presently available commercially (Hewlett-Packard, Tektronix, Digital Equipment PDP, Data General, Nicolet, Modular Computer Systems) offer comparable capabilities as regards speed, memory size, ease of programming, etc. They use BASIC or FORTRAN languages. We previously discussed the problem of memory size for such a system (14); a 24 to 32 K core is required to allow the use of high level programming languages and adequate facilities for data treatment. The input/output subroutines must however be written in Assembler language.

Such huge developments in mini- and micro-computer systems are now in progress that it seems impossible to give specific advice for making a definite choice. Microsystems such as Apple II, PET Commodore, Tandy, etc., have already reached capabilities comparable to those offered by the minicomputers of 5 to 10 years ago. The use of FORTRAN as the programming language is highly desirable since it allows appreciable improvements in execution times. This is particularly suitable for relaxation analysis on a large number of points and for curve fitting. Micro-computer systems generally use only BASIC as the programming language at the present time however.

We consider that the minimum composition of the system should be: 24 to 32 K of core memory; A/D and D/A interfaces for connecting transient recorder, oscilloscope and XY recorder; auxiliary memory (disc, magnetic tape or floppy disc); printing unit; video and keyboard unit.

Smoothing noisy electro-optical signals before treatment could be made by a wide variety of methods described in the literature for spectral data smoothing. In the case of simple amplitude calculations, smoothing does not appear to be useful. The steady-state amplitude determination with respect to the baseline preceding the electric pulse is better obtained by averaging a certain number of the real data point.

In the case of the relaxation curve analysis, the use of the 1,000 to 2,000 data points recorded on the transient recorder would make the running of the fitting program quite slow so that 100 to 150 data points are obtained by preprocessing the data to represent the decay or rise curves (see below under the Berkeley instrument) or the transistory signals are characterized by appropriate parameters (see below).

Using signal averaging procedures to improve the signal-to-noise ratio requires repetitive pulsing of the sample solution. The procedure has been applied to low field measurements (17,18). Care has to be taken to minimize electrode polarization effects and to avoid electrophoresis. Therefore, a train of sinewaves or of pulses of alternating signs, or accurately antisymmetric bipolar pulses should be used.

AN ELECTRO-OPTICAL INSTRUMENT AT LIEGE

We shall describe here the electric birefringence and dichroism apparatus (with optional fluorescence capability) at the Laboratoire de Chimie Physique of the University of Liège which put in concrete form choices related to those discussed above.

Optical Arrangement and Detector Circuits

Electric birefringence. The electric birefringence instrument is essentially similar to that previously described (7) except for the following improvements:

a) A 24V/150W tungsten-iodine lamp with a well-regulated power supply is used as light source in conjunction with a Zeiss prism monochromator.

b) The polarizer and analyzer are mounted on rotatable holders adjustable to within 0.1° and 0.01°, respectively, in order to allow proper adjustment of the analyzer angle (7), control of the instrument geometrical setting (see below) as well as simultaneous measurement of dichroism and birefringence on absorbing solutions (37). A $\lambda/4$ plate is present, mounted on the analyzer holder.

c) The response time has been lowered below 1 μs (see control of instrument setting); the amplifier circuit uses a 50J wideband differential FET amplifier (Analog Devices, Norwood, MA 02062). Relaxation time determinations are made from electric birefringence signals only, owing to the faster response of this detection circuit on our instrument.

   Electric dichroism. For the electric dichroism measurements
we now use a high sensitivity temperature-jump instrument (optical
part from DiaLog, 4000 Düsseldorf, Germany) with a 200W Xe-Hg light
source powered, after ignition, by a HP 6267B power supply, or by
an Electron Powerpacs Model 1152 power supply. The high luminosity
grating monochromator (Schoeffel model GM250) is usually set at input and exit slits of 3 and 1 mm, respectively, which gives a spectral bandwidth of the order of 6-8 nm.

   A rotatable Glan polarizer (orientation angle adjustable to
within 0.2°) has been installed after the beam splitter, in the
region of parallel beam geometry. The wavelength accessibility of
the instrument, with this polarizer, ranges from the visible to the
ultraviolet down to 230 nm. Even though the light beam from the 1
mm exit slit of the monochromator was focused (after polarization)
on the Kerr cell having a 1.5 mm separation between vertical electrodes, we did not notice any anomaly in the optical behaviour of
the system in the presence of the electric field (see below).

   Zeroing of the absorption photomultiplier signal was achieved
with the filtered reference signal ($\tau_c \approx$ 330 µs) from a second
photomultiplier receiving a fraction of the incident beam reflected
by the beam splitter. The two amplifier circuits of this apparatus
use the same 50J amplifiers as for the birefringence and allowed to
measure relative changes of absorbance $\Delta A/A$ down to 0.001 at absorbances up to 1-1.5; its response time is of the order of 3 µs.

Kerr Cell and Pulse Generator

   We are using the previously described Kerr cell assembly (7)
with full satisfaction; its Pt electrodes separation is 1.5 mm; the
sample volume required is around 0.5 ml. The rectangular quartz
cuvettes (Hellma 100 QS) are carefully selected with the aid of the
birefringence setting, in order to achieve strain birefringence not
exceeding 1° at 550 nm. A thermostatable cell holder is available
if required.

   Although an instrument setting for longitudinal dichroism would
have been simpler than that for transverse dichroism (23), it suffers
from several important inconveniencies: the optical pathlength is
equal to the electrode spacing unless a special cell design is used
(25) which however presents serious conductivity limitations; controls for perturbating effects of the electric field other than
orientation effects are not directly possible.

   The rectangular and bipolar electric pulses applied to our Kerr
cell are delivered by two high power generators (Cober Model 606 P;
see above) through a polarity reversing switch with facility for
impedance matching using 500Ω noninductive resistors.

Data Acquisition and Treatment

The photomultiplier signals are stored in a Biomation 8100 transient recorder (2048 words capacity; 8 bits accuracy (0.4%); fastest sampling rates available are 100 ns and 10 ns in dual and single channel operation, respectively) and displayed on a Tektronix 603 monitor. The attenuated ($\approx$ 2,000 times) electric pulse is recorded on the second channel of the transient recorder.

The transient recorder is interfaced to a mini-computer (Modular Computer Systems, Model II/10) with 24K of core memory and a number of peripheral devices: fast paper tape reader and punch, magnetic disc storage device (10^6 words, Diablo Model 4126), magnetic tape recorder (Perex Model Perifile 8041, Perex Ltd., Sintrom Group, Reading, Berks. RG2 OLS, U.K.), Teletype writer, Teletec Data-Screen video terminal and line printer with graphic capabilities (Printronix Model P150, Printronix Inc., Irvine, CA 92714). The mini-computer is also interfaced to an XY recorder allowing plotting of the signals (or any other calculated curve) in any desirable form after computer treatment.

The signals were analyzed by FORTRAN IV programs which call an Assembler Subroutine for the data transfer from the Biomation transient recorder. Although the Biomation 8100 can be set under computer control, we always use a manual setting of the various scales adjustments. Execution of FORTRAN programs is appreciably faster than BASIC programs which is an important factor when analyzing relaxation curves. A very schematic flow chart of such a program is shown in Fig. 2 and its complete listing is available upon request from C.H. (38). The successive important steps of the program are the following:
- choose program options and input required experimental parameters (wavelength, absorbance of sample, cell birefringence, solvent Kerr constant if correction desired).
- record light signal at rest, or input rest signal (read on a digital voltmeter); input analyzer angle (birefringence only) or absorbance (dichroism only).
- record electric pulse and photocurrent signal, and transfer to computer.
- localize beginning and end of electric pulse.
- calculate amplitude of electric field and steady-state amplitude of the photocurrent signal (averaging of the last 1/5th length of the signal).
- calculate amplitude of photocurrent signal for all points along the electro-optical signal (for relaxation analysis only).
- determine dichroism or birefringence steady-state value using the appropriate equations (see below).
- determine the birefringence values along the electro-optical signal, the rise and decay surfaces (giving the mean τ value) (for relaxation analysis only).

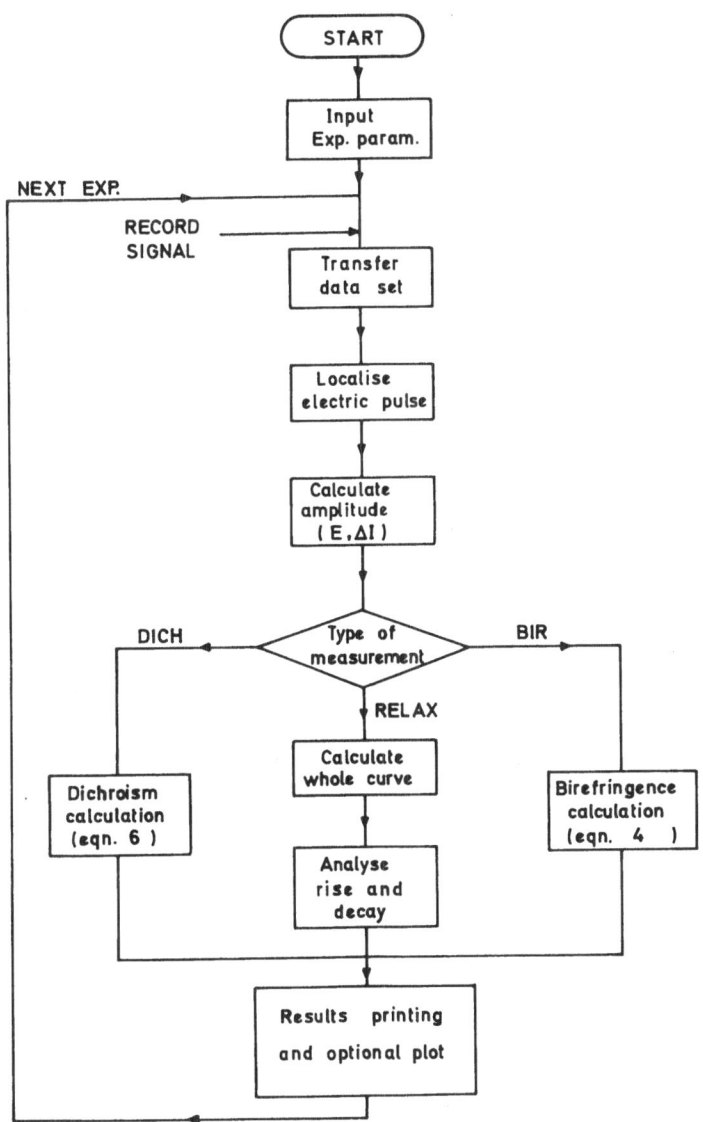

Fig. 2. Flow-chart of a schematic computer program for the treatment of electro-optical signals.

- print results (field strength and anisotropy, plus any other related quantities).
- optional plot of relaxation curve in direct or semi-log form.

Note: Until now we have analyzed the bipolar pulse signals manually from XY-recorder plots.

Optical relationships. For an optical arrangement with polar-izer at 45° and quarter-wave plate slow axis at 135° with respect to the field direction, the general expression for the calculation of the birefringence in regions of absorption where a dichroism contribution may be present, reads (37):

$$\left(\frac{\Delta I}{I_\alpha}\right)_{corr} = \frac{\exp(-\delta'/3) \{\cosh \delta' - \cos(\delta_t + 2\alpha)\}}{1 - \cos(\delta_o + 2\alpha)} - 1 \qquad (1)$$

where ΔI is the change of light intensity on the photomultiplier in the presence of the field with respect to the light intensity "at rest" I_α, for a rotation of the analyzer through an angle α from the crossed position. This relative change of light intensity is corrected for stray-light by the relation (7):

$$(\Delta I/I_\alpha)_{corr} = (\Delta I/I_\alpha)_{meas} \left\{1 + \frac{K_{SL}}{\sin^2\alpha}\right\} \qquad (2)$$

where $K_{SL} = I_{SL} \sin^2\alpha/(I_{t,\alpha} - I_{SL})$, where the stray-light constant K_{SL} (0.00025 for our present instrument setting) is determined in the absence of the cell, from the residual light intensity I_{SL} emerg-ing from the analyzer in the crossed position, and from the total light intensity $I_{t,\alpha}$ transmitted for a rotation α of the analyzer. Here, δ' is the differential absorption term: $\delta' = 2.303 \Delta A/2$ with $\Delta A = A_{||} - A_\perp$ and δ_t is the optical retardation which is the sum of solute sample (δ), solvent (δ_s) and cell (δ_o) contributions: $\delta_t = \delta + \delta_s + \delta_o$ with $\delta = 2\pi \ell\Delta n/\lambda$ and $\Delta n = n_{||} - n_\perp$.

For small δ_t, δ' and α values, equation (1) may be approxi-mated by (37):

$$(\Delta I/I_\alpha)_{corr} \sin^2\alpha = (\delta'/2)^2 + (\delta/2)^2 + (\delta/2) \sin 2\alpha \qquad (3)$$

In the absence of dichroism, equation (1) becomes:

$$(\Delta I/I_\alpha)_{corr} = \frac{\cos(2\alpha + \delta_o) - \cos(2\alpha + \delta_t)}{1 - \cos(2\alpha + \delta_o)} \tag{4}$$

In the absence of analyzer and quarter-wave plate, and for the polarizer set at an angle γ with respect to the field direction, the relative change of light intensity is given by (38-40):

$$(\Delta I_\gamma/I) = \cos^2\gamma \; \exp(-2.303 \; \Delta A/1.5)$$
$$+ \; \sin^2\gamma \; \exp(2.303 \; \Delta A/3) - 1 \tag{5}$$

If $\gamma = 0^o$, $\Delta I/I = \Delta I_{||}/I$ and the dichroism is deduced from the change of light intensity by the equation:

$$\Delta A = -1.5 \; \log \left(1 + \frac{\Delta I_{||}}{I}\right) \tag{6}$$

For $\gamma = 90^o$, we obtain:

$$\Delta A = 3 \; \log \left(1 + \frac{\Delta I_\perp}{I}\right) \tag{7}$$

Control of instrument setting. Equation (3) offers an easy means of checking the optical adjustment of the system with respect to the electric field direction. The absolute values of $(\Delta I/I_\alpha)_{corr} \; \sin^2\alpha$ are plotted as a function of $\sin 2\alpha$ for negative and positive α values. The two straight lines obtained should intercept for $\alpha = 0^o$ at an ordinate equal to $(\delta/2)^2 + (\delta'/2)^2$ (or $(\delta/2)^2$ outside the absorption bands or for non-dichroic samples). The application of this test to our instrumentation showed a deviation of less than 20' as compared to the orientation expected with respect to the electric field (Fig. 3). This is an instrument geometrical parameter which has not often been controlled.

The dichroism variation with the polarizer angle γ (eqn. (5)) also allows to control the overall optical arrangement of the system and to evidence the presence of perturbating effects such as chemical relaxation effects. Fig. 4 shows the variation observed with our adapted temperature-jump instrument using a DNA sample. The theoretical curve calculated on the basis of equation (5) perfectly fits the experimental points. The angle for which the orientation dichroism effect cancels (the so-called "magic angle") depends on the dichroism amplitude (39,40). It ranges from 62° to 47° for ΔA values ranging from -0.5 to +0.5.

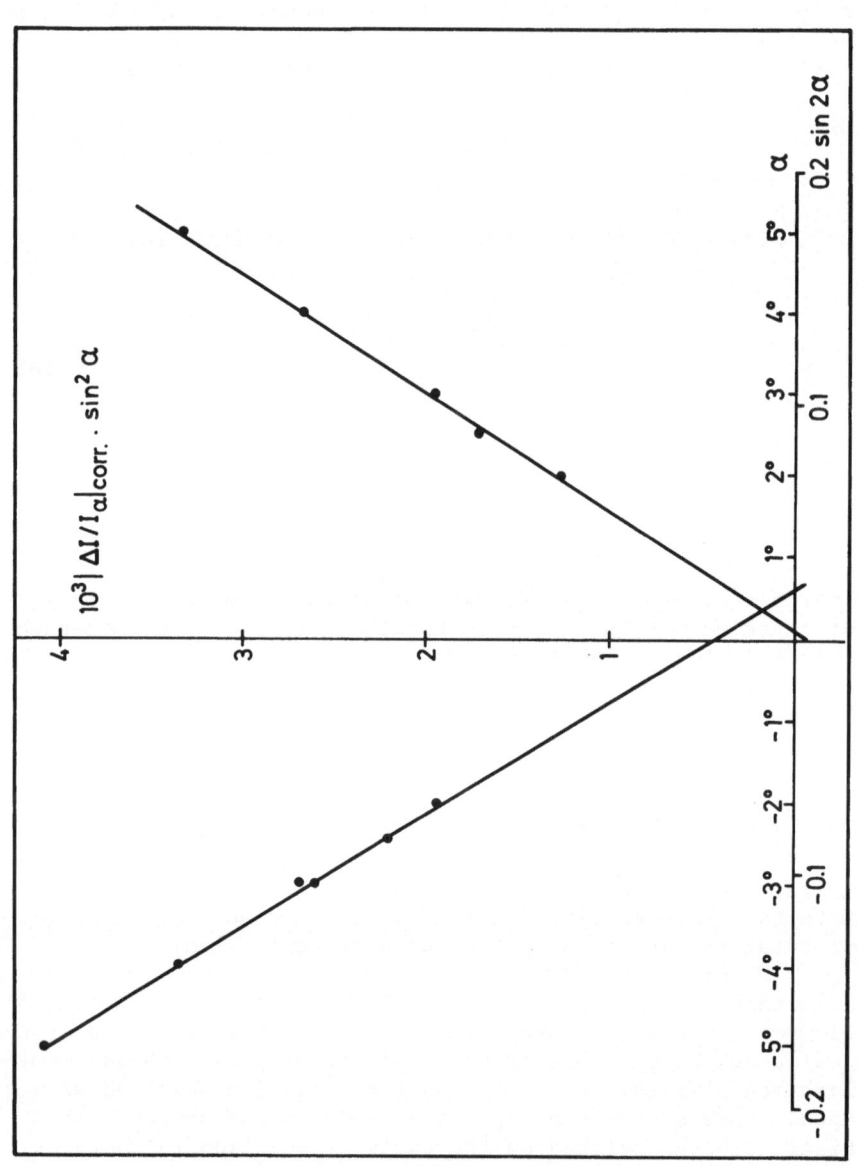

Fig. 3. Check of the optical arrangement for birefringence measurements, based on equation (3).

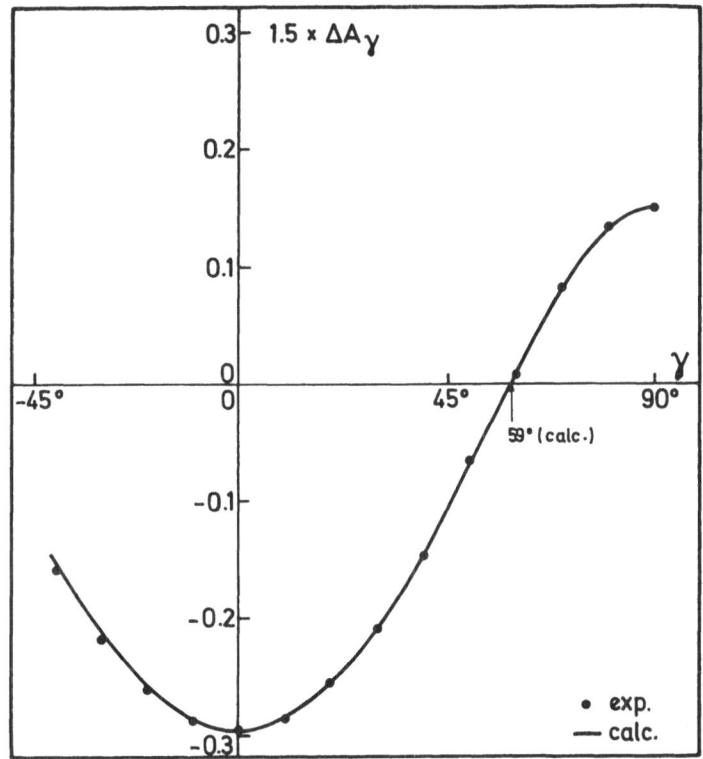

Fig. 4.   Variation of the relative change of absorbance with polar-
          ization direction γ for dichroism measurements on calf-
          thymus DNA (Sigma, purified; $M_w \approx 2.10^6$) in 1 mM NaCl at
          6.3 kV/cm; $\Delta A_\gamma = -\log(1 + (\Delta I_\gamma/I))$.

    For correct relaxation time determinations, the response time
of the detection circuit and the decay time of the electric pulses
must be several times smaller than the lowest τ values to be measured. When determined from the electric birefringence signal displayed by nitrobenzene, acetophenone or propylene carbonate, we
obtained a value of 0.9 - 1 μs, most probably partly due to the influence of the electric pulse decay. It was indeed not possible to
measure these non-conducting solvents in exactly the same output
impedance conditions for the pulse generator as those prevailing
for aqueous conducting solutions. We have, however, the following
indications that the response time is appreciably smaller than 1 μs:
(i) the τ values measured for nucleosomes and nucleosomal DNA, in
the range of 2 to 4 μs, were not modified by more than 0.1 to 0.2 μs
when the amplifier gain was changed by a factor of 2 to 5; (ii) measurements of τ for 150 to 170 b.p. DNA as a function of Dextran

Table 1. Stability and Reproducibility of Electro-Optical Measure-
          ments

---

1. Light signal "at rest"   :   less than 1% deviation over a period
                                   of 15 min.

2. Electric field strength
    (repeat pulsing on the
    same sample)   :   less than 0.5-1% deviation

3. Electric birefringence   :   less than 1-2% deviation for
                                   $\Delta n \geq 2 \times 10^{-7}$

4. Electric dichroism   :   less than 1-2% deviation for
                                   $\Delta A \geq 0.01$

5. Mean relaxation time   :   5 to 10% deviation for $\tau \geq$ 2-3 µs
                                   with $\Delta n \geq 2 \times 10^{-7}$

NB/ The standard deviations are given for about 10 measurements taken
    over a period of 15 min.

The signal-to-noise ratio was of the order of 5 to 10 for
$\Delta n \simeq 1 \times 10^{-7}$ (at less than 1 µs response time) with the birefringence
instrument setting, and for $\Delta A/A \simeq 0.01$ (at 3 µs response time) with
the dichroism and T-Jump measurements system.

Sensitivities down to 10^{-9}-10^{-10} in birefringence are achieved
with birefringence instrumentations equipped with lasers as light
sources, and in the range of 10^{-12} when signal averaging techniques
are used.

concentration gave extrapolated values at zero Dextran concentration not differing by more than 0.5 µs from the values measured in
the absence of Dextran.

The stability and reproducibility of the electro-optical data
obtained with this instrumentation are given in Table 1 (14,38).

Relaxation curves analysis. For a more critical evaluation of
the field-free decay, we have presently adopted the following procedure:

a) the initial slope of the field-free decay is determined by
    linear regression over the first 20 to 50 points (τ_{ini}), i.e.,
    on about 10% of the time range covered by the decay.

b)   characteristic decay times $\tau_x = t_x/x$ (with x = 0.5, 1, 1.5, 2,
     2.5, 3) are determined along the normalized decay curve $\Delta n_t/\Delta n_o$
     for relative drops to exp(-x) (1,7). For a monoexponential
     decay, all these decay times should be identical and equal to τ.

c)   the area under the normalized decay curve, τ_D, is estimated by
     numerical integration down to 0.02 amplitude.

     If the τ values obtained in a), b) and c) are similar, in
the limits of accuracy of the measurements (less than 20 to 50%
deviation) the relaxation curve is considered as monoexponential.
Then the relaxation time and its standard deviation are calculated
by linear regression analysis of the $\ln(\Delta n_t/\Delta n_o)$ versus time values,
over two exponential decay units. The semi-log plot of the relaxation curve is also used as a visual criterion for monoexponential
decays.

     For a more complete analysis of multiexponential relaxation
curves, 20 to 50 points are taken along the semi-logarithmic decay
curve after manual smoothing, and the data set is fitted with a sum
of exponential terms using a general multiparametric curve-fitting
program (41) available from L. Meites in the form of a new version
CFT4 (Clarkson College of Technology, Potsdam, NY). We consider
that not more than three exponential terms are extractable from
noisy decay curves, as it is generally admitted for any type of
relaxation data. The τ_x values may be used to choose the initial
guesses for the fitting.

     Field strength dependence analysis. Orientation functions
(ratio of the anisotropy An at a given field to its value An_s at
infinite field, i.e., at saturation of the orientation) useful for
the analysis of the field strength dependence of the birefringence
and dichroism have been derived (42-44) and are shown in Fig. 5.
These two anisotropic properties are related to $<\sin^2\theta>$ where θ is
the angle between the axis of the oriented particles and the electric field direction. In the case of the fluorescence, the various
measurable changes of intensity upon orientation of the particles
are expressed as complex functions of two averages $<\sin^2\theta>$ and
$<\sin^4\theta>$ and expressions for rod like particles at arbitrary fields
have been given by Sokerov and Weill (49).

     Fittings of a set of 15 to 30 data points of the anisotropy
($\Delta n/A$ or $\Lambda A/A$) versus field strength E are made using the same
general multiparametric curve-fitting program as that employed for
the relaxation analysis (see above; 38). Whenever the orientation
function cannot be put in analytical form, it is recommended to

1. ELONGATED PARTICLES WITH CYLINDRICAL SYMMETRY ($\Phi = An/An_s$)

GENERAL FORM (42) :

$$\Phi(\beta,\gamma) = \frac{3}{2} \frac{\int_{-1}^{+1} u^2 \exp(\beta u + \gamma u^2)du}{\int_{-1}^{+1} \exp(\beta u + \gamma u^2)du} - \frac{1}{2} \tag{1}$$

$$\Phi(\beta,\gamma) = \frac{3}{4\gamma}\left[\frac{e^{\beta^2/4\gamma} + \gamma \{\sqrt{\gamma}(e^\beta + e^{-\beta}) - (\beta/2\sqrt{\gamma})(e^\beta - e^{-\beta})\}}{\int_0^{(\beta/2\sqrt{\gamma})+\sqrt{\gamma}} e^{x^2}dx - \int_0^{(\beta/2\sqrt{\gamma})-\sqrt{\gamma}} e^{x^2}dx} + \frac{\beta^2}{2\gamma} - 1\right] - \frac{1}{2} \tag{2}$$

with $u=\cos\theta$, $\beta= \mu E/kT$, $\gamma= \Delta\alpha E^2/2kT$, $\beta^2/2\gamma = P/Q$

LIMITING FORMS (42) :

Pure permanent dipole :

$$\Phi(\beta) = 1 - 3(\coth\beta - 1/\beta)/\beta \tag{3}$$

Pure induced dipole :

$$\Phi(\gamma) = \frac{3}{4}\left[\frac{e^\gamma/\sqrt{\gamma}}{\int_0^{\sqrt{\gamma}} e^{x^2}dx} - \frac{1}{\gamma}\right] - \frac{1}{2} \tag{4}$$

Low fields :

$$\Phi(\beta,\gamma)= (\beta^2 + 2\gamma)/15 + \{(2\gamma)^2 + 2\beta^2(2\gamma) - 2\beta^4\}/315 \tag{5}$$

Very low fields (Kerr law region) :

$$\lim_{E\to o} \Phi(\beta,\gamma)= (\beta^2 + 2\gamma)/15 = \{(\mu/kT)^2 + (\Delta\alpha/kT)\} E^2/15 \tag{6}$$

Very high fields :

$$\lim_{E\to\infty} \Phi(\beta,\gamma)= 1 - 3/(\beta + 2\gamma) \tag{7}$$

SATURATING INDUCED DIPOLE :

Kikuchi and Yoshioka (46) :

$$\Phi(\kappa) = \frac{3}{2} \frac{\int_0^1 u^2(\sinh(\kappa u)/\kappa u)^n du}{\int_0^1 (\sinh(\kappa u)/\kappa u)^n du} - \frac{1}{2} \tag{8}$$

with $\kappa = Ze1E/2kT$ (n=number of weakly bound counterions)
at low fields : induced dipole with $\Delta\alpha = nZ^2e^21^2/12kT$
at high fields : permanent dipole with $\mu' = nZe1/2$

Fig. 5. Orientation functions for the analysis of the field
strength dependence of dichroism or birefringence of particles with cylindrical symmetry.

Orientation functions (cont.)

Hogan et al. (33) :

$$\phi(\gamma') \text{ identical to } \phi(\gamma) \text{ with } \gamma'= AE/kT$$

Shirai, Sokerov and Weill (47,48) :

$$\phi(\beta') = \frac{3}{2} \frac{\int_0^1 u^2 \exp(\beta'u)du}{\int_0^1 \exp(\beta'u)du} - \frac{1}{2} \tag{9}$$

$$= \frac{3}{2}\left[\frac{\exp(\beta') \{1 - \frac{2}{\beta}(1 - \frac{1}{\beta'})\}- (2/\beta^2)}{\exp(\beta') - 1} \right] - \frac{1}{2} \tag{10}$$

with $\beta' = \mu'E/kT$

at low fields : $\quad \lim_{E\to 0} \phi(\beta') = \beta'/8 \tag{11}$

at high fields : $\quad \lim_{E\to\infty} \phi(\beta') = 1 - (3/\beta') + (3/\beta'^2) \tag{12}$

2. DISK-SHAPED PARTICLES (44)   ($\phi= -An/2An_s$)

GENERAL FORM :

$$\phi(\beta,\gamma) = \frac{3}{2} \frac{\int_{-1}^{+1} u^2 \exp(\beta u - \gamma u^2)du}{\int_{-1}^{+1} \exp(\beta u - \gamma u^2)du} - \frac{1}{2} \tag{13}$$

$$\phi(\beta,\gamma) = \frac{3}{4\gamma}\left[\frac{e^{-(\beta^2/4\gamma +\gamma)}\{(\beta/2\sqrt{\gamma})(e^{-\beta}-e^{\beta}) -\sqrt{\gamma}(e^{-\beta}+e^{\beta})\}}{\int_0^{\sqrt{\gamma}-(\beta/2\sqrt{\gamma})} e^{-x^2}dx -\int_0^{-\sqrt{\gamma}-(\beta/2\sqrt{\gamma})} e^{-x^2}dx}+ \frac{\beta^2}{2\gamma} + 1 \right] - \frac{1}{2} \tag{14}$$

with β and γ as defined above (μ along symmetry axis ; $\gamma >0$, polarizability larger in disk plane)

LIMITING FORMS :

Pure permanent dipole :   same as eqn.(3)

Pure induced dipole :

$$\phi(\gamma) = \frac{3}{4}\left[\frac{1}{\gamma} - \frac{e^{-\gamma}/\sqrt{\gamma}}{\int_0^{\sqrt{\gamma}} e^{-x^2}dx} \right] - \frac{1}{2} \tag{15}$$

Low fields :

$$\phi(\beta,\gamma) = (\beta^2 - 2\gamma)/15 - \{(2\gamma)^2 - 3\beta^2\gamma - \beta^4\}/90 \tag{16}$$

Very low fields (Kerr law region) :

$$\lim_{E\to 0} \phi(\beta,\gamma) = (\beta^2 - 2\gamma)/15 \tag{17}$$

Fig. 5 (continued)

introduce the $\overline{\phi}$ values in table form in the Subroutine Function in order to accelerate the fitting process. The theoretical $\overline{\phi}$ values at each field corresponding to the data set are then calculated by linear interpolation. Tables with the values of $\overline{\phi}$ for different orientation mechanisms are available in the literature (42,43) or may be computed by numerical integration methods. For the calculation of the general orientation function derived by Holcomb and Tinoco (45) a modification of the computer program designed by these authors has been prepared (38).

When fitting with the $\overline{\phi}$ (β,γ) function (eqn. (2) in Fig. 5), we usually fix the $P/Q = \beta^2/2\gamma$ value (determined from bipolar pulse experiments) and a table of the Dawson integral

$$\exp(-q^2) \int_0^q \exp(x^2) dx$$

is used in the Subroutine Function to estimate the integral terms of this equation.

As has been frequently noticed, the field strength dependence of the anisotropy of polyelectrolytes is not satisfactorily accounted for by the $\overline{\phi}(\gamma)$ function, even though the bipolar pulse experiments and the ratio of rise and decay surfaces do reveal the dominance of an induced polarization mechanism (7,8). In all our previous studies we used a weighted sum of three orientation functions with different electric polarizability terms to fit the data, justifying this procedure by the presence of polydispersity and/or flexibility effects. We, however, recognize that such a procedure would enable one to account for any curvature of the experimental curve.

Several authors have considered that the anomalous shape of the field strength dependence of the anisotropy of polyelectrolytes has its origin in the progressive saturation of the ionic polarization (46-49). Kikuchi and Yoshioka (46) derived an orientation function (eqn. (8) in Fig. 5) which requires numerical integration for its calculation, making the fitting procedure with this function very slow, unless the n value (number of weakly bound counterions) is fixed which enables one to present the function in the form of a table. The function derived by Hogan, et al. (33) is identical to $\overline{\phi}(\gamma)$ except that $\gamma = AE/kT$ with A the equivalent permanent dipole corresponding to the saturated polarizability. The Subroutine Function used for the $\overline{\phi}(\gamma)$ function is thus directly applicable to this model with the appropriate change for the calculation of γ from the electric field strength. The reasoning followed by Shirai (48)

and by Sokerov and Weill (49) yields an orientation function similar
to $\bar{\phi}(\beta)$ except that integration in Eqn. (1) (Fig. 5) is performed
between 0 and 1 (this corresponds to considering that the electrostatic energy terms are identical for orientations of the saturated
induced dipole at Θ and $\pi - \Theta$ with respect to E). This function
may be put in analytical form (eqn. (10), Fig. 5) and is thus easily
calculated directly during the fitting procedure. Its predicts, as
the model of Hogan, et al. (33), a linear dependence of the anisotropy on the first power of the electric field at low E, which does
not correspond to the actual observations on short, monodispersed,
rigid polyelectrolytic molecules like nucleosomal DNA.

(Note: all the computer programs used for the analysis of the
electro-optical data have been listed elsewhere (38); updated ver-
sions are available from C. H. upon request.)

AN ELECTRO-OPTIC INSTRUMENT AT BERKELEY

Optics and Detector for Pulsed Electric Birefringence

The optical system employs a tungsten tight coil source, chosen
for stability and convenience, condensing lenses and a Glan-Thompson
polarizer, which is the last optical element before the cell, to
minimize depolarization due to anisotropic scattering and residual
birefringence of optical elements. After the cell, we use a fresnel
rhomb with optic axis set along the plane of polarization to introduce the $\lambda/4$ retardation before passing the transmitted beam through
another Glan-Thompson prism set near the crossed position. A fresnel
prism is superior to a conventional $\lambda/4$ plate because of its achro-
maticity.

It is advisable to rotate the analyzer in the direction, depending on the sign of the birefringence, so that the light intensity
increases with the signal. Our K_{SL} is normally around 1-2 x 10^{-4}
with selected cuvettes and with solution in place. Instrument settings are entered into the data record at the keyboard and a sub-
program is used to automatically compute the retardation values
during data analysis for any values up to $2(\pi/2 - \alpha_m)$, when the
light intensity reaches its maximum.

Several cell designs have been employed. We mainly have used
shiny platinum plane parallel electrodes which are mounted on teflon
or epoxy supports and immersed in the liquid samples held in square
spectro-photometer cuvettes (51,52,53). The cuvette is placed in a
thermostatted holder.

Electric dichroism optics also were employed in this laboratory (54,55). See the general considerations discussed above.

Pulsers, On-Line Computer and Data Handling Procedures

The computerized facility is shown in Fig. 6. The visible region optical system has been described previously (56). We use three kinds of high power pulse generators. These are a Cober 605-P, an 8kV, 50 ohm delay line thyratron pulser built here (53) and a 0-7kV vacuum tube power pulse amplifier with a linear pre-amplifier built here and which is driven by low amplitude wave-form

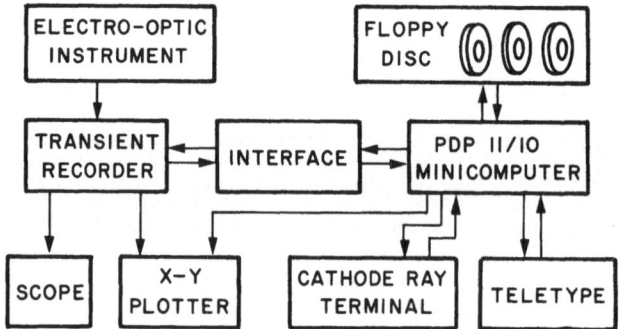

Fig. 6. Data acquisition facility using a transient recorder and a minicomputer system.

generators (Tektronix 161, 163 and a Hewlett-Packard 114). For sine waves we have employed an audio (20 Hz - 20 KHz) 1KW amplifier and a 100 watt RF surplus aircraft transmitter. A replaceable end-window photomultiplier tube and compactly constructed solid state preamplifier (dc to 100 MHz) is mounted on the optical bench. Output waveforms are stored in a digitizing transient recorder (Bio-mation 805 or ICE PTR 9200) which samples 2K times per transient. An interface allows rapid switching between two transient recorders. Normally we operate single pulse but a coadding program is also

available. The PDP 11/10 has 8K of core with 16-bit words, a
FORTRAN compiler and an RT-11 operating system, modified to allow
efficient use of flexible magnetic disc storage in an Advanced Electronics Design, Inc. Model 2500 triple drive unit. The Tektronix
Model 4010 CRT keyboard and a teletype may be used interchangably,
the latter when printed output is wanted. Graphics programs have
been written which permit copying of all graphical data on the X-Y
recorder (H-P Moseley 7000AM). Programs have been written using
various non-linear least squares methods for discrete relaxation
time analysis and a Fourier transform method for doing an inverse
Laplace transform to compute continuous distribution functions for
relaxation times (57).

The back-up system for long calculations and modelling studies
is the Campus Computer Center CDC-6400, used for batch operations.
To utilize our system as input/output for the tieline computers
available, we could add an acoustic coupler for telephone line use.

Transient data in digital form are transferred to the computer
and then stored on flexible discs. Sample intervals are adjustable
from 10 nsec in steps of 2, 5, 10, etc., to 50 sec, for a total of
2K raw data points. Kinetic analyses of relaxation processes are
accomplished with iterative nonlinear least squares computer programs. These have been extensively used for electro-optic data
analyses. Our main program was derived from a Los Alamos radio-
activity computational method ("Superfrenic") which is equivalent
to the "Nonlin" program written independently (R. Baer) at our
campus computer center. We have tested various weighting function
procedures and have also introduced modifications to do analyses of
signal buildups. For example, a least squares program employs
buildup exponential terms as well as decay terms, and a constant
term to fit a signal which first builds up, then "droops" to a lower
level. Signals like that due to a structural transition or "chemical field effect" are sometimes obtained during the electric field-
on period. Raw data are examined on the CRT with program PLOT which
connects consecutive data points. To process data files, the user
examines a display of numbers on the CRT corresponding to signal
amplitudes for consecutive samples and enters the "field-on" and
"field-off" data point indices and the manual instrument readings.
Program SMOOTH is used to compute prepulse and postpulse baseline,
amplitudes, standard deviations, and retardations from the raw data.
Program AVETAU is usable directly on the raw data to calculate the
average birefringence relaxation time. Program SMOOTH preprocesses
the data for nonlinear least squares fitting, producing "smoothed"
data files for both the pulse-on and the field-free decay periods.
It also computes the maximum retardation value and the value just
before removal of the field.

The smoothing procedure for decay signals is designed to reduce the number crunching on each file of around 1000 to 1500 useful
uniformly spaced data points, without significant loss of information in the recorded transients. The first 20 raw data points are
all taken to begin the file; then the next 60 are averaged three
at a time to give 20 more smoothed points; then the next 140 are
averaged seven at a time, etc., increasing the number of raw data
points used as $2^n - 1$ up to 31 at a time, which is the maximum
number that is averaged. This reduces the data files to around 100
to 120 points, typically. This greatly speeds the nonlinear least
square fitting procedure on our small minicomputer. Our strategy
of increasing the number of points averaged as the signal amplitude
falls exponentially (according to $2^n - 1$, where n is the index of
each set of 20 smoothed points) captures the rapidly breaking start
of a transient at the highest available resolution in any given
sweep; it also averages data points in the tail of the curve to
improve S/N at lower amplitudes.

Typically, 1-3 exponential terms (2 to 6 independent parameters)
are then fitted to the signal decay data with program NLSQ1, a non-
linear least squares fitting procedure which is based on a Taylor
expansion and utilizes matrix methods for solving the normal equations. Initial guesses are made automatically, based upon the 1/e
time of fall for a decay curve, or may be supplied by the user.
Variable damping is selected by the user to curtail oscillations
during iteration. Signal buildup and droop data are fit typically
with 5 or 7 parameters for a flexible macromolecule like DNA, for
example. Comparisons of the computed fitting functions are then
made graphically on the CRT with the smoothed data points on either
a linear signal scale (EOPLOT) or a logarithmic signal (LOGPLT)
display. The latter is far superior for mentally assimilating the
meaning of the experimental data, of course. Visual displays,
which permit viewing the data in a critical comparison with their
fitting function, are decisive in choosing the appropriate number
of fitting terms according to "Occam's Razor;" they also enable an
intelligent guess to be made as to how to choose the initial parameters for the first iteration when an additional term is added,
which often determines if convergence can be obtained. With 8K of
computer core, a single iteration required 12 seconds for six independent parameters; convergence (to within 1% for all parameters)
occurs in 5-7 iterations, typically. The fitted relaxation times,
τ_i, their percent contributions to the signal, s_i, the relaxation
rates, α_i, and their averages, $\bar{\tau}$ and $\bar{\alpha}$ are displayed on the CRT.
We also calculate $\bar{\tau}\,\bar{\alpha}$ which is a convenient criterion of the width
of the distribution of relaxation times. Statistical criteria of
the S/N ratio and of the fitting function are also presented with
the nonlinear least squares output (58).

A sample of the logarithmic plot of a 3-term exponential decay
fitting function (solid curve) and the smoothed data (on the CRT)
is given in Fig. 7. Another illustration, showing the buildup and
the high field intensity droop (birefringence signal) of a macromolecule undergoing a structural transition is shown in Fig. 8.

Several computer programs have been written for calculating
macromolecular parameters. An example, is a program to calculate
the length of a rod model for DNA by iteration of the Broersma
equation to fit a measured relaxation time. This facility may be
used for a wide variety of input signals, e.g., temperature and
concentration jump transients, charge-pulse transients, conductivity
and dielectric transients, electro-physiological recording and
analyses, i/V curves of planar membranes at various sweep rates,
etc. Many types of data can be analyzed with existing programs.
The magnetic disc storage and the RT-11 operating system permit new
ones to be introduced, and the graphics capabilities make this a
powerful system for innovative experimenters. A discussion of
mathematical methods and modelling for data analysis was published
earlier (59), together with a preliminary description of the present
system.

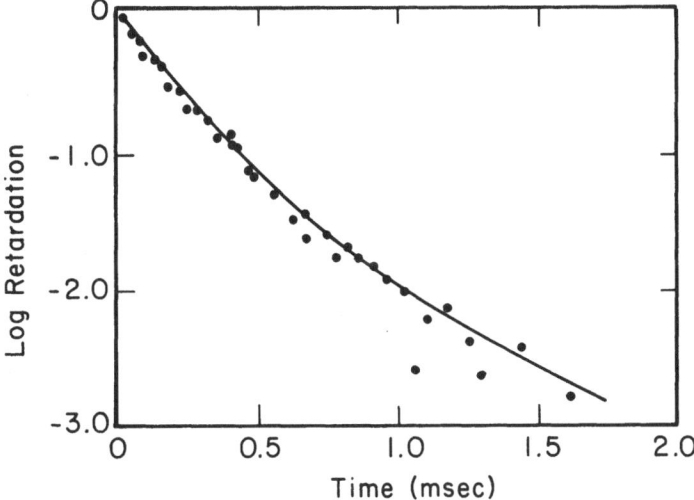

Fig. 7.   Log plot of smoothed data and fitting function for a
typical high molecular weight (ca. 6×10^6) DNA sample.

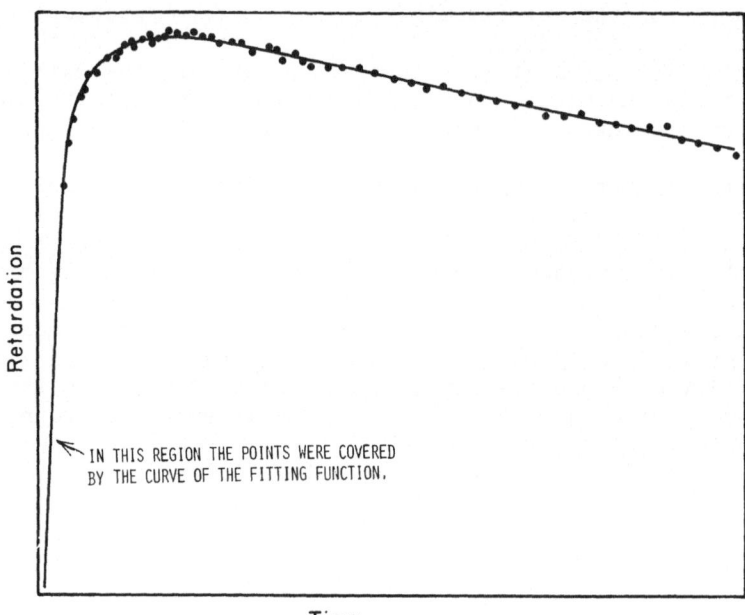

IN THIS REGION THE POINTS WERE COVERED
BY THE CURVE OF THE FITTING FUNCTION.

Time

Fig. 8. Plot of experimental buildup data and the fitted 3-term
curve for high field birefringence of calf thymus DNA.
DNA concn., 18 µg/ml. tris concn., 0.0005 M

The fitting function is:

-84.EXP(-T/3.3)-33EXP(-T/18)+70EXP(-T/465)+39
where T is the time in µsec.

ACKNOWLEDGEMENTS

    The expert technical assistance of Mr. Guy Bertin was of in-
valuable help in the setting up and running of the whole electric
dichroism, birefringence and computer system at the University of
Liège. C. H. also acknowledges the financial support of the Fonds
National de la Recherche Scientifique (Belgium). He also thanks
Prof. V. Desreux and Prof. E. Fredericq for constant interest in
this work and Dr. M. Tricot for fruitful collaboration. C. T. O.
acknowledges the support of the National Institute of Health, U.S.
Public Health Service, the collaboration of Drs. L. L. Mack, J. W.
Jost, D. I. Devore, and R. S. Farinato in setting up and improving
the system, and assistance from graduate students Michael Kwan,
Lloyd Shepard, John Zarkarian, Robert Purtell and programmers George
Wiley, Ernest Haberkern, Sam Mershon, and Dennis Reese. Thanks
are due to Dr. R. S. Farinato for useful comments on this manuscript.

REFERENCES

1. C. T. O'Konski, Encyclopedia Polym. Sci. Technol. 9:551-590 (1968).
2. K. Yoshioka, and H. Watanabe, in: "Physical Principles and Techniques of Protein Chemistry," Part A, S. J. Leach, ed., Academic Press, NY, pp. 335-367 (1969).
3. E. Charney, in: "Procedures in Nucleic Acids Research," Vol. 2, G. L. Cantoni and D. R. Davies, eds., Harper and Row, NY, pp. 176-204 (1971).
4. S. P. Stoylov, Adv. Colloid Interface Sci. 37:443-457 (1971).
5. M. Hanss, and B. Roux, Ann. Phys. Biol. et Méd. 3:133-163 (1972).
6. B. R. Jennings, in: "Light-Scattering from Polymer Solutions," Ch. 13, M. B. Huglin, ed., Academic Press, NY, pp. 530-579 (1972).
7. E. Fredericq, and C. Houssier, "Electric Dichroism and Electric Birefringence," Clarendon Press, Oxford, U.K. (1973).
8. M. Tricot, and C. Houssier, in: "Polyelectrolytes," K. C. Frisch, D. Klempner, and A. V. Patsis, eds., Technomic Publishing Co., Wesport, pp. 43-90 (1976).
9. C. T. O'Konski, "Molecular Electro-Optics. Part I. Theory and Methods," Marcel Dekker, NY (1976).
10. C. T. O'Konski, "Molecular Electro-Optics. Part II. Applications to Biopolymers," Marcel Dekker, NY (1978).
11. B. R. Jennings, Adv. Polym. Sci. 22:61-81 (1977).
12. B. R. Jennings, "Electro-Optics and Dielectrics of Macromolecules and Colloids," Plenum Press, NY (1979).
13. P. J. Rudd, and B. R. Jennings, Lab. Pract., pp. 535-537 (1973).
14. C. Houssier, Lab Pract. pp. 562-563 (1974).
15. J. W. Jost, and C. T. O'Konski, in: "Molecular Electro-Optics. Part II. Applications to Biopolymers," C. T. O'Konski, ed., Marcel Dekker, NY, pp. 529-564 (1978).
16. A. S. Verkman, A. A. Pandiscio, M. Jennings, and A. K. Solomon, Anal. Biochem. 102:189-195 (1980).
17. J. Newman, and H. L. Swinney, Biopolymers 15:301-315 (1976).
18. D. C. Rau, and V. A. Bloomfield, Biopolymers 18:2783-2805 (1979).
19. F. S. Allen, and K. E. VanHolde, Rev. Sci. Instrum. 41:211-216 (1970).
20. W. H. Rahe, R. J. Fraatz, L. K. Sun, D. R. C. Priore, and F. S. Allen, in: "Electro-Optics and Dielectrics of Macromolecules and Colloids," B. R. Jennings, ed., Plenum Press, NY, pp. 57-66 (1979).
21. B. R. Jennings, in: "Molecular Electro-Optics. Part I. Theory and Methods," C. T. O'Konski, ed., Marcel Dekker, NY, pp. 275-319 (1976).
22. P. J. Ridler, and B. R. Jennings, in: "Electro-Optics and Dielectrics of Macromolecules and Colloids," B. R. Jennings, ed., Plenum Press, NY, pp. 99-107 (1979).
23. E. D. Baily, and B. R. Jennings, Applied Optics 11:527-532 (1972).

24. E. D. Baily, and B. R. Jennings, J. Colloid Interface Sci. 45: 177-189 (1973).

25. A. R. Foweraker, and B. R. Jennings, Lab. Pract. 25:318-320 (1976).

26. H. J. Coles, Polymer 18:554-556 (1977).

27. H. Asai, N. Watanabe, and T. Okuyama, Rev. Sci. Instrum. 49: 236-237 (1978).

28. E. Neumann, in: "Topics in Bioelectrochemistry and Bioenergetics," G. Milazzo, ed., John Wiley & Sons, London (1979).

29. E. Neumann, and A. Katchalsky, Proc. Natl. Acad. Sci. U.S.A. 69:992-997 (1972).

30. M. Pollak, and H. A. Glick, Biopolymers 16:1007-1013 (1977).

31. G. B. Thurston, and R. S. Wilkinson, J. Phys. E. Sci. Instrum. 6:289-293 (1973).

32. J. Schweitzer, and B. R. Jennings, J. Phys. D. Appl. Phys. 5: 297-309 (1972).

33. M. Hogan, N. Dattagupta, and D. M. Crothers, Proc. Natl. Acad. Sci. U.S.A. 75:195-199 (1978).

34. R. Rigler, C. R. Rabel, and T. M. Jovin, Rev. Sci. Instrum. 45:580-588 (1974).

35. R. C. Williams, W. T. Ham, and A. K. Wright, Anal. Biochem. 73:52-64 (1976).

36. B. L. Brown, and B. R. Jennings, J. Phys. E. Sci. Instrum. 3: 195-198 (1970).

37. J. C. Ravey, and C. Houssier, in: "Electro-Optics and Dielectrics of Macromolecules and Colloids," B. R. Jennings, ed., Plenum Press, NY, pp. 67-76 (1979).

38. C. Houssier, Thèse d'Agrégation, Université de Liège (1977).

39. M. Dourlent, J. F. Hogrel, and C. Hélène, J. Am. Chem. Soc. 96: 3398-3406 (1974).

40. M. Dourlent, and J. F. Hogrel, Biochemistry 15:430-436 (1976).

41. T. Meites, and L. Meites, Talanta 19:1131-1139 (1972).

42. C. T. O'Konski, K. Yoshioka, and W. H. Orttung, J. Phys. Chem. 63:1558-1565 (1959).

43. M. Matsumoto, H. Watanabe, and K. Yoshioka, Scient. Pap. Coll. Gen. Educ. Tokyo 17:173-202 (1967).

44. M. J. Shah, J. Phys. Chem. 67:2215-2219 (1963).

45. D. N. Holcomb, and I. Tinoco, J. Phys. Chem. 67:2691-2698 (1963).

46. K. Kikuchi, and K. Yoshioka, Biopolymers 15:583-587 (1976).

47. M. Shirai, private communication referred to in Ref. 49.

48. S. Sokerov, and G. Weill, Biophys. Chem. 10:161-171 (1979).

49. S. Sokerov, and G. Weill, Biophys. Chem. 10:41-46 (1979).

50. C. T. O'Konski, and B. H. Zimm, Science 111:113-116 (1950).

51. C. T. O'Konski, and A. G. Haltner, J. Am. Chem. Soc. 78:3604-3610 (1956).

51a. C. T. O'Konski, and A. G. Haltner, J. Am. Chem. Soc. 79:5634-5649 (1957).

52. R. M. Pytkowicz, and C. T. O'Konski, Biochim. Biophys. Acta 36:466-470 (1959).

53. S. Krause, and C. T. O'Konski, J. Am. Chem. Soc. 81:5082-5088 (1959).

54. C. T. O'Konski, and K. Bergmann, J. Chem. Phys. 37:1573-1574 (1962).

55. C. M. Paulson, Ph.D. Thesis, Univ. of California, Berkeley; C. M. Paulson, and C. T. O'Konski, Polymer Preprints 7:1175 (1966); C. M. Paulson, Ch. 7, Electric Dichroism of Macromolecules, in: "Molecular Electro-Optics," Part 1, C. T. O'Konski, ed., Marcel Dekker, NY (1976).

56. C. T. O'Konski, and S. Krause, Electric Birefringence and Relaxation in Solutions of Rigid Macromolecules, Ch. 3, in: "Molecular Electro-Optics," Part 1, C. T. O'Konski, ed., Marcel Dekker, NY (1976).

57. M. Matsumoto, H. Watanabe, and K. Yoshioka, Biopolymers 6:929-938 (1968); same authors, Kolloid-Z. Z. Polymere 250:298-302 (1972).

58. A. E. Pritchard, and C. T. O'Konski, Ann. New York Acad. Sci. 303:159-169 (1977).

59. J. W. Jost, and C. T. O'Konski, Electro-Optic Data Acquisition and Processing, Ch. 15, in: "Molecular Electro-Optics. Part 2. Applications to Biopolymers," C. T. O'Konski, ed., Marcel Dekker, pp. 529-564 (1978).

THE ELECTRO-OPTICS OF PROTEINS

J. C. Bernengo

Laboratoire de Biophysique
Université de Nice
parc Valrose 06034 Nice, France

INTRODUCTION

Since the early work of Tinoco on fibrinogen in 1954, many
electro-optical investigations, especially birefringence experiments
have been made on protein solutions. The first experimenters were
particularly attracted by the simple spherical or rod shaped models
which fitted rather well with hydrodynamical data obtained a few
years before on some extensively studied proteins such as serum
albumin, fibrinogen, and collagen. Though their original purpose
was more to verify the validity of phenomenological or theoretical
equations on well defined, monodisperse molecules, they also had
the feeling that electro-optical techniques could bring some knowledge about associative, and more generally, functional properties
of many proteins. From a physico-chemical point of view, the initial purpose has been achieved fairly well, but the structural and
functional assumptions deduced from these experiments were often
later proved to be unsatisfactory or even wrong, as a result of
recent improvements of biochemical and structural techniques. The
main reasons for these discrepancies have to be sought in the
initial product preparation and identification and in the non-
physiological, low ionic strength solvents which had to be used for
electro-optics purposes. In the last decade, a new generation of
experiments has been carried out, not only on biological materials
(including proteins and nucleic acids, but also on higher order
particles, such as chromatin subfragments, mitochondria, membranes,
viruses, bacteriophages and even bacteria); these new experiments
deal with the most recent biochemical discoveries and try to comple-
ment their results.

This presentation will be focused on these recent works for

several reasons:

- Good reviews of protein electro-optics have already been
given, first in 1965 by Yoshioka and O'Konski in a general bibliography on molecular electro-optics[1] and later by Yoshioka and
Watanabe[2] with more details. As far as we know, the most complete
and up to date survey is due to Yoshioka in 1976, in a book entirely
devoted to the subject[3]; very helpful general considerations and
examples of the most significant experiments have also been given
by Kahn[4] and Houssier[5].

- It is impossible to summarize here all the results which have
been presented since 1954, and particularly the numerous exciting
debates which occurred, due to the discrepancies observed in the
same protein, either between two different electro-optical experiments, or when electro-optics and other physico-chemical methods
were compared.

- Our aim in this text is to point out the potentialities of
electro-optic measurements in biological studies. For that purpose,
we shall limit our review to a reasonable number of proteins, selected for their biological importance, and for which electro-optics
proved to be a valuable tool in structural and functional studies.

Some general considerations about hydrodynamical and electrical properties of proteins in solutions have to be discussed first,
as well as their consequences when applied to the results and interpretation of electro-optical measurements. Then, recent results
will be classified in three groups, according to the biological
origin of the molecules which are considered: connective tissue,
muscle, and blood. In each case, an attempt will be made to indicate the importance of the contribution of electro-optical experiments among the various other methods used to further our understanding of the biological problem.

Some Remarks on Hydrodynamical and Electrical Properties of Proteins.

Electro-optics has been widely used to measure the rotary
diffusion constant D_r of proteins, mainly through decay signals

obtained after the electric field is cut off. Since each relaxation
process contributes to the decay by adding an exponential term, it
is in principle possible to separate several relaxation mechanisms,
corresponding for instance to several molecular species, from one
single experiment. As an example, aggregation properties of a given
protein could be followed that way without disturbing the equilibrium (at least if limiting low fields are applied), since, contrary to synthetic polymers or many nucleic acids, pure protein
solutions should not be polydisperse except if aggregation occurs.
As a matter of fact, the polyelectrolytic nature of proteins

facilitates the formation of aggregates, particularly at low concentrations and low ionic strength. It is often possible to find
adequate solvents in which the protein is truly monomeric, but
unfortunately, high salt and low solute concentrations are often
needed. (Consequences of this point will be examined later.) Since
additional relaxation processes originate from flexibility and
reorientation around the different axes of a molecule, great care
has to be taken to interpret birefringence decay curves in terms of
hydrodynamical molecular data.

     Even in the simple case where a single rotary relaxation time
can be determined, a model has to be assumed in order to evaluate
a molecular dimension. Except for a spherical shape, the complete
hydrodynamical characterization of an equivalent ellipsoid is only
possible if at least a second parameter is available: it can be
other rotational data obtained for instance from dielectric constant
measurements (as in Moser et al. work on serum albumin[6]) or from a
two exponential decay curve analysis, but in that case it has to be
proved that the two observed relaxation times correspond to two
different orientations of the same molecule (see for instance
Wright et al. (48)).

     As a matter of fact, this characterization is of very poor
accuracy for weakly elongated and oblate ellipsoids, even if the
extra unknown due to protein solvation is temporarily ignored, as
first shown by Daune et al.[7] Fig. 1 represents the function $t^3.r^{-1}$
(curve A), the value of which is calculated from experimentally
determined translational and rotational diffusion constants, versus
the axial ratio $\rho = \dfrac{b}{a}$, according to Perrin's equations. The
uncertainty of that junction, directly related to experimental
errors, has to be multiplied by the factor $E(\rho)$ (curve B) in order
to get the actual relative uncertainty over ρ, and it is clear that
for $\rho > 0.8$ very poor (or even valueless) calculations of molecular
dimensions are achieved through hydrodynamical measurements. If
solvation phenomena are considered, this type of determination may
be totally impossible, as for instance for a highly hydrated molecule such as fibrinogen (more details will be given below).

     From the foregoing considerations, it appears that the most
significant results are likely to be obtained on long rod shaped
molecules, weakly hydrated, such as myosin or collagen: they
present the advantages of a suitable relaxation time range (10 to
500 μs), rather high optical and electrical polarizabilities
involving high specific birefringence and hence, good accuracies
on relaxation time determinations. End to end aggregation is in
this case easy to characterize, but side by side association is
quite undetectable up to axial ratios of about 0.1.

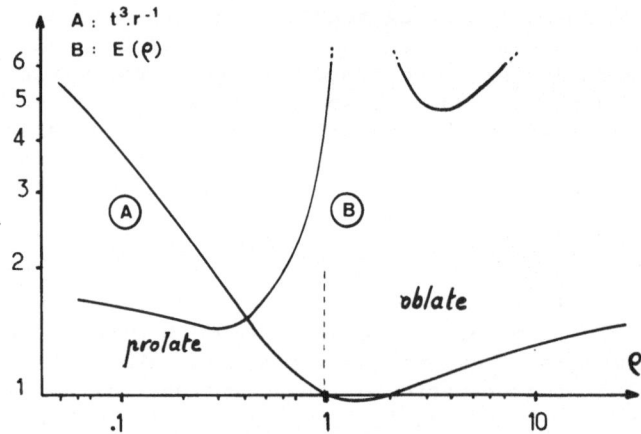

Figure 1.   The Perrin function $t^3 r^{-1}$ (ρ) versus the ellipticity
$\rho = \dfrac{b}{a}$ and the error function $E(\rho)$ defined by:

$$\frac{\Delta\rho}{\rho} = E(\rho) \frac{\Delta\,(t^3 r^{-1})}{t^3 r^{-1}}$$

If we are now interested in the electrical properties of
proteins, we have to deal with the origin of the orienting torque
acting on the molecule, either due to a permanent or to a field
induced dipole moment, or even more likely to both together. Two
components contribute to the permanent dipole moment: the vector
addition of the elementary moments of amino-acids and peptide
bonds and the dissymmetry of the charge separation on the molecular
surface, the latter being preponderant on elongated proteins. The
induced polarization includes an electronic contribution; but, as
for any polyelectrolyte, the modification of the equilibrium
between the polyion and the counterions cloud is dominant, particularly at low frequencies. Many theories have been proposed to
model the ionic polarizability of polyelectrolytes, and a good
example of experimental agreement obtained through dielectric
constant measurements is given by the Minakata and Kobayashi results
on heavy meromyosin[8].

Since permanent dipole moments are of great interest for structural and interaction studies, several methods have been proposed
to evaluate their contribution (with respect to induced polarization)
to the total birefringence. When applicable, the best one seems to
be the use of alternate pulses, first suggested by Tinoco[9] in 1958,
and later applied to TMV proteins, fibrinogen, myosin, collagen, and
recently, histones. Unfortunately, this method does not easily

discriminate between permanent and slow induced dipole moments,
though an attempt has been made by Tinoco himself to derive reverse
pulse equations in the case of a first order induced relaxation
process. Better discrimination could be achieved by combining
electro-optical and dielectric measurements, as described by Moser
et al.[6] for bovine serum albumin or by Hanss et al. for DNA[10].
Another more practical limitation comes from the difficulty encoun-
ered in generating the high voltage, high current alternate pulses
at frequencies suitable for small proteins. In that case, an
alternative (and in principle equivalent) technique consists of
applying sine wave bursts to the solution, taking advantage of the
high signal to noise ratio of a tuned detector to minimize the
required electric field. However, the interpretation of experimental dispersion curves involves polynomial curve fitting according to
the equations of Thurston and Bowling[11], and a straightforward
estimation of the polarization origin is not possible, contrary to
the reverse pulse method.

The absolute determination of dipole moments and electrical
anisotropies by means of an electro-optical technique requires a
measure of the optical anisotropy of the molecules. O'Konski et
al.[12] have proposed applying saturating electric fields to the
solution. Though theoretically attractive, this method, developed
later by Holcomb and Tinoco[13] does not seem realistic when dealing
with proteins for at least two reasons:

- on fibrous proteins, deformations and internal motion become
more and more important when the electric field is increased, and
even if saturation is reached, it affects more than just the molecular orientation process;

- for many globular proteins, for which the above effect is
unlikely to occur, saturating fields are too high to be easily
generated by standard laboratory equipment (around 100 Kv/cm) and
an extrapolation to limiting high fields from values obtained at
the beginning of the saturation is always hazardous.

A good way to get an accurate value of the optical anisotropy
of proteins is to carry out flow birefringence experiments, such as
in the recent work by Miyahara and Noda on myosin[(29)], but this
technique is not sensitive enough to get results on weakly anisotropic globular proteins. Dynamic depolarized light scattering is
also in principle capable of measuring optical anisotropies and
rotary diffusion constants, as shown for instance by Fletcher on
collagen[14]. The main advantage of this latter technique is to be
able to separate several molecular species by a mathematical analysis of the light intensity autocorrelation function, and to obtain
the optical anisotropy factor of each. Combined with an electro-
optical technique such as electric birefringence, it should provide

a powerful tool for obtaining hydrodynamical, electrical, and optical parameters at the same time, for example, during the early phases of an aggregation process.

This chapter will end with the main limitation of electro-optics with regard to protein studies, i.e. the rise of temperature occurring during the pulse when working on conductive solutions at high fields. To avoid this effect, and also because high power pulse generators were not currently available, many experimenters have looked for low conductivity media in which the protein to be studied might not be damaged. To solubilize long rod shaped proteins such as myosin or collagen at low ionic strength, denaturing agents are often added (urea, acetic acid). Even globular proteins, though much more soluble at low ionic strength, are not sufficiently stable to avoid intermolecular interacting, and extrapolations at infinite dilutions have to be carried out to find the actual rotary diffusion constant corresponding to the isolated molecule. Moreover, a comparison of hydrodynamic data obtained from electro-optics with sedimentation or viscosity measurements is often impossible, since the experimental ionic strengths are very different.

If biological processes such as collagen fibrillogenesis myosin association, or glutamate dehydrogenase multimerisation are now considered, it becomes absolutely necessary to get electro-optical methods to work at ionic strength up to 1M. Generators capable of handling such high powers have been built (see for instance Bernengo et al.[15]) and are now commercially available, but because of heating effects, can only be used for short pulses, and therefore are only suitable for orientating small proteins. More precisely, we have represented in Figure 2 the electric field for which an increase of 5°C occurs in the cell, versus the relaxation time τ of the molecules to be studied, assuming a pure permanent dipole orientation. In that case, the pulse duration T has to be at least equal to 12τ in order to reach a steady state value within 2% (see for instance ref. 9). As an example, a myosin solution in .5M KCl equivalent ionic strength has been considered: for the monomer of 40μs relaxation time, the maximum field is around 800V/cm, but if the experimenter wants to orient both monomer and dimer ($\tau \simeq 200$μs), the maximum usable field decreases down to 350V/cm, and the amplitude of the optical signal is reduced by a factor of 5, together with the signal to noise ratio. Therefore, this ratio has to be considerably improved, for instance by means of solid state photodetectors when the incoming light is intense enough (as in birefringence measurements), low noise video amplifiers and finally by using averaging techniques or synchronous sine wave detection. The use of low electric fields not only has the advantages of fulfilling the heat dissipation requirements at high ionic strength, but also avoids flexibility effects and field induced phenomena, which have been demonstrated by many authors; see for example Gerber et al. on bacterial flagellar protein filaments[16],

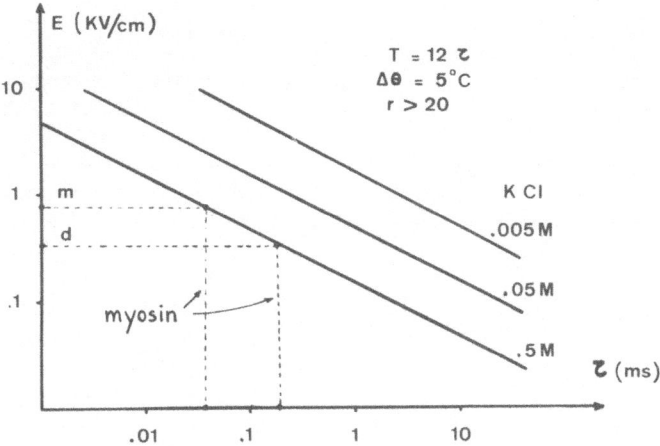

Figure 2.   The maximum field E which can be applied to a solution
of permanent dipolar molecules of relaxation time τ for an increase
of temperature of 5°C at three KCl concentrations.

Kahn and Witnauer on soluble collagen[17] and Riddiford[18] on several
proteins.

The preceeding general considerations apply to any electro-
optical technique, but in fact, due to its high sensitivity, only
electric birefringence has been widely employed in protein studies.
As a matter of fact, the number of aromatic and heterocyclic residues in most proteins is too low to get a measurable signal at
280nm in absorption techniques such as electric dichroism or in
emission processes such as fluorescence in an electric field.   The
only exceptions deal with hemo-proteins which present a strong
absorption band in the near U.V. range and several weaker visible
bands.   A typical example of an electric dichroism study on haptoglobin-hemoglobin complexes is given by Makinen et al.[45] and will
be presented later in the blood proteins section.   Nevertheless,
the recent technological improvements might make possible such
measurements on other proteins, and in that case a valuable new
tool would be provided to study the binding on specific cyclic
amino-acid sites, for instance during enzymatic reactions.

Connective Tissue Proteins

Acid soluble collagen molecules are long and rod shaped, 300 nm

in length, 1.5m in diameter. X-ray scattering reveals a rigid triple
helix structure, with two short non-helicoidal ends of high importance in the fibril formation process since the covalent bonds are
situated in these regions. This molecule is more than an excellent
physico-chemical model, since it has the capability of forming fibers "in vitro" by simply increasing the pH of the solution. In
fact, true physiological fiber formation has to be studied by
starting from neutro-soluble collagen, which possesses many less
covalent intramolecular bonds and is only soluble at high ionic
strength.

Acid soluble collagen from rat tail tendon was first studied
by Yoshioka and O'Konski[19] in 1966. They found a two exponential
birefringence decay, with a long relaxation time attributed to the
molecular reorientation and leading to a diffusion constant around
1000 s^{-1}, corresponding to a rod length of 270nm. The orientation
was mainly due to a permanent dipole moment, estimated at 1500 D
through a rather questionable method involving extrapolation at
infinitely high fields. Later, Kahn and Witnauer[20] found a monoexponential decay for calf skin collagen dissolved in citrate
buffer and showed anomalous behaviors in highly concentrated
solutions[21] and even field induced effects[22]. Ananthanarayanan
and Veis[23] made the first attempt to characterize aggregates in
dilute acetic acid solutions by combining intrinsic viscosity and
electric birefringence and concluded that the observations were
due to the formation of an end to end dimer.

In 1974, the reverse pulse technique, applied to collagen by
Bernengo et al.[24], proved the orientation to be due to permanent
dipoles at low fields and low concentrations (down to 50 mg/1)
in .1 M acetic acid. The same authors[25] studied the birefringence
decay curves in solutions of various pH down to very low residual
birefringences and found a well defined long relaxation process
independent of pH between 2.5 and 4, which was attributed to a
700nm long aggregate. Its electrostatic nature was attested by
the concentration dependence of the corresponding birefringence
amplitude as shown on curve a, b, c, d, of Figure 3 at various
pH; dotted line (e) corresponds to cartilage collagen aggregation
confirmed later to be of covalent nature, as it had been assumed
from birefringence results. Recently, chemically substituted
collagens have been studied in order to look at the importance
of lateral changes in the molecular association mechanisms. (Bernengo et al.[26]): aspartic acid carboxyl groups have been blocked by
methylation and lysine amino-groups by acetylation. The electrical properties of both monomers and aggregates have been determined
by combining the numerical analysis (in terms of two relaxation
mechanisms) of decay, alternate and sine wave responses. From this
work it was concluded that two different types of aggregates of
equal dimensions appeared in the collagen solutions: a non polar

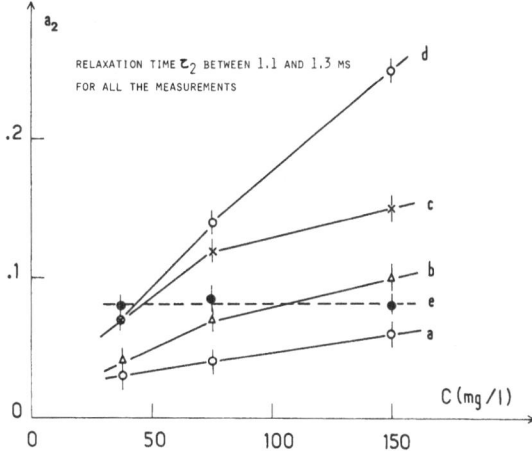

Figure 3.   The contribution a_2 of the aggregates to the total normalized birefringence, versus collagen concentration. Pepsin treated
acid soluble skin collagen, in 10 mM acetic acid (a) 1 mM (b),
.2 mM (c) and .1 mM (d).   Curve (e) corresponds to a pepsin
extracted cartilage collagen in 100 mM acetic acid.   (Bernengo et
al.[25])

form of high pH native collagen and of low pH methylated collagen,
and a polar association of acylated solutions.   Therefore, the
absence of negative charges facilitates a non polar aggregation
(possibly anti-parallel), though the neutralization of positive
charges is in favor of polar aggregates.

     The first steps of fibrillogenesis are not up to now well
understood, since no intermediate aggregates have been actually
observed before the fibrillae become visible under the electron
microscope. In 1978, Piez proposed the existence of intermediates
from turbidity results, but all attempts to characterize these
intermediaries by physico-chemical methods have failed.   To apply
electro-optical techniques to this problem was attractive, but
the experiment had to be carried out at high ionic strength ($\sim$ 2M
NaCl) in neutral salts.   Averaging birefringence techniques have
been used together with dynamic light scattering, but without
detection of intermediate aggregates which could be considered as
the "nucleus" proposed by Piez during the early steps of fibrillogenesis (Veis, Bernengo et al. in press).

     Since, in connective tissue collagen interacts with proteogly-
cans, attempts have been made to characterize such processes in

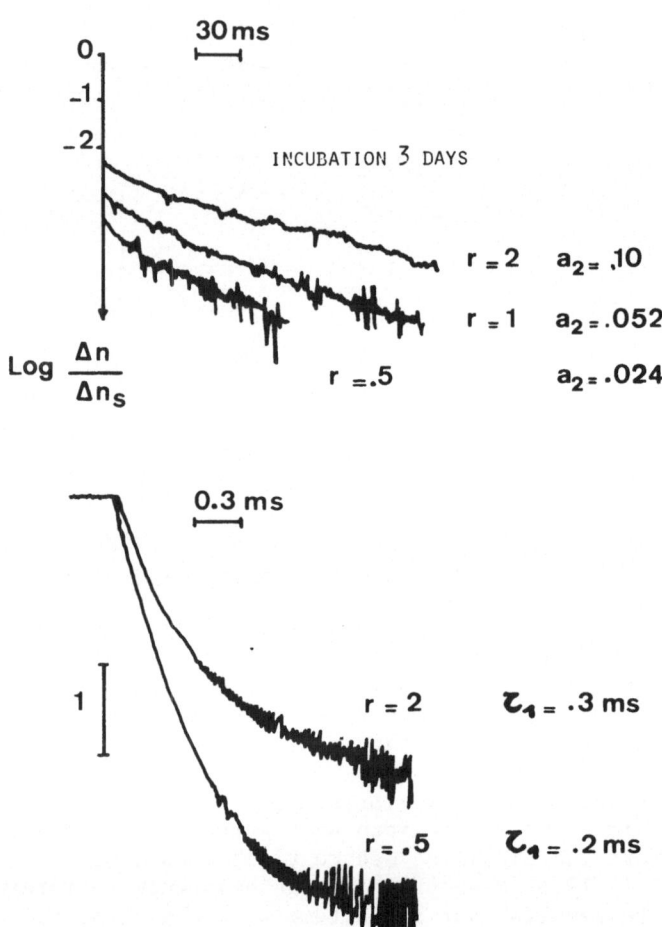

Figure 4. Birefringence decays obtained on collagen solutions
incubated 3 days with chondroitin-6-sulfate (r is the number of
disaccharides per 100 amino acid residues). (Bernengo et al.[27])

vitro - for instance, a study of collagen. Chondroitin-6-sulfate
(a proteoglycan component) interactions have been carried out, following the aggregation by means of electric birefringence (Bernengo
et al.[27]); chondroitin-6-sulfate molecules associate slowly with
collagen and the process stops when no free C-6-S remains in the
solution. The initial speed of aggregate growth depends on the
ratio of C-6-S to collagen, but the final size of the aggregates
does not. (The aggregates have relaxation times of around 80 ms
corresponding to a sphere of 100 nm diameter.) The aggregate
concentration is proportional to the quantity of C-6-S, and an average of 3 polysaccharide molecules are bound on each collagen molecule
to form the aggregates which are non polar in nature, as shown by
alternate pulse experiments. Figure 4 shows birefringence decays
obtained in this type of experiment. (Refer to caption for more
details.)

Proteoglycans have been studied by Isles et al.[28] in a work
which shows the helpful contribution that can be offered to biochemical problems by electro-optical measurements. Proteoglycans
dissolved in water at pH7 have a positive birefringence with a
multi-exponential decay from the initial slope of which a 3.6 ms
relaxation time is determined, corresponding to a length of 580 nm.
After the addition of hyaluronic acid, the positive birefringence
vanishes, while a negative, much slower birefringence appears, with
a relaxation time of 650 ms. From these results, a model of association between proteoglycans and hyaluronic acid has been proposed,
as shown on Figure 5.

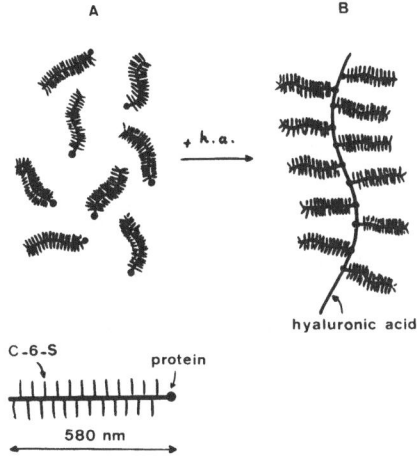

Figure 5. Proteoglycan aggregation model proposed by Isles et
al.[28]   A proteoglycan

       B hyaluronic acid - proteoglycan complex.

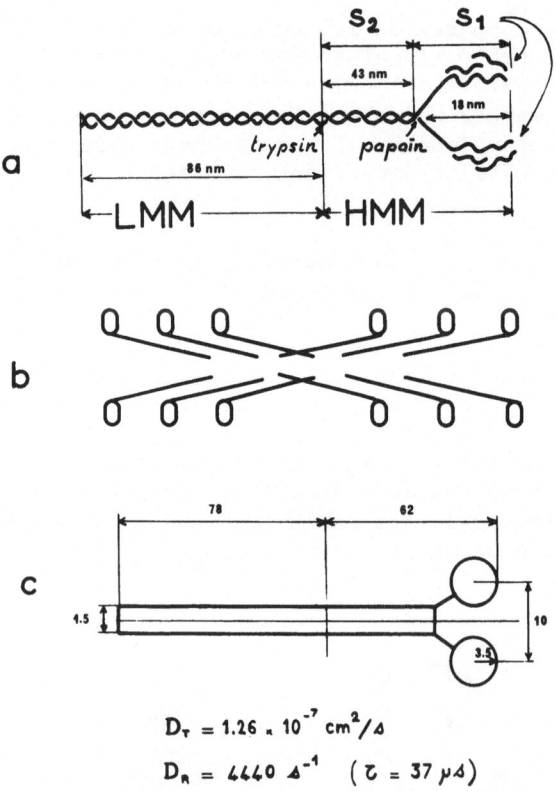

Figure 6.   (a) schematic representation of the myosin molecule, as it is seen in electron microscopy.   LMM: Light meromyosin; HMM heavy meromyosin;
            (b) the arrangement of molecules in myosin fibers;
            (c) modelization by Nakajima and Wada[33].

Muscle Proteins

The contractile machinery of striated muscle cells consists of
a small number of proteins aggregated together in filaments. Myosin
and actin are the two major components which interact to produce
contraction. If we first consider myosin, it appears as a non-
symmetric protein with a rod shaped helicoidal part and a rather
globular head composed of two subfragments S1 (Figure 6a). Trypsin
digestion dissociates the molecule into two meromyosins (at a place
of the rod which has been suggested to be a flexing point): LMM, rod
shaped with a length of 80 nm and HMM containing the two previously
mentioned subfragments S1 and the remaining helical part S2 (length
around 45 nm). S1 is responsible for the ATP-ase activity of the
myosin, and binds to actin fibers; S2 has been proposed as the
location of the mechanical effects. Figure 6b shows Huxley's model
of the molecular arrangement in a myosin filament, and it can be
seen that the association is parallel except for the central bare
zone, the content of which could depend on the anti-parallel interaction and its competition with the parallel one. That is to say
that the length of the myosin filament could be regulated "in vivo"
by the relative importance of anti-parallel and parallel interactions.
Consequently, the characterization of the dimers which were first
observed by sedimentation and viscosity measurements, then confirmed
by dynamic light scattering (Herbert and Carlson[30]) is of great
interest, as well as the interaction between S1 heads and ATP or
actin.

Unfortunately, myosin is only soluble at high ionic strength
(.2M KCl + .5M phosphate buffer), and mainly for this reason it
was not studied by means of electric birefringence until 1979.
As a matter of fact, Miyahara and Noda[29] have worked in 1M urea to
escape from the high conductivity problems, but under these conditions, it is not certain that they were studying the native protein;
above 3M urea, they reported a visible denaturation effect. Figure
7a shows a typical birefringence decay curve obtained in a 1mM·
tris 1M urea solution (pH 7.7): the short relaxation time corresponds to an apparent length of 240nm, greater than the 210 nm value
of Herbert and Carlson[30]. Flow birefringence has been used to
measure the optical anisotropy factor of myosin, but a strong velocity dependence of the birefringence was observed, and the extrapolation to high velocity gradiant was difficult to achieve, leading
to an underestimate of this optical factor. The electric polarizability and the permanent dipole moment were then calculated from
the steady state birefringence value and the alternate pulse response of the solutions. The values of the total permanent and
apparent dipole moment versus myosin concentration in different
solvents are shown in Figure 7b, from which the authors deduced
that the dimers should be anti-parallel, since the total apparent
dipole moment was decreasing when the concentration, i.e. the

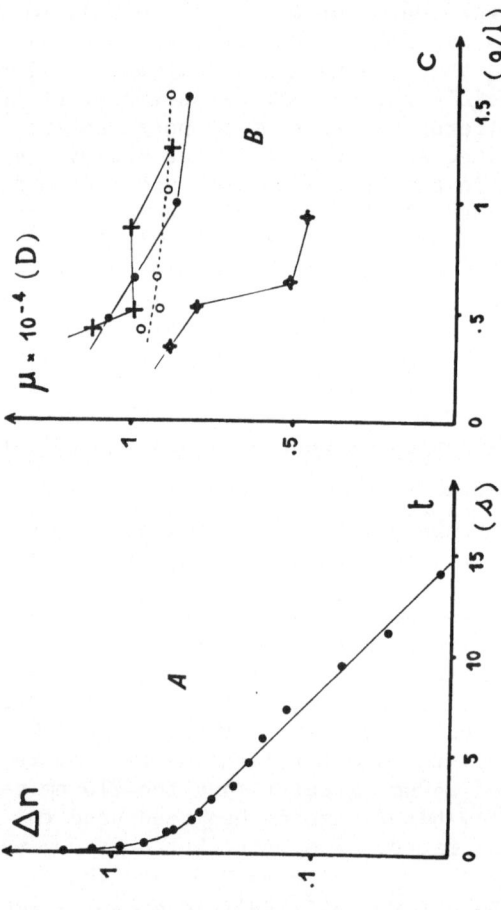

Figure 7. A. A typical birefringence decay obtained on a myosin solution in 1 M urea + 1 mM Tris
               (c = 0.6 g/l) τ_1 = .23 ms, τ_2 = 6.2 ms.

           B. Apparent dipole moment of myosin as a function of concentration:
               + 2M urea + 1 mM Tris pH 7.5.
               • 3M urea + 1 mM Tris pH 7.5.
               ✦ .3 mM solution pyrophosphate pH 7.2.

Circles correspond to the permanent dipole moment for the 2 M urea solution
(Miyahara and Noda 1979) (29).

quantity of dimers, increased. This assumption does not seem reasonable, because the permanent dipole moment, on the same figure (circles) does not change significantly with myosin concentration, as it should behave when dealing with anti-parallel association. In fact, we really think that this type of experiment has to be carried out in high ionic strength solvent, in which it is possible to cover the whole range of dimer concentrations, without any risk of protein saturation.

The different myosin subfractions, soluble at low ionic strength, have been extensively studied by electrical techniques (Minakata and Kobayashi[31]) already mentioned, and electric birefringence measurements. Kobayashi and Totsuka[32] have found two well separated relaxation times, which vary with concentration (Figure 8) for HMM subfragments. The zero-concentration extrapolation of 5 µs for the long relaxation time corresponds to a length of 55 nm, (assuming an axial ratio of 10) and the fast mechanism is attributed to the orientation of subfragment S1, found to present a relaxation time of .25 µs when measured alone (length of 24 nm, axial ratio of 7). The orientation of HMM was attributed to a slow induced dipole moment in good accord with dielectric constant measurements[31], by using the Tinoco and Yamaoka[9] method, though the authors could not be sure of the absence of a permanent dipole. A particularly noticeable result of this study is the strong Kerr constant decrease of HMM and S1 when ATP interacts with the latter, even after hydrolysis to ADP and inorganic phosphate. On the other hand, the observation that the relaxation times of these two subfragments are not significantly changed is in favor of the absence of shape modification

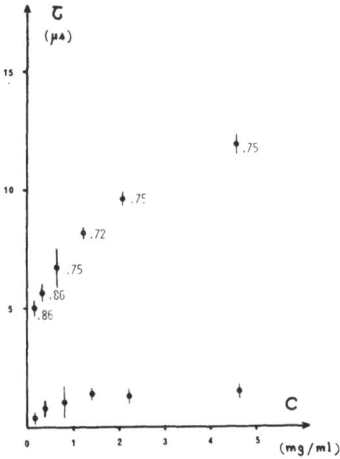

Figure 8.   Concentration dependence of the relaxation times of heavy meromyosin (Kobayasi and Totsuka[32]).

during the interaction. To conclude about myosin studies, a remark
has to be made about the hydrodynamical model of the molecule: the
application of the Perrin formula to an ellipsoidal model has to be
replaced by a more up to data and appropriate model such as that of
Nakajima and Wada[33] or Garcia De la Torre and Bloomfield's[34]
assembly of spheres. As an example, Figure 6c shows the results of
the Nakajima and Wada[33] calculation for myosin, and the computed
values of D_t and D_r are very close to the experimental values
obtained with dynamic light scattering and electric birefringence
in high ionic strength solvents.

        Actin filaments are depolymerized into globular monomers by
reducing the ionic strength (G-actin, 5.5 nm diameter). Kobayashi
et al.[35] have studied the polarizability of G-actin and of actin
filaments using the reverse pulse technique. G-actin and small
polymers gave a positive birefringence (either streaming or electric)
and a permanent dipole moment of 75 Debye per monomer. On the
other hand, actin filaments displayed a large negative electric
birefringence and a large positive flow birefringence, showing they
were oriented in a direction perpendicular to the applied field by
a transverse dipole moment. This negative electric birefringence
disappeared when HMM was added to the filaments, probably due to
the cancellation of the permanent dipole moment. An ingenious
confirmation was given by the same author[36], by measuring the
birefringence changes of F-actin oriented by flow under the application of an electric field parallel or perpendicular to the flow.
Moreover, the HMM action on F-actin was explained by a strong interaction between their two permanent dipole moments. Similar
conclusions were obtained from ultraviolet dichroism experiments
which also suggested an F-actin conformational change in the presence
of HMM, possibly involved in muscle contraction; Taniguchi[37].

        Because of the connection with a special type of contraction
called the "catch contraction", the aggregation properties of a
molluscan protein, paramyosin, have recently been examined by Krause
and Delaney[38] using transient electric birefringence. Paramyosin
is a rod shaped, positively birefringent molecule with a two chain
helical structure and a length of about 130nm (relaxation time of
about 20µs.). Two types of soluble aggregates occur depending on
the pH range: at pH 6-7, big disk shaped, negatively birefringent
aggregates which are assumed to be large lateral stacks of monomers
modeled as an oblate ellipsoids of 200 to 300nm diameter. At pH
7-10, the aggregates are positively birefringent close-packed rods
of length 200 nm in which monomers overlap each other for 20 to 60
nm of their 120 nm length, in agreement with electron micrographs of
paramyosin crystals. In this work, a comparison of pH dependence
of the relaxation time of the positively birefringent aggregates and
of the protein titration curve shows that their dissociation takes
place in the pH range where tryosine and lysine residues are

titrated. This important observation, as pointed out by the
authors, should be confirmed by measuring the relaxation times
under the ionic strength conditions (.3M KCl) which correspond to
the titration curves, and compared to some similar results obtained
on other proteins (see for instance ref. 26 on collagen).

Blood Proteins

        Since the early work of Tinoco on fibrinogen, carried out just
after the determination of its molecular weight and diffusion
constants (Katz, 1952 and Shulman 1953), an extraordinary large
number of publications have appeared on the physico-chemical and
biochemical properties of this protein, as well as on the fibrin
formation mechanism, of obvious medical importance. Unfortunately,
as pointed out by Doolittle[39] in an excellent review of this subject:
"there is no accord on the general shape of the fibrinogen molecule,
conceptions ranging from long rigid rods to globules or even symmetrical sponges" (Figure 9). Due to a strong hydration, it is impossible to decide from hydrodynamic data alone between a spherical and an elongated shape, and also to relate these data to electron micrographs (which themselves show very strong discrepancies
according to preparation). Haschemeyer and Tinoco[40] and later
Haschemeyer alone[41] extensively studied the electric birefringence
of fibrinogen and fibrin intermediates. Fibrinogen was found to be
an anisotropic molecule oriented through an induced dipole moment
at low pH and through both induced and permanent effects at high

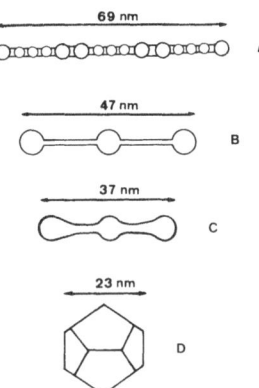

Figure 9.   Different fibrinogens models derived from electron
microscopy:
             A.   Kay and Cuddigan (1967);
             B.   Hall and Slayer (1959);
             C.   Bang (1964);
             D.   Köppel (1967).

pH (8-9). An ellipsoidal shape was assumed, with an axial/ratio
of at least 5 and a length of about 50nm. Partial conversion to
fibrin by thrombin lead to the following results:

- during fibrin formation, a polar transient intermediary
occurs, from the 4300 D dipole moment value of which it is possible
to deduce that fibrino-peptide A units are situated near the
opposite ends of an elongated molecule.

- the fibrino-peptides A are located on the same side of the
parent molecule, and the fibrino-peptides B are situated equatori-
ally.

- association between polar intermediate and fibrin monomer
corresponds to an end to end dimer, with parallel orientation.

These conclusions were widely accepted by the investigators up
to the revelation that all six amino-terminal segments are bound
together by a series of disulfide bonds presumably near the center
of the molecule. Asked by Doolittle to repeat her birefringence experiments on well characterized fibrinogen, Haschemeyer[42] did not
find any of the previously observed properties and had to recognize
that a plasma component may have been at the origin of the reported
effects. As a matter of fact, these last experiments showed only
a little birefringence at low fields, consistent with a nearly
spherical shape, and no polar intermediate was observed after
thrombin treatment. New attempts have recently been made to
characterize fibrin monomers and intermediates, for instance through
dynamic light scattering (Nemoto et al.[43] and Brosstad et al.[44]),
and we think it worthwhile to re-examine birefringence measurements
in the light of new information now available such as the formation
of soluble complexes of fibrinogen - fibrin monomer.

Electro-optics of hemoglobin have been extensively studied by
Orttung (see for instance ref. 3 for a list of references), who
showed the presence of a permanent dipole moment along the two-fold
axis from the optical dispersion of electric birefringence. In the
last paper of the series[45] the heme contribution to the polarizability anisotropy has been computed and found to be very sensitive
to the heme orientation and to the proton fluctuation anisotropy.
Electro-optics has also been used by Makinen et al.[46] to look at
hemoglobin-haptoglobin, investigated as a model of specific inter-
ations of globular protein surfaces. Figure 10 shows the permanent
and induced dipole moments of the complex Haptoglobin-Hemoglobin as
a function of saturation of Hb binding sites at pH 6.8, determined
through the Tinoco and Yamaoka method[9]. The solid curve is the
concentration of the half saturated complex, and these results
suggest that first half of the Hb molecule fits on one side of
haptoglobin, followed by the other half symmetrically cancelling

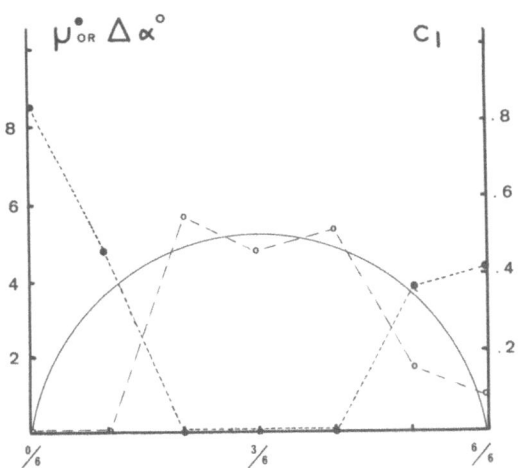

Figure 10. Dependence of the permanent dipole moment (·) and anisotropy of polarizability (0) of haptoglobin as a function of the saturation of hemoglobin binding sites at pH 6.8. (From Makinen et al. 46).

the permanent dipole which had appeared. From the increase in haptoglobin length when Hb was added, it was deduced that the subunits of Hb were bound with each pair near the terminal regions of the haptoglobin molecule.

This presentation will end with two examples of globular protein studied in a denaturing medium widely used for biochemistry experiments: Rowe and Steinhart[47] have tried to relate the relaxation time of some globular proteins in sodium dodecyl sulfate to their molecular weight. These data were re-examined some months later by Wright et al.[48], and these authors showed that the logarithmic relation found between the relaxation time and the molecular weight was in fact more accurate if, instead of considering one orientation mechanism, the decay was decomposed in two relaxation processes, corresponding to the rotation around two perpendicular symmetry axes of the molecules. Moreover, they determined the length of these two axes, assuming an ellipsoidal model. Figure 11 shows the two sets of results, compared to electrophoretic gel determinations, and it can be seen that the dispersion of relaxation time values is not much higher than obtained on electrophoresis, for a much faster experiment. There are many other valuable works on globular proteins, and the electro-optics studies of their associative properties seems highly promising, particularly if high ionic strength experiments can be achieved by improving instrumentation.

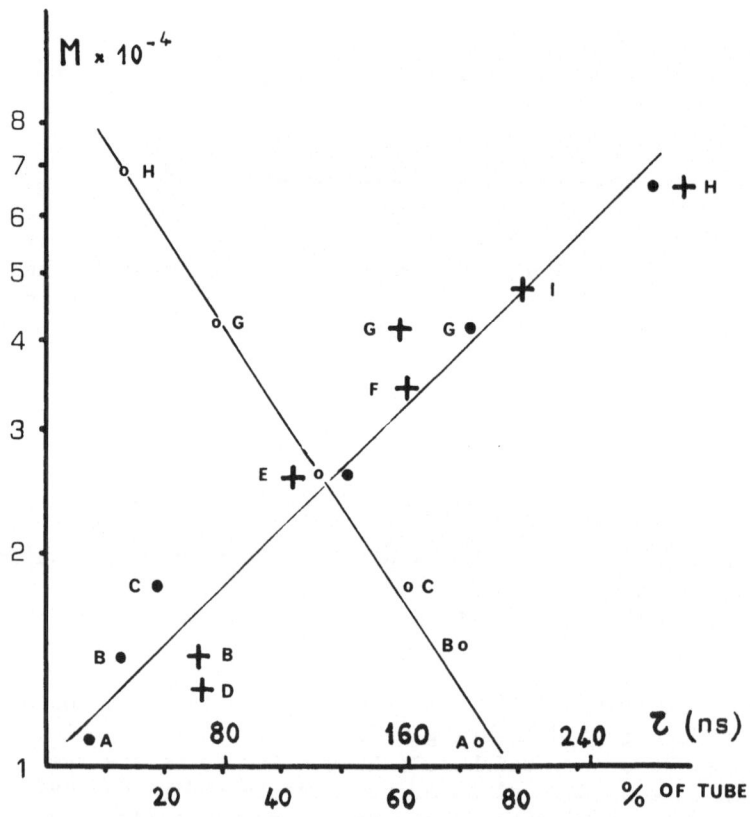

Figure 11.   Comparison between molecular weight determination by
gel electrophoresis and transient electric birefringence:
          0   10% acrylamide gel electrophoresis;
          ·   slow relaxation time (Wright et al.[48]);
          +   mean relaxation time (Rowe and Steinhardt[47]).
A: cytochrome C, B: lysozyme, C: lactoglobulin, D: myoglobin, E:
chymotrypsinogen, F: pepsin, G: ovalbumin, H: BSA, I: fumarase.

1. K. Yoshioka and C. T. O'Konski, Nat. Inst. of Gen. Med. Sciences, technical report n°4, (1965).

2. K. Yoshioka and H. Watanabe, Phys. Principles Tech. Protein Chem., A, 335-367, (1969), Academic Press, New York.

3. K. Yoshioka, Molecular Electro-optics, I, 2, 601-644, Marcel Dekker, New York (1978).

4. L. D. Kahn, Methods. Enzymol., 26, 323-337, Academic Press, New York (1972).

5. E. Fredericq and C. Houssier, Electric dichroism and electric birefringence, 167-180, Clarendon Press, Oxford (1973).

6. P. Moser, P. G. Squire, and C. T. O'Konski, J. Phys. Chem., 70, 3, 744-755 (1966).

7. M. Daune, L. Freund and G. Spach, J. Chim. Phys., 59, 485-491 (1962).

8. A. Minakata and S. Kobayashi, Biopolymers, 12, 2623-2631 (1973).

9. I. Tinoco and K. Yamaoka, J. Phys. Chem. 63, 423-432 (1959).

10. M. Hanss and J. C. Bernengo, Biopolymers, 13, 2151-2162 (1973).

11. G. B. Thurston and D. I. Bowling, J. of Coll. and Interface Sci. 30, 1, 34-45 (1969).

12. C. T. O'Konski, K. Yoshioka and W. H. Orttung, J. Phys. Chem., 63, 1558 (1959).

13. D. N. Holcomb and I. Tinoco, J. Phys. Chem., 67, 2691-2698 (1963).

14. G. Fletcher, Biopolymers, 15, 2201-2217 (1976).

15. J. C. Bernengo, B. Roux and M. Hanss, Rev. Sci. Instr. 44, 1083-1087 (1973).

16. R. Gerber, A. Minakata and D. Kahn, J. Mol. Biol. 92, 507-528 (1975).

17. L. D. Kahn and L. P. Witnauer, Biochim. Biophys. Acta, 393, 247-252 (1975).

18. C. L. Riddiford, J. Amer. Chem. Soc., 11, 2 427-429 (1978).

19. K. Yoshioka and C. T. O'Konski, Biopolymers, 4, 499-507 (1966).

20. L. D. Kahn and L. P. Witnauer, J. Amer. Leather Chem. Assoc., 64, 12-18 (1969).

21. L. D. Kahn and L. P. Witnauer, J. Applied Polymer Sci., 13, 141 (1969).

22. L. D. Kahn and L. P. Witnauer, Biochim. Biophys. Acta, 243, 388 (1971).

23. S. Ananthanarayanan and A. Veis, Biopolymers, 11, 1365-1377 (1972).

24. J. C. Bernengo, B. Roux and D. Herbage, Biopolymers, 13, 641-647 (1974).

25. J. C. Bernengo, D. Herbage, C. Marion, and B. Roux, Biochim. Biophys. Acta, 532, 305-314 (1978).

26. J. C. Bernengo, B. Roux and D. Herbage, Electro-optics and dielectrics of macromolecules, 219-230, Jennings ed., Plenum Press, New York (1979).

27. J. C. Bernengo, B. Roux and D. Herbage, Ber. Bunsen Ges. Phys. Chem., 80, 246-249 (1976).

28. M. Isles, A. R. Forewaker, B. R. Jennings, T. Hardingham and H. Muir, Biochem J., 173, 237-243 (1978).

29. M. Miyahara and H. Noda, J. Biochem. 86, 239-248 (1979).

30. T. J. Herbert and F. D. Carlson, Biopolymers, 10, 2231-2252 (1971).

31. A. Minakata and S. Kobayashi, Biopolymers, 12, 2623-2630 (1973).

32. S. Kobayashi and T. Totsuka, Biochim. Biophys. Acta, 376, 375-383 (1975).

33. H. Nakajima and Y. Wada, Biopolymers, 16, 875-893 (1977).

34. De la Torre and V. A. Bloomfield, Electro-optics and dielectrics of macromolecules, 183-196, Jennings ed., Plenum Press, New York (1979).

35. S. Kobayashi, H. Asai, and F. Oosawa, Biochim. Biophys. Acta, 88, 528 (1964).

36. S. Kobayashi, Biochim. Biophys. Acta, 88, 541 (1964).

37. M. Taniguchi, Electro-optics and dielectrics of macromolecules, 203-210, Plenum Press, New York.

38. S. Krause and D. E. Delaney, Biopolymers, 16, 1167-1181 (1977).

39. R. F. Doolittle, Adv. Protein Chem. 27, 1-109 (1973).

40. A. E. V. Hashemeyer and I. Tinoco, Biochemistry, 1, 996 (1962).

41. A. E. V. Hashemeyer, Biochemistry, 2, 851 (1963).

42. A. E. V. Hashemeyer, Thromb. Res. 7, 1, 59-65 (1975).

43. N. Nemoto, F. H. M. Nestler, J. L. Schrag, and J. D. Ferry, Biopolymers, 16, 1957-1969 (1977).

44. F. Brosstad, P. Kierulf and H. C. Godal, Thromb. Res., 14, 705-712 (1979).

45. W. H. Orttung, J. Phys. Chem., 73, 2908-2915 (1969).

46. M. W. Makinen, J. B. Milstien and H. Kon, Biochemistry, 11, 21, 3851-3860 (1972).

47. E. S. Rowe and J. Steinhardt, Biochemistry 15, 12, 2579-2585 (1976).

48. A. K. Wright, M. R. Thompson, and R. L. Miller, Biochemistry, 14, 14, 3224-3228 (1975).

INVESTIGATING NUCLEIC ACIDS, NUCLEOPROTEINS, POLYNUCLEOTIDES, AND

THEIR INTERACTIONS WITH SMALL LIGANDS BY ELECTRO-OPTICAL METHODS

Claude Houssier

Laboratoire de Chimie Physique
Université de Liège (B6)
Sart-Tilman, B-4000 LIEGE (Belgium)

INTRODUCTION

Nucleic acids and nucleoproteins molecules are particularly suitable for investigation through electro-optical methods: (i) they form relatively rigid polyelectrolytic particles which are sufficiently asymmetric and electrically polarizable to be significantly oriented in an electric field; (ii) they hold planar heterocyclic chromophores (the purine and pyrimidine bases) absorbing in an easily accessible wavelength region (the 260 nm band); (iii) many ligands of biological interest interacting with these molecules are also planar heterocyclic dye rings, absorbing in the visible or near ultraviolet range. These considerations are also applicable to some extent to polynucleotides although some of them usually form less rigid structures.

In this review we shall focus our attention on the information which can be gained by combining the observations obtained by several of these methods. Namely: (i) the conformation and conformational changes exhibited by these molecules in various experimental conditions; (ii) the arrangements of the bound ligand molecules attached to the macromolecular backbone.

We shall only consider here the effects resulting from the orientation of the particles, leaving the question of electric field effects on equilibrium and structural properties of these molecules to the other appropriate sections of this conference (see chapter by Neumann, Interaction of Electric Fields).

Interpreting the results of electro-optical studies of this type requires a knowledge of several independent pieces of

information such as the orientation mechanism of the particles
(allowing the determination of their electrical characteristics),
the assignment of the chromophore absorption bands and the determination of the direction of their corresponding transition moments.
The latter information is obtained experimentally from fluorescence polarization, linear dichroism on crystals, on stretched
films and in liquid crystal matrices. This problem has been extensively discussed in recent reviews (1,2).

The electro-optical behaviour of nucleic acids, nucleoproteins,
polynucleotides and their interactions with small ligands has been
the subject of several earlier reviews (3-5). We shall not repeat
here the discussion of the early literature but rather shall concentrate our attention on the most recent results, trying to show
the present state of the conclusions which can be derived from
these data.

We shall mainly consider here electric dichroism and birefringence measurements which have been the most widely used methods
for these studies. The fluorescence polarization in electric fields
will only be briefly mentioned since it is covered in detail in
another chapter (Weill, Polarized Fluorescence).

PARAMETERS ACCESSIBLE FROM ELECTRO-OPTICAL MEASUREMENTS

Any optical anisotropy effect resulting from the orientation
of particles in an electric field may be expressed by the general
relation:

$$\Delta n = \Delta n_s \, \Phi \tag{1}$$

where Δn is the steady-state anisotropy measured at a given amplitude of the field E, Δn_s its value at saturation of the orientation
(infinite field strength) and Φ an orientation function ($0 \leq \Phi \leq 1$)
which reflects the degree of orientation of the particle at the
field E. The knowledge of Δn_s gives access to structural parameters such as the orientation of chromophoric groups. Φ depends
on the electric factors responsible for the orientation, thus
allowing their determination from the analysis of the field strength
dependence of Δn with the aid of various theoretical orientation
functions (see Stoylov, Rotational Diffusion; and Houssier and
O'Konski, Electro-Optical Instrumentation).

When using electric pulses to orient the particles, transitory
states for Δn are generated from which the particle dimensions (from
the decay or relaxation of the signals) and the electric parameters
(from the rise and reversed phases of the signals) can be deduced.

Structural Parameters - Chromophore Orientations

The measurement of the dichroism offers the most direct means of obtaining the chromophore orientation with respect to the orientation axis of the particles. Transition moment directions are in fact obtained, and inferring chromophore orientation requires the measurement of the dichroism in at least two transitions with different and known polarizations. The transitions of the nucleotide bases and of many compounds involved in dye-interaction studies have been determined and are given in Fig. 1.

The reduced dichroism at saturation is related to the mean angle α of the transition moment μ^o with respect to the orientation axis of the particle by the expression:

$$(\Delta A/A)_s = 3(3<\cos^2\alpha> - 1)/2 \tag{2}$$

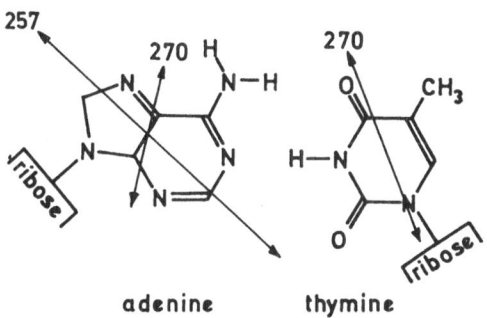

adenine   thymine

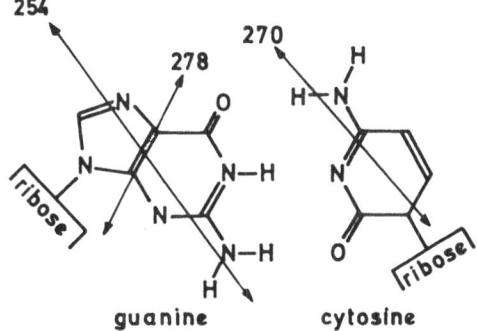

guanine   cytosine

Fig. 1a. Chemical structures and transition moment directions of nucleic acids bases (taken from Ref. 2).

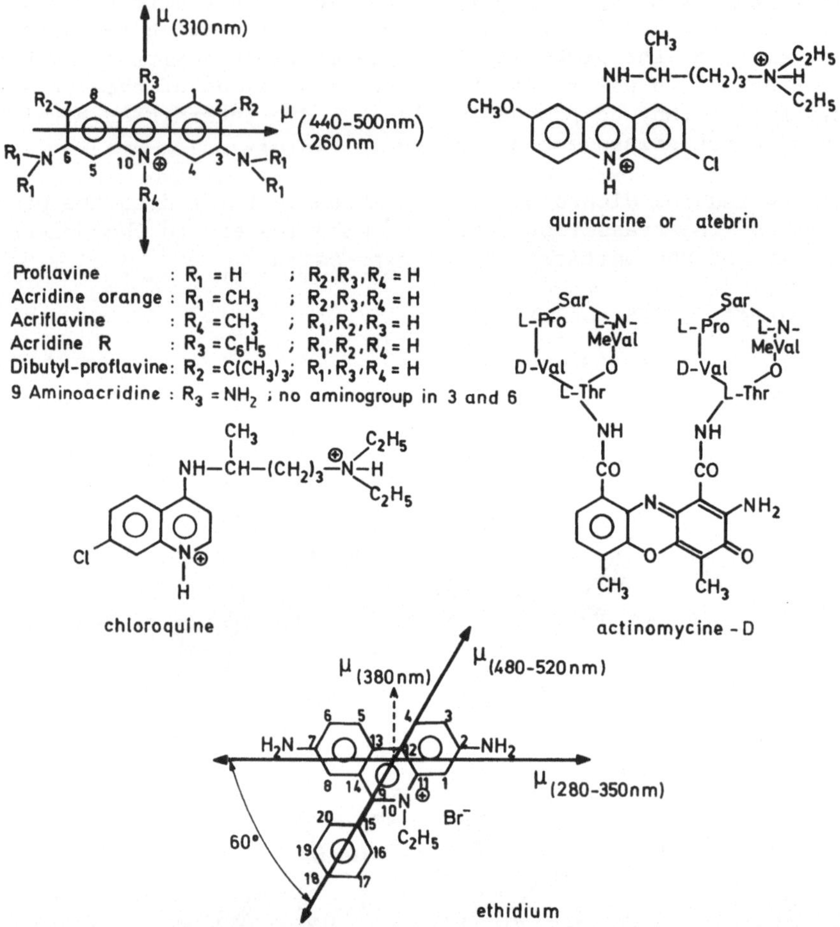

Fig. 1b. Chemical structures and transition moment directions of several important dye molecules interacting with nucleic acids and nucleoproteins (taken from Refs. 6, 7, & 49). For 9-aminoacridine, the long wavelength absorption band would be short-axis polarized (81).

where $\Delta A = A_{\parallel} - A_{\perp}$ is the difference of absorbance for the polarization directions parallel and perpendicular to the field, and $A = (A_{\parallel} + 2A_{\perp})/3$ the absorbance "at rest," i.e., in the absence of the field.

If the chromophoric groups are arranged in superhelical structures, the reduced dichroism at saturation depends on the superhelix pitch angle θ, with $\tan\theta = \pi d/p$ (with d the diameter and p the pitch of the superhelix) and on the direction of orientation of the superhelix axis with respect to the orienting force (8,9b,10). For <u>orientation parallel</u> to the superhelical axis, and irrespective of the number of superhelical turns, one obtains:

$$(\Delta A/A)^{\parallel}_{s,sh} = 3/4 (3\cos^2\theta - 1)(3<\cos^2\alpha> - 1)$$

$$= (\Delta A/A)_{s,h} (3\cos^2\theta - 1)/2 \qquad (3)$$

where the underscripts sh and h denote the superhelical and helical structures, respectively. A similar relationship holds for the birefringence (11).

When the <u>superhelix axis</u> is <u>oriented perpendicularly</u> to the field with orientation axis along the C_2 axis, the reduced dichroism will, in addition, depend on the total number n of superhelical turns and on the integer number N of complete superhelical turns according to the expression:

$$(\Delta A/A)^{\perp}_{s,sh} = - \frac{(\Delta A/A)_{s,h}}{4} [(3\cos^2\theta - 1)$$

$$+ \frac{3}{2\pi n} \sin(2\pi(n-N)) \cdot \sin^2\theta] \qquad (4)$$

The following limiting expression of equation (4):

$$(\Delta A/A)^{\perp}_{s,sh} = - (\Delta A/A)_{s,h}(3\cos^2\theta - 1)/4$$

$$= - 3(\Delta A/A)^{\parallel}_{s,sh}/2 \qquad (5)$$

is found for any total number of superhelical turns equal to an integer multiple of a half-turn ((n-N) = 0, 0.5 or 1) or for n → ∞. It is however very improbable that a long superhelix would be oriented orthogonally to the field.

Fig. 2 represents the variation of $(\Delta A/A)_{s,sh}$ with the number of turns for a few typical situations which could be encountered with DNA superhelices.

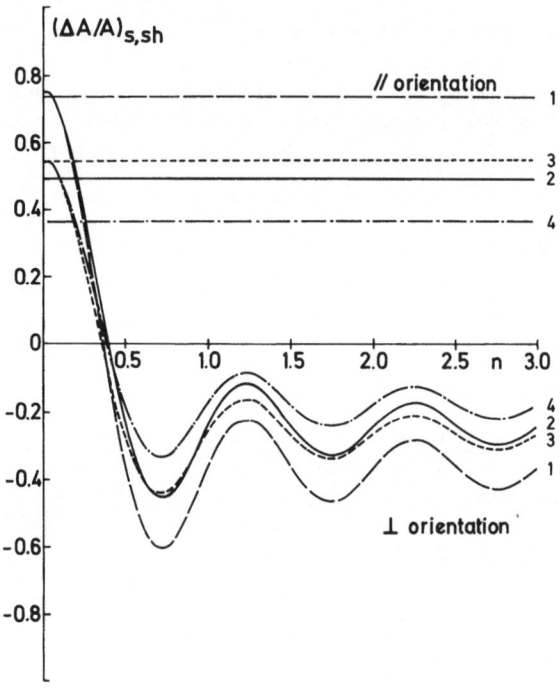

Fig. 2. Calculated reduced dichroism at saturation for superhelical
         arrangements of DNA, according to Eqn. (3) (superhelical
         axis parallel to orienting force), and Eqn. (4) as a func-
         tion of the number of superhelical turns n (superhelical
         axis perpendicular to orienting force). Curves (1,2) and
         (3,4) respectively refer to $(\Delta A/A)_{s,sh}$ = -1.5 and -1.1. The
         pitch angle values θ are 85° (curves 1 and 3) and 70°
         (curves 2 and 4).

Relations for higher order superhelices have been derived for the birefringence and for the dichroism by Maestre and Kilkson (11) and by Crothers, et al. (10), respectively.

Structural analysis on the basis of the above relationships requires the knowledge of the saturation anisotropy which constitutes the major difficulty of these studies (see below). In the case of dye-interaction studies, a comparison of the reduced dichroism observed in the dye and polymer absorption bands can give their relative orientation in a qualitative way without any requirement for extrapolation to infinite field; we shall discuss this approach in the analysis of the literature results below.

The measurement of fluorescence polarization in an electric field would in principle give more detailed information on these structural parameters. However, owing to the complexity of the relationships involved (12), only qualitative comparisons of the respective signs of the changes of fluorescence intensity for the various polarization directions have usually been made (13-15).

The influence of flexibility on the linear dichroism of helices like DNA has been analyzed by Wilson and Schellman (16) and Callis and Davidson (17). The treatment of electro-optical data for flexible polymer chains is the subject of the chapter by Jernigan and has also been reviewed previously (18). It is clear that the electric field can perturb the relative abundances of the various configurations of a flexible chain, i.e., produce distortions of the chain.

Wavelength Dependence of the Dichroism and Birefringence

The dichroism and birefringence vary with wavelength according to typical absorption and dispersion curves. If the absorption band originates from a single transition, the reduced dichroism should remain constant throughout the band; this is also true for absorption bands which are broadened by the presence of vibrational components which do not usually affect the transition moment direction. The birefringence dispersion in dichroic absorption bands shows the characteristic S-shaped curve with positive or negative troughs at high wavelength depending on the corresponding sign of the dichroism, to which it is related by the well-known Kramers-Kronig relationships (19,20). Under the assumption of Gaussian shape for the dichroic band components, the expression for Δn versus $\tilde{\nu}$ may be expressed as:

$$\Delta n(\tilde{\nu}) = \sum_i k_i \, \Delta A_{max,i}/\ell \tag{6a}$$

with

$$k_i = (2.303/2\pi\sqrt{\pi}\tilde{\nu}) \left[e^{-q_i^2} \int_0^{q_i} e^{u^2} du - \frac{\Delta\tilde{\nu}_{1/2,i}}{3.33(\tilde{\nu}_{max,i} + \tilde{\nu})} \right] \quad (6b)$$

where

$$q_i = 1.6651(\tilde{\nu}_{max,i} - \tilde{\nu})/\Delta\tilde{\nu}_{1/2,i};$$

$\tilde{\nu}_{max,i}$ is the position of the dichroism maximum (in wavenumbers) and $\Delta\tilde{\nu}_{1/2,i}$ the width of the dichroic band component at half maximum.

The question of the wavelength dispersion of the birefringence and dichroism is further discussed in the chapter by Charney.

Hydrodynamics

The hydrodynamic characteristics of the oriented rigid particles may be obtained for monodispersed systems from the field-free relaxation of the anisotropy which has the form:

$$An_D(t) = An_o \exp(-t/\tau) \quad (7)$$

where An_o is the steady-state anisotropy and $\tau = 1/6\Theta$ the relaxation time related to the rotational diffusion coefficient Θ .

The rise of the electro-optical signals depends, in addition to τ, on the electrical parameters through the relation:

$$An_R(t)/An_o = 1 - \frac{3R}{2(R+1)} \exp(-t/3\tau) + \frac{R-2}{2(R+1)} \exp(-t/\tau) \quad (8)$$

where R is the ratio of permanent over induced dipole terms:

$$R = P/Q = \beta^2/2\gamma \text{ , with } \beta = \mu E/kT \text{ and } \gamma = \Delta\alpha E^2/2kT$$

The initial slope of the rise signal is given by:

$$\left(\frac{d(An(t))}{dt} \right)_{t \to o} = \Delta\alpha E^2 An_s/15\tau kT \quad (9)$$

A very useful way of analyzing the orientation mechanism is to measure the frequency dispersion of the anisotropy which follows the relation:

$$\frac{An_{av}}{An_o} = \frac{1}{1+R}\left(1 + \frac{R}{1 + (3\omega\tau)^2}\right) \tag{10}$$

where An_o is the anisotropy in d.c. field and An_{av} is the steady-
state component of the signal at the frequency ω. The relaxation
time may be deduced from the critical frequency ω_c at which the dispersion curve reaches its inflection point ($\omega_c = 1/3\tau$). Equations
(8) and (10) are limiting forms which are only applicable in the
Kerr law region (low fields where An varies linearly with E^2), while
Equation (9) has been shown to hold at any field strength (21).

Even though the above analysis could be performed from any
electro-optical signal, it is most frequently made from birefringence measurements which usually offer the highest sensitivity. The
advantage of using the dichroism signal for this determination is
that it does not usually contain any solvent contribution which may,
however, yield complex birefringence signals. We also consider it
preferable to determine the hydrodynamic parameters from the field-
free decay because it is independent of the orientation mechanism.

It is not possible to infer the shape of anisotropic particles
from electro-optical relaxation data alone. Even though three rotational diffusion coefficients could be expected for monodispersed
suspensions of particles of general ellipsoidal shape, only two are
generally sufficiently different to be reasonably distinctly observable in practical cases (22). This is of course the case for revolution ellipsoids (23). In addition, in the case of long particles
with cylindrical symmetry, only one relaxation decay will be observable, either because rotation about the symmetry axis is much too
fast or because it does not result in changes of the recorded optical property.

For flat particles (disks or oblate ellipsoids) the three relaxation times will be usually so close to one another (22, 23) that
they will not be distinguishable. Experimentally, a relaxation decay consisting of two exponential terms with τ values not differing
by more than a factor 2-3 will not be distinguishable from a monoexponential decay if the two terms have comparable amplitudes. The
presence of two distinct relaxation times were however claimed to
be detected on monodispersed proteins and tRNA samples by electric
birefringence decay measurements with an instrument of fast response
(24, 25).

The rotational diffusion coefficient Θ of particles with
cylindrical symmetry is related to their long (a) and short (b)
semi-axes through the general Perrin equation:

$$\Theta = \frac{kT}{\pi\eta_o} \frac{1}{8\ a\ b^2}\ f(p) \tag{11}$$

where kT is the Boltzmann energy term, η_o the solvent viscosity
and f(p) a function of the axial ratio p (=a/b) which has the
following forms for different particle shapes (only rotation of the
longest axis has been considered):

spheres: $a = b\ ;\ p = 1\ ;\ f(p) = 1$ \hfill (12a)

prolate ellipsoids: $a > b\ ;\ p \geq 2\ ;\ f(p) = \frac{3}{p^2}\ (\ln\ 2p - 0.5)$ \hfill (12b)

oblate ellipsoids: $b > a\ ;\ p < 1\ ;\ f(p) = \frac{3p^4}{4(1-p^4)}\left[\ -1\right.$

$$\left. + \frac{2 - p^2}{p^3(1-p^2)^{1/2}}\ \arctan\ \frac{(1-p^2)^{1/2}}{p}\right] \tag{12c}$$

long rods: $a > b\ ;\ p \geq 3.5\ ;\ f(p) = \frac{3}{p^2}\ (\ln\ 2p - \gamma)$ \hfill (12d)

with $\gamma = 0.8$ (Burgers, see Ref. 26)

or $\gamma = 1.57 - 7(\frac{1}{\ln\ 2p} - 0.28)^2$ (ref. 26)*

At 20°C in water ($\eta_o \simeq 0.01$ Poise = 0.001 kg m^{-1}s^{-1}), the term kT/$\pi\eta_o$
amounts to 1.288 x 10^{-18} m^{3}s^{-1} so that the field-free relaxation time
is given by:

$$\tau(\mu s) = 0.129\ x\ 10^{-6}\ (8ab^2)/f(p) \tag{13}$$

with a and b in Å.

The treatment of dynamic aspects of electro-optical properties
of flexible polymer chains has been previously reviewed (18) and is
treated in the chapter by Jernigan. According to the

---

* New coefficients have been recently given for the γ
expression (96, 97) which should read:

$$\gamma = 1.45 - 7.5\ (\frac{1}{\ln\ 2p} - 0.27)^2.$$

An analysis according to the rigid rod model of Tirado and Garcia
de la Torre (98) can also be made.

Zimm-Rouse model, the τ_k value for the kth mode of relaxation of a non-free draining chain is given as (see Ref. 17):

$$\tau_k = M_w \eta_o [\eta]/0.586 \, RT \, \lambda_k \qquad (14)$$

with λ_k = 4.04, 12.8, 24.2 and 37.9 for k = 1, 2, 3 and 4, respectively. Knowing the dependence of the intrinsic viscosity of the polymer on its molecular weight gives by substitution the molecular weight M_w dependence of τ. The fraction of the signal decaying with the longest relaxation time τ_1 has been estimated to 87% (17).

The analysis of the relaxation behaviour of polydispersed systems is a very complex problem, especially when this effect is superimposed on a flexibility effect. In principle the length distribution should be obtainable from an analysis of the multiexponential electro-optical decays. The noise in the recorded signals, however, hinders this analysis and more than three relaxation components are not usually extractable from the decays. Various relaxation time averages are obtained from determinations at different field strengths and from different characteristics of the decay curves (initial slope, total area). These are helpful for getting several important characteristic parameters of the distribution curve. We shall not further discuss this question here and we advise the reader to consult the appropriate literature (27-32).

Electric Parameters

The analysis of the field strength dependence of the measured anisotropy can give the intrinsic anisotropy at saturation (indispensable for the analysis in absolute structural terms; see above) and the electric parameters of the oriented particles. This treatment is performed through a fitting of the experimental data with the aid of theoretical orientation functions $\overline{\Phi}$; these have been presented in other chapters (Stoylov, Rotational Diffusion Function; O'Konski, Kerr Effect; Houssier & O'Konski, Electro-Optical Instrumentation).

It is better to have a preliminary knowledge of the orientation mechanism in order to make the appropriate choice of $\overline{\Phi}$ rather than to try to infer this mechanism from the shape of the field strength dependence of the anisotropy. Three experimental parameters are useful to determine this mechanism: 1) the ratio of the rise over decay areas of the normalized electro-optical signals which is given by:

$$S_R/S_D = (4R + 1)/(R + 1) \qquad (15)$$

and thus ranges from 1 to 4 for R varying from 0 (pure induced
polarization mechanism) to ∞ (pure permanent dipole mechanism);
2) the depth of the minimum under bipolar pulse which also yields
R from the relation:

$$R = [1 - (An_{min}/An_o)]/[0.1547 + (An_{min}/An_o)] \qquad (16)$$

where An_{min} is the minimum of the transient reached immediately
after the field reversal; 3) the ratio of the anisotropy in d.c.
field to the limiting anisotropy at high frequency (where the permanent dipoles cannot follow the alternating field anymore) giving R
by:

$$R = \frac{An_o}{\lim\limits_{\omega \to \infty} (An_{av})} - 1 \qquad (17)$$

Of the three above methods, the third one certainly contains
the largest wealth of information. When carried out with light-
scattering data, this analysis directly allows the unambiguous determination of the electric parameters since the optical term is
obtainable from a measurement in the absence of orienting field (33).
However, the treatment of light-scattering data is more difficult
and the analysis of the frequency dispersion a bit tedious, so that
we consider the second procedure probably is the easiest and most
direct way to obtain the ratio of permanent over-induced dipole
contributions to the orientation mechanism.

Relation (16) (as well as (15) and (17)) is only strictly
applicable at low fields (Kerr law region) to particles showing
cylindrical symmetry (34) such as revolution ellipsoids, rods, flat
disks. Matsumoto, et al. (35) extended the theory to intermediate
fields for the same type of particle symmetry. The depth of the
minimum becomes deeper with increasing field at a given R value.
This means that, in the worst situation, one would overestimate the
permanent dipole contribution by measuring the reversed transients
at intermediate or high fields.

The permanent dipole component appearing in relation (16)
through the R parameter (see Eqn. (8)), lies along the symmetry
axis of the particle. If transverse components μ_b, μ_c have to be
included, then the term $\beta^2/2\gamma$ has to be replaced by (34):

a) $\beta_a^2/[2\gamma - (\beta_b^2/2)]$ for elongated particles (long rods or prolate
    ellipsoids with relaxation time much larger along symmetry
    axis, $\tau_a \gg \tau_b$), and by

b) $[\beta_a^2 - (\beta_b^2/2)]/2\gamma$ for flat disks or oblate ellipsoids (with

$\tau_a \approx \tau_b$), with $\beta_a^2 = (\mu_a E/kT)^2$ and $\beta_b^2 = (\mu_b^2 + \mu_c^2)E^2/2(kT)^2$.

Complex shapes of the electro-optical signals may be observed in these cases (34).

SURVEY OF THE LITERATURE RESULTS

DNA and Chromatin

Orientation mechanism and field strength dependence of the anisotropy. Apart from the still-unexplained anomalous behaviour observed with sonicated DNA samples (36, 37), it seems now well-established that fully-deproteinized DNA orients in an electric field by an almost pure induced polarization mechanism as revealed by bipolar pulse experiments and by the rise over decay areas of the electro-optical signals under rectangular pulses (3, 37-40). The report of Ding, et al. (41) appears to be definitively due to an experimental artifact (37, 39). However, the frequency dispersion showed a low frequency dispersion more consistent with a permanent dipole orientation mechanism, which made it necessary to invoke special mechanisms not yet fully understood (42-45). The question of the electric polarizability of polyelectrolytic macromolecules being the subject of another section of this conference (see the chapter by Mandel), we shall not discuss it further here.

It is also noticeable that the shape of the field strength dependence of the DNA anisotropy is not satisfactorily described by the orientation function for a pure induced dipole $\overline{\underset{I}{\phi}}$ (γ).

Several new functions have been presented (46-48) which derive from the idea that the induced dipole would saturate at high fields. These functions, however, do not appear to satisfactorily fit the data: (i) the Kijuchi and Yoshioka function (46) yields a too pronounced curvature at high fields, and too low values of the number of weakly bound counterions have to be chosen to obtain the best fit (49); (ii) the functions derived by Hogan, et al. (47) and by Sokerov and Weill (48) predict a linear dependence on E at low fields which is in fact observed for high molecular weight DNA (3, 50) but not for sonicated (49) or nucleosomal DNA (51) (see Fig. 3).

The electric parameters of DNA deduced from electro-optical data are relatively consistent with one another, despite the above-mentioned difficulties. Values of the polarizability term $\Delta\alpha$ range from (10 to 20) x 10^{-32} F m^2 ((1 to 2) x 10^{-15} cm^3) for

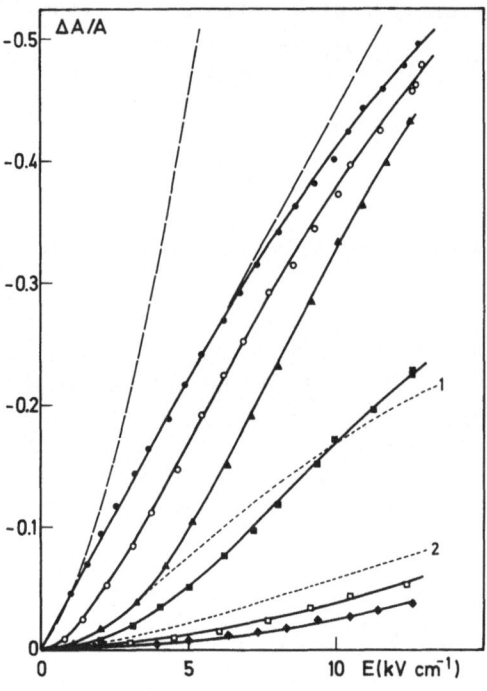

Fig. 3. Field strength dependence of the electric dichroism of DNA
         and chromatin subunits.

         (---): sonicated calf-thymus DNA (M_w = 100,000 to 300,000)
                 at (1-4) x 10^{-4} ionic strength (taken from Refs. 41
                 and 66).

         DNA in 1 mM NaCl: calf-thymus ($\bullet$; M_w = 2 x 10^6); chicken
                 erythrocytes (0; sonicated sample: M_w $\approx$ 300,000); mono-
                 nucleosomal ($\blacktriangle$; 170 b.p.).

         nucleosomes from chicken erythrocytes: H1/H5 depleted
                 sample ($\blacksquare$), native mononucleosomes ($\square$) and core par-
                 ticles ($\blacklozenge$; EXOnuclease treatment); all samples at
                 about 0.001 ionic strength (Ref. 51).
         oligonucleosomes from chicken erythrocytes (5 to 10 mono-
                 nucleosomes in the chain): curves 1 and 2 at 0.5 x 10^{-3}
                 and 2 x 10^{-3} ionic strength, respectively (Houssier,
                 Lasters, Muyldermans and Wijns, to be published).

         Electric birefringence studies of the electric polariza-
tion and relaxation behaviour for restriction fragments of DNA
(91, 92) and for fractionated sonicated DNA (93) have been presented during this Advanced Study Institute and will probably be
of definite importance in this field of research.

sonicated and native DNA, respectively (3, 13, 37, 49, and 52), and $(2-4) \times 10^{-32}$ F m^2 for nucleosomal DNA (51). If reinterpreted in terms of an induced dipole saturating at high fields (above 6 kV/cm), "saturating-dipole moments" of 6000 D (2×10^{-26} C m) (Ref. 48) to 9000-13000 D (47) are found.

Thus, although the question of the orientation of polyelectro-lytic macromolecules in an electric field, and of DNA in particular, has been reinvestigated in several recent works, there does not seem to exist any totally consistent interpretation which would satis-factorily account for all the available experimental observations. A thorough investigation by dielectric and electro-optical methods on monodispersed short fragments of DNA would be useful since pre-vious investigations were dealing either with high molecular weight DNA samples with important polydispersity and flexibility, or with sonicated DNA which displayed an anomalous behaviour in pulsed elec-tric fields (36, 37) as well as in the frequency dispersion of its Kerr constant (45).

In the case of chromatin, early observations on sonicated samples using bipolar-pulse experiments revealed the predominence of an induced polarization mechanism (37, 38) but with a slightly larger permanent dipole contribution than for DNA, however. The electric polarizability values obtained from the fitting of the field strength dependence of the birefringence were of the order of $(3-5) \times 10^{-32}$ F m^2 (($(3-5) \times 10^{-16}$ cm^3), and the permanent dipole term of the order of 3000 D (10^{-26} C m). It is now known from elec-tron microscopic observations that sonicated or sheared chromatin contains fragments with long internucleosomal segments resulting from disruption of nucleosomes. The question of the chromatin elec-tro-optical properties is now being reinvestigated with nuclease digested samples in which the beads-on-a-string structure is pre-served.

Chromatin subunits have been recently investigated by the elec-tric birefringence (53, 54) and dichroism (55-58) methods. In the study of Marion and Roux (54) the ratio of the rise over decay areas of the birefringence signals for H1-containing mononucleosomes yielded an R value of the order of 0.5, indicating a predominent induced-dipole interaction mechanism. Crothers, et al. (55, 56) concluded from the absence of variation of the dichroism with ionic strength in the 2-5 mM range, and from an analysis of the field strength dependence of the dichroism for nucleosomal core particles (containing 140 and 175 b.p. and the octamer of histones H2a, H2b, H3, H4, but no H1) that they were oriented by a permanent dipole of about 1100 D (3.7×10^{-27} C m) lying in the plane of the core-particle disk plane, along the C_2 axis. In the light of the above

considerations on DNA, we consider that the shape of the field
strength dependence of the anisotropy is not sufficiently characteristic to allow the inference of the orientation mechanism from
its analysis only.

Recent studies in our laboratory (51) on mononucleosomes subjected to bipolar electric pulses yielded R values ranging from
0.3 to 0.7 depending on the H1/H5 content of the samples. The
shape of the birefringence transients were interpreted by a disk-
shaped structure having a negative electric polarizability term and
a permanent dipole contribution in the disk plane (with a predominence of the induced dipole term, however).

Dichroism, birefringence and structural implications. Large
discrepancies appear between the observed electric dichroism amplitudes reported in the literature for the 1 to 15 kV/cm field-strength
range and for various DNA samples (3, 41, 47, 59, and 60). Some
reconciliation of these results may be achieved if the ionic strength
conditions, sample molecular weight and concentration are taken into
account. In order to avoid denaturation at room temperature, DNA at
concentrations lower than 10 mg/dl ($A_{260nm}^{10} \leq 2.$) must be kept at
ionic strengths not lower than 0.001, for which the melting temperature is of the order of 50-60°C, depending on the residual salt
present in the solid DNA samples and on the G-C content. Ionic
strengths as low as 0.0001 may be used at higher pH (e.g., pH 8).
The effect of ionic strength on the electric birefringence and dichroism of high molecular weight DNA has already been reported (3).
In our latter studies, DNA was most frequently studied in unbuffered
solutions at 0.001 ionic strength. We never noticed any significant
influence of the nature of different buffered media on the birefringence and dichroism in this ionic strength range (cacodylate,
citrate, phosphate or Tris-HCl). The electric dichroism versus field
strength curves given in Fig. 3 are representative of numerous measurements done in our laboratory on many different DNAs from various
sources (calf-thymus, chicken erythrocytes, micrococcus). This
figure compares the behaviour of high-molecular-weight DNA with
sonicated and nucleosomal DNA, and to nucleosomes. We never extended
our measurements above 12-14 kV/cm and hope in this way to have
avoided possible perturbations of the double-stranded structure by
the electric field (61). The variation of the dichroism with the
polarization angle supports this assertion (see Houssier & O'Konski,
Electro-Optical Instrumentation).

An even more controversial question is that of the saturation
dichroism of DNA from which the inclination of the base pairs with
respect to the helix axis is deduced. Our analysis on the basis
of an induced polarization mechanism yields values as low as -(0.6
to 0.8) at 0.001 ionic strength (37, 59). When analyzed in terms
of a permanent dipole orientation mechanism at high fields, much
higher negative values are found (-1.1 (Ref. 47) at about 0.002

ionic strength, and $-(1.3$ to $1.4)$ at very low ionic strength (41)
or for DNA samples alleged to be in a "bundled" state (60)). From
their limiting dichroism value observed on short DNA fragments,
Hogan, et al. (47) concluded, using Equation (2), that the base
planes are inclined at an angle of $(73 \pm 3)^{\circ}$ with respect to the
helix axis, a result in agreement with the revised B-form DNA model
presented by Levitt (62). Although we do not feel very much motiv-
ated by this kind of controversy, our opinion is that even very
short DNA fragments retain some internal flexibility (they can be
wound in a superhelix in the nucleosomes after all). The slight
bending of the chain, clearly apparent in the electron microscopic
pictures of DNA and dependent on the experimental conditions, would
be responsible for the lower dichroism values as compared to the
limiting dichroism expected for a perfect perpendicularity of the
base planes $((\Delta A/A)_s = -1.5)$. Lowering the ionic strength, and
perhaps also increasing the field strength could produce a straight-
ening of the chain which would finally give the expected limiting
dichroism value.

Birefringence versus field strength curves show, of course, the
same shape as the corresponding dichroism curves; the $\Delta n/A_{260nm}^{10}$ at
12.5 kV/cm for DNA at 0.001 ionic strength is of the order of
$-(18$ to $20) \times 10^{-7}$. The optical anisotropy factor $g_a - g_b$ obtained
from the analysis of these curves is of the order of -2×10^{-2} (Ref.
3, 63) in agreement with other reports (4, 48).

Rill and Van Holde (9a) early reported the very low electric
dichroism of nuclease-resistant chromatin fragments obtained by
limited digestion of calf-thymus chromatin with micrococcal nuclease.
They have already suggested a compact, perhaps superhelical, arrangement of DNA in these subunits. In more recent and extensive studies
on nucleosome core particles, Crothers, et al. (56) interpreted the
negative dichroism observed for the 175 b.p. $((\Delta A/A)_s = -0.49)$ and
for the 140 b.p. $((\Delta A/A)_s = -0.29)$ particles by a disk-shaped model
oriented with the DNA superhelix axis perpendicular to the field
and with 100 b.p. per turn (see Fig. 2 and Equation (4)); the presence of a permanent dipole along the C_2 axis was assumed in the
analysis of the results. A model was also proposed for the spacer-
less dinucleosome (55). Electric dichroism was also used to study
a reconstituted nucleosomal particle whose condensation into a
native structure was induced by the addition of lysine-rich histones
(57). The structural transition encountered by nucleosome core
particles in the 1 mM salt-concentration range was found to result
in a large increase of electric dichroism, and of assigned permanent
dipole contribution. This was interpreted as being due to an unfolding of the nucleosome as the ionic strength decreases, to a 140 b.p.
segment for each 0.9 superhelical turn (58). Cross-linking of

nucleosomal proteins by dimethylsuberimidate inhibited this transition.

The electric birefringence of rat liver oligonucleosomes containing H1 displayed a negative birefringence for less than 6
particles in the chain, with a maximum amplitude for the dinucleosome. The absence of birefringence for the hexanucleosome was
interpreted on the basis of a symmetric helical structure which
could be one of the repeating unit of chromatin in some conditions
(54).

From a comparison of the nucleosome fractions obtained by
sucrose gradient and by Sepharose 6B chromatography, we observed
that prolonged chromatographic separation produces a release of
histone H1, yielding a larger negative anisotropy (64). Sucrose
gradient fractions of mononucleosomes from calf-thymus displayed an
almost zero or slightly positive birefringence (64). A more thorough reinvestigation of the role of H1/H5 on the DNA coiling in the
nucleosome using electric dichroism and birefringence techniques
recently led us to consider the effect of nucleosome tails unfolding
as well as core-DNA unwinding in the interpretation of the results
(51).

Early observations on gel-forming nucleohistone (3) and sonicated chromatin (37, 38) showed a lower negative anisotropy as
compared to DNA. This is now easily understandable on the basis of
the nucleosomal subunit structure. However, this question will have
to be reinvestigated using long chromatin fragments retaining an
intact nucleosomal structure. Preliminary measurements indicated
that the general trends of the observations are preserved, with
of course the condensation effect of divalent cations (63, see
below).

Wavelength dependence of the dichroism and birefringence. The
reduced dichroism spectrum of DNA shows a slight decrease above
280 nm. (41, 50). No definite interpretation of this observation
has been given: the presence of out-of-plane $n \rightarrow \pi*$ transitions is a
possible explanation as well as the effect of transitions from
guanylic and cytidylic residues (50). The dichroism spectrum that
we previously reported (3) shows larger decreases of $\Delta A/A$ above
280 nm and below 240 nm than reported in any other work; this possibly arises from the lower accuracy of the measurements in these
wavelength regions with our early instrumentation. It seems that
measurements using Xe-Hg light sources may show anomalous variations
of the dichroism·at the peaks characteristic of these arc lamps. It
is advisable to determine the absorption spectrum of the samples
with the dichroism instrument setting in order to check for such
possible artifacts. One should also be very attentive to stray-light
perturbations with these high intensity sources.

The birefringence measured in the visible region for nucleic
acids and nucleoproteins arises from ultraviolet transitions, predominantly the in-plane $\pi \rightarrow \pi^*$ transitions of the purine and pyrimidine
bases. The 260 nm band is not mainly responsible for this effect
but rather other transitions of higher intensity located below 200
nm as revealed by the electric birefringence dispersion curve of
DNA determined by Yamaoka and Charney (50) in the 220-320 nm range.
A rough estimation of the contribution arising from the 260 nm band
using Equation (6) with a reduced dichroism of -0.5 (at 12.5 kV/cm)
(see Fig. 3) yields for $\Delta n_{550}/A_{260nm}^{10}$ a value of the order of
-2.5×10^{-7}, i.e., about 7-8 times lower than the actually measured
value at this field strength. In the case of nucleoprotein complexes, the ratio of the birefringence in the visible range Δn_{550}
to the dichroism at 260 nm, ΔA_{260}, drops from about 43×10^{-7} (for
DNA) to $(23-25) \times 10^{-7}$. Several factors may be responsible for this
effect: (i) a positive contribution of the protein moiety to the
birefringence; (ii) a decrease of the intrinsic negative birefringence and an increase of the form birefringence contribution
as a result of the structural alteration of the DNA in the nucleoprotein structure. The ratio $\Delta n_{550}/\Delta A_{260}$ may in fact be used as a
relatively sensitive probe of the protein content and structure of
chromatin samples (51).

Hydrodynamic behaviour. Long DNA chains are flexible and
follow quite well the bead-and-spring model of Rouse and Zimm. At
very low electric fields, the longest relaxation time obtained from
the birefringence decay was found independent of field strength and
was considered as reflecting the orientation of the macromolecule
as a whole (40b). Such analyses are, however, complicated by polydispersity effects unless viral DNA samples are used (41, 65).
Sonicated samples approach the behaviour of rigid rods. This behaviour is perfectly followed by short DNA fragments extracted from
nucleosomal subunits or obtained by restriction enzyme treatment
(47). Fig. 4 summarizes the relaxation time values reported for
DNA samples with molecular weights ranging from 100,000 to 125×10^6.
The persistence length of DNA has been estimated of the order of
500-650 Å (16, 65, 66) and the problem of the rigidity of double-
stranded polynucleotides structures has been thoroughly discussed
(66).

We never noticed with our DNA samples (even those extracted
from nucleosomal particles) any anomalous increase of τ below
2 mM salt, as reported by Hogan, et al. (47), provided that the
cautions to avoid denaturation have been taken, as discussed above.
We never observed either any effect comparable to that attributed
by Mandelkern, et al. (60) to a bundling of short DNA fragments into
specific aggregates of about 7 parallel molecules as the ionic

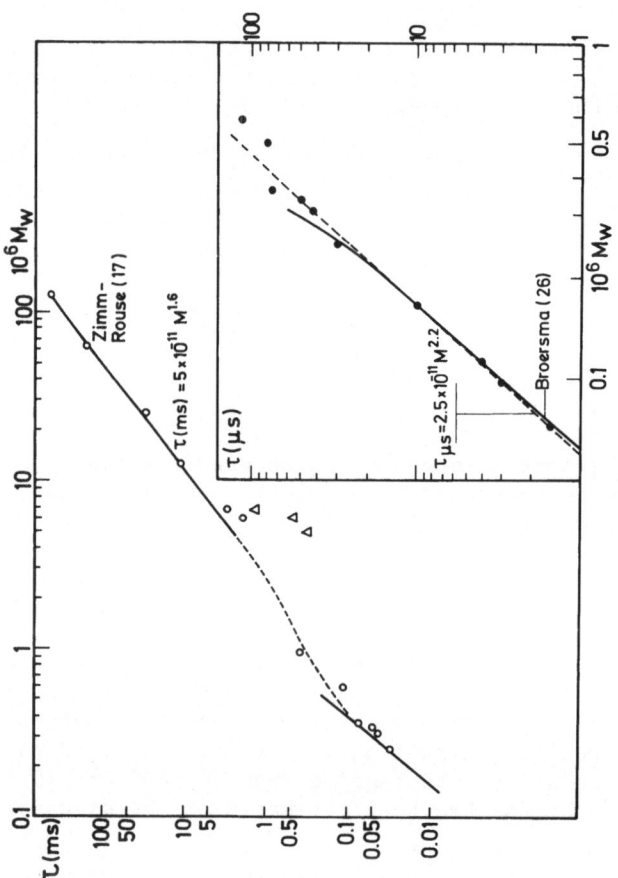

Fig. 4. Molecular weight dependence of DNA relaxation times (data from Ref. 17, 41, 47, and 65). The circles represent the longest relaxation times, the triangles mean relaxation times. Monoexponential decays were observed with the lowest molecular-weight samples.

Electric birefringence studies of the electric polarizability and relaxation behaviour for restriction fragments of DNA (91, 92) and for fractionated sonicated DNA (93) have been presented during this Advanced Study Institute and will probably be of definite importance in this field of research.

strength is decreased, or the concentration increased above 10^{-5}-10^{-4} M in base pairs. Measurements in our laboratory on 160-180 b.p. nucleosomal DNA in 1 mM NaCl at absorbances of 0.5 to 1.0 yielded relaxation times of the order of 3.4-3.6 μs in perfect agreement with a rigid rod behaviour (51). It is possible that the rise in temperature caused by the condenser discharge method or the use of the rise of the dichroism signals to determine τ , as carried out by the Crothers group may have some responsibility in these findings.

The relaxation data of nucleosome particles have been analyzed on the basis of models of oblate ellipsoids or spheres (Eqs. (11) to (13)). From the rise of the dichroism signal ($\tau_r \sim 0.6$ μs) and assuming a permanent dipole orientation mechanism ($\tau_r = 1/2\Theta$ for the long time portion of the rise curve; see Eqn. (8)), Crothers, et al. (56) estimated a value of 130 $\overset{o}{A}$ for the longest distance across the 140 b.p. particle, which is compatible with the disk model of Finch, et al. (67). For this calculation, the lowest re-laxation time above the 1 mM-salt structural transition was used and the measurements in viscous Dextran solutions were extrapolated to zero Dextran concentration.

The field-free relaxation data of Marion and Roux (54) on H1 containing nucleosomes ($\tau_d = 0.6$ μs) would be compatible with an oblate ellipsoid of larger dimensions than calculated by these authors (namely $\sim 100 \times 200 \times 200$ $\overset{o}{A}$ overall dimensions for an axial ratio of 0.5, or $\sim 68 \times 205 \times 205$ $\overset{o}{A}$ for an axial ratio of 1/3, using Equation (12c)). Let us recall that the shape of the particles cannot be inferred from electro-optical relaxation data and that the field-free relaxation is related to the rotational diffusion coef-ficient by the relation $\tau_d = 1/6\Theta$ irrespective of the orientation mechanism. The hydrodynamic behaviour of rat liver chromatin sub-units has been recently reinterpreted (94) on the basis of models developed by Garcia-de-la-Torre and Bloomfield (95).

The sensitivity of the relaxation times to conformational changes has also been used to follow the effect of gamma-rays on superhelical PM2 and linear calf-thymus DNA (68).

RNA, Ribosomes

Early studies on the electro-optical behaviour of RNA and ribo-somes have been previously reviewed (3, 4). Few recent reports have appeared on this subject. This mainly arises from the low degrees of orientation attainable with these molecules and from their more

irregular structures as compared to the double helical and superhelical structures of DNA and chromatin.

The determination of the dimensions of a purified yeast tRNA^{Phe}
sample has been recently reported (25). Two relaxation times were
observed (18 and 44 ns) corresponding to a prolate ellipsoid of
(54.1 x 18.7 x 18.7) Å. Above 20°C, the relaxation time and the
negative birefringence decreased sharply and the Δn became positive
above 40°C, as a result of thermal denaturation of the molecule. A
decrease of τ with increasing tRNA concentration and upon Mg^{2+}
addition was attributed to the formation of a more compact structure
(25).

Ribosomal RNA showed a negative dichroism at 260 nm (amounting
to about -0.11 at 13 kV/cm; Ref. 69) while no measurable signal
could be detected with ribosomes in the same experimental conditions
and in the presence of Mg^{2+}. Mg^{2+} complexation by EDTA, and dye interactions produce an unfolding of the ribosome structure.

Polynucleotides

The electric dichroism of poly(A) has been thoroughly investigated recently by Charney and Milstien (66). The influence of
molecular weight revealed an increased flexibility with increasing
chain length. The base stacking was found to be retained in single-
stranded poly(A) which behaved as a rod-like structure similar to
the double-stranded polymer. The amplitude of the negative reduced
dichroism at saturation was however unexpectedly low (-0.6 at
2.5 x 10^{-4} M buffer concentration), which yielded these authors to
envisage the possibility of perturbations in the electronic absorption properties giving rise to out-of-plane components of the transition moments.

One of the important aims in studying the electro-optical properties of synthetic polyelectrolytes such as polynucleotides, poly-
styrenesulfonates, polyvinylpyridinium is to test the validity of
the counterion polarization theory postulated for their orientation
mechanism. A thorough discussion of this important question is
presented in the chapter by Mandel, and will not be repeated here.

Poly(C) has also been studied by electric dichroism and birefringence (70,71). The double-stranded structure present in solution at low pH showed a considerable tilting of the base planes
with respect to the helical axis and much less rigidity than poly(A).
The influence of molecular weight has been examined and the electro-optical behaviour analyzed in terms of the counterion polarization mechanism (71).

Interactions with Cationic Dyes and Other Small Ligands

Not much attention has been paid to the possible perturbations
of the orientation mechanism upon binding of dyes to nucleic acids
and nucleoproteins. In most cases, in fact, the absence of such
perturbations has been assumed. The appearance of a permanent
dipole component has, however, been observed by bipolar pulse experiments in the case of the interaction of chromatin with ethidium
bromide and proflavine. This was also observed by increase of salt
concentration (37). This effect was attributed to partial release
of histones, or segments of their chain; histone H1 which is bound
to the internucleosomal segments could be at the origin of this
effect.

Dichroism at low binding ratios. In the context of these interactions studies, the dichroism has been used as one of the most
definitive proofs of intercalation. This is the case when a negative dichroism of amplitude close to that of the purine and pyrimidine bases was observed. Also, departure from the parallelism with
respect to the bases can be deduced from such a comparison. There
are two different approaches for analyzing such dichroism data.
The first one consists in extrapolating the dichroism to infinite
field (with or without fitting with an appropriate orientation function) and using Equation (2) to determine the angle of the dye
transition moment. This is the procedure adopted by Hogan, et al.
(72). This type of analysis often suffers from the absence of
clear approach to saturation of the effect and of adequate orientation function for the fitting of the data (see above).

We personally prefer a second method based on a comparison of
the reduced dichroism at a given field for the bases and for the dye
chromophore in their respective absorption bands. If the degree of
orientation in the electric field and the macromolecular structure
are not affected by the binding (a reasonable assumption at low
amounts of binding), the ratio of the two reduced dichroism values
may be expressed as:

$$\frac{(\Delta A/A)_{dye}}{(\Delta A/A)_{bases}} = \frac{3\cos^2\beta - 1}{3\cos^2\alpha - 1} \tag{18}$$

where α and β are the angles between the transition moments of the
bases and of the dye chromophores, respectively, and the orientation
axis of the particles. This is the approach followed in several
other works (37, 73-75).

A comparison of the ligand orientation deduced from these two
types of analyses is given in Table 1 for several intercalating and
non-intercalating dyes bound to DNA. A few results for chromophores
covalently bound to DNA are also presented; in this case,

Table 1. Electric Dichroism of DNA Complexes with Various Ligands (limiting dichroism at low binding ratios)

Ligands	$(\Delta A/A)_s$	(λ_{nm})	β(deg)	$\dfrac{(\Delta A/A)_{dye}}{(\Delta A/A)_{bases}}$	(λ_{nm})	β(deg) with $\alpha = 90°$	β(deg) with $\alpha = 73°$	Ref.
Intercalating dyes								
Proflavine	-1.42	(430)	82	1.09	(450)	90	75	72
	-1.25	(450)	76					
Acridine Orange	--	--	--	1.02	(460)	90	73.5	3
	--	--	--	0.95	(460)	90	72	59
				1.09	(502)	90	75	77
				0.73	(310)	72.5	67	77
10-Methyl Acridine Orange	--	--	--	1.	(410–530)	90	73	50
Ethidium	-1.	(520)	71	0.73	(520)	72.5	67	72
	-0.81	(320)	67	0.68	(320)	71	66	72
	--	--	--	0.81	(510)	75.5	68.5	7
	--	--	--	0.67	(310)	70.5	66	7
Actinomycin-D	-0.6	(470)	63	0.52	(440)	66.5	63	72
	--	--	--	0.36	(460)	62.5	60.5	37
9-aminoacridine	-1.13	(407)	73	0.755	(407)	73.5	67.5	72
	--	--	--	1.	(365)	90	73	72
1-Methyl Phenyl Neutral Red	-1.45	(530)	84	---	---	--	--	72

Table 1 (continued)

Ligands	$(\Delta A/A)_s$	(λ_{nm})	β(deg)	$\dfrac{(\Delta A/A)_{dye}}{(\Delta A/A)_{bases}}$	(λ_{nm})	β(deg) with $\alpha = 90°$	β(deg) with $\alpha = 73°$	Ref.
Non-intercalating compounds								
Dibutylproflavine	---	---	--	-0.109	(500)	52.5	53	78
				0.09–0.18	(435)	57–59	56–57.5	78
33258 Hoechst	---	---	--	-0.545	(377)	44	47	79
Distamycin	---	---	--	not given	(315)	39	--	80
Stilbene derivative (OHSA)	---	---	--	-0.05*	(334)	54	54	12
Covalently bound chromophores								
N-2 acetylaminofluorene derivatives								
N-acetoxy	---	---	--	0.25	(300)	60	59	73
	---	---	--	0.606	(302)+	69	65	74
7-fluoro	---	---	--	0.39	(313)+	63	61	74
7-iodo	---	---	--	0.41	(313)	64	61	74
Benzopyrene derivative	---	---	--	-0.96	(310–350)	36	41	75

*Ratio with respect to acridine orange as intercalating dye.

+DNA structure perturbed; value calculated with respect to free DNA.

perturbations of the DNA structure have been sometimes evidenced.
When the angle α of the bases planes is set at 73° (Ref. 47), the
variation in the orientation angle of the ligand chromophores is
confined to the $60-70^\circ$ range for most intercalating dyes.

The lower dichroism values observed with dyes having bulky
substituents (actinomycin-D, ethidium) could be interpreted by the
appearance of kinks in the macromolecular structure as in the models
of Sobell, et al. (76). However, as recently noticed by Hogan,
et al. (72), the absence of variation of the reduced dichroism with
the amount of binding at low binding ratios (see below) argues
against this suggestion. These authors rather proposed a model of
intercalation between propeller-twisted bases with a reduction of
base stacking and a significant deviation of the dye chromophore
orientation from the perpendicularity with respect to the DNA helix
axis (72).

The interaction of sonicated DNA with diacridines in which
flexible methylene chains of variable lengths join the amino-groups
of their heterocyclic 9-amino-acridine rings has been investigated
by electric dichroism in the long-wavelength visible absorption band
of the dye (81). A comparison of the reduced dichroism of the ligand
chromophore with that of DNA alone indicated that the dye transition
moment was parallel to the base planes for chains containing more
than 6 carbon atoms or less than 3. An average twist of 10 to 17°
was observed for chains with 3 to 6 carbon atoms. Parallel measurements of viscosity and sedimentation allowed the interpretation of
the observations as a monofunctional intercalation for one to four
methylene groups in the linker chain and bifunctional intercalation
for chains with more than 6 carbons.

In the study of the 2-hydroxy-4,4'-diamidinostilbene - DNA
interaction (12) a comparison of fluorescence and dichroism effects
under pulsed electric fields, revealed the existence of non-fluorescent sites with an orientation of the dye chromophore at larger
angles with respect to the helix axis.

Complexes of chromatin with cationic dyes have also been studied
by the electric dichroism method (37). In this case, some unfolding
of the chromatin structure could be present as was recently noticed
upon binding of ethidium bromide to nucleosomes (82; see below).

Variation of the dichroism with the binding ratio. The study
of the variation of the reduced dichroism with the binding ratio r
(= molar bound-dye concentration over molar mononucleotide concentration) is indispensable to fully characterize the binding process.
In the strong binding range, the reduced dichroism does not appreciably change with r for intercalating dyes bound to DNA and
sonicated chromatin (37, 59, 72; see Fig. 5). In the latter case,
the binding of the first dye molecules probably takes place in the

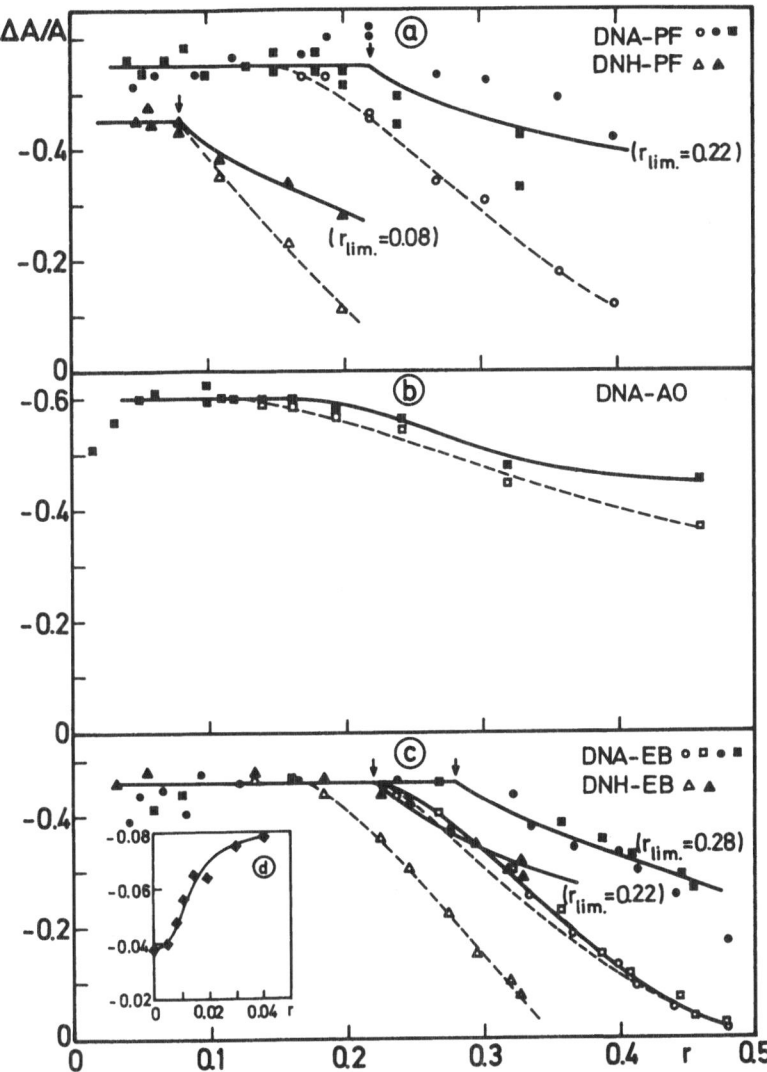

Fig. 5.   Influence of binding ratio r on the reduced electric dichroism, in 1 mM NaCl at 12-13 kV/cm, for complexes of DNA
and chromatin (DNH) with proflavine (PF; part a), acridine
orange (AO; part b) and ethidium (EB; part c). The full
lines in parts a and c are calculated on the basis of Eqn.
(19) with the appropriate adjustment of the r_{lim} value
(indicated by the vertical arrow). Open and filled symbols
respectively refer to measured dichroism and dichroism corrected for free dye contribution to the absorption. Part d:
chromatin core-particle interaction with ethidium (data from
Ref. 81).

long unfolded internucleosomal segments which were revealed in these samples (see above). Some slight decrease of $\Delta A/A$ is often noticed at very low r values but this is most probably a concentration effect since these measurements are made at constant dye concentration and variable macromolecular concentration.

Above a certain limiting value of r, r_{lim}, due to exclusion of adjacent binding sites (in the case of DNA, $r_{lim} \approx 0.25$), the dichroism decreases as a result of the increase of free-dye concentration in solution, and of the presence of externally-bound dye molecules which do not contribute to the measured dichroism. When corrected for the free-dye contribution to the absorption, the reduced dichroism in the isosbestic region of absorption may be expressed by the relation (37, 49) for $r > r_{lim}$:

$$(\Delta A/A)_{corr} = (\Delta A/A)_{int} \cdot r_{lim/r} \tag{19}$$

where $(\Delta A/A)_{int}'$ is the plateau value of the dichroism at $r \leq r_{lim}$. The calculated curves shown in Fig. 5 after proper adjustment of the r_{lim} value satisfactorily account for the experimental data.

The behaviour of nucleosome-ethidium complexes is very different, showing an increase of dichroism with r (Inset d in Fig. 5) which reflects the unfolding of the nucleosome (82). The reduced dichroism at saturation, for r = 0.03 (where an ill-defined plateau is reached; see Fig. 5d), measured at 265 and 320 nm, was found to be larger than the value in the visible absorption band of the dye (530 nm) and than the corresponding free-nucleosome dichroism in the ultraviolet absorption band. Wu, et al. (82) thus concluded that ethidium is bound in clusters along the DNA superhelix in nucleosomes, through one of the DNA grooves. From an analysis of the ionic strength dependence of the dichroism, the same authors deduced that the nucleosomes unfolded by ethidium are oriented with their superhelix axis parallel to the field, in opposition to free nucleosomes (see above); this would result from the lengthening of the superhelix axis evidenced by the increase of τ .

RNA-dye complexes also show a progressive increase of dichroism with r at low binding ratios (Fig. 6). This was interpreted by a stiffening of the double-helical regions of these molecules (49, 69, 83).

For the complexes with non-intercalating dyes, complicated variations of $\Delta A/A$ with r, dependent on the wavelength of the measurement, were encountered (78, 79); they have been accounted for by the occurrence of two external binding processes with very different orientation of the dye chromophores.

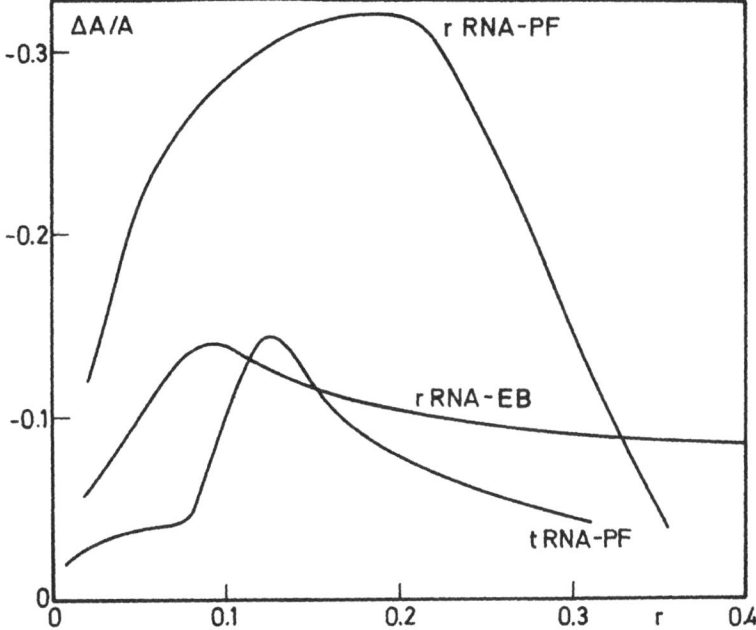

Fig. 6. Influence of binding ratio r on the reduced electric
         dichroism, at 0.001 ionic strength and 12-13 kV/cm, for
         complexes of tRNA with proflavine (PF; Grosjean, Wérenne
         and Houssier, unpublished observations) and of ribosomal
         RNA with proflavine (PF; from Ref. 82) and ethidium (EB;
         from Ref. 69).

Birefringence. The birefringence of dye complexes Δn_r is a
composite of the birefringence of the macromolecular backbone Δn_o
plus the birefringence Δn arising from the presence of new dispersion curves in the dichroic absorption bands of the dye. This is
true even if Δn_r is measured outside the absorption band owing to
the tails of the dispersion curves. The contribution Δn may be estimated from the measured dichroism in each band ΔA_i and from the
separation between the dichroic bands and the wavelength where the
birefringence is measured. Assuming a Gaussian shape for all the
components of the absorption bands and using Equation (6), it can
be shown (37, 49) that the relative increase of birefringence should
vary linearly with r according to the relation:

$$\frac{\Delta n_r - \Delta n_o}{\Delta n_o} = \frac{r}{\Delta n_M} \sum_i k_i \, (\Delta A/A)_{max,i} \cdot \epsilon_{max,i} \tag{20}$$

where $(\Delta A/A)_{max,i}$ is the reduced dichroism at the maximum of the Gaussian absorption component i with extinction coefficient $\epsilon_{max,i}$, and k_i is defined above (Eq. (6b)).

Comparing the birefringence variation upon dye binding with that predicted by Equation (20) is a useful additional criterion for possible structural alterations of the macromolecular backbone resulting from the ligand fixation. This analysis has been performed on DNA and sonicated chromatin complexes with proflavine and ethidium (37, 49). Some departure from the expected variation was noticed for the ethidium complexes but the accuracy in the determination of the birefringence variation with r was low owing to the relatively small extinction coefficient of this dye in the visible region.

Other electro-optical effects. The use of fluorescence polarization measurements under pulsed electric fields for macromolecular complex characterization has been mentioned above. In several instances the method was applied to DNA-intercalating-dye complexes for the determination of the electric polarizability (13, 48, 52) and of the rotational relaxation time (13). Polynucleotides and DNA interactions with acridine orange were also investigated (14) and some characteristics of the binding geometry were deduced. For the poly(A)-acridine orange complex, the absorption transition azimuth was found to be of the order of $66°$, i.e., close to the value deduced for the adenine bases from electric dichroism measurements (66); the value obtained for the emission transition moment orientation was found to be anomalously low ($54°$). Except for the poly(U)-acridine orange complexes, the sign of the fluorescence intensity changes were the same for all the polynucleotides investigated (poly(A), poly(C), poly(G), poly(I)). A strong influence of the concentration was noticed in the case of poly(A), poly(C), and poly(U) but not for poly(G) and poly(I). This was attributed to the formation of more compact, flexible structures in the former cases.

Hydrodynamic properties. The intercalation of planar heterocyclic dye rings into DNA-like structures is expected to produce a lengthening equivalent to a base-pair separation per ligand molecule bound. This means that the cube root of the relaxation time $\overline{\tau}_r$ should increase linearly with r according to the equation:

$$(\overline{\tau}_r/\overline{\tau}_o)^{1/3} = 1 + 2 \, r \tag{21}$$

where $\overline{\tau}_o$ is the relaxation time of the macromolecule in the absence

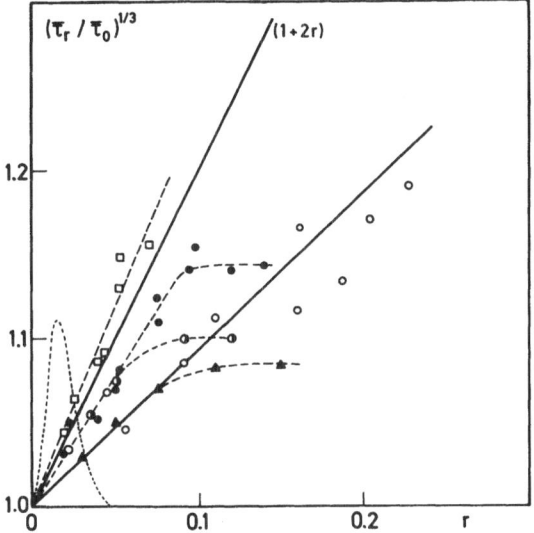

Fig. 7. Variation of relaxation time with binding ratio r for
the complexation of sonicated DNA with proflavine (0; Ref.
59; ❶: Ref. 72), ethidium (●: Ref. 72), 9-aminoacridine
(▲: Ref. 72) and actinomycin-D (■: Ref. 72). DNA sample
fractionated on Sepharose 4B for the data from Ref. 72.
(-----): interaction of distamycin with calf-thymus DNA.

of ligand. Rigid, short DNA molecules have to be used in order to
conform to this idealized limiting behaviour. However, in most
cases so far studied, the increase of length observed was lower
than expected from this equation (Fig. 7). This indicates a change
in the flexibility or structure of the chain perhaps related to a
reduction of the base stacking as in the model of dye intercalation
between propeller twisted DNA (72).

The intercalation of ethidium in the DNA of nucleosomal particles produces a large increase of relaxation time which is an additional indication of the unfolding of the DNA superhelix in these
structures, upon dye binding (82).

Complexation of the diaminosteroid irehdiamine-A to DNA produces a length decrease at low binding ratios followed by a length
increase of 5-6% at saturation for three bacterial DNAs samples,

while two eukaryotic DNAs (calf-thymus and human placenta) showed only a length increase of about 13%. In addition, the binding of the drug induced a tilt of about 31° of the bases with respect to the perpendicularity to the orientation axis, as deduced from the reduced dichroism amplitude at saturation (84). These results are consistent with the β-kinked conformation proposed by Sobell, et al. (76).

The binding of the oligopeptide antibiotic distamycin to DNA from various sources (85) revealed a quite unique behaviour of calf-thymus DNA. This is characterized by a cooperative binding and a large increase of length (determined from the electric dichroism rise time) reaching a maximum of 11% lengthening for one drug molecule per 30 b.p. (Fig. 7). This was interpreted by the induction of an allosteric transition into an altered form with increased affinity for the drug and modified arrangement of the intercalation sites accessible to ethidium and for which the visible transition moment becomes perpendicular to the DNA helix axis.

Interactions with Metal Cations and Coordination Complexes

The importance of metal-nucleic acids' interactions in biological systems has often been emphasized (86, 87). Strong alterations of the nucleic acids and nucleoprotein structures upon metal interactions have been evidenced by electric birefringence and dichroism (63, 88, 89). A large decrease of the birefringence in the case of DNA and reversals of its sign for chromatin were observed in the presence of sufficient amounts of Mn^{2+} and platinum coordination complexes (88). DNA-Hg^{2+} and DNA-Ag^{+} interactions yielded an important decrease of ultraviolet dichroism (89). Two interpretations may be put forward for these observations and have been previously discussed in detail (88): (i) the formation of compact more rigid structures; (ii) a modification of the optical properties of the base chromophores resulting in the appearance of out-of-plane components of their transition moments. Although the second alternative has received substantial experimental support from the dichroism spectrum for the DNA-Ag^{+} complexes, there are several experimental evidence of structural alterations (sedimentation, electron microscopy) which are in favour of the occurrence of compact structures, particularly in the case of platinum-chromatin interactions.

The covalent binding of cis-$Pt(NH_3)_2Cl_2$ to the double-stranded concatenary structure of poly(I) · poly(C) produces a drastic decrease of its negative birefringence and of its relaxation time (90). These effects are related to the dissociation effect of cis-$Pt(NH_3)_2Cl_2$ evidenced by circular dichroism, poly(I)-agarose affinity chromatography, density equilibrium sedimentation and acid titrations.

ACKNOWLEDGEMENTS

    This work has received the financial support of the Fonds
National de la Recherche Scientifique. Thanks are also due to Prof.
E. Charney and Prof. D. M. Crothers for making manuscripts of their
most recent works available to me. The results on nucleosomes,
oligonucleosomes and nucleosomal DNA summarized in Fig. 3 were
obtained on samples prepared and characterized at the Vrije Univer-
siteit Brussel by I. Lasters, S. Muyldermans and L. Wijns whose
collaborations are very greatly appreciated. The fruitful collaborations of Dr. J. Ramstein, Dr. C. Gatti, Dr. H. Grosjean, Dr. J.
Werenne, Dr. J. Bontemps, Dr. X. Edmonds-Alt, Dr. R. Hacha, M. C.
DePauw-Gillet and D. Hermann are gratefully acknowledged. I am also
very grateful to Prof. E. Fredericq for his invaluable help and
for his constant interest in this work.

REFERENCES

1.   J. Hofrichter, and W. A. Eaton, Ann. Rev. Biophys. Bioeng.
     5:511 (1976).
2.   B. Norden, Appl. Spectr. Rev. 14:157-248 (1978).
3.   E. Fredericq, and C. Houssier, "Electric Dichroism and Elec-
     tric Birefringence," Clarendon Press, Oxford, U.K. (1973).
4.   N. C. Stellwagen, in: "Molecular Electro-Optics. Part II.
     Applications to Biopolymers," C. T. O'Konski, ed., Marcel
     Dekker, NY, pp. 645-683 (1978).
5.   M. F. Maestre, in: "Molecular Electro-Optics. Part II.
     Applications to Biopolymers," C. T. O'Konski, ed., Marcel
     Dekker, NY, pp. 713-741 (1978).
6.   G. Weill, and M. Calvin, Biopolymers 1:401-417 (1963).
7.   C. Houssier, B. Hardy, and E. Fredericq, Biopolymers 13:1141-
     1160 (1974).
8.   R. Rill, Biopolymers 11:1929-1941 (1972).
9a. R. Rill, and K. E. VanHolde, J. Biol. Chem. 248:1080-1083
     (1973).
9b. R. Rill, and K. E. VanHolde, J. Mol. Biol. 83:459-471 (1974).
10. D. M. Crothers, N. Dattagupta, M. Hogan, L. Klevan, and K. S.
     Lee, Biochemistry 17:4525-4533 (1978).
11. M. F. Maestre, and G. Kilkson, Biophys. J. 5:275-287 (1965).
12. G. Weill, and J. Sturm, Biopolymers 14:2537-2553 (1975).
13. B. R. Jennings, and P. J. Ridler, Chem. Phys. Letters 45:550-
     555 (1977).
14. P. J. Ridler, and B. R. Jennings, in: "Electro-Optics and
     Dielectrics of Macromolecules and Colloids," B. R. Jennings,
     ed., Plenum Press, NY, pp. 99-107 (1979).
15. P. J. Ridler, and B. R. Jennings, SPIE 164:94-99 (1978).
16. R. W. Wilson, and J. A. Schellman, Biopolymers 16:2143-2165
     (1977).

17.   P. R. Callis, and N. Davidson, Biopolymers 8:379-390 (1969).
18.   R. L. Jernigan, and D. S. Thompson, in: "Molecular Electro-
      Optics. Part I. Theory and Methods," C. T. O'Konski, ed.,
      Marcel Dekker, NY, pp. 159-206 (1976).
19.   H. G. Kuball, Z. Naturforsch. A22:1407-1412 (1967).
20.   C. Houssier, and H. G. Kuball, Biopolymers 10:2421-2433 (1971).
21.   K. Nishinari, and K. Yoshioka, Kolloid-Z und Z. fur Polymere
      235:1189-1192 (1969).
22.   E. W. Small, and I. Isenberg, Biopolymers 16:1907-1928 (1977).
23.   S. H. Koenig, Biopolymers 14:2421-2423 (1975).
24.   A. K. Wright, and M. R. Thompson, Biophys. J. 15:137-141 (1975).
25.   M. R. Thompson, R. C. Williams, and C. H. O'Neal, Biophys. J.
      24:264-266 (1978).
26.   S. Broersma, J. Chem. Phys. 32:1626-1631 (1960).
27.   J. Schweitzer, and B. R. Jennings, Biopolymers 12:2439-2441
      (1973).
28.   A. R. Foweraker, and B. R. Jennings, Adv. Mol. Relxn. Intn.
      Processes 8:103-110 (1976).
29.   A. R. Foweraker, V. J. Morris, and B. R. Jennings, in: "Parti-
      cle Size Analysis," M. J. Groves, ed., Heyden Publishers,
      pp. 147-154 (1977).
30.   M. M. Judy, and S. R. Bernfeld, in: "Non Linear Systems and
      Applications," Academic Press, NY, pp. 577-590 (1977).
31.   E. Y. Hawkin, A. R. Foweraker, and B. R. Jennings, Polymer
      19:1233-1236 (1978).
32.   V. J. Morris, A. R. Foweraker, and B. R. Jennings, Adv. Mol.
      Relxn. Intn. Processes 12:201-210 and 211-220 (1978).
33.   B. R. Jennings, Adv. Polym. Sci. 22:61-81 (1977).
34.   I. Tinoco, Jr., and K. Yamaoka, J. Phys. Chem. 63:423-427
      (1959).
35.   M. Matsumoto, H. Watanabe, and K. Yoshioka, J. Phys. Chem.
      74:2182-2188 (1970).
36.   P. Colson, C. Houssier, E. Fredericq, and J. Bertolotto,
      Polymer 15:396-397 (1974).
37.   C. Houssier, J. Bontemps, X. Emonds-Alt, and E. Fredericq,
      Ann. N.Y. Acad. Sci. 303:170-189 (1977).
38.   P. Colson, C. Houssier, and E. Fredericq, Biochim. Biophys.
      Acta 340:244-261 (1974).
39.   J. Greve, and M. E. DeHeij, Biopolymers 14:2441-2443 (1975).
40a. J. C. Bernengo, B. Roux, and D. Herbage, Ber. Bunsenges. Phys.
      Chem. 80:246-249 (1976).
40b. B. Roux, J. C. Bernengo, C. Marion, and M. Hanss, J. Coll.
      Interface Sci. 66:421-427 (1978).
41.   D. Ding, R. Rill, and K. E. VanHolde, Biopolymers 11:2109-2124
      (1972).
42.   M. Hanss, and J. C. Bernengo, Biopolymers 12:2151-2159 (1973).
43.   F. Van der Touw, and M. Mandel, Biophys. Chem. 2:218-230 (1974).
44.   M. Sakamoto, H. Kanda, H. Reinosuke, and Y. Wada, Biopolymers
      15:879-892 (1976).

45. R. S. Wilkinson, and G. B. Thurston, Biopolymers 15:1555-1572 (1976).
46. K. Kikuchi, and K. Yoshioka, Biopolymers 15:583-587 (1976).
47. M. Hogan, N. Dattagupta, and D. M. Crothers, Proc. Natl. Acad. Sci. U.S.A. 75:195-199 (1978).
48. S. Sokerov, and G. Weill, Biophys. Chem. 10:161-171 (1979).
49. C. Houssier, Thèse d'Agrégation, Université de Liège (1977).
50. K. Yamaoka, and E. Charney, Macromolecules 6:66-76 (1973).
51. C. Houssier, I. Lasters, S. Muyldermans, and L. Wijns, in preparation (1980).
52. C. Hornick, and G. Weill, Biopolymers 10:2345-2358 (1971).
53. B. Roux, C. Marion, and J. C. Bernengo, in: "Electro-Optics and Dielectrics of Macromolecules and Colloids," B. R. Jennings, ed., Plenum Press, NY, pp. 163-173 (1979).
54. C. Marion, and B. Roux, Nucleic Acids Res. 5:4431-4449 (1978).
55. L. Klevan, M. Hogan, N. Dattagupta, and D. M. Crothers, Cold Spring Harbor Symp. Quant. Biol. 42:207-214 (1977).
56. D. M. Crothers, N. Dattagupta, M. Hogan, L. Klevan, and K. S. Lee, Biochemistry 17:4525-4533 (1978).
57. L. Klevan, N. Dattagupta, M. Hogan, and D. M. Crothers, Biochemistry 17:4533-4540 (1978).
58. H. M. Wu, N. Dattagupta, M. Hogan, and D. M. Crothers, Biochemistry 18:3960-3965 (1979).
59. J. Ramstein, C. Houssier, and M. Leng, Biochim. Biophys. Acta 335:54-68 (1974).
60. M. Mandelkern, N. Dattagupta, and D. M. Crothers, in preparation (1980).
61. M. Pollak, and H. A. Glick, Biopolymers 16:1007-1013 (1977).
62. M. Levitt, Proc. Natl. Acad. Sci. 75:640-644 (1978).
63. X. Emonds-Alt, C. Houssier, and E. Fredericq, Biophys. Chem. 10:27-39 (1979).
64. C. Houssier, R. Hacha, M. C. DePauw-Gillet, J. L. Pieczynski, and E. Fredericq, submitted for publication (1980).
65. D. C. Rau, and V. A. Bloomfield, Biopolymers 18:2783-2805 (1979).
66. E. Charney, and J. B. Milstien, Biopolymers 17:1629-1655 (1978).
67. J. T. Finch, L. C. Lutter, D. Rhodes, R. S. Brown, B. Rushton, M. Levitt, and A. Klug, Nature 269:29-36 (1977).
68. C. T. O'Konski, and R. S. Farinato, in: "Electro-Optics and Dielectrics of Macromolecules and Colloids," B. R. Jennings, ed., Plenum Press, NY, pp. 133-142 (1979).
69. C. Gatti, C. Houssier, and E. Fredericq, Biochim. Biophys. Acta 407:308-319.
70. H. H. Chen, and E. Charney, Biopolymers, in press (1980).
71. E. Charney, and H. H. Chen, in preparation (1980).
72. M. Hogan, N. Dattagupta, and D. M. Crothers, Biochemistry 18:280-288 (1979).
73. C. T. Chang, S. J. Miller, and J. G. Wetmur, Biochemistry 13:2142-2148 (1974).

74. R. P. P. Fuchs, J. F. Lefevre, J. Pouyet, and M. P. Daune,
     Biochemistry 15:3347-3351 (1976).
75. N. E. Geacintov, A. Gagliano, V. Ivanivic, and I. B. Weinstein,
     Biochemistry 17:5256-5262 (1978).
76. H. M. Sobell, C. C. Tsai, S. C. Jain, and S. G. Gilbert, J.
     Mol. Biol. 114:333-365 (1977).
77. E. Fredericq, and C. Houssier, Biopolymers 11:2281-2308 (1972).
78. J. Bontemps, C. Houssier, and E. Fredericq, Biophys. Chem.
     2:301-315 (1974).
79. J. Bontemps, C. Houssier, and E. Fredericq, Nucleic Acids Res.
     2:971-984 (1975).
80. A. K. Krey, Polymer 18:495-499 (1977).
81. L. P. G. Wakelin, M. Romanos, T. K. Chen, E. S. Glaubiger,
     E. S. Canellakis, and M. J. Waring, Biochemistry 17:5057-5063
     (1978).
82. H. M. Wu, N. Dattagupta, M. Hogan, and D. M. Crothers, Bio-
     chemistry 19:626-634 (1980).
83. M. Schoentjes, and E. Fredericq, Biopolymers 11:361-374 (1972).
84. N. Dattagupta, M. Hogan, and D. M. Crothers, Proc. Natl. Acad.
     Sci. USA 75:4286-4290 (1978).
85. M. Hogan, N. Dattagupta, and D. M. Crothers, Nature 278:521-
     524 (1979).
86. M. Daune, Metal Ions Biol. Syst. 3:1 (1974).
87. T. G. Spiro, Nucleic Acids - Metal Ions Interactions, in:
     "The Metal Ions in Biology Series," Vol. 1 (1980).
88. C. Houssier, and M. Tricot, in: "Electro-Optics and Dielec-
     tric of Macromolecules and Colloids," B. R. Jennings, ed.,
     Plenum Press, NY, pp. 247-257 (1979).
89. D. Ding, and F. S. Allen, in: "Electro-Optics and Dielec-
     trics of Macromolecules and Colloids," B. R. Jennings, ed.,
     Plenum Press, NY, pp. 143-147 (1979).
90. D. Hermann, C. Houssier, and W. Guschlbauer, Biochim. Biophys.
     Acta 564:456-472 (1979).
91. J. G. Elias, and D. Eden, submitted for publication (1980).
92. N. C. Stellwagen, submitted for publication (1980).
93. R. S. Farinato, and C. T. O'Konski, in preparation (1980).
94. J. C. Bernengo, P. Bezot, C. Bezot, B. Roux, and C. Marion,
     submitted for publication (1980).
95. J. G. Garcia-de-la-Torre, and V. M. Bloomfield, Biopolymers
     16:1765-1778 (1977).
96. J. Newman, H. L. Swinney, and L. A. Day, J. Mol. Biol. 116:
     593-603 (1977).
97. F. C. Chen, G. Koopmans, R. L. Wiseman, L. A. Day, and H. L.
     Swinney, Biochemistry 19:1373-1376 (1980).
98. M. M. Tirado, and J. G. Garcia-de-la-Torre, J. Chem. Phys.,
     in press (1980).

ELECTRO-OPTICS OF VIRUSES AND BACTERIOPHAGE

Fritz S. Allen

Department of Chemistry
University of New Mexico
Albuquerque, NM 87131

Bacteriophage and viruses have been given much attention by
workers in the area of electro-optics. There are several reasons for
this activity. The first is that these particles have the right
sizes, shapes, and electrical anisotropies to exhibit considerable
electro-optic effects. The second is that these particles offer a
progression of structural complexity. It is possible to find very
simple particles consisting of only a cylinder, to very complex
particles which possess many structural components with movable and
extendable portions. The third reason to study these materials is
that they are simple systems which embody many of the essential
genetic processes of life. Whether virus particles are to be considered alive or not, they are very near the threshold of life. They
provide an ideal means to study many of the steps important in molecular biology. The area of electro-optics of bacteriophage and
viruses has been previously reviewed by Houssier and Fredericq (1)
and by Maestre (2). In this survey we shall deal mainly with
developments which have appeared since the time of the previous contributions.

Let us begin our review with tobacco mosaic virus (TMV). This
is perhaps the most well studied of all viral material. An electron
micrograph of this particle is shown in Figure 1. TMV is a simple
cylinder of length 3000 Å and diameter 180 Å. TMV is an RNA virus
with the single strand of viral RNA contained within helix of coat
protein.

Brownsey and Jennings (3) studied TMV by means of electrophoretic
light scattering. They have determined the rotational and translation
diffusion constants as 285 sec^{-1} and 4.5 x 10^{-8} cm^2/sec, respectively.
By comparing the known properties of the TMV particle with these

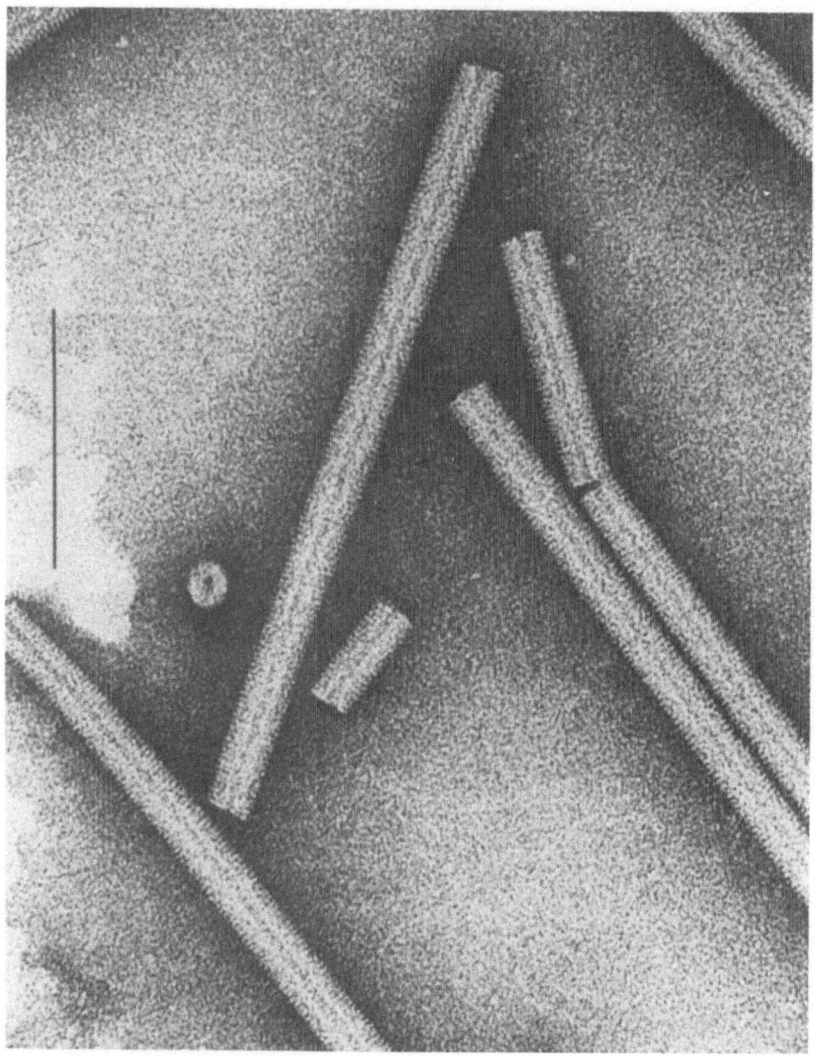

Fig. 1.   An electron micrograph of tobacco mosaic viruses. The line
          in the figure is 100 nm in length. We are grateful to
          Charles C. Thomas, Publishers for permission to reprint this
          figure.

experimental results the authors concluded that these were very reasonable values. They did note that by knowing both the rotational
and translational diffusion constants one could calculate the dimensions of the particle, assuming cylindrical symmetry, from the
Broersma equations. They also indicated that the value for the
translational diffusion constant decreased when the sample was
allowed to stand for two days. They ascribed this to aggregation.

In a paper which has since become classic, Newman and Swinney (4) applied the techniques of transient electric birefringence to tobacco mosaic virus. In this paper the authors introduce the use of repetitively pulsed, laser, birefringence instruments. The sample of virus which they used was extraordinarly well characterized. Newman and Swinney determined the rotational diffusion constant for TMV to be 318 sec^{-1} at 20°C. They did note that upon standing for four months their sample exhibited an increase in rotary diffusion constant. They ascribed this to degradation. By assuming a diameter of 150 Å for the TMV particle from electron microscopy. These workers solved the Broersma equation and determined the length to be 2925 Å. This is in good agreement with the electron microscopy length. Stoylov (5) has presented a review of the electro-optical properties of TMV and Newman and Swinney (4) have presented a compilation of TMV rotational diffusion constants. In the process of preparing this diffusion constant table, several values for the rotational diffusion constant were altered to account for changes in solution temperature. It is worth noting that the changes for 5°, that is from 293-298°K are dramatic and amount to the order of 11-12%. Perhaps a good deal of the scatter in the tabulated values can be accounted for by differences in temperature of the samples. This serves to underscore the importance of accurate temperature control when one is determining hydrodynamic parameters from electro-optic measurements.

Rahe, et al. have recently published a dichroism spectrum for TMV. The presence of seven chromophores are clearly indicated in their data. Anisotropic light scattering was accounted for by the methods of Allen and Van Holde (6). It was assumed there was no form dichroism.

The T bacteriophages are among the most complicated of the bacterial viruses. A T-even bacteriophage is shown in figure 2 in a schematic fashion. An electron micrograph is shown in figure 3 which demonstrates the morphology of these organisms. The particle consists of a protein head which contains the DNA, a tail segment which is comprised of a collar, tail core, a contractile tail sheath and a tail plate to which are attached six kinked tail fibers. The host for the T series bacteriophage is the common bacterium E. coli. the T phages are called virulent phages because an attack by a bacteriophage always destroys the host bacterium. The cycle of phage attack begins by adsorption of phage by the tail fibers onto the surface of the bacterial membrane. The tail fibers draw the body of the organism down to the membrane surface, the surface of the membrane is either penetrated or dissolved away, the contraction of the tail sheath occurs and the DNA is injected into the cytoplasm of the bacterial cell. Within a few minutes the metabolic activity of the cell is redirected toward the synthesis of new viral particles. In a period of approximately 45 min, the cell lyses and a new generation of bacteriophage are released to the solution. The T even phages, i.e., T-2, T-4, and T-6 are all morphologically similar and have been

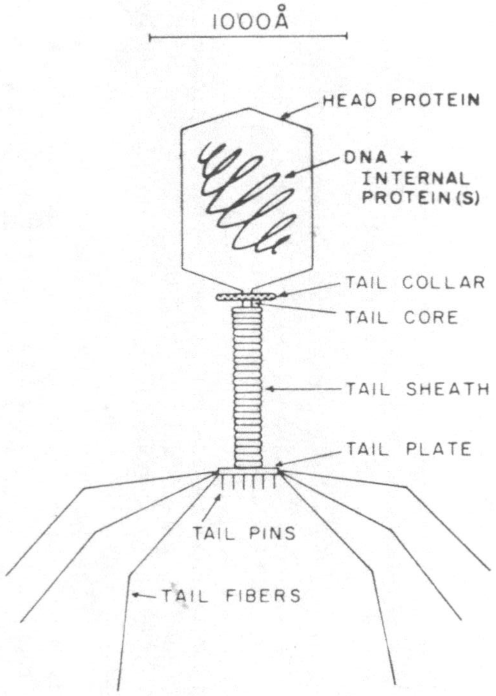

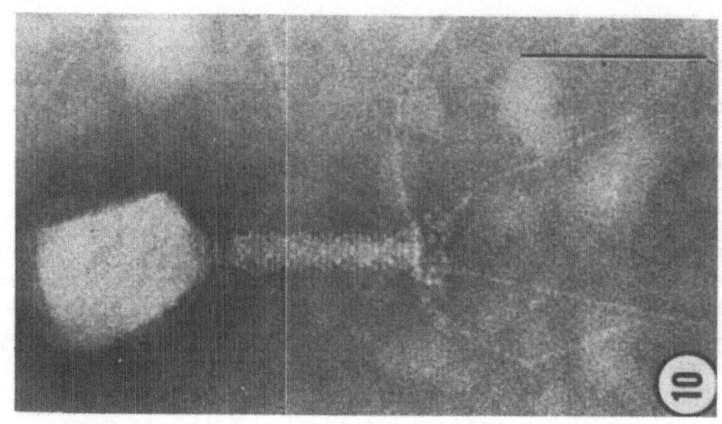

Fig. 3.   An electron micrograph of the
          T-4 particle. The line in the
          figure represents 100 nm. We
          are grateful to Academic Press
          for permission to reprint this
          figure.

Fig. 2. A schematic drawing showing
        the structural components
        of the T-even bacteriophage
        particle.

studied by electric birefringence (7-12) and electrophoretic and
dynamic light scattering methods (13-18). T-7 has been examined by
electric birefringence (9) and electric dichroism (19). T-7 has a
much abbreviated tail structure attached to a head which is similar
to the T-even phages. The most extensive electro-optic work characterizing the T-even phages is that of Greve and Blok (8-12). These
workers have studied the bacteriophage, T-2, T-4, and T-6, in both
the slow and the fast forms. The slow form of a bacteriophage occurs
when the tail fibers are extended downward into the solution. The
fast form of the phage results when the tail fibers are retracted
upward along the tail so that the kink approaches the collar.

In their series of papers, Greve and Blok have determined the
solution conditions which cause the tail fibers to reorient from the
fast to the slow form. They have also determined the dipole moments
and rotary diffusion constant for these materials. A summary of their
results is given in Table 1. They have concluded from these results
that for the phage particles in the fast form with retracted fibers
T-2 and T-6 exhibit very similar rotary frictional properties.
Whereas, T-4 is considerably different. Similarly for the slow form
where the tail fibers are extended, all of the T-phage particles are
experimentally different. This indicates that the tail fibers are
oriented in a particular fashion for the different phages. The phage
particles were found to orient by a permanent dipole mechanism. The
magnitude of the dipole was found to depend upon ionic strength. As
the ionic strength was increased the dipole moment was decreased.
This was explained by shielding in a manner similar to the Debye
Huckle Theory. Below ionic strength of 0.01 the dipole moments were
found to be essentially constant. Large changes in dipole moment can
be measured when the phages undergo the fast-slow transition. The
change is nearly an order of magnitude for T-4, that for T-2 approximately doubles, and the dipole of T-6 remains approximately the same.
The authors ascribe the changes in dipole moment to the effects of
fiber reorientation and changes in charge distribution. Calculations
show that as few as three positive charges on the tips of the fibers
would account for the large changes in dipole moment observed. It
should be noted that it is also possible to account for apparent
changes in dipole moment for charged particles by a movement of the
hydrodynamic center for rotational reaction.

The experiments described above have been correlated with
parallel experiments involving phage adsorption to the bacterial
membrane. A number of conclusions have emerged. The first is, as
the rotary diffusion constant is decreased, more of the phage adsorbs.
This is reasonable since the extended fibers make first contact with
the bacterial membrane. A second conclusion is that as the ionic
strength is increased at constant rotational diffusion constant, less
adsorption is obtained. This indicates that some electrical screening of charge patterns on the membrane surface or at the fiber tips
prevents adsorption. A recent publication by this group involving

Table 1. Electro-Optic Summary

Organism	Dipole Moment μ (D)	Rotational Diffusion Constant D_R (sec^{-1})	Translational Diffusion Constant $D_R \times 10^8$ (cm^2 sec^{-1})	Reference
T4B fiberless	95,000	295	3.35	8; 13; 13
T4B fast	24,000	280	~ 3.35	8; 11; 22
T4B slow	200,000	133	2.86	9; 12; 16
T2L fast	125,000	325	3.53	11; 11; 16
T2LO slow	280,000	120	3.05	12; 12; 16
T6 fast	200,000	322		11; 11;
T6 slow	170,000	157		12; 12
T7	5,600; 5100	5,356; 5,290		19,10; 19,10
TMV		291-318	4.5	4; 5; 3
Xf		17.3	2.53	26
fd		20.9	2.58	25; 25
λ head			6.5	29

depolarized, forward, inelastic light scattering (18) has allowed
the determination of the rotational diffusion constant for T-4 as
258 sec^{-1}. This is reasonably consistent with the values obtained
by means of the techniques of electro-optics and gives assurance that
there has been no distortion of the bacteriophage structure by the
electric field in the electro-optic measurements. In the same paper,
the rotational diffusion constant for T-7 was determined to be 4528
sec^{-1}.

Another very active group in bacteriophage electro-optics is
that of Victor Bloomfield. This laboratory has constructed an
inelastic light scattering instrument where the sample scattering
area is a band electrophoresis column. This allows separation of
samples, while at the same time one can determine the rotational and
translational diffusion parameters. This instrument was calibrated
with red blood cell ghosts and hemoglobin and good agreement with
literature value diffusion constants was obtained. As an example of
the ability of this instrument to separate samples with different
electrophoretic mobilities, a 1:1 mixture of whole and fiberless T-4
phage particles was studied (13). The results are shown in figure 4.

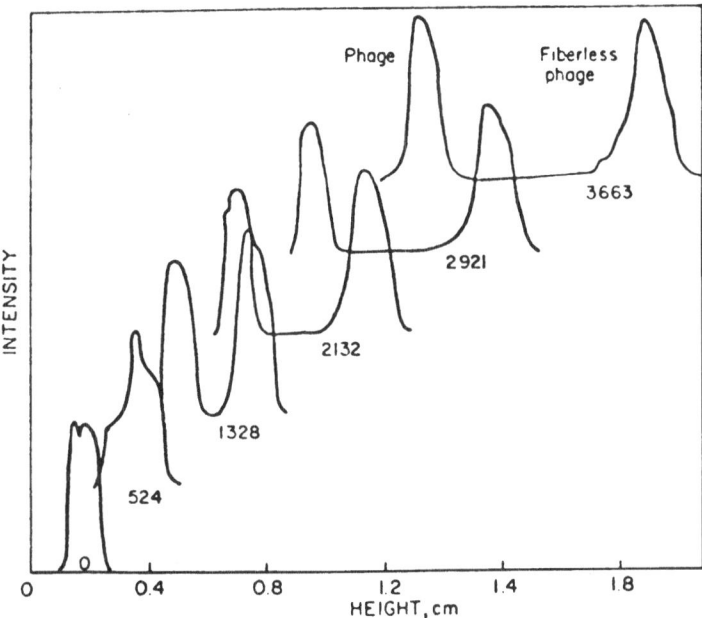

Fig. 4.   Electrophoresis of a 1:1 mixture of intact and fiberless T4D
           particles. The numbers below each scattering profile give
           the length of application of the electric field in seconds.
           This figure is from reference 12 and we are grateful to John
           Wiley and Sons for permission to reprint this figure.

The data show the scattered intensity as a function of height in the
electrophoresis column for a variety of times of application of
electric field. Very clearly, the fiberless phage and the intact
phage are resolved, with the fiberless or high diffusion constant
form of the phage moving much more rapidly. As in the more usual
electrophoretic light scattering experiment, the electrophoretic mobility can be obtained from this instrument as the slope of the plot of
the distance moved by a sample peak as a function of the time of field
application. Baran and Bloomfield (14) have studied the attachment
of tail fibers to the fiberless T-4 particles. The scattered intensity data for this system is given in figure 5. The seven forms of
the bacteriophage with zero to six tail fibers attached can be
resolved. Increasing the concentration of free tail fibers, shifts
the equilibrium to higher fiber configurations. The translational
diffusion constant measured by Bloomfield and coworkers for T-4 and
T-2 phages are given in Table 1. These numbers are to be compared
with the calculations of Bloomfield and Garcia de la Torre (20-22).
The calculations are in excellent agreement for the fiberless particle
and not quite as good agreement for intact bacteriophage with tail
fibers extended.

Welch and Bloomfield (15) have studied the thermodynamics of
the slow to fast transition in bacteriophage T-2. For a series of
equilibrium mixtures of the slow and fast form they have determined
translational diffusion constants and sedimentation coefficients.
This has allowed the writing of an apparent equilibrium constant for
the slow to fast form. By making these determinations as a function
of temperature and making a Van't Hoff plot the thermodynamics of the
transition can be investigated. Welch and Bloomfield found ΔH to be
-135 ± 18 kcal/mol, ΔS was determined to be -230 ± 45 cal/mol°K and
ΔG was determined to be 66 ± 22 kcal/mole at 293°K. They also determined that the order of 8 to 9 protons were involved in the retraction
of the tail fibers. This reaction was also studied as a function of
concentration (16).

Kosturko, et al. (19) have studied the electric dichroism of
bacteriophage T-7. An electron micrograph of this bacteriophage is
shown in figure 6. These collaborators worked at 265 nm and measured
a positive dichroism which, when extrapolated to saturating field,
gave a value of 0.12. Since at 265 nm they are principally examining
the DNA chromophores, a positive dichroism indicates a general tendency of the helix backbone to be across the orientation axis of the
bacteriophage. If one assumes that the symmetry axis of the bacteriophage is the orientation axis, then several models can be examined
to see which is the most realistic. Kosturko, et al. determined that
the model where the DNA was wound into a solenoidal winding which was
tilted at 43° to the symmetry axis of the bacteriophage, gave the
best agreement with the data. In their model calculations, they
assumed that the bases made a 73° tilt angle with respect to the
helix axis of the DNA itself. This is consistent with the

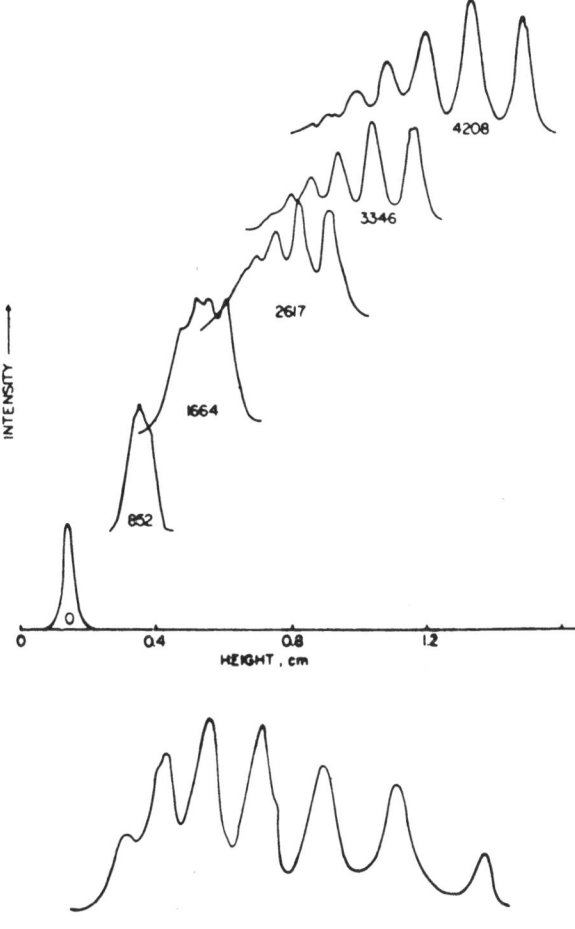

Fig. 5. Electrophoresis of a mixture of reconstituted T4 particles
         showing phage with zero to six fibers attached. The numbers
         below each scattering profile give the length of application
         of electric field in seconds. The lower portion of the
         figure shows the effect of increasing the ratio of tail
         fibers per phage particles. This figure is from reference
         13 and we are grateful to John Wiley and Sons for permission
         to reprint this figure.

assumptions and findings of Levitt (23), as has been suggested by
Hogan, et al. (24). They have also determined the rotational diffusion constant for T-7 to be 5356 ± 200 sec^{-1}.

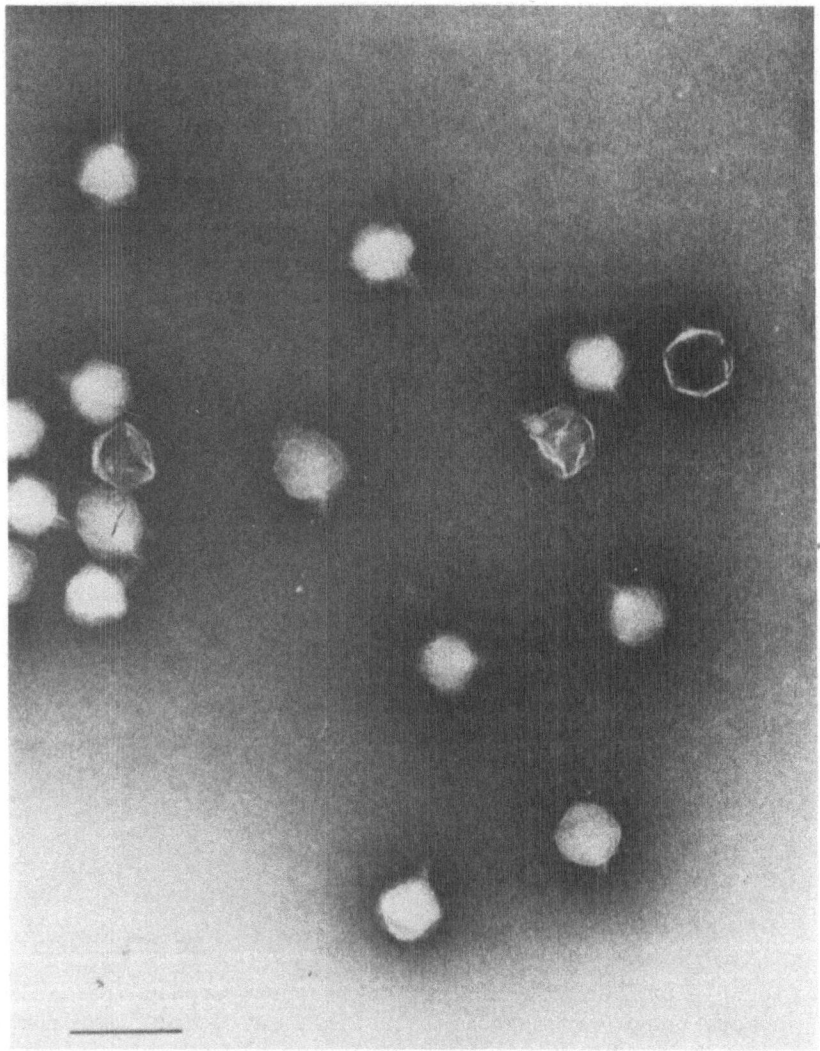

Fig. 6.   An electron micrograph of bacteriophage T-7. The line in
         the photo represents 100 nm. We are grateful to Charles C.
         Thomas Publishers for permission to reprint this figure.

      Newman, et al. (25) have studied the bacteriophage fd. fd is a
temperature bacteriophage meaning that it does not kill the host
cell. fd is a filamentous particle 9400 Å in length and 50 Å in
diameter. It has a single strand closed circle of DNA. It is very
simple bacteriophage having only two coat proteins and perhaps only
a total of eight in the phage genetic material. An electron micrograph is shown in figure 7. Newman, Swinney, and Day (25) studied

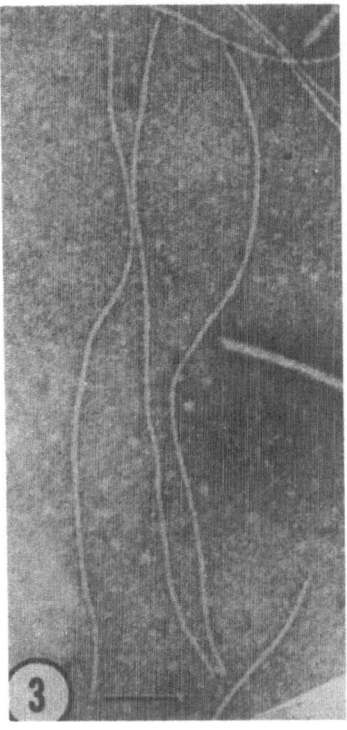

Fig. 7. An electron micrograph of bacteriophage fd. The line in
        the photo represents 100 nm. We are grateful to Academic
        Press for permission to reprint this figure.

this virus particle by means of transient electric birefringence and
quasi elastic light scattering. From the electric birefringence
measurements they determined the rotational diffusion constant and
from the quasi-elastic light scattering measurements they evaluated
the translational diffusion constant. The rotational diffusion
constant was determined to be $20.9 \pm .3$ sec^{-1}. The translational
diffusion constant was given as $2.58 \pm .04 \times 10^{-8}$ cm^2/sec. These
data are included in Table 1. Assuming that the fd virus particle
is a rigid cylinder it is possible to apply the Broersma equations
for rotational and translational diffusion constants along with the
known measured values for these parameters and determine the diameter and length of the particle. This was done and the reported
values are 8950 ± 200 Å for the length and 90 ± 10 Å for the diameter. These are in reasonable agreement with the literature values.
The most recent form of the Broersma equations was employed in this
work. They are given (25) as

$$D_R = (3kT/\pi\eta L^3)(\delta - \xi)$$

where $\delta = \ln(2L/d)$ and $\xi = 1.45 - 7.5 \, (\frac{1}{\delta} - 0.27)^2$. L is the particle
length, d is the diameter and η is the solvent viscosity.

$$D_T = (kT/3\pi\eta L)(\delta - 1/2(\gamma_{\shortparallel} + \gamma_{\perp}))$$

where $\gamma_{\shortparallel} = 1.27 - 7.4 \, (1/\delta - 0.34)^2$ and $\gamma_{\perp} = 0.19 - 4.2 \, (1/\delta - 0.39)^2$.

A very similar analysis was carried out by Chen, et al. (26) for
the filamentous virus Xf. Xf is very similar to fd. The particle
length is given as 9770 Å, and the diameter as 80 Å by electron microscopy. An electron micrograph of this particle is given in figure 8.
Essentially the same techniques as have been described for fd were
carried out on the Xf particle by these workers. Transient birefringence was used to determined the rotational diffusion constant
and quasi-elastic light scattering was used to determined the translational diffusion constant. Simultaneous solution of the Broersma
equations lead to length of 9800 Å and a diameter at 70 Å. These are
in good agreement with the electron microscopy measurements reported.
A novel way of presenting the rotational and translational diffusion
constant data is given in this paper. The two Broersma equations
are plotted in length diameter space yielding a surface for D_R and a
surface for D_T. Where the contours for the proper values on the two
surfaces intersect the length and diameter can be determined. The
advantage of this method is that it is quite easy to see the sensitivity of the length and diameter result to the errors in the measured
diffusion constants. The plot of Chen, et al. is shown in figure 9.

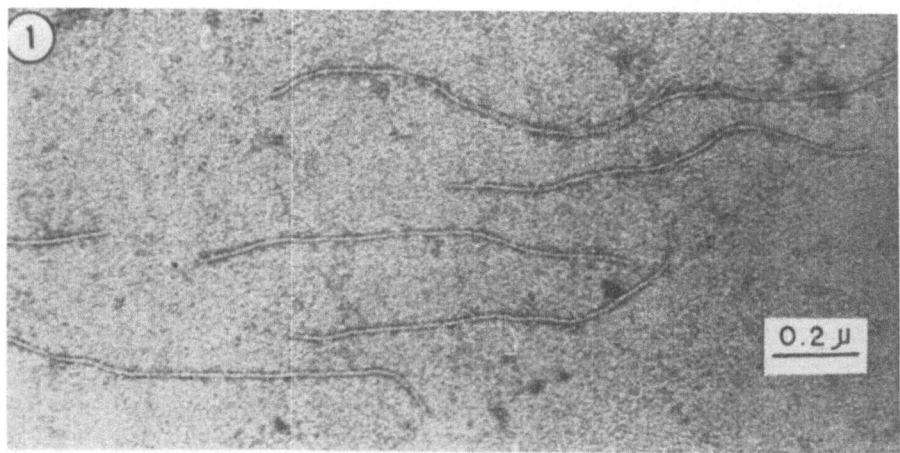

Fig. 8.   An electron micrograph of bacteriophage Xf. The line in the
          photo represents 200 nm. We are grateful to Academic Press
          for permission to reprint this figure.

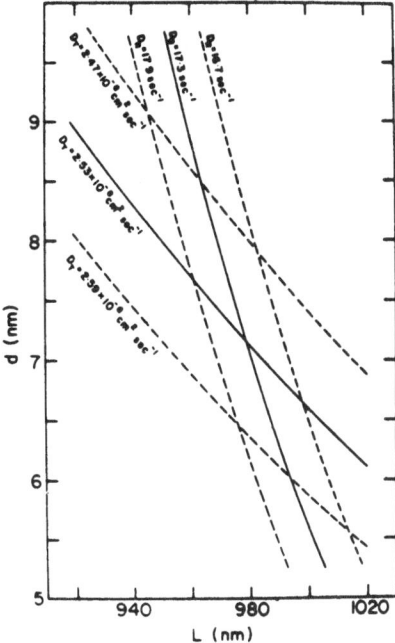

Fig. 9. A plot of rotational diffusion constant D_R and translational
diffusion constant D_T in length-diameter space for cylindrical particles. This figure is from reference 25. We are
grateful to the American Chemical Society for permission to
reprint this figure.

The filamentous bacteriophage fl has been studied by Rossomando
and Milstein (27-28). An electron micrograph of this material is
given in figure 10. There are two proteins in the coat or capsid of
the intact fl bacteriophage. Protein A is present in only one copy
whereas protein B is replicated many times and is the main structural
material. Rossomando and Milstein worked with three mutants of fl
and the native particle itself. The mutants served to substitute
different amino acids for a tryptophane in the A protein of the native
fl particle. Mutant 1 has a serine replacing the tryptophane. Mutant
2 has a glutamine replacing the tryptophane and Mutant 3 has a tryosine replacing the tryptophane. When Rossomando and Milstein made
electro-optic measurements they observed that the native particle and
the tryosine modified particle behaved similarly in both electric
birefringence and electric dichroism. Whereas both the serine and
glutamine modified particles were significantly different. They were
able to correlate these differences with differences in particle
stability as measured by resistance to ultransonic vibration. Hence,

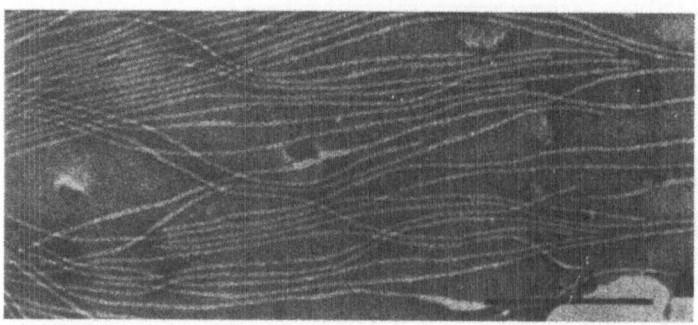

Fig. 10.   An electron micrograph of bacteriophage f1. The line in
           the photo is 200 nm in length. We are grateful to Academic
           Press for permission to reprint this figure.

they concluded that the A protein determines the arrangement of the
subunits of the B protein which form the capsid. Thus, it appears
that the A protein is a nucleation or scaffolding protein for the B
protein.

The last work to be described is that of Kuntzler and Hohn (29)
and their quasi-elastic light scattering measurements on the head of
λ bacteriophage. λ is a very interesting particle. It is a lysogen-
ous bacteriophage. This means that the phage λ can include it's own
genetic material into the genetic material of the host bacterium.
This grants immunity to further λ infection to the bacterium but
daughter cells of that bacterium will carry with them the information
to reproduce λ phage. When conditions are appropriate, the daughter
cells may fill with λ phage and lyse, releasing the phage particles
to the solution. Kuntzler and Hohn have examined a whole series of
developmental stages in the formation of the head of the λ particle.
An electron microphage of the λ phage is given in figure 11. The
head of the phage particle has a diameter of 500 Å and the tail is
somewhat flexible and 1500 Å in length. The translational diffusion
constant determined by the quasi-elastic light scattering measurement
is 6.5×10^{-8} cm^2/sec. This is the most rapidly diffusing particle
we have encountered but it is also the smallest and most symmetric.

There has been a definate increase in the interest and activity
in the electro-optics of viruses and bacteriophage. Electric birefringence is now a commonly applied means of characterization. With
the advent of improved electric dichroism instrumentation, it will
be possible to utilize the small dichroism intrinsic in most viruses
and phage to obtain structural information about these interesting
particles. Clearly the next few years will result in many more
investigations.

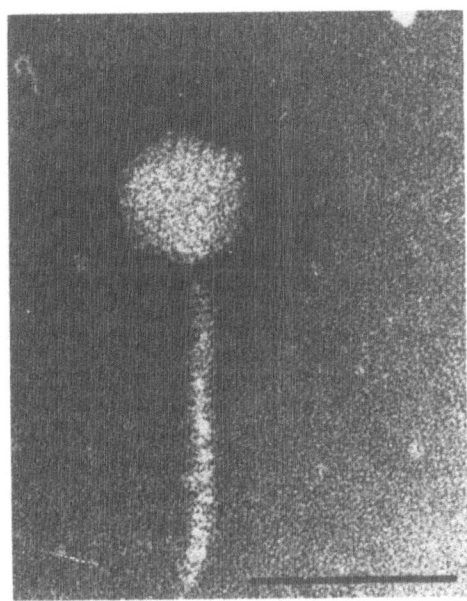

Fig. 11.   An electron micrograph of bacteriophage λ. The line in the
           figure is 100 nm in length. We are grateful to Charles C.
           Thomas Publisher for permission to reprint this figure.

REFERENCES

1. E. Fredericq and C. Houssier, "Electric Dichroism and Electric
    Birefringence," Clarendon Press, Oxford (1973).
2. M. Maestre, Electro-Optics of Nucleoproteins and Viruses, in:
    "Molecular Electro Optics, Part 2," C. T. O'Kowski, ed.,
    Marcell Dekker, New York (1978).
3. C. J. Brownsey and B. R. Jennings, Electrically Induced Varia-
    tions of Rayleigh Scattering Simultaneous Monitoring of the
    Intensity and Spectral Width from Macromolecular Solutions,
    J. Chem. Phys., 68:926 (1978).
4. J. Newman and H. L. Swinney, Length and Dipole Moment of TMV by
    Laser Signal-Averaging Transient Electric Birefringence,
    Biopolymers, 15:301 (1976).
5. S. P. Stoylov, Colloid Electro-Optics Electrically Induced Optical
    Phenomena in Disperse Systems, Advan. Colloid Interface Sci.,
    3:45 (1971).
6. W. H. Rahe, R. J. Fraatz, L. K. Sun, and F. S. Allen, A New
    Instrument to Measure Solution Electric Dichroism, Rev. Sci.
    Instr., in press.
7. M. F. Maestre, Transient Electric Birefringence of T2 Bacteri-
    ophage and T2 Ghost, Biopolymers, 6:415 (1968).
8. J. Greve and J. Blok, Transient Electric Birefringence of T-Even

Bacteriophages. I. T4B in the Absence of Tryptophan and
Fiberless T4 Particles, Biopolymers, 12:2607 (1973).

9. J. Greve and J. Blok, Transient Electric Birefringence of T-Even
Bacteriophages. II. T4B in the Presence of Tryptophan and
T4D, Biopolymers, 14:139 (1975).

10. G. deGroot, J. Greve, and J. Blok, Transient Electric Birefringence of the Bacteriophages T3 and T7, Biopolymers, 16:639
(1977).

11. W. Boontje, J. Greve, and J. Blok, Transient Electric Birefrin-
ence of T-Even Bacteriophages. III. T2L and T6 with
Retracted Fibers Compared with T4B, Biopolymers, 16:551 (1977).

12. W. Boontje, J. Greve, and J. Blok, Transient Electric Birefringence of T-Even Bacteriophages. IV. T2Lo and T6 with Extended
Tail Fibers, Biopolymers, 17:2689 (1978).

13. T. K. Lim, G. J. Baran, and V. A. Bloomfield, Measurement of
Diffusion Coefficient and Electrophoretic Mobility with a
Quasielastic Light-Scattering-Band-Electrophoresis Apparatus,
Biopolymers, 16:1473 (1977).

14. G. J. Baran and V. A. Bloomfield, Tail-Fiber Attachment in
Bacteriophage T4D Studied by Quasielastic Light Scattering-
Band Electrophoresis, Biopolymers, 17:2015 (1978).

15. J. B. Welch and V. A. Bloomfield, Thermodynamics of the Slow-
Fast Transition in Bacteriophage T2L, Biopolymers, 17:1987
(1978).

16. J. B. Welch and V. A. Bloomfield, Concentration-Dependent Isomeri-
zation of Bacteriophage T2L, Biopolymers, 17:2001 (1978).

17. J. Aksiyote-Benbasat and V. A. Bloomfield, Joining of Bacteriophage T4D Heads and Tails: A Kinetic Study by Inelastic
Light Scattering, J. Mol. Biol., 95:335 (1975).

18. P. C. Hopman, G. Koopmans, and J. Greve, Influence of Double
Scattering in Determination of Rotational Diffusion
Coefficients by Depolarized Dynamic Light Scattering: Application to the Bacteriophages T7 and T4B, Biopolymers, 19:1241
(1980).

19. L. D. Kosturko, M. Hogan, and N. Dattagupta, Structure of DNA
within Three Isometric Bacteriophages, Cell, 16:515 (1979).

20. J. Garcia De La Torre and V. Bloomfield, Hydrodynamic Properties
of Macromolecular Complexes. I. Translation, Biopolymers,
16:1747 (1977).

21. J. Garcia De La Torre and V. Bloomfield, Hydrodynamics of Macromolecular Complexes. II. Rotation, Biopolymers, 16:1765
(1977).

22. J. Garcia De La Torre and V. Bloomfield, Hydrodynamics of Macromolecular Complexes. III. Bacterial Viruses, Biopolymers,
16:1779 (1977).

23. M. Levitt, How Many Base-Pairs per Turn Does DNA have in Solution
and in Chromatin? Some Theoretical Calculations, Proc. Natl.
Acad. Sci., 75:640 (1978).

24. M. Hogan, N. Dattagupta, and D. M. Crothers, Transient Electric
Dichroism of Rod-Like DNA Molecules, Proc. Natl. Acad. Sci.,

75:195 (1978).

25. J. Newman, H. L. Swinney, and L. A. Day, Hydrodynamic Properties and Structure of fd Virus, J. Mol. Biol., 116:593 (1977).

26. F. C. Chen, G. Koopmans, R. L. Wiseman, L. A. Day, and H. L. Swinney, Dimensions of Xf Virus from Its Rotational and Translational Diffusion Coefficients, Biochemistry, 19:1373 (1980).

27. E. F. Rossomando and J. B. Milstein, Electro-Optic Evidence for the Control of the Structure of Bacteriophage fl by a Minor Coat Protein, J. Mol. Biol., 58:187 (1971).

28. J. B. Milstein and E. F. Rossomando, Electro-Optics Studies on the Effect of Heat Treatment on Structure in Bacteriophage fl, Virology, 46:655 (1971).

29. P. Kunzler and T. Hohn, Stages of Bacteriophage Lambda Head Morphogenesis: Physical Analysis of Particles in Solution, J. Mol. Biol., 122-191 (1978).

APPLICATION OF ELECTRO-OPTICS TO THE STUDY OF COLLOIDS

S. P. Stoylov

Institute of Physical Chemistry
Bulgarian Academy of Sciences
1040 Sofia, Bulgaria

The first steps in electro-optics were made in the field of colloid science. This is not surprising, since for many of the optical phenomena, both the largest effects and the slowest effects can be achieved most easily by colloids. This greatly simplifies the experimental problems. Therefore, it is astonishing to see that at present the application of electro-optics to colloids is relatively unutilized with respect to its application to biological macromolecules. However, we might expect that in the near future the relative importance of the electro-optics of colloids, biological colloids being largely included, could grow considerably.

NATURE OF THE ELECTRO-OPTIC EFFECT

The nature of the electro-optic effect by colloids is not very different from that by large macromolecules. Besides orientational effects, aggregational, deformational (structural or conformational), chemical, permeational and other phenomena are possible.

In Fig. 1 an attempt is made to give a rough scheme of the different possible electro-optical effects, of the interrelation of these effects, and of some of the criteria by which the different electro-optical effects can be distinguished. Of considerable interest, although not being studied currently, is the aggregational electro-optic effect. It could in principle be both electro-aggregational and aggregational, the association being induced by some non-electric agent. The electro-aggregational electro-optic effect (to some extent also true for the pure aggregational electro-optic effect) could be either reversible or irreversible. A demonstration of both possibilities has been made in Refs. 1 and 2.

417

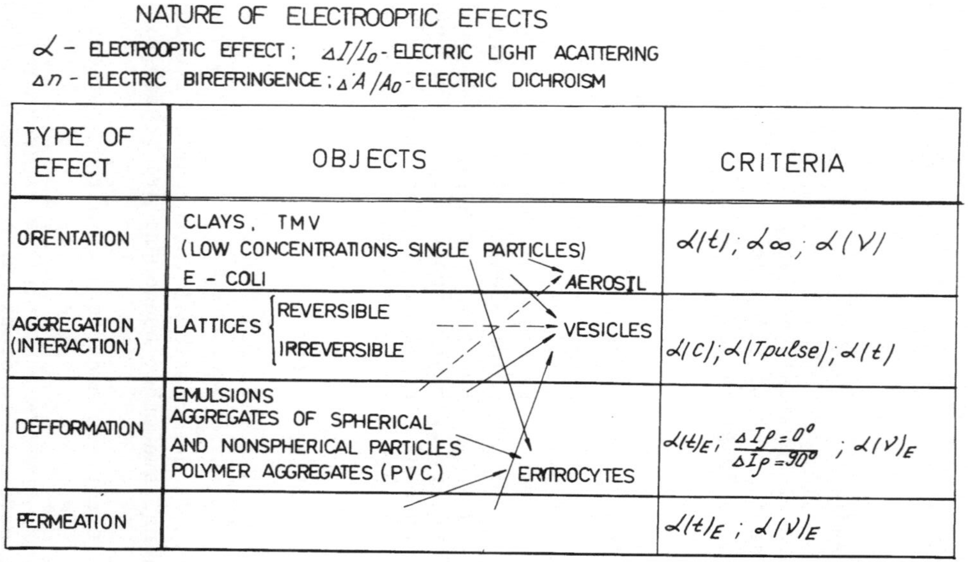

Fig. 1. The nature of electro-optical effects.

Deformable reversible or irreversible aggregates are of considerable interest in colloid science. This is a supplementary
source of information on the structure and properties of these
aggregates. In some cases, aggregates of well-defined simple
spherical single particles, e.g., aerosils, should be an interesting
and useful model for the study of electro-optic effects.

Effects of a mixed nature can be observed in natural and model
membranes. For example, in charged and non-charged lipid vesicles,
observed electric light scattering effects are connected with deformational and orientational effects (3) although permeational effects
are not excluded. Upon addition of calcium ions or cytochrome C,
aggregational (non-reversible) effects have been also detected (4).
With erythrocytes, it seems that permeational effects are definitely
present and are accompanied by both deformational and orientational
effects (5).

One must not forget that in many cases the application of
electro-optics is to particles of considerable dimensions (e.g.,
the erythrocytes), for which the optical approximations of the theory
are scarcely followed. However, there are electro-optical expressions and ways of interpretation which are not very sensitive to
optical approximations, for example, equations derived for low
degrees of orientation of various particles. For axially symmetric
particles, a large proprotion of the existing expressions is useable
without worrying about optical properties of the particles. Another
independent way of estimating the electrical properties of the
particles is the utilization of the value of the applied electric
field at which deviation from the square field dependence observed
at low degrees of orientation begins. For this field, the orientation energy has to be approximately equal to the energy of the
thermal motion of the particles. It is then not difficult to make
a rough estimate of the electric moments of the particles.

In the case of orientational electro-optic effects, the interaction of the electric field with the electric moments (permanent
and induced) of the colloidal particles is not the only possible
origin of orientation. The extensive analysis of Wilfried Heller
(9) (made as early as 1942) showing other possible sources for
orientation like electro-osmosis, electrophoresis, sedimentation,
convection, electric field inhomogeneities and interfacial polarizability is still pertinent. It should be noticed that this work
(9) practically was the first to suggest the interfacial nature of
the electric polarizability of colloid particles, which in recent
years has gained considerable recognition. For high fields, big
particles, highly conducting systems, long pulses, etc., the other
orientational effects also could become of considerable importance.
In some cases this makes visual control of the electro-optic effect
very helpful.

CHARACTERIZATION POSSIBILITIES FOR ELECTRO-OPTICAL EFFECTS

In Table I an attempt is made to summarize the possibilities
which the electro-optic methods provide for the characterization of
the geometric, electrical and optical properties of colloid particles. Electric birefringence, electric dichorism and electric light
scattering are the only methods considered herein.

For the geometrical properties, there are mainly two types of
electro-optic measurements that can be used: determination of the
dimensions of particles of known form from the electric light scattering effect at full particle orientation, or determination of the
dimensions of particles of known form from the decay of one or
another electro-optic effect, where this decay for monodisperse,
rigid, axially symmetric particles with circular cross-section is
monoexponential (10). The polydispersity of the samples can also
be deduced by these two methods. However, one must note that when
using the time dependence of the electro-optic effect at saturating
electric fields, non-monoexponential decays are possible (1,2) with
highly assymetric and optically polarizable colloid particles. A
unique possibility which has been used recently is the dependence of
the decay of an electro-optic effect on the electric pulse length
and the electric field strength. Thus, particles of varying dimension (and form) become effective at different times and fields in
the electro-optic effect. A direct possibility for studying the
rigidity of the colloid particles which could be quite important in
the case of aggregates, emulsions, membranes, etc., is provided by
the ratio of the light scattering effects at different angles from
the direction of the applied electric field. These possibilities
have been applied in the study of a number of colloids: clays,
viruses, bacteria, etc. (6,12).

The optical properties of colloid particles, e.g., the optical
anisotropy and the optical asymmetry (in the case of absorbing
particles) can be measured by various electro-optical effects at
saturating electric fields. If we have Rayleigh particles, for
which the dimensions of the particles are much smaller than the wavelength of the incident light, both electric light scattering and
electric birefringence can lead to calculation of the optical anisotropy. Electric birefringence gives greater sensitivity, especially with small particles. For particles comparable to the wavelength of light, particles with high geometric anisotropy but low
optical (intrinsic) anisotropy, greater sensitivity is obtained with
electric light scattering. In the case of "beyond Rayleigh" particles with high geometric and optical (intrinsic, not due to form)
anisotropies, all the electro-optical effects at full particle
orientation depend on the particle dimensions. In this case, for
the calculation of any optical property (here the optical asymmetry
has to be included) it will be necessary either to know the dimension by independent methods or to have information on the dimensions

from electro-optic decay analysis. It should be clear that measurements are highly complicated in the case of polydisperse particles.

    The determination of the electric properties of the colloid
particles is of great interest. There are three parameters which
can be determined. One is the electric polarizability, the second
is its dispersion, which could contain more than one parameter (more
than one dispersion) and the third is the permanent dipole moment.

    For the determination of the electric polarizability, one can
enumerate at least four electro-optic methods (see Table I): (1)
the dispersion dependences of the electro-optic effect which for low
degrees of orientation (energy of orientation much lower than the
energy of the Brownian motion) seems to be little influenced by the
particle geometry and optics; (2) the initial slope of a plot of the
electro-optic measurement versus the square of the electric field at
a frequency at which the permanent dipole moment (or slow polarizabilities) have no time to play a role in particle orientation; (3)
the full electric field dependence; and (4) the time dependence
(the build-up) of the electro-optic effect at saturating electric
fields. In the three last cases, when the particles exceed the
Rayleigh approximation, it should be kept in mind that considerable
deviations can occur which may be quite different for the different
electro-optic effects. Special attention has to be paid when combining or comparing data for high and low degrees of orientation.
The study of the dispersion of the electro-optic effect is of importance for the correct determination of the electric polarizability
of the colloidal particles, and in practically all cases it is preferable that such a study is made before the determination of the
electric polarizability. This could be done in two ways. The
first way, and by far the most used at present, is to study the
frequency dependence of the electro-optic effect, always at low
degrees of orientation. The other is to study the build-up of
the electro-optic effect where the different dispersions have to
manifest themselves in the time-domain. This should also be done
at low degrees of orientations.

    The determination of the value of the permanent dipole moment
is also useful. However, this presents considerable difficulties
as it is difficult to distinguish a dipole moment from a slow electric polarizability. The principal methods are the same as the
methods for the determination of the electric polarizability, the
essential difference being that the dipole moment is measured at
low frequencies (ordinarily much lower with the limiting case of
the d.c. field) at which the molecule rotates freely. The value of
the dipole moment is obtained by comparison with measurements at
frequencies where the dipole moment cannot act at all, i.e., where
the molecule cannot rotate and where the electric polarizability is
determined. The study of slow electric polarizability employs the

Table 1. Electro-optic Characterization of Geometric (Mechanic), Optic and Electric Parameters of Colloid Particles

α = electro-optic effect; $\Delta I/I_o$ - electric light scattering; Δn - electric bire-fringence; $\Delta A/A_o$ - electric dichroism

Type of Parameter	Parameters	Type of Electro-optic Measurements
Geometric (Mechanical)	Dimension	$\left(\dfrac{\Delta I}{I_o}\right)_\infty$, $\alpha(t)$
	Form	$\left(\dfrac{\Delta I}{I_o}\right)_\infty$, $\alpha(t)$
	Polydispersity	$\alpha(t)_\infty$, $\alpha(E^2)$, $\left(\dfrac{\Delta I}{I_o}\right)_\infty$, $\alpha(t) = f\left(E, T_{pulse}\right)$
	Rigidity	$\dfrac{\Delta I(\beta = 90^\circ)}{\Delta I(\beta = 0^\circ)}$, $\alpha(t)$
Optical	Optical Anisotropy	Δn_∞, $\left(\dfrac{\Delta I}{I_o}\right)_\infty$
	Optical Asymmetry or Angle Absorption/Symmetry Particle Axes	$\left(\dfrac{\Delta A}{A_o}\right)_\infty$

Table 1 (continued)

Type of Parameter	Parameters		Type of Electro-optic Measurements
Electrical	Electric Polarizability		$\left(\dfrac{d\alpha}{dE^2}\right) \approx$; $\alpha_E(t)$; $\alpha(\nu)$; $\alpha(E^2)$
	Dispersion of Electric Polarizability	Volume	$(d\alpha/dE^2)$ $\nu > 10^8$ Hz; $\alpha(\nu)$
		Interfacial	$(d\alpha/dE^2)$ $\nu < 10$ MHz; $\alpha(\nu)$
		Slow	$(d\alpha/dE^2)$ $\nu < 100$ Hz; $\alpha(\nu)$; reverse; $\alpha_E(t)$
	Permanent Dipole Moment		$\alpha(\nu)$; $\left(\dfrac{d\alpha}{dE^2}\right) =$; reverse; $\alpha_E(t)$; $\alpha(E^2)$

same methods as those for the permanent dipole moment. A slow
electric polarizability can be distinguished from a permanent dipole
moment via the dispersion curve, the build-up curve and the decay
curve of the appropriate electro-optic effect.

THE INTERFACIAL NATURE OF THE ELECTRIC POLARIZABILITY OF COLLOID PARTICLES AND ITS IMPACT ON SOME BASIC PROBLEMS IN COLLOID SCIENCE

The first to describe interfacial polarizability in terms of
deformation of the electric double layer were Errera, Overbeek and
Sack (13) as early as 1935. Later in 1956-57, O'Konski and Haltner
(14) in a detailed study of the electric birefringence of tobacco
mosaic virus (TMV) gave for the first time firm evidence for the
reality of the interfacial electric polarizability. They showed a
strong dependence of the electric polarizability on the frequency
of the applied electric field below the MHz range. The polarizability had an experimental value some 50 times greater than that
calculated from volume polarizability theory. Later, Stoylov (15)
indicated a still greater difference between the measured and calculated electric polarizabilities for a typical colloid sample,
the clay palygorskite, and showed the existence of a strong frequency dependence in the kHz range. In the following years, Stoylov
and Petkanchin gave new evidence for the importance of the interfacial electric polarizability for clays (16) showing a pronounced
ionic strength dependence also varying with the frequency of the
electric field in the kHz range. In addition, it was shown that
for TMV there is a considerable interfacial electric polarizability
(large value and frequency dependence in the kHz range) at the isoelectric point (17). At the beginning of the seventies, the interfacial nature of the electric polarizability in the case of clays
was further confirmed by the demonstration of the great effect of
ionic surface active substances on their electric polarizability
(18,19).

In colloid science it is generally accepted that the stability
of colloid suspensions is one of the most important theoretical
and experimental current questions. The large value of the electric
polarizability and its interfacial nature not only directly influences the magnitude of the polarizability interactions between
the colloid particles (the electrodynamic attraction and the electrostatic repulsion (20)), but also influences the kinetics of
the interactions, which might be much slower than are currently
accepted.

Another problem is the electrokinetic effect. This is connected with the fact that the interfacial polarizability is due to
the asymmetry of the double electric layer induced by the external
electric field. This is the physical basis for the need of a
relaxation correction in electrophoresis. Electro-optics give the

possibility for both an experimental and a theoretical approach to
the electrokinetic phenomena. Up to now, the theory of the electrokinetic phenomena has been extensively used in the elaboration of
the theories of the interfacial electric polarizability (6). It may
be expected that further experimental and theoretical developments
in the fields of electro-optics and electrokinetics will be closely
interrelated. In principle, one can speak of orientational electrokinetics in which details in the distribution of the interfacial
charge became important.

ELECTRO-OPTICS AND THE AGGREGATIONAL STATE OF COLLOID SYSTEMS

Another important problem in colloid science where electro-
optics could gain considerable importance is the problem of the
aggregational state of colloid systems and the structure of aggregates. It should not be forgotten that electro-optics are perhaps
the most sensitive method for detecting the presence of very small
fractions of aggregates. This provides the possibility of following
the initial stages of aggregation where theory is most reliable.
Generally, one might expect that the aggregation of a small number
of particles is geometrically anisotropic. Therefore, in the most
favorable case of spherical primary particles, electro-optics will
be able to observe the expected nonspherical aggregates.

A case in which electric light scattering clearly demonstrated
the potential of the electro-optical methods to detect aggregates
of small shperical particles is that of aerosils (7). Unfortunately,
it is very difficult to control the number of spheres in the aggregates, which although monodisperse are composed of a large number
of spheres (over twenty when the diameter of the primary particle
was 0.05 nm). Future experiments with small diameter latexes could
be useful.

As far as the structure of the aggregates is concerned, an
early and still very good example is the electric light scattering
study of the structure of montmorillonite aggregates made by
Schweitzer and Jennings (21). The complicated structure they proposed corresponds well to that predicted by clay colloid chemists.
Structures of aggregates have been obtained mostly from decay
curves of electro-optics effects, and, to some extent, from the
electric light scattering effect at full orientation of the particles.

Another possibility which is seldom used is the utilization of
electric properties such as dipole moment for the study of aggregate
structures. There is evidence for a number of objects (TMV, clays)
that a considerable transverse permanent dipole moment often exists
in the aggregates. With aerosils, this adds feasibility to the
suggestion that formation of the aggregates is accompanied by redistribution of the interfacial charge which leads to the formation

of a permanent dipole moment. In the case of aerosil aggregates,
there is no ground to assume that the spherical primary particles
are electrically anisotropic, hence they scarcely can have a permanent dipole moment before aggregation.

Another important problem for the future development of electro-optics is connected with the problem of colloidal aggregates.
That is, what type of objects can be studied so that the theory can
be set up with the lowest number of assumptions? This is particularly important for the study of the interfacial electric polarizability. It seems that in many respects, the rigid, linear, aggregates of very small spheres are good candidates. The theory of the
decay curve for such aggregates is well developed (22) and there are
no reasons to expect that the elaboration of the theory for the
interfacial electric polarizability, for the orienting torque, and
for the build-up curve should be more complicated than for other
nonspherical forms.

SOME NEW DEVELOPMENTS IN THE ELECTRO-OPTIC DETERMINATION OF THE
ELECTRIC PROPERTIES OF COLLOID PARTICLES

The determination of the transverse permanent dipole moment of
aerosil aggregates should be reliable since it is made at sufficiently
low degrees of orientation. However, there exists a doubt about the
transverse permanent dipole moments which have been determined up
until now. This doubt has been expressed by Morris and Jennings
(23) in their electro-optic study of bacterias.

The extent to which the deviation from the requirements for
low degrees of orientation could lead to misleading interpretations
has been experimentally demonstrated on palygorskite suspensions (24).
The comparison of field dependence measurements made with d.c. fields,
bursts of sinusoidal fields and bursts of square reverse pulsed
fields of the same frequency as the sinusoidal one (ordinarily a
frequency at which the permanent dipole moment cannot act) showed
that, above the electric fields for which there are low degrees of
orientation, the electro-optical effects observed no longer correspond to the effective field strength of the sinusoidal field.
Rather, they correspond to half the peak-to-peak value of the electric field, i.e., the real orienting electric field could be larger
by as much as 40% than one might think. Fig. 2 shows the deviation
of the real field strength dependence of the electro-optic effect
(the continuous curve) from that which could be drawn when the difference of orienting field strength from the effective field strength
beyond the range of low degrees of orientation is not taken into
account (the dotted curve). In Fig. 3, the experimental data for
palygorskite are shown. It can be seen that with the increase of
the value of the electric field at which the dispersion curve (i.e.,
the frequency dependence) is measured, it changes from a falling

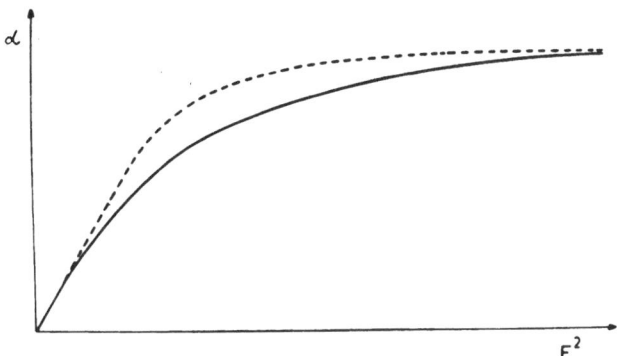

Fig. 2. Dependence of an arbitrary electro-optic effect on the
         electric field strength (in arbitrary units). The dotted
         line, presenting the high frequency effect is related to
         an effective field, deviates from the real (continuous)
         curve.

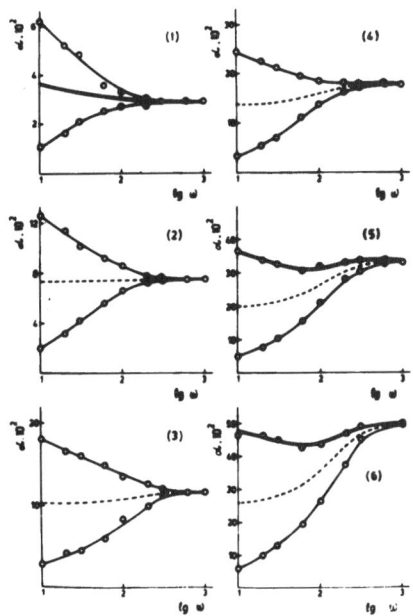

Fig. 3. Experimental dependences of the instantaneous minimum and
         maximum values and the average value of the electric light
         scattering on field frequency for palygorskite: (1) E =
         64 V/cm; (2) E = 96 V/cm; (3) E = 128 V/cm; (4) E = 160
         V/cm; (5) E = 256 V/cm; and (6) E = 320 V/cm.

frequency dependence to a rising one. A longitudinal permanent
dipole moment (or low frequency electric polarizability) at low
fields and a transverse one at high fields can be seen. It has to
be noted that only the lowest field is in the linear part of the
square field dependence of the electro-optic effect when it is corrected for the deviations presented with Fig. 2. It is important
to remember that this correction has the effect of shortening the
linear section of the field dependence curves.

So one must be careful when studying dispersion dependences
and one should not forget that with the rise of the frequency and
the electric field strength beyond those necessary for low degrees of
orientation, the real orienting electric field shifts from the effective field strength to half the peak-to-peak value. This becomes
more important in the case of large particles where the dispersion
dependency is least influenced by deviations from the optical approximations (11,25). Finally, it should be realized that some of
the reported data giving the presence of a transverse permanent
dipole moment could be due to this effect.

Another development which concerns the permanent dipole moment,
the electric polarizability, and the screening of the permanent
dipole moment by the ions in the double layer is the investigation
of aerosols at different humidities. It is possible to eliminate
completely the action of the diffuse double layer. This approach
is limited to particles which do not change their interface appreciably when dispersed in air instead of water. The first results
do not give evidence for strong screening or for interaction between
the two types of polarizability (26). Quite recently, the method
of electric light scattering for aerosols has been considerably
improved (27,28) and should prove useful for future work in this
field.

ELECTRO-OPTIC STUDIES OF COLLOIDAL STABILITY

When discussing the impact of the concept of interfacial electric polarizability on colloid science, the impact on our ideas
about colloidal stability should be mentioned first. Further on in
this part, this question will be discussed in more detail, with
more attention being paid to the correlations found among the values
of electric polarizability, electrophoretic mobility (the electrokinetic potential) and the stability of the suspensions.

Practically in all cases studied, correlations between the
variation of the values of the electric polarizability and the electrophoretic mobility with the addition of cationic surface active
substance have been found. In some respects this could be considered as a further proof for the predominantly interfacial electric polarizability. In Fig. 4, an attempt is made to summarize

OBJECT	INITIAL ELECTRIC POLARIZABILITY (cm^3)	ISOELECTRIC ELECTRIC POLARIZABILITY (cm^3)	C_{isoel} CONCENTRATION OF ISOELECTRIC POINT	C_{rextr} CONCENTRATION OF EXTREMUM ELECTRIC POLARIZABILITY	REMARKS
PALYGORSKITE-I	$0,9.10^{-12}$	$0,59.10^{-12}$	CETYL PYRIDINIUM CHORIDE (CPC) $1,5.10^{-2}$ M/1	10^{-5} M/1	MINIMUM STABILITY AT C_{rextr}
PALYGORSKITE-II	$0,85.10^{-12}$	$0,50.10^{-12}$	$Th(NO_3)_4$; 10^{-4}M/1	7.10^{-5} M/1	MINIMUM STABILITY AT C_{rextr}
BENTONITE	1.10^{-12}	$1,30.10^{-12}$	CPC; $6,5.10^{-5}$ M/1	$6,5.10^{-5}$ M/1	MINIMUM STABILITY AT C_{rextr}
ILLITE CLAY	$1,3.10^{-12}$	$0,8.10^{-12}$	CPC; $1,5.10^{-4}$ M/1	10^{-6} M/1	MINIMUM $\frac{\partial \psi}{\partial C_{cpc}}$ MINIMUM $\frac{\partial r}{\partial C_{cpc}}$
TOBACCO MOSAIC VIRUS	$0,12.10^{-12}$	$0,03.10^{-12}$	pH= 3,3	10^{-4} M/1	NO MINIMUM OF STABILITY OBSERVED
KAOLINITE	$8,45.10^{-12}$	$4,5.10^{-12}$	CPC, 5.10^{-4} M/1 ionic strength 5.10^{-4}	$5,7.10^{-6}$ M/1	MINIMUM STABILITY AT C_{rextr}
AgI	$1,8.10^{-13}$	$\gamma_{min} = 7,2.10^{-4}$	DECYL DIMETHYL AMMONIUM OXIDE (DDAO) NO γ=0 FOR pH=6	10^{-4} M/1	MINIMUM STABILITY AT C BUT FOR pH=7

Fig. 4. Colloidal Stability, Electric Polarizability. and the Isoelectric Point

the greater part of the existing studies on the correlations among
electric polarizability, electrophoretic mobility and the stability
of the colloid systems studied. The electric polarizability is
calculated from the slope of the electro-optic effect versus the
square of the applied electric field at a frequency previously determined from the dispersion curves to correspond with orientations
not perturbed by the action of the permanent dipole moment (frequencies above the dipole dispersion frequency). The electrophoretic
mobility is determined by microelectrophoresis, and stability is
followed in most cases both by light scattering intensity measurements and by the decay of the electro-optic curves. Except for the
measurements of kaolinite and AgI which were made by electric birefringence, all other measurements were made by the electric light
scattering method.

The common trend observed for all objects presented in Fig. 4
is the marked correlation between the electric polarizability and
the electrophoretic mobility variations with the addition of ionic
surfactant. This is well observed in Fig. 5 where the measurements
with the clay palygorskite are presented. Similar clear correlations have also been observed on other objects (sepiolite (29) and
biocolloid suspensions (30)) by Jennings and coworkers.

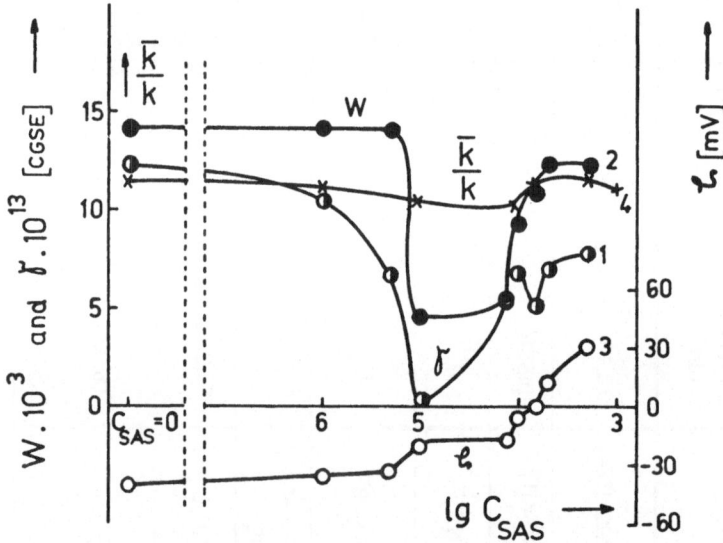

Fig. 5. Dependence of the electric polarizability γ (curve 1), the
         stability w (curve 2), the ζ-potential (curve 3) and the
         relative conductivity $\bar{k}/k$ (curve 4) on the concentration
         of the surface active substance cetylpyridinium chloride
         C_{SAS} (in m/ℓ).

In all cases studied, a clear correlation is observed between the variation of the electric polarizability and the stability on the addition of an ionic surfactant. Different types of correlations are observed: a correlation of minimum stability with a minimum in the value of the electric polarizability (palygorskite, kaolinite, AgI) with a maximum in the value of the electric polarizability (bentonite) and a correlation of minimum slope in the dependence of the electric polarizability on surfactant concentration with a minimum slope in the dependence of the stability on surfactant concentration (illite clay).

Quite surprisingly, there are cases of correlation of the electric polarizability extreme with the minimum stability concentration of the surfactant. In some cases, these correlations occur away from the minimum value of the electrophoretic mobility which is the isoelectric point. There are a number of cases known in colloid science where the minimum stability does not coincide with the isoelectric point (6) and this always has been a problem considered suitable for deeper analysis of the interaction phenomena in colloid suspensions. At present, this difference is best seen in the case of palygoskite. Data for AgI and TMV also indicate such differences.

It is logical to expect similar correlations between colloidal stability and permanent dipole moment. This has been demonstrated in the case of AgI and the data are presented in Figure 6. In this case the correlation is simple to explain by a higher dipole moment, therefore, greater particle interactions (attractions) and lower stability. A more complicated situation is the explanation of the correlation between electric polarizability and colloid stability. It is clear that not only is the electrostatic repulsion important, as it is related to the electrophoretic mobility, but also the electrodynamic interaction should be considered. This could have a considerable influence on the interfacial electric polarizability (20). Furthermore, the interfacial nature of this interaction will be closely connected with the repulsion as the same ions are responsible for both effects. In the one case their dynamic and in the other their electrostatic properties should be taken into account.

Before trying to derive some conclusions about the connection of the interfacial electric polarizability with the stability of the colloid systems, it is necessary to introduce further refinements in the measuring procedure. In the first place it is recommended to use methods for following the electric polarizability which are less influenced by association phenomena than those used in existing studies (29,30). At the moment, it seems that this requirement is best met in dispersion dependences at low degrees of orientation. Another refinement is related to the study of stability. It might be useful to make parallel studies of interactions for more concentrated systems where faster and larger effects may give more detail on the stability/surfactant concentration curves.

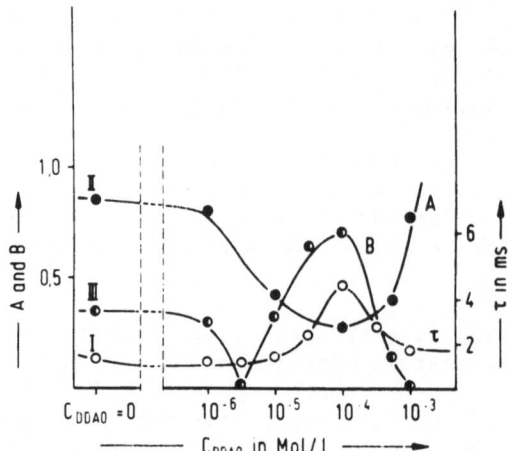

Fig. 6. Dependence of the relaxation time τ (curve I), the initial slope (A) of the curve $\Delta n/E_{\sim}^2$ (Eq. (3), curve II) and
the difference (B) of the initial slopes of the curves
$\Delta n/E_{=}^2$ and $\Delta n/E_{\sim}^2$ (curve III) on the DDAO concentration.*

CONCLUSIONS

Although quite limited at present, the application of electro-
optics to colloids seems to have evolved from its infancy. It is becoming more and more recognized as a new tool for the characterization of the geometrical, optical and electric properties of colloidal
particles. Further refinement of the concept of the interfacial nature of electric polarizabilities of the colloid particles and the
application of electro-optics to the study of colloid stability,
aggregation and structure of aggregates of colloid particles can be
expected to bring new findings to the field of colloid science. In
addition, among colloid systems there exist good candidates for model
objects for electro-optic studies. Therefore it is not impossible
that in the near future the application of colloids to electro-optics
will gain in importance.

*Dodecyldimethylamine-oxide

REFERENCES

1. S. Sokerov, T. **Vorobeva,** and S. P. Stoylov, <u>J. Polym. Sci.</u> 44:
    147 (1974).
2. S. Sokerov, and T. Vorobeva, "Proc. Intern. Conf. on Colloid and
    Interface Science," Vol. 1, Budapest, Acad. Sci., p. 391 (1975).
3. B. G. Tenchov, S. Kh. Sokerov, and S. P. Stoylov, <u>Studia Bio-</u>
    <u>physica</u> 77:109 (1979).
4. B. G. Tenchov, and S. P. Stoylov, unpublished results.
5. S. P. Stoylov, Studia Biophysica 72:193 (1978).
6. S. P. Stoylov, <u>Adv. Interface Coll. Sci.</u> 3:45 (1971); S. P.
    Stoylov, V. N. Shilov, S. S. Dukhin, S. Sokerov, and I.
    Petranchin, "Electro-Optics of Colloids," Nankova Dumka, Kiev,
    (in Russian) (1979).
7. M. Buleva, and S. P. Stoylov, in preparation.
8. A. Andonovski, S. Sokerov, and S. P. Stoylov, Intern. Polym.
    Symposium, Mainz (1979).
9. W. Heller, <u>Rev. Mod. Physics</u> 14:390 (1942).
10. M. Stoimenova, Ph.D. Thesis, Sofia (1979).
11. S. P. Stoylov, this volume.
12. B. R. Jennings, <u>in</u>: "Molecular Electro-Optics," C. T. O'Konski,
    ed., Vol. 1, Part 1, Ch. 8, Marcel Dekker, NY, p. 275 (1976).
13. J. Errera, J. Th. C. Overbeek, and H. Sack, <u>J. Chim. Phys.</u> 32:
    681 (1935).
14. C. T. O'Konski, and A. J. Haltner, <u>J. Am. Chem. Soc.</u> 78:3604
    (1956); ibid. 79:5634 (1957).
15. S. P. Stoylov,"Proc. Intern. Congr. Surface Activity," 4th,
    Brussels, p. 171 (1964).
16. S. P. Stoylov, and J. Petkanchin, Izv. Inst. Fizikokhim., <u>Bulg.</u>
    <u>Akad. Nauki</u> 5:73 (1965).
17. S. P. Stoylov, and J. Petkanchin, <u>God. Sof. Univ.</u>, 61:227 (1966-
    1967).
18. S. P. Stoylov, and J. Petkanchin, Dokladi BAN, L4:487 (1971)
19. G. L. Brownsey, V. J. Morris, and B. R. Jennings, <u>in</u>: "Electro-
    Optics and Dielectrics of Macromolecules and Colloids, " B. R.
    Jennings, ed., Plenum Press, London, p. 337 (1979).
20. J. Mahanty, and B. W. Ninham, "Dispersion Forces," Acad. Press,
    London (1976).
21. J. Schweitzer, and B. R. Jennings, <u>J. Coll. Interface Sci.</u> 37:
    443 (1971).
22. J. Garcia de la Torre, in this volume.
23 V. J. Morris, and B. R. Jennings, <u>J. Chem. Soc. Trans. II</u> 71:
    1948 (1975).
24. M. Stoimenova, Ts. Radeva, and S. P. Stoylov, <u>Coll. Polym. J.</u>
    257:1226 (1979).
25. M. Stoimenova, Thesis, Sofia (1979).
26. S. P. Stoylov, Y.' S. Lyubovtseva, N. V. Melnikov, and J.
    Petkanchin, "Proc. VII Colloid Conference," USSR (1977).
27. V. N. Kapustin, G. V. Rosenberg, N. C. Ahlquist, D. S. Covert,
    A. P. Waggonner, and R. J. Charlson, <u>Appl. Optics</u> 19:1345 (1980).

28. V. N. Kapustin, and D. S. Covert, <u>Appl. Optics</u> 19:134 (1980).
29. S. P. Stoylov, "Proc. Intern. Conf. Coll. Interface Sci.,"
     Budapest, p. 279 (1975).
30. S. P. Stoylov, I. Petkanchin, and S. Sokerov, "Proc. Vth Con-
     gress on Surface Activity," p. 163 (1969); I. Petkanchin, S.
     Sokerov, and S. P. Stoylov, "Proc. VIth Congress on Surface
     Activity," p. 671 (1973).

ELECTRO-OPTICAL PROPERTIES OF LIQUID CRYSTALS

D. A. Dunmur

Department of Chemistry
The University
Sheffield, S3 7HF, U.K.

1. INTRODUCTION

It is generally believed that Reinitzer's microscopic observations[1] on a cholesterol derivative marked the beginning of scientific interest in a range of new phases of matter known as liquid crystals. These thermodynamically stable phases are often referred to as mesophases since they occur between the solid crystalline phase and the fluid liquid phase, and they possess some of the physical characteristics of both solids and liquids. Although many fascinating properties of liquid crystals had been identified and characterized in the period 1900 to 1940, interest in these materials was renewed in the mid-1960's prompted in part by the publication of G. W. Gray's book "Molecular Structure and the Properties of Liquid Crystals"[2] which reviewed the earlier work. The major stimulus to recent research on liquid crystals arose from the realization in the late 1960's and early 1970's that many of the unusual properties of liquid crystals, especially their electro-optical properties, could have technological applications. These applications have been the subject of a number of reviews,[3,4,5] and it is intended in this chapter to concentrate on the application of electro-optical techniques to the investigation of the molecular aspects of the liquid crystalline phase.

Liquid crystals are disordered phases in which partial orientational order results in the nematic phase, or if the constituent molecules are optically active the cholesteric phase. If some degree of translational order is also present then smectic (soap-like) phases may be formed in which the density is spatially modulated giving rise to a variety of layered structures. A simplified view of some liquid crystalline phases is given in Figure 1. The

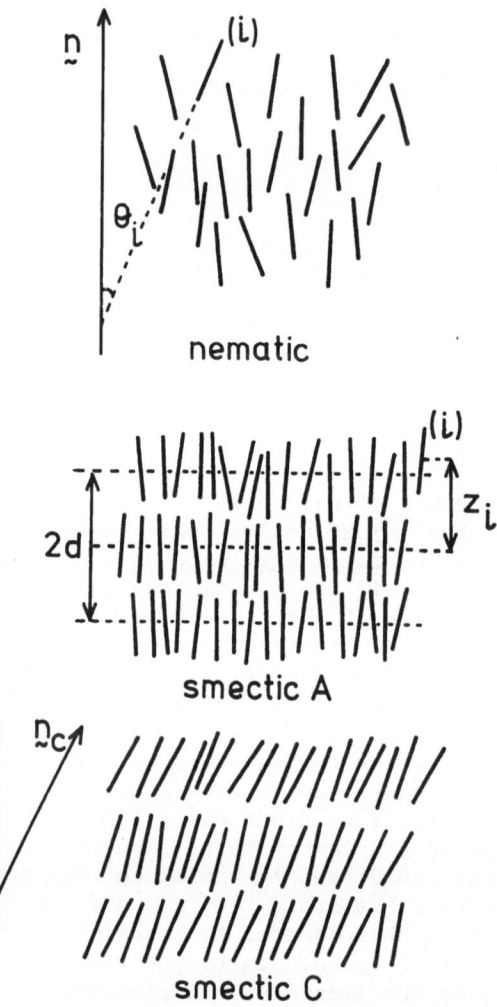

Figure 1.   Schematic Diagram of Liquid Crystal Structures

feature common to all liquid crystal phases is that of partial
orientational ordering of the molecules (mesogens), and so these
mesophases are formed by anisometric molecules which are usually
lath-like; recently some liquid crystal phases have been identified[6]
in which the constituent molecules are disc-like. Orientational
ordering of the molecules in liquid crystals gives rise to a local
anisotropy in some physical properties such as electrical and thermal conductivity, diffusivity, electric permittivity, magnetic
susceptibility, and aligned liquid crystal samples can be produced
which exhibit a macroscopic anisotropy in their physical properties.
Liquid crystals may be formed by pure compounds on heating (thermotropic) or by dissolving an amphiphilic molecule in a polar or
non-polar solvent (lyotropic)[7]; we shall only be concerned with
thermotropic liquid crystals in this chapter, although the electro-
optical properties of lyotropic systems present a challenging new
area of research.

The local orientation order in a liquid crystal can be characterized by an order parameter S defined by:

$$S = <\frac{1}{2} (3\cos^2\theta_i - 1)> \tag{1}$$

where θ_i is the angle between a unique axis in the molecule i
(termed the long molecular axis) and a direction $\underset{\sim}{n}$ which is the
local average direction of the long molecular axis. Angular brackets indicate an ensemble average. The direction $\underset{\sim}{n}$ is termed the
director, and the distribution in space of this unit vector determines the macroscopic anisotropy of a liquid crystal sample. In
a fully aligned liquid crystal $\underset{\sim}{n}$ has a uniform direction throughout
the sample, except in cholesteric liquid crystals where the distribution of $\underset{\sim}{n}$ follows a regular helix. It is sometimes convenient
to introduce a macroscopic order parameter defined in terms of some
anisotropic property of the liquid crystal: we shall denote this
order parameter as Q, and the relationship between S and Q will be
considered in the next section.

Molecular statistical theories can be used to calculate S.
Maier and Saupe[8] developed a theory of the nematic phase based on
a mean-field potential of the form:

$$u(\theta) = -\frac{\varepsilon S}{2} (3\cos^2\theta - 1). \tag{2}$$

This contains a single parameter ε which was interpreted as measuring the strength of the anisotropic dispersion forces believed to
be responsible for the stability of the nematic phase. The value
of ε can be obtained from the equilibrium temperature (T_{NI}) for the
nematic and isotropic liquid phases ($\varepsilon = kT_{NI}/0.2203$), and the
order parameter is predicted to be a universal function of the reduced temperature T/T_{NI}. The order parameter is also predicted to

decrease through the nematic phase to a value of 0.43 at T_{NI}; the
discontinuity in S being consistent with the first order nature of
the nematic to isotropic transition. In practice real nematic liquid crystals have order parameters in remarkably good agreement
with the predictions of the Maier-Saupe theory, and this agreement
has been further improved by Luckhurst et al.[9] by adding extra terms
to the potential of eq. (2). The Maier-Saupe-Luckhurst theory of
nematics only models the attractive part of the intermolecular potential, but Onsager showed[10] that repulsive interactions alone
could give rise to an orientationally ordered phase. Unfortunately
the Onsager approach is only valid for molecules with molecular
length-to-breadth ratios greater than 100, and the predicted order
parameter is very high (~ 0.8).

   Recently[11,12] a van der Waals theory of nematics has been
developed that includes both repulsions and attractions in the
intermolecular potential. The repulsive interactions are treated
by scaled particle theory, while the attractive part of the potential is replaced by its mean field average. Predictions of this
theory for the transitional order parameter lie between those of
the Onsager approach and the Maier-Saupe value, but one interesting
conclusion of the theory is that coupling between isotropic attractions and anisotropic repulsions are primarily responsible for the
stability of the liquid crystalline phase; i.e. the shape of
molecules is much more important in stabilizing the ordered phase
than the anisotropy in dispersion forces. The importance of molecular shape has been further demonstrated by Gelbart and Barboy[13],
who have shown that departures from axial symmetry can account for
the observed S_{NI} being lower than the Maier-Saupe value of 0.43.
Their theory can also account for the existence of discotic phases
of plate-like molecules having negative order parameters.

   Although there is now a reasonable understanding of the factors
that contribute to nematic phase stability, the situation with
smectics is much less clear[14]. A number of smectic phases have
been identified (at least eight and possibly ten) which have varying degrees of translational order in addition to the orientational
order characteristic of liquid crystals. To characterize the
additional order new parameters are introduced which describe the
ordering of smectic layers: these are defined as

$$\rho_m = \langle \cos(2\pi m z_i/d) \rangle$$

where d is the layer spacing and z_i is the position of molecule i
in a direction perpendicular to the layer plane. Unfortunately
measurements of ρ_m have not been made, although X-ray scattering
may be a route to them. Optically active molecules give rise to
chiral smectic phases, and some of the phases are biaxial in which
the molecules are tilted with respect to the layer normal. Since
smectic phases are layered they must be stabilized by increased

transverse interactions that act perpendicular to the director or molecular axis. McMillan[15] has suggested that dipolar interactions are responsible for smectic C phase formation, the location of the polar groups in the molecule determining whether or not the phase is tilted. While dipolar interactions may contribute to smectic phase stability it is clear that other factors are also important[16, 17], and much more information is necessary before many of the subtleties of the variety of smectic phase behaviour are understood.

2. ORIENTATIONAL ORDER IN LIQUID CRYSTALS

The distinguishing feature of liquid crystal phases is orientational ordering of the molecules, and so the characterization and measurement of the degree of order is of great importance. Ideally the orientational distribution function $f(\Omega)$ is required; this may be represented as an expansion in angular functions that relate the molecular orientation to some laboratory fixed axis system. For example $f(\Omega)$ can be written[18] as:

$$f(\Omega) = \frac{1}{8\pi^2} \{1 + 5S_{\alpha\beta}n_\alpha n_\beta + 9S_{\alpha\beta\gamma\delta}n_\alpha n_\beta n_\gamma n_\delta + \ldots\} \tag{3}$$

where n_α are direction cosines of the director of the liquid crystal phase (assumed uniaxial) with respect to a molecule-fixed axis system. Repeated Greek suffixes imply a summation over all possible values of the suffixes. $S_{\alpha\beta}$ and $S_{\alpha\beta\gamma\delta}$ are ordering tensors, and the second rank ordering matrix is defined as:

$$S_{\alpha\beta} = \frac{1}{2} <3n_\alpha n_\beta - \delta_{\alpha\beta}> . \tag{4}$$

If the molecule possesses some elements of symmetry, or more particularly if the property being used to probe local order has some symmetry, then the expression for $\underset{\approx}{S}$ is simplified: for axial symmetry $\underset{\approx}{S}$ is diagonal and $S_{xx} = S_{yy} = -\frac{1}{2}S_{zz} = -\frac{1}{2}S$.

Many techniques have been used to probe the orientational order in a liquid crystal. Spectroscopic techniques such as NMR, ESR, Raman scattering and fluorescence depolarization measure the order of a segment of a molecule or a particular chromophore, and different parts of a non-rigid molecule may have different order parameters. Statistical theories of liquid crystals are based on a model for the molecule, and the order parameter that emerges from the theory refers to the model mesogen. Comparison of calculated with measured order parameters therefore has to be made with some care.

Microscopic ordering of molecules leads to a local anisotropy in various physical properties, and the magnitude of the anisotropy depends on the degree of local order. Thus measurement of the

anisotropy of physical properties can in principle lead to values
for the order parameter. The local axis of the anisotropy is
parallel to the director, and ordering of the director results in a
macroscopically ordered sample. An alternative definition of
orientational order can be introduced via the macroscopic ordering
tensor $Q_{\alpha\beta}$:

$$Q_{\alpha\beta} = \frac{3}{2} (\Delta A^{(o)})^{-1} \{A_{\alpha\beta} - \frac{1}{3} A_{\gamma\gamma} \delta_{\alpha\beta}\} \tag{5}$$

$A_{\alpha\beta}$ is any second rank tensor property of the liquid crystal, and
$\Delta A^{(o)} = A_{||}^{(o)} - A_{\perp}^{(o)}$ is the hypothetical anisotropy in A for the
perfectly ordered liquid crystal (assumed to be uniaxial with S =
1). It is not usually possible to measure $\Delta A^{(o)}$ and estimates may
be made from measurements on crystals, or by extrapolation[19] from
measurements on liquid crystals.

To relate Q to S it is necessary to develop appropriate statistical theories for the tensor property that defines Q. We shall
consider three such properties, which are of importance in the description of the electro-optics and magneto-optics of liquid crystals:
electric permittivity, refractive index and magnetic permeability.
The last of these is the simplest, and one can write:

$$\kappa_o^{-1}(\kappa_{\alpha\beta} - \delta_{\alpha\beta}) = \sum_i \chi_{\alpha\beta}^{(i)} \equiv A_{\alpha\beta} \quad . \tag{6}$$

$\kappa_{\alpha\beta}$ is the permeability tensor, κ_o the magnetic permeability of
free space and $\chi_{\alpha\beta}^{(i)}$ is the diamagnetic susceptibility of molecule
(i); the summation is over all molecules in a unit volume. For
paramagnetic molecules a temperature dependent term would contribute to eq. (6). This particularly simple relation results because
$(\kappa - 1)\kappa_o^{-1}$ is very small ($\sim 10^{-6}$) and the internal magnetic field
is therefore equal to the macroscopic magnetic field. If the
molecular susceptibility tensor is axially symmetric, then:

$$\chi_{\alpha\beta}^{(i)} = \frac{1}{3} \chi_{\gamma\gamma} \delta_{\alpha\beta} + \frac{1}{3} \Delta\chi(3\ell_\alpha^{(i)} \ell_\beta^{(i)} - \delta_{\alpha\beta}). \tag{7}$$

$\ell^{(i)}$ is the molecular symmetry axis. Replacing the summation in
eq. (6) by N times the average of eq. (7) gives:

$$Q_{\alpha\beta} = S_{\alpha\beta} = \frac{3(N\Delta\chi)^{-1}}{2} (\kappa_{\alpha\beta} - \frac{1}{3} \kappa_{\gamma\gamma} \delta_{\alpha\beta}) \kappa_o^{-1} \tag{8}$$

where N is the number of molecules per unit volume and $\Delta\chi$ is the
anisotropy in the molecular susceptibility. It is more difficult
to relate S and Q via the electric permittivity tensor or refractive indices because of the complications which arise from the internal electric field. There have been a number of attempts[20,21]
to define the internal electric field in a liquid crystal, and
studies of the order in liquid crystals may help to clarify the
situation. Using the Vuks relation for the internal field in a

non-polar dielectric leads to expressions relating the refractive indices of a uniaxial liquid crystal to the molecular polarizability of the mesogen:

$$(n_{\parallel}^2 - 1) = \frac{N(n^2+2)}{3\varepsilon_o} \left\{ \bar{\alpha} + \frac{2\Delta\alpha S}{3} \right\}$$

$$(n_{\perp}^2 - 1) = \frac{N(n^2+2)}{3\varepsilon_o} \left\{ \bar{\alpha} - \frac{\Delta\alpha S}{3} \right\} .$$

(9)

$n_{\parallel}$ and $n_{\perp}$ are principal refractive indices parallel and perpendicular to the director, n is the average refractive index ($n^2 = \frac{1}{3}(n_{\parallel}^2 + 2n_{\perp}^2)$), $\bar{\alpha}$ and $\Delta\alpha$ are respectively the isotropic and anisotropic polarizabilities for a molecule assumed to be axially symmetric. Equations (9) give:

$$(n_{\parallel}^2 - n_{\perp}^2) = \frac{N(n^2+2)\Delta\alpha S}{3\varepsilon_o}$$

(10)

so that the independent component of the macroscopic ordering matrix becomes (z corresponds to the parallel direction):

$$Q_{zz} = S = \frac{3\varepsilon_o(n_{\parallel}^2 - n_{\perp}^2)}{N\Delta\alpha(n^2+2)}$$

(11)

Measurement of the principal refractive indices and a knowledge of $\Delta\alpha$ should therefore provide a value for S.

Maier and Meier[22] derived equations relating the principal components of the permittivity tensor to the order parameter:

$$(\varepsilon_{\parallel} - 1) = NLF\varepsilon_o^{-1} \left\{ \bar{\alpha} + \frac{2\Delta\alpha S}{3} + \frac{F\mu^2}{3kT}(1 + 2S) \right\}$$

$$(\varepsilon_{\perp} - 1) = NLF\varepsilon_o^{-1} \left\{ \bar{\alpha} - \frac{\Delta\alpha S}{3} + \frac{F\mu^2}{3kT}(1 - S) \right\} .$$

(12)

μ is the molecular dipole moment and L and F are cavity and reaction field factors for a spherical cavity in an isotropic dielectric continuum. They have been derived for the general case of an ellipsoidal cavity in an anisotropic dielectric, and the consequences of introducing anisotropy into the internal field factors for the solution of the Maier and Meier equations has been considered[23]. For simplicity here L and F are assumed to be the same for both parallel and perpendicular directions, giving

$$(\varepsilon_{\parallel} - \varepsilon_{\perp}) = NLF\varepsilon_o^{-1} \left\{ \Delta\alpha + \frac{F\mu^2}{kT} \right\} S$$

(13)

and hence:

$$Q_{zz} = S = \frac{(\varepsilon_{\parallel} - \varepsilon_{\perp})}{NLF\varepsilon_o^{-1}} \left\{ \Delta\alpha + \frac{F\mu^2}{kT} \right\}^{-1}$$

(14)

L and F for the isotropic case are given by:

$$L = \frac{3\varepsilon}{(2\varepsilon+1)} \text{ and } F = \frac{(n^2+2)(2\varepsilon+1)}{3(n^2+2\varepsilon)} .$$

Equations (11) and (14) maintain the equivalence of S and Q, but if the internal field factors depend on the anisotropy of the medium and hence S, then Q_{zz} will no longer equal S.

The homologous series of alkyl-cyanobiphenyl liquid crystals has been extensively studied by various workers, and values have been reported for refractive indices[23,24], magnetic permeabilities[25] and electric permittivities[23,26]. Values for the order parameters for the homologous series have been derived using equations (8), (11) and (14), and are listed in Table I. In obtaining S from eq. (14) some allowance has been made for dipole-dipole correlation which is known[21] to be important in cyano-biphenyl liquid crystals. It is clear from the table that different macroscopic properties yield

Table I. Order Parameters for Alkyl Cyano-Biphenyl
Liquid Crystals at a Reduced Temperature T_R = 0.98

$R \cdot \phi \cdot \phi \cdot CN$	eq. (8) (i)	eq. (11) (ii)	eq. (14) (iii)
$R = C_5H_{11}$	0.55	0.64(0.41)	0.43
C_6H_{13}	0.51	0.70(0.40)	0.43
C_7H_{15}	0.55	0.80(0.46)	0.48
C_8H_{17}	0.52	0.84(0.46)	0.44

(i)   magnetic permeabilities from ref. (25); molecular suscep-
       tibility anisotropy obtained by bond additivity calcula-
       tion.

(ii)   refractive indices from ref. (24) and polarizability ansio-
       tropies from ref. (23). Values in brackets are order para-
       meters given in ref. (24) which were deduced using estimated
       molecular polarizabilities.

(iii) dielectric permittivities and molecular parameters from
       ref. (23). An average dipole correlation factor ($g_1 \simeq 0.45$)
       has been included such that $\mu^2 = g_1\mu_0^2$, where μ_0 is the
       free molecule dipole moment - see ref. (21).

different values for the order parameters. If the microscopic equation for S is used, then accurate values for the molecular properties are necessary, while if a macroscopic order parameter is to be
calculated then the evaluation of the extrapolated macroscopic anisotropy presents a problem. Order parameters derived from magnetic
susceptibility measurements should be the most reliable, but only
if accurate values for the molecular susceptibility are available.

In the isotropic phase of a liquid crystal the order parameter
is zero, but close to the transition temperature there are large
fluctuations in the order parameter which give rise to intense
light scattering. Application of an electric or magnetic field to
the isotropic phase in the pretransitional region can result in
field-induced order giving anomalously large magnetic or electric
birefringence. The latter will be discussed in detail in a later
section, and in anticipation of that the Landau theory of phase
transitions will be discussed.

3. LANDAU THEORY OF THE ISOTROPIC NEMATIC PHASE TRANSITION

Field free fluids - When a crystal melts to a nematic or disordered smectic (A or C) there is considerable release of energy
as latent heat of fusion ($\sim$ 10 kJ mol^{-1}), but at the liquid crystal
to liquid transition the latent heat is much smaller ($\sim$ kJ mol^{-1})
indicating that most of the lattice energy of a crystal is due to
translational ordering. The near vanishing value of the enthalpy,
entropy and volume change at the nematic to isotropic transition
means that this transition is close to being second order: i.e.
the first derivative of the free energy is almost continuous at T_{NI}.
The behaviour of the Gibbs free energy close to first and second
order phase transitions is illustrated in Figure 2. Certain physical properties diverge as the temperature approaches the critical
temperature associated with a second order transition, and there
is also a dramatic increase in the amplitude of fluctuations in some
quantities[27]. For liquid crystals the important quantity is the
order parameter, and although this is discontinuous at the liquid
crystal to isotropic transition, the weak first order nature of the
transition has encouraged the application[28,29] of theories that
were originally developed to describe second order phase transitions.

At a second order transition the order parameter goes continuously to zero, so Landau[30] proposed that around the transition
temperature the free energy can be expanded as a Taylor series in
the order parameter. It is appropriate to use the macroscopic
order parameter Q since Landau's theory is a macroscopic theory.
The free energy density may be written as

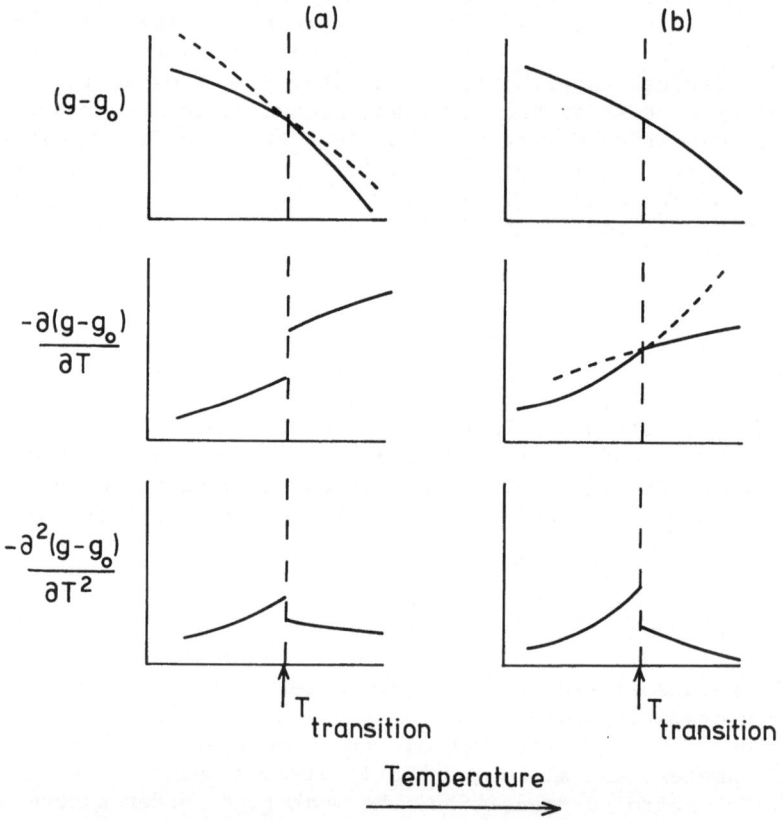

Figure 2.   Discontinuities in Free Energy Derivatives at First
Order (a) and Second Order (b) Phase Transitions

$$g = g_o + \left(\frac{\partial g}{\partial Q_{\alpha\beta}}\right)_{Q=0} Q_{\alpha\beta} + \frac{1}{2}\left(\frac{\partial^2 g}{\partial Q_{\alpha\beta}\partial Q_{\gamma\delta}}\right)_{Q=0} Q_{\alpha\beta}Q_{\gamma\delta} + \cdots \tag{15}$$

Retaining only rotationally invariant terms, the first order term is identically zero, and the expansion may be simplified to[29]:

$$g = g_o + \frac{a'}{2} Q_{\alpha\beta}Q_{\beta\alpha} + \frac{b'}{3} Q_{\alpha\beta}Q_{\beta\gamma}Q_{\gamma\alpha} + \frac{c'}{4}(Q_{\alpha\beta}Q_{\gamma\delta})^2 + 0(Q^5) \tag{16}$$

If $\underset{\sim}{Q}$ has uniaxial symmetry it may be written in terms of the director orientation as:

$$Q_{\alpha\beta}(\underset{\sim}{r}) = \frac{Q}{2}(3n_\alpha(\underset{\sim}{r})n_\beta(\underset{\sim}{r}) - \delta_{\alpha\beta}) \tag{17}$$

where $Q = \Delta A/\Delta A^{(o)}$ is the ratio of the macroscopic anisotropy at the appropriate temperature to the extrapolated anisotropy. $\underset{\sim}{n}(\underset{\sim}{r})$ is the director orientation at the point $\underset{\sim}{r}$ in the fluid. Assuming a uniformly aligned sample and using equation (17), the free energy density becomes:

$$g = g_o + \frac{a}{2} Q^2 - \frac{b}{3} Q^3 + \frac{c}{4} Q^4 + 0(Q^5) \quad . \tag{18}$$

The coefficients in this expansion differ by simple numerical factors from those appearing in eq. (16).[†] When $(g - g_o)$ is zero the orientational free energy vanishes, and stable states are represented by minima in $(g - g_o)$ plotted as a function of Q. a, b and c are functions of temperature and pressure, but for simplicity a is set equal to $a_o(T - T^*)$ and b and c are assumed independent of temperature: the pressure remains constant. It should be remembered however that the coefficients b and c probably vary more with temperature than a, and higher order coefficients of the free energy expansion may have singularities as functions of temperature and pressure[30]. Thus the expansion (18) is only reliable close to the transition and when Q is small.

The general form of the free energy is illustrated in Figure 3 as a function of order parameter; typical values have been assumed for a_o, b, c and T^*. Expressions for the first and second derivatives of the free energy with respect to order parameter are:

$$\frac{\partial(g-g_o)}{\partial Q} = aQ - bQ^2 + cQ^3 + 0(Q^4) \tag{19}$$

$$\frac{\partial^2(g-g_o)}{\partial Q^2} = a - 2bQ + 3cQ^2 + 0(Q^3) \tag{20}$$

[†] $a' = 2a/3$, $b' = -4b/3$, $c' = 4c/9$

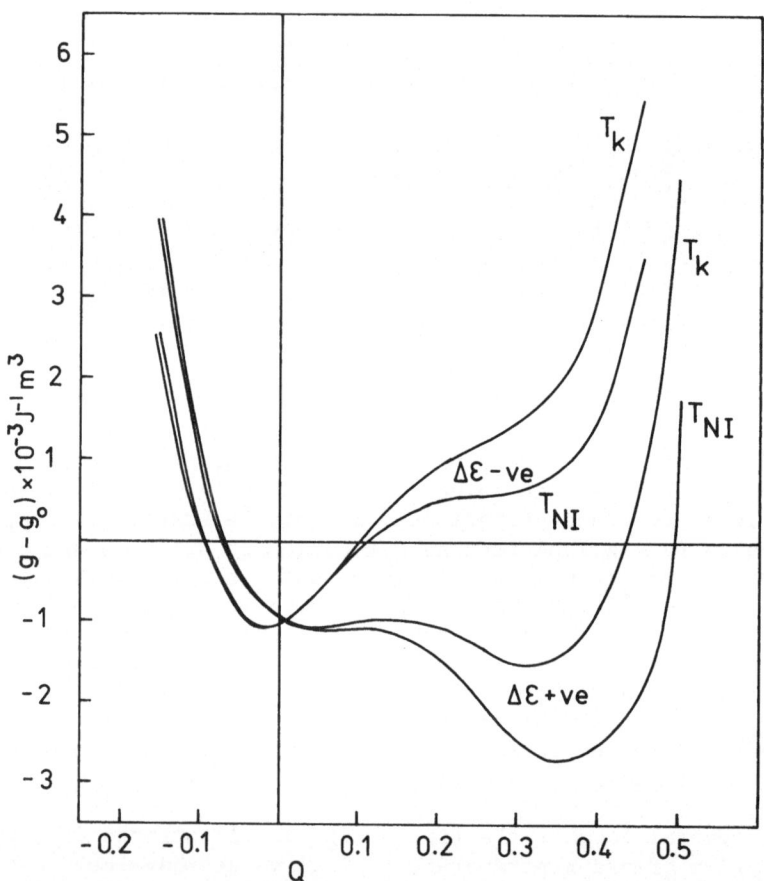

Figure 3.   Free Energy as a Function of Order Parameter
according to eq. (18). a_o = 0.2 x 10^6 Jm^{-3} K^{-1}, b = 1.5 x 10^6 Jm^{-3},
c = 3 x 10^6 Jm^{-3}, and T^* ≃ 314 K.   These values correspond approximately to those for n-heptyl cyano-biphenyl (ref. 31)

There is always a turning point in g at the point Q = 0, and this
is a maximum if a <0 and a minimum if a >0, i.e. the isotropic phase
can only be stable for T > T*. Above the temperature T_k (see Fig.
3) eq. (19) has only one real root at Q = 0, and a stable ordered
phase cannot exist; from eq. (19) T_k is given by:

$$(T_k - T^*) = \frac{b^2}{4a_o c} . \tag{21}$$

Between T_k and T* two free energy minima are predicted, but only at
T_{NI} are these of equal energy corresponding to a phase equilibrium
between the isotropic phase (Q = 0) and the ordered phase with
Q_{NI} = 3a/b = 2b/3c. The corresponding transition temperature is:

$$(T_{NI} - T^*) = 2b^2/9a_o c \tag{22}$$

and the enthalpy change at the transition can be evaluated from

$$\Delta h_{NI} = - T_{NI} \frac{\partial(g-g_o)}{\partial T} = \frac{2a_o b^2}{9c^2} T_{NI} . \tag{23}$$

By making estimates of the various parameters of the Landau free
energy expansion, it is possible to calculate values for the order
parameter as a function of temperature below T_{NI}. The calculated
values[31,32] for two typical liquid crystals are in reasonable agreement with the rather uncertain experimental measurements, but in the
nematic phase contributions from neglected higher order terms to
the free energy expansion may be significant[28]. If the coefficient
of Q^3 in the expansion is zero then the transition becomes second
order and the order parameter goes continuously to zero as T
approaches T*.

The coherence length - If the director orientation is inhomo-
genous, then spatial gradient terms of Q should be added to the expression for g (18). This leads to the fundamental equation of the
static continuum theory of nematics[33,34]:

$$g = g_o' + \frac{k_{11}}{2} (div\underset{\sim}{n})^2 + \frac{k_{22}}{2} (\underset{\sim}{n} \cdot curl\underset{\sim}{n})^2 + \frac{k_{33}}{2} (\underset{\sim}{n} \times curl\underset{\sim}{n})^2 \tag{24}$$

g_o' is the free energy density for a nematic having a uniform director
orientation and a given Q and includes all the terms of eq. (18),
which may not be convergent in the nematic phase. The elastic constants k_{11}, k_{22} and k_{33} describe splay, twist and bend deformations,
which can be supported by a nematic liquid crystal. In an isotropic
liquid the instantaneous magnitude of Q and the orientation of n
will depend on position, and additional terms are added to eq. (16):

$$g = g_o + \frac{a'}{2} Q_{\alpha\beta}Q_{\beta\alpha} + \frac{\ell}{2} (\nabla_\alpha Q_{\beta\gamma})^2 . \tag{25}$$

For simplicity only the term quadratic in Q has been included. The constant ℓ may be regarded as a combination of two elastic constants for the isotropic phase, and provides the definition of a coherence length ξ [28]:

$$\xi^2 = \ell/a' = 3\ell/2a = \xi_o^2 \ T^*(T - T^*)^{-1}. \qquad (26)$$

The significance of ξ is that it indicates the range over which the local order in a liquid crystal is correlated, and the spatial correlation function may be written as [28]:

$$<Q(0)Q(r)> = const \ r^{-1} \ exp - r/\xi \ . \qquad (27)$$

Field effects - The effects of external fields on the isotropic to nematic phase transition, and the associated pretransitional behaviour can be investigated by adding field dependent terms to the free energy expressions: we shall consider only electric fields but the treatment of magnetic fields is analogous. Assuming that the electric polarization is linear in the applied field $(\underset{\sim}{E})$, eq. (16) becomes:

$$g = g_o + \frac{a'}{2}Q_{\alpha\beta}Q_{\beta\gamma} + \frac{b'}{3}Q_{\alpha\beta}Q_{\beta\gamma}Q_{\gamma\alpha} + \frac{c'}{4}(Q_{\alpha\beta}Q_{\gamma\delta})^2 -$$

$$\frac{\varepsilon_o}{2} (\varepsilon_{\alpha\beta} - \delta_{\alpha\beta}) E_\alpha E_\beta . \qquad (28)$$

From the definition of Q we can write:

$$\varepsilon_{\alpha\beta} = \frac{2\Delta\varepsilon^{(o)}}{3} Q_{\alpha\beta} + \frac{1}{3} \varepsilon_{\gamma\gamma}\delta_{\alpha\beta} \qquad (29)$$

and so the expression for g now contains a term linear in the order parameter. Neglecting for the moment terms of order Q^3 and higher gives that the state of minimum free energy is ordered with:

$$Q_{\alpha\beta}^{(E)} = \frac{\varepsilon_o \Delta\varepsilon^{(o)}}{9a'} (3E_\alpha E_\beta - E^2 \delta_{\alpha\beta}) \qquad (30)$$

and so the field induced order varies as $(T - T^*)^{-1}$. If terms of order Q^3 and higher are included then Q is predicted [29] to vary more rapidly than $(T - T^*)^{-1}$ close to T_{NI}. The effect of applying an electric field to a nematic liquid crystal may be understood from Figure 4, where eq. (18) is plotted with the same values of a_o, b, c and T^* as in Fig. 3 but including an electric field term. The value of this term corresponds to a dielectric anisotropy $\Delta\varepsilon^{(o)}$ of approximately 20 and a mean permittivity of 10; the field strength is 10^5 V cm^{-1}. When $\Delta\varepsilon^{(o)}$ is positive the ordered phase is stabilized, and T_{NI} is increased; however if $\Delta\varepsilon^{(o)}$ is negative the transition may become second order as the minima in the free

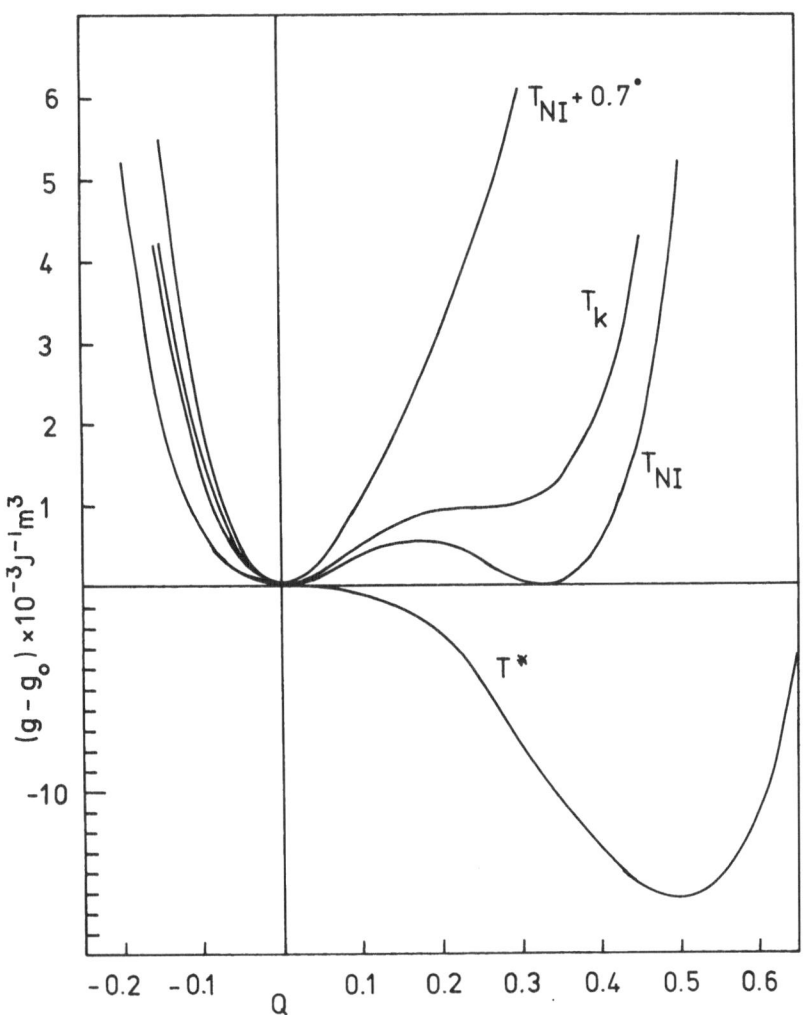

Figure 4. Free energy as a function of order parameter in the presence of an electric field (see text): T_k and T_{NI} are the field-free temperatures, i.e. the same as in Fig. 3.

energy curve coalesce. The possibility of a second order transition to a biaxial nematic phase also appears when $\Delta\varepsilon^{(o)}$ is negative[29].

4. ELECTRO-OPTICAL EXPERIMENTS

It is possible to induce macroscopic changes in the alignment and optical properties of liquid crystal films by electric fields, and these effects form the basis of a number of electro-optical devices. However measurements of electro-optical properties such as Kerr constants and their relaxation times are difficult in the liquid crystalline phase because of its strong turbidity below the isotropic transition temperature. In this section electro-optical experiments performed on the isotropic phase of liquid crystals in the pretransitional region will be discussed.

The birefringence induced in liquid crystals in the pretransitional region by electric fields, strong optical fields or magnetic fields is very large compared with ordinary liquids, and this may be attributed to the ease with which macroscopic order can be induced in the isotropic phase by external fields. In the previous section we have seen that the induced order is inversely proportional to the mean field parameter a: using the same typical values for a_o and $\Delta\varepsilon^{(o)}$ as in Figure 3 we find that the order induced 3° above T^* by an electric field of strength $10^4 Vcm^{-1}$ is $Q_{zz}^{(E)} = 1 \times 10^{-4}$. For nitrobenzene the order induced by the same field strength is approximately 7×10^{-6} i.e. fourteen times smaller.

The static Kerr effect - It has already been shown that application of an electric field to the isotropic phase of a liquid crystal results in field-induced order. For a field in the z-direction of E_z, the order is from eq. (30)

$$Q_{zz}^{(E)} = \frac{\varepsilon_o \Delta\varepsilon^{(o)} E_z^2}{3a} \tag{31}$$

and the corresponding induced optical anisotropy becomes:

$$(n_z^2 - n_x^2) = \frac{\varepsilon_o \Delta\varepsilon^{(o)} E_z^2}{3a} (n_z^{(o)2} - n_x^{(o)2}) \tag{32}$$

x is a direction perpendicular to the electric field direction and n_z and n_x are the appropriate principal refractive indices. The superscript (o) refers to the fully aligned state i.e. $Q_{zz} = 1$. Using eq. (32) and the usual definition of the Kerr constant B gives:

$$B = \frac{(n_z - n_x)}{\lambda E_z^2} = \frac{\varepsilon_o \Delta\varepsilon^{(o)} (n_z^{(o)2} - n_x^{(o)2})}{3a\lambda (n_z + n_x)} \tag{33}$$

which can be approximated to:

$$B = \frac{\varepsilon_o \Delta\varepsilon^{(o)} \Delta n^{(o)}}{3a\lambda} \tag{34}$$

Assuming the mean field result for $a = a_o(T - T^*)$, B is predicted to vary with temperature as $(T - T^*)^{-1}$, and hence diverges as the critical temperature T^* is approached.

Measurements have been reported of the static Kerr constant for a number of liquid crystals in the pretransitional region[35→41], and most recent work confirms the mean field temperature dependence of B. By plotting B^{-1} against T the value of T^* may be determined, and if $\Delta\varepsilon^{(o)}$ and $\Delta n^{(o)}$ are known the mean field parameter a_o can be obtained. Some typical results are shown in Figure 5. Deviations from a simple inverse temperature dependence have been noted by some authors[42,43], who have analyzed their results according to an empirical expression of the form:

$$B = B_o(T - T^*)^{-\gamma} \tag{35}$$

Values for γ between 0.5 and 1.0 have been reported in the literature, but later work[44] suggests that the mean field result is obeyed over a temperature range of between $5°$ and $20°$ for most substances. Departures from a $(T - T^*)^{-1}$ law are observed at temperatures within about $1°$ of the nematic/isotopic transition[40,45], and at high temperatures[44].

From equation (34) the sign of B is determined by the sign of $\Delta\varepsilon^{(o)}$, since $\Delta n^{(o)}$ is positive (except possibly for discotic materials), so that materials having a negative dielectric anisotropy should have a negative B, while positive materials will have a positive B. However some liquid crystals such as PAA (para-azoxy-anisole)[46] show a change in sign of the Kerr constant as the temperature is raised. Such behaviour is not explicable in terms of eq. (34) in which only a is temperature dependent. Within the context of the Landau-deGennes theory the factor $\Delta\varepsilon^{(o)}$ is a constant, independent of temperature, which equals the dielectric anisotropy of the perfectly ordered nematogen. Thus an attempt[47] to explain the change in sign of B in terms of the temperature dependence of $\Delta\varepsilon^{(o)}$ is questionable.

The optical Kerr effect - This is similar to the static Kerr effect, except that the aligning field is at optical frequencies from a laser, and so only electronic polarization properties are probed. Molecules can reorient in the laser field through their polarisability anisotropy, and there is no contribution to the induced birefringence from molecular dipoles. For an aligning field of frequency (ω'), the Kerr constant measured at a frequency ω is given by:

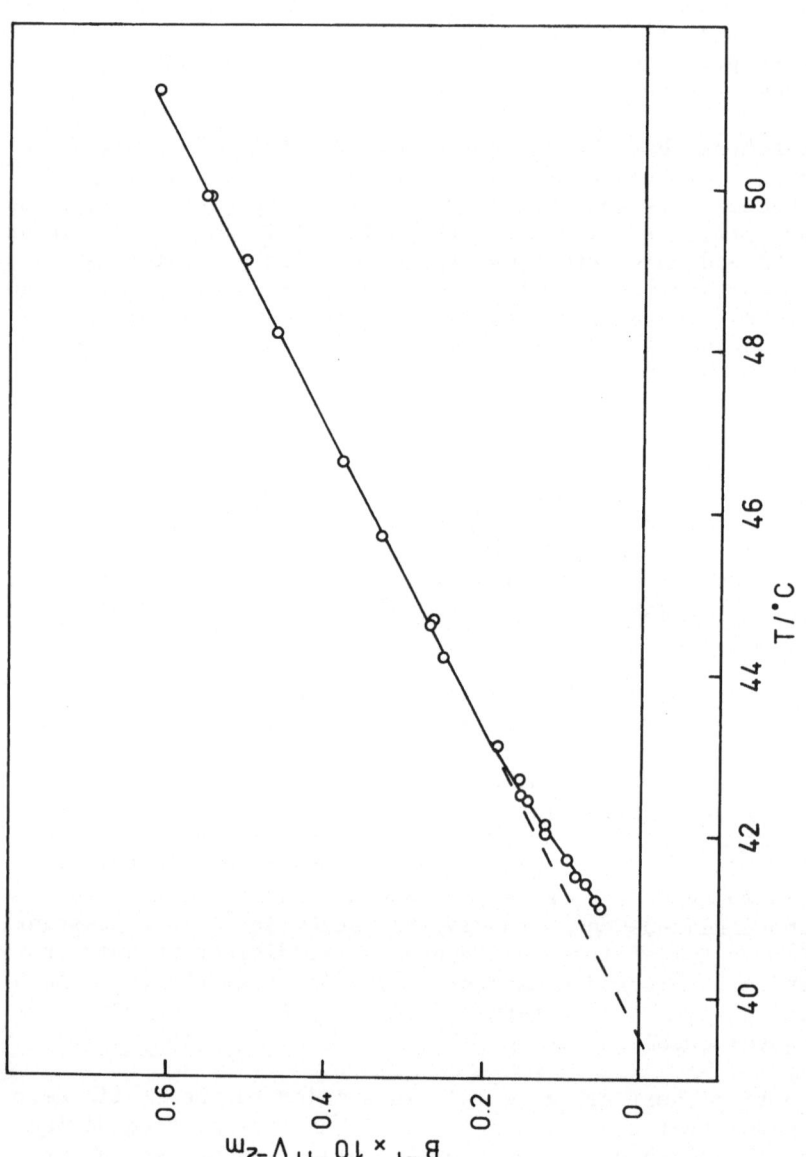

Figure 5.   Static Kerr constant as a function of temperature for n-octyl cyano-biphenyl
(D. A. Dunmur and R. Parker previously unpublished.)

$$B_{opt} = \frac{2n(\omega')\varepsilon_o}{3a\lambda} \Delta n(\omega)^{(o)} \Delta n(\omega')^{(o)} \tag{36}$$

The optical Kerr constant is often defined in terms of optical susceptibilities[48], such that

$$B_{opt} = \frac{8\pi^2 \varepsilon_o \Delta\chi(\omega')\Delta\chi(\omega)}{3an(\omega)\lambda} \tag{37}$$

where $\Delta\chi = (n_z^{(o)2} - n_x^{(o)2})/4\pi$. Neglecting any dispersion in the refractive indices, B_{opt} may be written in terms of the molecular polarizability anisotropy as:

$$B_{opt} = \frac{N^2(n^2 + 2)^2 \Delta\alpha^2}{54\varepsilon_o an\lambda} \tag{38}$$

The value of a is proportional to the number density N, and the temperature dependence of B_{opt} is determined by the temperature dependence of a. This result should be contrasted with the molecular statistical result quoted in the following section.

Closely related to the optical Kerr effect is the azimuthal rotation of incident elliptically polarized light by a medium subjected to a high intensity light beam. This effect was discovered in 1964 by Maker, Terhune and Savage[49], and results from changes in the intensity dependent refractive indices as the elliptically polarized beam traverses the sample. If the incident beam travelling along the y-direction is characterized in terms of its left and right circularly polarized components $E_\pm = \frac{1}{\sqrt{2}}(E_z \pm iE_x)$, then it has been shown[50] that the field-induced changes in the left and right circular refractive indices may be written as:

$$\delta n_\pm = C|E|^2 + D|E_\mp|^2 \tag{39}$$

where $|E|^2 = |E_+|^2 + |E_-|^2$. The corresponding ellipse rotation for a path length y is:

$$\theta = \frac{\pi}{\lambda} (\delta n_+ - \delta n_-)y = \frac{\pi Dy}{\lambda}(|E_-|^2 - |E_+|^2) \tag{40}$$

This equation is valid for small values of θ[51], and predicts that the ellipse rotation is proportional to the ellipticity of the incident light: i.e.,

$$\theta = \frac{\pi Dey}{\lambda}|E_+ + E_-|^2 \tag{41}$$

e is the ratio of the axes of the polarization ellipse. Equation (41) is not valid for incident left or right circularly polarized light (e = ±1 and $E_\mp$ = 0), since the effect of intensity dependence

in the circular refractive index would be only to change the phase
velocity of the propagating circularly polarized wave.

If hyperpolarizability effects are neglected it has been
shown[52] that the ellipse rotation constant is directly proportional
to the optical Kerr constant i.e. $B_{opt} = 2D/\lambda$. The mean field theory
predicts[48] that D varies as $(T - T^*)^{-1}$ and measurements[48,53,54]
of the ellipse rotation in the pretransitional region of liquid
crystals has confirmed this.

It is more difficult to obtain absolute values for B_{opt} than
for the static Kerr constant because of the difficulty of accurately
measuring the laser power and its distribution across the beam.
Furthermore there is often evidence of saturation[39,55], so the
measured effect may not always be proportional to the square of the
laser field strength. In spite of these difficulties measurements
have been reported of the optical Kerr constant for a number of
different nematogens using either the direct method[39,48,55,56] or
the ellipse rotation technique[53,54] described above. The results
all confirm the mean field prediction that $B_{opt} \alpha (T - T^*)^{-1}$, and
the measurements have been used to obtain values for a_o and T^*.

Kerr effect relaxation times – The time response of the field-
induced birefringence in a fluid may be probed using pulsed fields.
Such experiments have been performed on nematic liquid crystals and
have led to some interesting results for both static and optical
frequency aligning fields. According to the Landau-deGennes theory,
the relaxation time for the Kerr constant will be determined by the
dynamical behaviour of the macroscopic order parameter $\underset{\approx}{Q}$. De-
Gennes[28] has formulated a hydrodynamic theory for the isotropic
phase of a nematogen which couples time fluctuations in $\underset{\approx}{Q}$ with
spatial fluctuations in the local fluid velocity ($\underline{v}$). However if
the fluctuations in $\underset{\approx}{Q}$ are much faster than the frequencies associated with $\underline{v}$, the time dependence of $\underset{\approx}{Q}$ may be simply written as:

$$\nu \frac{\partial Q_{\alpha\beta}}{\partial t} = - \frac{\partial g}{\partial Q_{\alpha\beta}} \tag{42}$$

From eq. (28), and only including terms linear and quadratic in
$\underset{\approx}{Q}$, we obtain:

$$\nu \frac{\partial Q_{\alpha\beta}}{\partial t} = -a'Q_{\alpha\beta} + \frac{\varepsilon_o \Delta\varepsilon^{(o)}}{3} (E_\alpha E_\beta - \frac{1}{3}E^2\delta_{\alpha\beta}) \tag{43}$$

Integrating eq. (43):

$$Q_{\alpha\beta}(t) = \frac{\varepsilon_o \Delta\varepsilon^{(o)}}{3\nu} \int_{-\infty}^{t} (E_\alpha(t')E_\beta(t') - \tfrac{1}{3}E(t')^2\delta_{\alpha\beta}) \exp - \tau^{-1}(t - t')dt' \qquad (44)$$

where $\tau = \nu/a' = 3\nu/2a$. If $\underset{\sim}{E}$ is constant until $t = 0$, when it
becomes equal to zero i.e. it represents a pulse the decay time of
which is short compared with the relaxation time τ, the ordering
tensor becomes:

$$Q_{\alpha\beta}(t) = \frac{\varepsilon_o \Delta\varepsilon^{(o)}}{3\nu} (E_\alpha(0)E_\beta(0) - \tfrac{1}{3}E(0)^2\delta_{\alpha\beta})\tau \exp -t/\tau \qquad (45)$$

Thus the induced order and hence the Kerr constant decays exponentially with a relaxation time equal to $3\nu/2a$[48,53].

According to macroscopic theory the relaxation times for both
the static and optical Kerr effects should be equal depending only
on the energy parameter a and the viscosity ν. A comparison of
the optical and static relaxation times can therefore provide some
indication of the applicability of mean field theory. Some results
taken from the literature for Kerr relaxation times are presented
in Figure 6, where it is seen that correspondence between the
dynamic response of the static and optical Kerr effects is fair.
A number of other nematogens have been studied[55,57,41], and in all
cases the relaxation time has been found to vary with temperature
as $(T - T^*)^{-1}$. A dispersion experiment to measure the relaxation
of the static Kerr effect has been described[40] in which the frequency dependence of the Kerr constant was measured through the
range 0 to 10 MHz. The results were analyzed in terms of two
relaxation times, although it was found that a single relaxation
dominated and exhibited a $(T - T^*)^{-1}$ temperature dependence.

By using very short laser pulses ($\sim 10^{-12}$ secs) and fast detection techniques it has proved possible to show the existence of
an additional fast relaxation process in the optical Kerr effect
of some nematogens[58]. This has been attributed to individual
molecule reorientation, and is not expected to show the critical
behaviour that has been observed for the collective or macroscopic
relaxation time.

5. MOLECULAR STATISTICAL ASPECTS

A molecular interpretation of liquid crystalline behaviour
has been, and remains the goal of much research. Early attempts
to explain the properties of liquid crystals were based on a molecular theory: the swarm theory[59] suggested that liquid crystals
contained swarms or domains of parallel molecules which interacted
collectively with external fields. The great achievements of

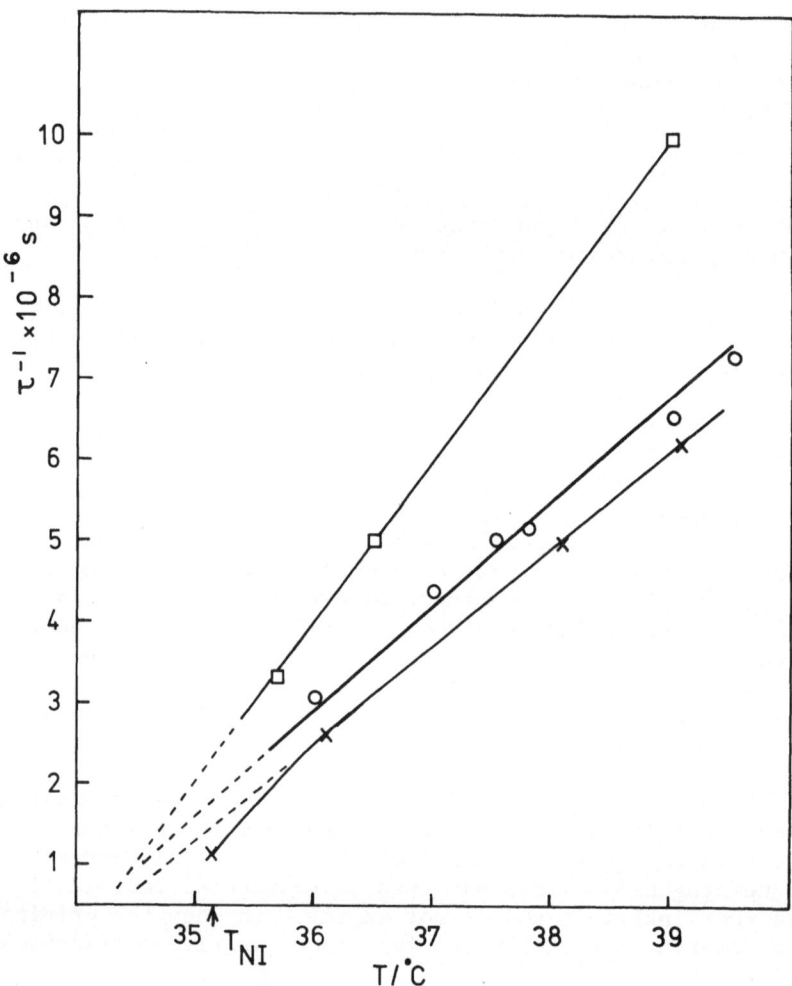

Figure 6.   Kerr relaxation times for n–pentyl cyano–biphenyl: X =
static rise time (ref. 57),   □ = optical decay time and ₒ = static
    decay time (H. J. Coles and B. R. Jennings, Mol. Phys., 36,
                           1661 (1978)).

continuum theory led to the discreditation of the swarm theory, but we now know that the basic idea behind the latter is correct. Of course swarms do not have any actual existence, but they can be statistically defined in terms of distribution functions and coherence lengths. The molecular picture of the liquid crystalline phase is groups of molecules whose orientations (and positions in the case of smectics) are correlated over distances which may equal many molecular diameters. Despite this, the highly successful mean field theory of Maier and Saupe[8] neglects correlation between molecules. We will show in this section that electro-optical measurements can provide a very valuable probe of molecular correlation in liquid crystals.

The theory of electro-optical behaviour in liquid crystals outlined so far has been essentially macroscopic, and the experiments described have been interpreted in terms of the macroscopic theory. A link may be made with molecular properties through the statistical expressions given in §2 for the order parameter, since these allow in principle the calculation of $\Delta n^{(o)}$ and $\Delta \varepsilon^{(o)}$ which appear in the macroscopic equations for the Kerr constant. Unfortunately difficulties associated with the description of the internal electric field make the estimates of $\Delta n^{(o)}$ and $\Delta \varepsilon^{(o)}$ subject to considerable uncertainty. There is a further problem with the evaluation of $\Delta \varepsilon^{(o)}$ from equation (13), which concerns its apparent temperature dependence through the dipolar term. In fact $\Delta \varepsilon^{(o)}$ should be independent of temperature since it refers to the hypothetical dielectric anisotropy of a perfectly ordered liquid crystal. Such a system has an infinitely high barrier to dipole reorientation, and so it might be expected that the orientational polarization will be 'frozen out' of the dielectric anisotropy leading to $\Delta \varepsilon^{(o)}$ equalling $2n\Delta n^{(o)}$. Alternatively the state S=1 might be achieved by extrapolating $\Delta \varepsilon$ to 0 K, where according to eq. (13) $\Delta \varepsilon \to \infty$. Both of these limiting values for $\Delta \varepsilon^{(o)}$ lead to unphysical results, and more careful consideration of the statistical theory is necessary.

Molecular statistical theories of the Kerr effect - As explained in §4, measurements of the Kerr effect in liquid crystalline phases are not yet available, and we shall consider the application of statistical theories of the Kerr effect in isotropic liquids to the interpretation of electro-optical measurements in the pretransitional region.

The starting point for any molecular theory must be the description of the electro-optical properties of a system of noninteracting molecules; i.e. the low density gas. Defining a molar Kerr constant as:

$$_m K = \lim_{E \to o} \frac{N_A 6n(n_z - n_x)}{N(n^2+2)^2(\varepsilon+2)^2 E^2} = \frac{6nN_A \lambda B}{N(n^2+2)^2(\varepsilon+2)^2} \tag{46}$$

where N_A is Avogadro's number, it has been shown[60] that this may be expressed in terms of molecular properties by the equation:

$$_m K = \frac{N_A}{81\varepsilon_o} \left\{ \gamma + \frac{1}{kT} \left(\frac{2\mu\beta}{3} + \frac{9\bar{\alpha}^{(o)}}{5} \bar{\alpha}\kappa(\alpha)^2 \right) + \frac{3\mu^2}{10k^2T^2}(\alpha_{33} - \bar{\alpha}) \right\} \tag{47}$$

β and γ are related to the components of the first and second hyper-polarizabilities, while $\bar{\alpha}$ and $\bar{\alpha}^{(o)}$ are the mean optical and static polarizabilities. The anisotropy factor $\kappa(\alpha)$ is defined by:

$$\kappa(\alpha)^2 = (\alpha_{\alpha\beta}\alpha_{\alpha\beta}^{(o)} - 3\bar{\alpha}\bar{\alpha}^{(o)})/6\bar{\alpha}\bar{\alpha}^{(o)} \tag{48}$$

and becomes equal to $\Delta\alpha/3\bar{\alpha}$ if the molecule has axial symmetry. α_{33} is the component of the polarizability tensor along the dipole axis. N.B. the 3-axis is not necessarily a principal axis of the polarizability. For the optical Kerr effect the terms involving the molecular dipole are zero, and the corresponding expression for the optical molar Kerr constant is:

$$_m K_{opt} = \frac{N_A}{81\varepsilon_o} \left\{ \gamma + \frac{3}{10kT} (\alpha_{\alpha\beta}\alpha_{\alpha\beta}^{(o)} - 3\bar{\alpha}\bar{\alpha}^{(o)}) \right\} \tag{49}$$

Little is known about β and γ for large molecules[61], but for the highly anisotropic molecules that form liquid crystalline phases it is probably reasonable to neglect contributions to $_m K$ from hyper-polarizabilities.

In order to develop a molecular theory of the Kerr effect in condensed phases three factors must be considered. Firstly the internal electric field must be correctly described, and while this is a difficult problem for anisotropic disordered systems[21,62], we shall assume that the Lorentz description implicit in eq. (46) is adequate for liquid crystals in the pretransitional region. A second consideration is how the molecular parameters such as α and μ are modified by interactions with other molecules, and finally the effect of angular correlation between molecules through an angle-dependent intermolecular potential must be accounted for. The separation of these last two effects can be difficult, and relies on obtaining accurate values for the effective molecular parameters. Dilute solution measurements[23] can yield estimates for the effective values of α and μ, but to be reliable the measurements should be made for a solvent having a refractive index and permittivity equal to those of the pure solute. For liquid crystal molecules this is almost impossible, and in the subsequent discussion effective molecular parameters obtained from measurements on benzene or carbon tetrachloride solutions will be used. Our primary objective in analyzing Kerr effect data for liquid crystals is to obtain information on angular correlation, although recently[63] some progress has been made in the evaluation of effective polarizabilities for simple liquids.

Essentially similar theories of the Kerr effect in liquids have been presented by Buckingham and Raab[64] and Kielich[65], and both these theories explicitly include the effects of angular correlation. Restricting attention for the moment to molecules having axial symmetry, these theories give that:

$$_mK = \frac{N_A}{405\varepsilon_o kT} \left\{ \Delta\alpha\Delta\alpha^{(o)} g_2 + \frac{\Delta\alpha\mu^2}{kT} \left(2(g_1-1) + g_2 + g_3 \right) \right\} \qquad (50)$$

where $g_1 = 1 + \sum_{j\neq1} <\cos\theta_{1j}>$

$g_2 = 1 + \sum_{j\neq1} \frac{1}{2}<3\cos^2\theta_{1j} - 1>$

and $g_3 = \sum_{j\neq1}\sum_{k\neq j\neq1} \frac{1}{2}<3\cos\theta_{1j}\cos\theta_{1k} - \cos\theta_{jk}>$.

In writing this equation certain internal field factors[64] have been neglected in the dipolar term, and μ represents an effective dipole moment for the molecular in its local environment in the presence of the field. The correlation factors g_1 and g_2 refer to pairs of molecules, but g_3 involves the triplet distribution function[18] and is therefore difficult to obtain. When there are no interactions present $g_1 = g_2 = 1$ and $g_3 = 0$. The Kerr constant for non-polar molecules and the optical Kerr constant is given by the first term of eq. (50), and this may be compared with the result of the general theory of Ladanyi and Keyes[66]:

$$_mK = \frac{N_A}{405\varepsilon_o kT} \Delta\alpha_o(\omega)\Delta\alpha_o(\omega')[1+ \pounds(\omega)][1+\pounds(\omega')]g_2 \qquad (51)$$

where $\Delta\alpha_o(\omega)$ is the free molecule polarizability anisotropy at frequency ω, and ω and ω' are the frequencies of the probe field and the aligning field. The factor (ω) relates the effective polarizability anisotropy to the free molecule value ($\Delta\alpha_o(\omega)$), and is given by[66]:

$$\pounds(\omega) = \tau_{20}[9\bar{\alpha}_o(\omega)^2 + 3\bar{\alpha}_o(\omega)\Delta\alpha_o(\omega) + 2\Delta\alpha_o(\omega)^2]/9\Delta\alpha_o(\omega) \qquad (52)$$

τ_{20} is a parameter of the liquid structure and for a simple model fluid of anisotropic molecules it has been calculated[67] to be between -3 and -4 times $N(4\pi\varepsilon_o)^{-1}$.

Applications – From equation (50) the divergence of the optical Kerr constant of liquid crystals as the temperature approaches the nematic/isotropic transition can be explained in terms of a divergence in the pair correlation factor g_2. If values are available for the effective polarizability anisotropy g_2 may be calculated from $_mK_{opt}$, or if dilute solution measurements of the optical Kerr effect are available then,

$$g_2 = {_mK_{opt}} / {_mK^\infty_{opt}}$$ (53)

where $_mK^\infty_{opt}$ is the molar Kerr constant extrapolated for an infinitely dilute solution[68]. Combining equations (53), (46) and (38) gives:

$$g_2 = \frac{45NkT}{(n^2+2)^2 a_o (T-T^*)}$$ (54)

and shows that as the temperature approaches T^* the pair correlation factor g_2 diverges according to $(1 - T^*/T)^{-1}$. Using published values for the optical Kerr constants of pure MBBA (p-methoxy-benzylidene-p'-butyl aniline) and its solutions in carbon tetrachloride we have calculated the molar Kerr constants and hence derived values for g_2 using eq. (53). The results are given in Figure 7 where they are compared with values obtained from a light scattering study[69].

The extraction of information on molecular correlation from the static or low frequency Kerr constants where there is a significant contribution from the molecular dipole is more difficult. However if values are available for the optical and static molar Kerr constants measured at infinite dilution, then equation (5) may be re-arranged to give:

$$\frac{(_mK - {_mK_{opt}}) {_mK^\infty_{opt}} - (_mK^\infty - {_mK^\infty_{opt}}) {_mK_{opt}}}{_mK^\infty_{opt} (_mK^\infty - {_mK^\infty_{opt}})} = 2(g_1-1) + g_3$$ (55)

g_1 values may be available from dielectric measurements, so information on the triplet correlation factor g_3 can be obtained.

Additional complications arise if the molecule is not axially symmetric. For example, the electro-optical properties of MBBA and its solutions have been studied by various workers[37,39], and in Figure 8 the molar Kerr constants are plotted against temperature. The static Kerr constant is negative and strongly temperature dependent in the pretransitional region, while $_mK^\infty$ is positive and only weakly dependent on temperature. It has been shown[37] that in solutions of MBBA the static Kerr constant may change sign from negative to positive with increasing temperature, and it seems possible that at a sufficiently high temperature MBBA would have a positive Kerr constant, i.e. it would then parallel the behaviour of PAA (p-azoxyanisole) which shows a similar change of sign as

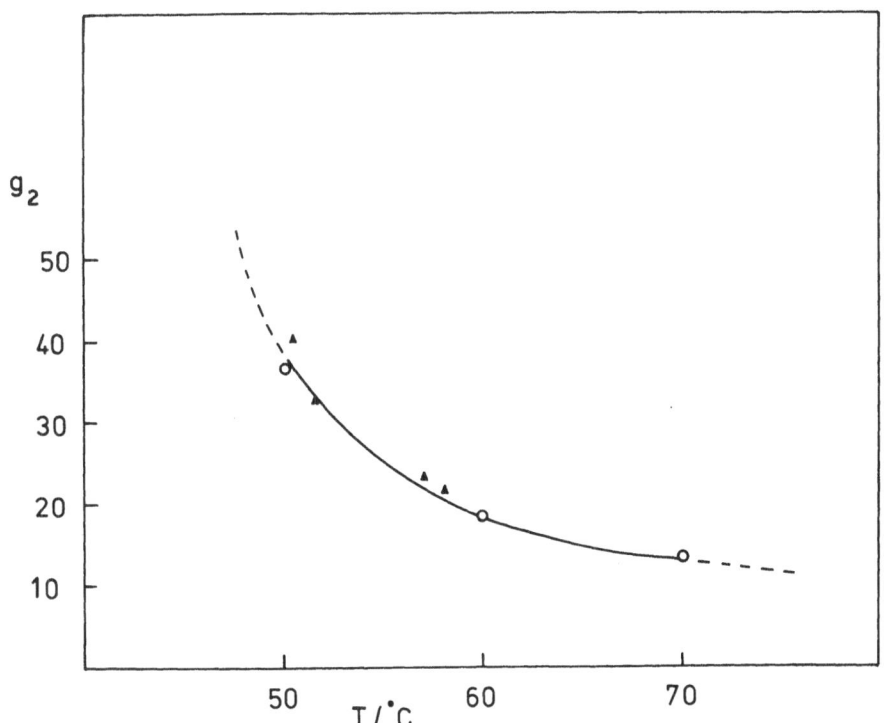

Figure 7. Pair Correlation Factor (g_2) for MBBA
Δ = optical Kerr effect (ref. 39)
o = light scattering (ref. 69)

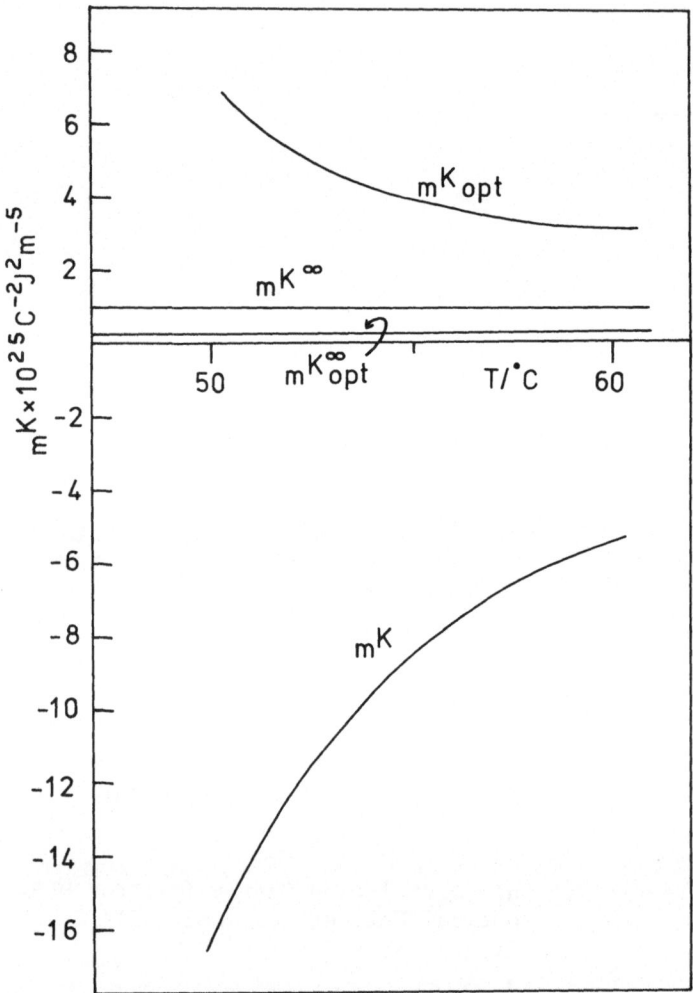

Figure 8.   Molar Kerr Constants for MBBA

the temperature is raised $5°$ above T_{NI}[46]. The fact that $_mK^\infty$ is positive and greater than $_mK^\infty_{opt}$ means that the dipolar term in eq. (47) is also positive. The change of sign of the Kerr constant of solutions of MBBA cannot therefore be a consequence of the intrinsic T^{-1} dependence of the dipolar term, although such a mechanism has been proposed to explain similar behaviour with PAA[70,47]. The generalization of eq. (50) for non-axially symmetric molecules contains a number of extra terms and correlation factors, and is not expressible in a compact form. However the negative divergence of $_mK$ for MBBA can only result from negative correlation factors in the dipolar term of eq. (50). Anti-parallel correlation of components of the molecular dipole along the axis of maximum polarizability will result in a reduction of the polarizability along the effective dipole axis. Thus the dipolar term in equation (50) can change sign as a result of molecular association.

Any theory that only includes pair correlations is unlikely to be successful in explaining the electro-optical behaviour of liquid crystals. Close to the nematic/isotropic transition, the values for g_2 are ~ 10, which implies that many particle effects are likely to be important. A cluster model has been proposed for nematic liquid crystals[70,71], and used to interpret the pretransitional electric and magnetic birefringence in PAA. The predicted T^* is itself temperature dependent, and the factor corresponding to g_2 is only weakly dependent on temperature and not divergent at T_{NI} as deduced experimentally.

There have been a number of attempts to calculate the properties of liquid crystals by computer simulation[72-75]. Lattice models in which the molecular positions are fixed, but the molecules have orientational freedom, undergo first order phase transitions to an orientationally ordered phase[72,74], and various parameters of the one and two particle distribution function have been calculated[75]. It is also possible to obtain values for the pair correlation factor in the pretransitional region above the nematic to isotropic transition temperature[75,76,77]. The results of such calculations using a simple pair potential of the form $u_{ij} = -\varepsilon_{ij}P_2(\cos\theta_{ij})$ and including only nearest neighbor interactions show a divergence in g_2 as T_{NI} is approached from the isotropic phase, and the values obtained are in remarkably good agreement with experiment: see Table II. This agreement may be fortuitous, but suggests that there is much to be learned from the computer simulation of liquid crystals.

6. DISCUSSION AND CONCLUSIONS

The information that is available from electro-optical measurements on liquid crystals in the pretransitional region falls into two categories. If the Landau-de Gennes theory is accurately obeyed

Table II.   Calculated and Experimentally Deduced
Correlation Factors for MBBA (g_2) and
n-pentyl cyano-biphenyl (5CB)(g_1)

	$T-T_{NI}$	calc.(ref. 77)	expt.
MBBA	5	g_2 = 25	38 (ref. 69)
	15	18	18 "
	25	13	13 "
5CB	2	g_1 = 0.68	0.53 (ref. 21)

then various macroscopic parameters may be determined from a study
of the temperature dependence of the Kerr constant or Kerr relaxation
time. Molecular properties can be extracted from the macroscopic
order parameter if the statistical theory can be relied upon, however a better route to molecular information is to combine electro-
optical measurements on pure materials with those made on dilute
solutions. Such studies can yield values for correlation factors
which are sensitive indicators of the local structure in the isotropic liquid crystal phase.

A question that has stimulated much research is, "What features
of the molecular structure are responsible for liquid crystal formation?" The most important feature seems to be the shape, i.e.
the repulsive potential that surrounds the molecule, but other factors such as the number and location of dipolar groups, the distribution of polarizability and the flexibility of the molecule also
contribute. In fact liquid crystal molecules behave in dilute solution as other molecules that do not form mesophases, so the formation
of the liquid crystalline phase is not directly determined by molecular properties, but by interactions between molecules. All molecules that are anisometric can form an orientationally ordered liquid
crystal phase, but for most materials such phases occur in a region
of thermodynamic phase space where the free energy of the crystalline phase is lower[78]. The extrapolated transition temperatures
for such 'hidden' phases are sometimes called virtual transition
temperatures. Supercooling below the liquid/crystal transition
temperature can sometimes reveal metastable liquid crystal phases,
and a useful indication of the proximity of a liquid crystal phase
can be pretransitional behaviour. For example if a mesophase transition lies just a few degrees below the melting point, then the
liquid above the melting point may exhibit divergence in light
scattering and electro-optical properties characteristic of a second

order transition, despite the fact that a strong first order transition (freezing) intervenes before the mesophase transition is reached. Similar behaviour can be observed in solutions of molecules: those having strong liquid crystal tendencies may show departures from a simple mixture law. Measurements of the Kerr effect in solutions of MBBA[37] and structurally related molecules[79] (see Figure 9) showed that BBA behaved in a similar fashion to MBBA, while in solutions of MBA and BB liquid crystal tendencies were not apparent. The depolarized light scattering intensity in the pretransitional region of liquid crystals diverges and has also been used[80] to probe the tendencies towards liquid mesophase formation of different molecules.

The mean field parameter a_o has been obtained from optical and static Kerr constant measurements for a number of mesogens. For the alkyl and alkoxy cyano-biphenyl homologous series a_o shows an odd even effect[41] that parallels many other properties such as T_{NI}[81], transitional volume change[82] etc., but such an alternation was not observed for a series of alkyl azoxy-benzenes[53]. It should be emphasized that the a_o values determined from Kerr and light scattering measurements rely on an accurate knowledge of $\Delta n^{(o)}$ and $\Delta \varepsilon^{(o)}$, and as explained in §3 these are not readily accessible quantities. Relaxation times can give values for the ratio ν/a_o and if a_o is available from other measurements the viscosity ν can be extracted.

Macroscopic interpretation of the Kerr measurements relies on the validity of the Landau theory. Experiments indicate that close to T_{NI} there are divergences of the pretransitional behaviour from a simple $(T - T^*)^{-1}$ temperature dependence. Extending the mean field theory to include pair correlations, it has been shown that such behaviour is predicted[83] and the induced birefringence varies as $(T - T^*)^{-\gamma}$ where $\gamma > 1$ and is temperature dependent. At temperatures much higher than T^* there are also deviations from the predictions of the Landau-de Gennes theory. The coherence length varies as $(T - T^*)^{-\frac{1}{2}}$, but when it becomes comparable to molecular dimensions the macroscopic theory is no longer applicable and we must turn to the molecular statistical theory.

It has been demonstrated that all the peculiarities of liquid crystal behaviour must be a result of molecular correlation. While it is possible to formulate theories which take account of pair correlation, the many body effects that are undoubtedly important in liquid crystals are much more difficult to describe, and the only correlation factors that we have reliable information on are g_1 and g_2. The factor g_1 is usually less than one indicating an anti-parallel arrangement of dipoles. This is consistent with the fact that ferro-electric nematic phases are not formed, although chiral smectic C phases are ferro-electric[84]. A pair correlation function $G_2(r)$ may be defined as:

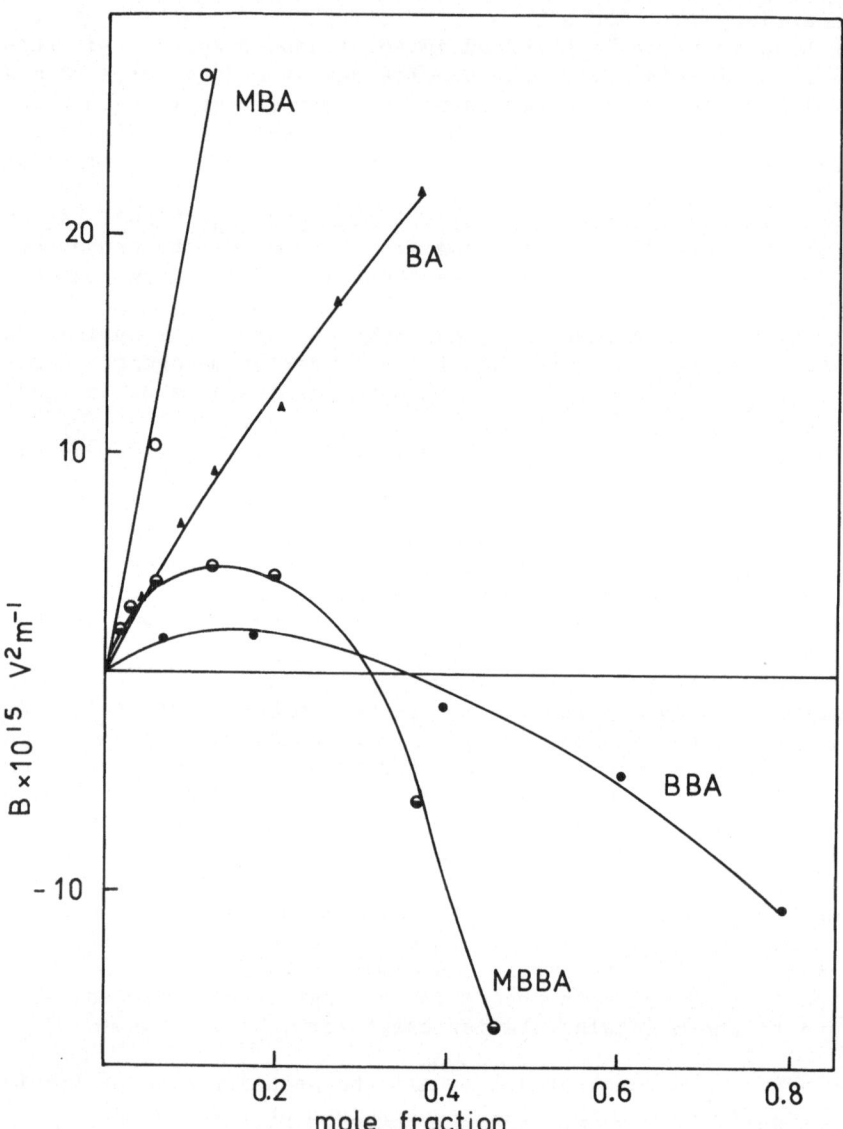

Figure 9.   Kerr constants for solutions in carbon tetrachloride of
MBBA and structurally related molecules (refs. 37 and 79).
        MBBA – p–methoxy–benzylidene p'–n–butyl aniline
        MBA – p–methoxy–benzylidene aniline
        BBA – benzylidene p–n–butyl aniline
        BA   – benzylidene aniline

$$G_2(r) = < \frac{1}{2}(3\cos^2\theta_{ij}(r) - 1) > \tag{56}$$

where $\theta_{ij}(r)$ is the angle between molecules i and j separated by a distance r. $G_2(r)$ is related to the pair correlation factor g_2 by:

$$g_2 = \int_0^\infty r^2 G_2(r)\,dr \tag{57}$$

Monte Carlo calculations have been made[85] of $G_2(r)$, and in both the nematic and isotropic phases $G_2(r)$ falls monotonically from one to S^2 as r increases (N.B. in the isotropic phase S = 0). The large value for g_2 at the nematic/isotropic transition is therefore not a result of closely correlated nearest neighbours, but due to the correlation extending over many molecular diameters.

Combining the results of the molecular and Landau-de Gennes theories leads to the result, eq. (54), that in the pretransitional region $g_2 = AT(T - T^*)^{-1}$, where A is a constant approximately independent of temperature. It has been suggested[69] that A must be unity, since as $T\to\infty$ $g_2\to 1$. However, as has been explained, it is not appropriate to use eq. (54) at temperatures far from T^*, and so the value of g_2 close to the nematic/isotropic transition is a function of the strength of the transition as measured through a_0.

The coherence length ξ is also a measure of molecular correlation and it also diverges as T^* is approached. A correlation volume can be defined which is proportional to ξ^3, and this varies with temperature as $V_c \propto (T - T^*)^{-3/2}$; g_2 varies as $(T - T^*)^{-1}$ and so is not directly proportional to the correlation volume or number of correlated molecules.

The dynamics of the Kerr effect can be studied by either measuring the Kerr constant as a function of frequency of the aligning field, or by measuring the time dependence of the birefringence induced by an electric field pulse. Analysis of the former can be complicated, and only a single such experiment has been reported[40] for liquid crystals. Pulse experiments have been performed to observe the rise or decay (or both) of the optical and static Kerr effects, and the corresponding relaxation times all show pretransitional effects, becoming longer as T_{NI} is approached.

There are two relaxation times that must be considered in the Kerr effect[86]: one refers to the dipole (τ_1) and the other to the polarizability (τ_2). Only the latter contributes to the optical Kerr effect and the rise and decay times are expected to be equal and characteristic of the reorientation dynamics of the polarizability ellipsoid. For the static Kerr effect however τ_1 and τ_2 contribute to the rise time, while only τ_2 determine the rate of decay. Dielectric relaxation on the other hand probes dipole orientation and

and the corresponding rise and decay times should be equal to τ_1.

The Kerr relaxation experiments reported for liquid crystals in the pretransitional region indicate that decay times for the optical and static Kerr effects are approximately equal and lie in the range $\tau_2 = 50 \rightarrow 500$ nanoseconds. Measurements of the rise time[57] for the static Kerr constant also give times in this range suggesting that the transient response is dominated by the relaxation of the induced dipole. Further evidence for this comes from the Kerr dispersion experiment[40].

Dielectric relaxation experiments in the pretransitional region[38,87,88] give relaxation times τ_1 that do not diverge as T_{NI} is approached and the values ($\tau_1 = 0.1 \rightarrow 5$ nanoseconds) are much smaller than τ_2. This suggests that dipole relaxation is a result of individual molecular reorientation, while relaxation of the induced dipole occurs through a collective mechanism involving many molecules. The additional fast relaxation process observed[58] in the dynamic response of the optical Kerr effect in MBBA has been attributed to individual molecule reorientation with a relaxation time $\tau_{2f} = 3 \times 10^{-12}$s at 50°C. Measurements[87] of the dipole relaxation in pure MBBA give $\tau_1 = 138 \times 10^{-12}$s at 50°C and a value for $\tau_1 = 30.6 \times 10^{-12}$s for MBBA in benzene solution. The diffusion model predicts that $\tau_1 = 3\tau_{2f}$, but the results are not in accord with this, and suggest that the fast process detected in the optical Kerr effect relaxation is not molecular reorientation.

Electro-optical studies of liquid crystals have already provided much valuable information on the nature of the liquid crystal phase and the subtleties of cooperative interactions and molecular correlation. In this chapter we have concentrated on information that is available from measurements on the isotropic phase, but application of the various electro-optical techniques to the ordered liquid crystal phases would greatly increase our understanding of these phases and the molecular interactions responsible for their stability.

Liquid crystals are undoubtedly very complex materials, and if we are to achieve more than a superficial knowledge of them all the physical techniques at our disposal must be used and their results coordinated. In the area of electro-optics this means careful measurements of optical and static Kerr constants of the pure materials and their dilute solutions. Similarly we can learn a great deal from comparative measurements of relaxation times obtained from different techniques including the Kerr effect and dielectric relaxation. From the point of view of the macroscopic theory, careful measurements at temperatures close to the transition will reveal divergences from the Landau theory which may be correlated with more exact theories of the phase transition. Also of interest would be measurements far from the transition where the coherence length is of the order of molecular dimensions and the macroscopic theory might also be expected to fail.

It is a pleasure to thank Professor Kivelson for hospitality received at the Department of Chemistry, University of California, Los Angeles, where this chapter was written. Financial support is also gratefully acknowledged from NATO, the Fulbright Commission, the Royal Society and the Science Research Council.

REFERENCES

1. F. Reinitzer, Montash Chem. 9:421 (1888).
2. G. W. Gray, "Molecular Structure and the Properties of Liquid Crystals", Academic Press, London 1962.
3. "Introduction to Liquid Crystals" (Ed. E. B. Priestley, P. W. Wojtwiez and Ping Sheng), Plenum Press, New York 1974.
4. "Non-emissive Electro-optic Displays" (Ed. A. R. Kmetz and F. K. von Willisen), Plenum Press, New York 1976.
5. T. J. Scheffer and H. C. Gruler, "Molecular Electro-optics", (Ed. Chester T. O'Konski), Vol. 1, part 2, Marcel Dekker, New York 1976, p. 761.
6. S. Chandrasekhar, B. K. Sadashiva, K. A. Suresh, N. V. Madhusudana, S. Kumar, R. Shashidhar and G. Venkatesh, J. de Phys., 40:C3-120 (1979).
7. P. A. Winsor, "Liquid Crystals and Plastic Crystals", (Ed. G. W. Gray and P. A. Winsor), Vol. 1, Ellis Horwood, Chichester 1974, p. 199.
8. W. Maier and A. Saupe, Z. Naturforschg., 13a:564 (1958).
9. R. L. Humphries, P. G. James and G. R. Luckhurst, J. Chem. Soc. Faraday Trans. II, 68:1031 (1972).
10. L. Onsager, Ann. N.Y. Acad. Sci., 51:627 (1949).
11. M. A. Cotter, J. Chem. Phys., 66:1098 (1977).
12. B. A. Baron and W. M. Gelbart, J. Chem. Phys., 67:5795 (1977).
13. W. M. Gelbart and B. Barboy, Mol. Cryst. Liq. Cryst., 55:209 (1979).
14. G. W. Gray, "Molecular Physics of Liquid Crystals", (Ed. G. W. Gray and G. R. Luckhurst), Academic Press, London 1979, p. 263.
15. W. L. McMillan, Phys. Rev., A8:1921 (1973).
16. B. W. van der Meer and G. Vertogen, J. de Phys., 40:C3-222 (1979).
17. A. Wulf, Phys. Rev., A11:365 (1975).
18. A. D. Buckingham, Disc. Faraday Soc., 43:205 (1967).
19. I. Haller, H. A. Huggins, H. R. Lilienthal and T. R. McGuire, J. Phys. Chem., 77:950 (1973).
20. W. H. de Jeu and P. Bordewijk, J. Chem. Phys., 68:109 (1978).
21. D. A. Dunmur and W. H. Miller, Mol. Cryst. Liq. Cryst., 60:281(1980).
22. W. Maier and G. Meier, Z. Naturforschg., 16a:262 (1961).
23. D. A. Dunmur, M. R. Manterfield, W. H. Miller and J. K. Dunleavy, Mol. Cryst. Liq. Cryst., 45:127 (1978).
24. P. P. Karat and M. V. Madhusudana, Mol. Cryst. Liq. Cryst., 36:51 (1976).
25. D. A. Crellin and P. L. Sherrell, J. de Phys., 40:C3-211 (1979).

26. B. R. Ratna and R. Shashidhar, Pramana, 6:278 (1976).
27. H. E. Stanley, "Introduction to Phase Transitions and Critical Phenomena", Oxford University Press, Oxford and New York 1971.
28. P. G. de Gennes, Mol. Cryst. Liq. Cryst., 12:193 (1971).
29. C. Fan and M. J. Stephen, Phys. Rev. Lett., 25:500 (1970).
30. L. D. Landau and E. M. Lifshitz, "Statistical Physics", Pergamon Press, London 1958, p. 430.
31. Y. Poggi, J. C. Filippini and R. Aleonard, Phys. Lett., 57A:53 (1976).
32. Y. Poggi, P. Atten and J. C. Filippini, Mol. Cryst. Liq. Cryst., 37:1 (1976).
33. C. W. Oseen, Trans. Faraday Soc., 29:883 (1933).
34. F. C. Frank, Disc. Faraday Soc., 25:19 (1958).
35. A. R. Johnston, J. Appl. Phys., 44:2971 (1973).
36. J. C. Filippini and Y. Poggi, J. de Phys. Lett., 35:L99 (1974).
37. M. S. Beevers, Mol. Cryst. Liq. Cryst., 31:333 (1975).
38. M. Davies, A. H. Price, R. F. S. Moutran, M. Beevers and G. Williams, J. Chem. Soc. Faraday Trans. II, 72:1447 (1976).
39. H. J. Coles and B. R. Jennings, Mol. Phys., 31:571 (1976).
40. M. Schadt, J. Chem. Phys., 67:210 (1977).
41. R. Yamamoto, S. Ishihara, S. Hayakawa and K. Morimoto, Phys. Lett. A, 276 (1978).
42. M. Schadt and W. Helfrich, Mol. Cryst. Liq. Cryst., 17:355 (1972).
43. J. C. Filippini, Compte, Rend. Acad. des Sci., B275:349 (1972).
44. J. C. Filippini and Y. Poggi, J. de Phys. Lett., 37:L17 (1976).
45. B. Malraison, Y. Poggi and J. C. Filippini, Sol. St. Comm., 31: 843 (1979).
46. V. N. Tsvetkov and E. I. Ryumtsev, Sov. Phys. Cryst., 13:225 (1968).
47. T. Bischofberger, R. Yu and Y. R. Shen, Mol. Cryst. Liq. Cryst., 43:287 (1977).
48. G. K. L. Wong and Y. R. Shen, Phys. Rev., A10:1277 (1974).
49. P. D. Maker, R. W. Terhune and C. M. Savage, Phys. Rev. Lett., 12:507 (1964).
50. P. D. Maker and R. W. Terhune, Phys. Rev., 137:A801 (1965).
51. A. Owyoung, IEEE J. Quant. Elec., QE9:1064 (1973).
52. R. W. Hellwarth, A. Owyoung and N. George, Phys. Rev., A4:2342 (1971).
53. E. G. Hanson, Y. R. Shen and G. K. L. Wong, Phys. Rev., A14:1281 (1976).
54. K. Inoue and Y. R. Shen, Mol. Cryst. Liq. Cryst., 51:190 (1979).
55. H. J. Coles, Mol. Cryst. Liq. Cryst. Lett., 41:231 (1978).
56. J. Prost and J. R. Lalanne, Phys. Rev., A8:2090 (1973).
57. J. C. Filippini and Y. Poggi, Phys. Lett., 65A:30 (1978).
58. J. R. Lalanne, B. Martin and B. Pouligny, Mol. Cryst. Liq. Cryst., 42:153 (1977).
59. L. S. Ornstein and W. Kast, Trans. Faraday Soc., 29:931 (1933).
60. A. D. Buckingham and J. A. Pople, Proc. Phys. Soc., A68:905 (1955).

61. M. P. Bogaard and B. J. Orr, Int. Rev. Sci. Series 2, (Ed. A. D. Buckingham), Vol. 2, Butterworths, London 1975, p. 149.

62. P. Bordewijk, Physica, 75:146 (1974).

63. M. R. Battaglia, T. I. Cox and P. A. Madden, Mol. Phys., 37:1413 (1979).

64. A. D. Buckingham and R. E. Raab, J. Chem. Soc., 1957:2341.

65. S. Kielich, Mol. Phys., 6:49 (1963).

66. B. M. Ladanyi and T. Keyes, Mol. Phys., 34:1643 (1977).

67. B. M. Ladanyi and T. Keyes, Mol. Phys., 33:1063 (1977).

68. C. G. Le Fèvre and R. J. W. Le Fèvre, Rev. Pure and Appl. Chem., 5:261 (1955).

69. T. D. Gierke and W. H. Flygare, J. Chem. Phys., 61:2231 (1974).

70. N. V. Madhusudana and S. Chandrasekhar, Pramana, 1:12 (1973).

71. N. V. Madhusudana, K. L. Savithramma and S. Chandrasekhar, Pramana, 8:22 (1977).

72. G. Lasher, Phys. Rev., A5:1350 (1972).

73. J. Viellard-Baron, Mol. Phys., 28:809 (1974).

74. H. Meirovitch, Chem. Phys., 21:251 (1976).

75. C. Zannoni, "Molecular Physics of Liquid Crystals", (Ed. G. W. Gray and G. R. Luckhurst), Academic Press, London 1979, p. 191.

76. G. R. Luckhurst, unpublished.

77. W. H. Miller, thesis, University of Sheffield (1979).

78. W. M. Gelbart and B. Barboy, Acc. Chem. Res., in press.

79. M. S. Beevers and G. Williams, J. Chem. Soc. Faraday Trans. II, 72:2171 (1976).

80. C. Destrade, H. Gasparoux and F. Guillon, Mol. Cryst. Liq. Cryst., 40:163 (1977).

81. See refs. in S. Chandrasekhar, "Liquid Crystals", Cambridge University Press, Cambridge 1977, p. 51.

82. D. A. Dunmur and W. H. Miller, J. de Phys., 40:C3-141 (1979).

83. Lin Lei, Phys. Rev. Lett., 43:1604 (1979).

84. R. B. Meyer, L. Liebert, L. Strzelecki and P. Keller, J. de Phys. Lett., 36:L69 (1975).

85. Ref. 75, p. 208, but see also T. E. Faber, Proc. Roy. Soc., A370:509 (1980).

86. M. S. Beevers, J. Crossley, D. C. Garrington and G. Williams, J. Chem. Soc. Faraday Trans. II, 72:1482 (1976).

87. P. Maurel and A. H. Price, J. Chem. Soc. Faraday Trans. II, 69:1486 (1973).

88. D. Lippens, J. P. Parneix and A. Chapoton, J. de Phys., 38:1465 (1977).

AN INTRODUCTION TO MAGNETIC BIREFRINGENCE OF LARGE MOLECULES

G. Weill

Centre de Recherches sur les Macromolecules and
Universite L. Pasteur
Strasbourg, FRANCE

Magneto-optics covers a wide field of phenomena, the majority
of which concerns the properties of solid state magnetic materials.
Let us quote as an example the magneto-optic Kerr effect discovered
in 1876, one year after the electro-optic Kerr effect. Another
area of considerable activity is that of magneto-optics of paramagnetic atoms or molecules with phenomena such as the observation
of the population of Zeeman sublevels of paramagnetic states by
optical detection of paramagnetic resonance, or the effect of magnetic field on triplet triplet recombination in organic crystals
as followed by the change in the delayed fluorescence. There are
two effects which are present in any diamagnetic molecule and are
of help in the study of large molecules.

- The Faraday effect (1846), i.e., the induction of optical
activity by a magnetic field along the direction of propagation.
Although this effect is now very easily measurable, especially
magnetic circular dichroism (MCD) in an absorption band using
superconducting coils in a circular dichroism spectrometer, and
has been used for conformational studies of biopolymers, its theoretical expression contains many terms, so that a precise knowledge
of the magnitude and relative positions of electric and magnetic
dipole transition moments is required for its calculation and
interpretation (1). Calculation of the MCD of helical polymers
with excitonic bands has been performed by Tinoco (2). One interesting suggestion relies on the fact that the Faraday effect has
the symmetry of a mixed product so that it is identical for the two
enantiomers of a naturally optically active substance. In the
case of isotactic synthetic polymers such as polystyrene (3), the
racemic mixture of possible helical structures of opposite sense

473

for L and D residues can present an increased MCD when compared
to the atactic polymer.

    - The Cotton-Mouton effect (1912), i.e., the magnetic bire-
fringence or dichroism resulting from the orientation of the molecules in a transverse magnetic field B, is the only effect which
will be considered in what follows with particular emphasis placed
on the study of the stiffness of semi-flexible chain molecules,
and on the observation of orientational correlation in macromolecular aggregates or polymeric amorphous liquids or solids.

MAGNETIC BIREFRINGENCE OF SOLUTIONS OF NON-INTERACTING MOLECULES

    The orientation of molecules in a magnetic field can be treated
in exact analogy with the orientation in an electric field, starting from the angular dependence of the magnetic energy

$$u = \frac{1}{2} [(\chi_{\parallel} - \chi_{\perp}) \cos^2\theta + \chi_{\perp}] B^2 - \vec{\mu} \cdot \vec{B} \tag{1}$$

The diamagnetic susceptibility tensor χ (considered here as
cylindrically symmetric with two principal values $\chi_{\parallel}$ and $\chi_{\perp}$) plays
the role of the electronic polarizability tensor α in the corresponding electrical energy. Even in paramagnetic molecules, the
quantum nature of the electronic spin and the very weak spin-orbit
coupling makes the contribution of the second term, formally
analogous to the permanent electric dipole moment in the electrical
case, negligible.

    Due to the small magnitude of χ $(10^{-28} - 10^{-27}$ J T$^{-2})$ for
small molecules, extremely high magnetic fields are required to
reach values of U high enough with respect to kT ($\sim 4 \ 10^{-21}$ J at
room temperature). With the highest steady state magnetic fields
currently available (10 - 20 T), U/kT remains of the order of
10^{-5}. The degree of orientation remains so low that only birefringence studies (far from the absorption bands or in the region
of anomalous dispersion) are currently performed. The situation
improves, however, in the case of large stiff molecules or molecular aggregates with a large orientational correlation.

    Using (1) and transposing from the Kerr effect for the case
of orientation due to polarizability anisotropy, one gets the specific Cotton-Mouton constant C_m for a solution of molecules with
concentration c and molecular weight M

$$C_m = \frac{\Delta n}{\lambda \, c \, B^2} = \frac{2\pi}{15 \, n \, kT \, \lambda} \left(\frac{n^2 + 2}{3} \right)^2 (\chi_{\parallel} - \chi_{\perp}) (g_{\parallel} - g_{\perp}) \, \bar{v}$$

$$= \frac{2\pi}{15 \, n \, kT \, \lambda} \left(\frac{n^2 + 2}{3} \right)^2 (\chi_{\parallel} - \chi_{\perp}) (\alpha^o_{\parallel} - \alpha^o_{\perp}) \frac{N}{M} \qquad (2)$$

λ is the wavelength of light in vacuum,

$g_{\parallel} - g_{\perp}$ is the difference of optical polarizabilities per unit volume of the solute

$\bar{v}$ is the solute partial specific volume

$\alpha^o_{\parallel} - \alpha^o_{\perp}$ is the difference in optical polarizability of one solute molecule

N is Avogadro's number

n is the solvent index of refraction and an internal field correction $(n^2 + 2/3)^2$ has been introduced in the formula.

Relation (2) assumes that the magnetic and optical suscepti-bility tensors have the same axis. Since $(\chi_{\parallel} - \chi_{\perp})$ and $(\alpha^o_{\parallel} - \alpha^o_{\perp})$ are electronic properties of the molecule, a band additivity scheme is supposed to hold, at least if there is no large delocalization along the polymer backbone. We can therefore foresee that magnetic birefringence will have the same basic advantages as the optico-optical Kerr effect. In addition, the orientation should not be dependent upon the complicated ionic polarization mechanism in the case of polyelectrolyte solutions. Also, the magneto-optic effect will be much simpler to measure.

INSTRUMENTATION

With the sensitivity of modern birefringence measuring devices, including signal averaging ($\Delta n \sim 10^{-12}$), the magnetic birefringence of solutions can be measured with conventional electromagnets, such as are currently used in high resolution NMR with B = 1.5 - 2 T. Such an instrument has been described by an English group (4) work-ing mostly with synthetic polymers in organic solvents. It is, however, very desirable in the case of polyelectrolyte systems in water to reach much lower concentrations and therefore to use the higher fields available from a Bitter solenoid. This is the case in the set-up at the High Field Institute of the German Max Planck

Institute for Solid State Physics located at Grenoble, France (5).
The use of a compensating Pockel cell (6) allows continuous recording of the birefringence, proportional to the applied voltage on the
solenoid, as a function of increasing field, as obtained from the
current intensity. The overall sensitivity is $\Delta n \sim 10^{-10}$. A
possible difficulty in the enhancement of the sensitivity can arise
from the reflections required to put the laser beam perpendicular
to the solenoid axis. The coil diameter (5 cm) would then also
limit the possible path length. A direct radial access allowing
a 3 cm sample has been developed recently. It is evident that no
transient experiments can be carried out because self-induction
puts a severe limitation on the magnetic field rise and decay times.

THE CASE OF FLEXIBLE OR SEMI-FLEXIBLE MOLECULES

Theory

Measurements have been carried out on a number of chain molecules, both uncharged synthetic polymers such as polyisobutene (7),
polystyrene (8), polycarbonate (9), polybenzamides (10,11), and
charged biopolymers such as polynucleotides and DNA (12,13). They
have, however, very often been limited to one molecular weight, and
have been interpreted in a very crude way, the ratio of the Cotton-
Mouton constant to that of a small model molecule being taken as
the number of monomers in the statistical element. Having in mind
the fact, outlined above, that the expression for magnetic birefringence exactly matches that for electric birefringence, pro-
vided $(\chi_{||} - \chi_\perp)$ is replaced by the electric polarizability aniso-
tropy $(\alpha_{||} - \alpha_\perp)$ the formulae first developed by Nagai (14), then
by Jernigan (15,16) for the latter case are directly applicable.
Utilizing the correspondence of the magnetic and polarizability
anisotropy, Wilson (33) derived a general expression for the magnetic dichroism with an explicit calculation restricted to the case
of a worm-like chain of infinite length. In what follows we shall
outline the derivation and extend it to the case of a worm-like
chain of arbitrary length.

For fields not high enough to disturb the field-free conformation, the result of Nagai and Jernigan for no permanent dipole
moment is

$$\Delta n = \frac{2\pi}{n} \left(\frac{n^2 + 2}{3}\right)^2 \frac{Nc}{M} \frac{1}{15} [3\langle Tr(\alpha^o \alpha)\rangle - \langle Tr\alpha^o Tr\alpha\rangle] \frac{E^2}{kT}$$

(3)

where $\mathrm{Tr}\underset{\sim}{\alpha}^o$, $\mathrm{Tr}\underset{\sim}{\alpha}$ and $\mathrm{Tr}\underset{\sim}{\alpha}_o\underset{\sim}{\alpha}$ are the traces of the tensors and the brackets indicate averaging over all conformations. Substituting $\underset{\sim}{\alpha}$ by $\underset{\sim}{\chi}$, E^2 by B^2 and denoting by $\hat{\underset{\sim}{\alpha}}^o$ and $\hat{\underset{\sim}{\chi}}$ the traceless tensors, (3) becomes:

$$\Delta n = \frac{2\pi}{n} \left(\frac{n^2 + 2}{3}\right)^2 \frac{Nc}{M} \frac{1}{10} <\mathrm{Tr}\ \hat{\underset{\sim}{\alpha}}^o\ \hat{\underset{\sim}{\chi}}> \frac{B^2}{kT} \tag{4}$$

Therefore, if $\hat{\underset{\sim}{\alpha}}^o_i$ and $\hat{\underset{\sim}{\chi}}_i$ are the traceless tensors associated with one segment in the chain, we can calculate the average by summing over all bonds. This can be performed using the rotational isomer Ising model (16) for stereoregular polymers or in a simpler way in the framework of the worm-like chain model.

If we assume that $\hat{\underset{\sim}{\alpha}}^o$ and $\hat{\underset{\sim}{\chi}}$ are cylindrically symmetric with their axes along the chain, we can directly derive an expression for the product of the magnetic and optical anisotropies in relation (2)

$$(\chi_{||} - \chi_{\perp})\ (\alpha^o_{||} - \alpha^o_{\perp}) = \Delta\chi_i\ \Delta\alpha_i\ \sum_i \sum_j \left\langle \frac{3\cos^2\theta_{ij} - 1}{2} \right\rangle \tag{5}$$

where $\Delta\alpha_i$ and $\Delta\chi_i$ are the segment anisotropies and the double sum extends over all N segments in the chain. The same sum is found when dealing with the optical anisotropy as measured by depolarized scattering and we have calculated it in a previous paper (17) for any value of the contour length L as compared to the persistence length q with the result

$$(\chi_{||} - \chi_{\perp})\ (\alpha^o_{||} - \alpha^o_{\perp}) = (\Delta\chi_i \cdot \Delta\alpha_i)\ N^2\ \frac{2q}{L}\ [1 - \frac{q}{3L}$$

$$(1 - \exp - \frac{3L}{q})] \tag{6}$$

For L << q expansion of the exponential gives

$$(\chi_{||} - \chi_{\perp})\ (\alpha^o_{||} - \alpha^o_{\perp}) = (\Delta\chi_i \cdot \Delta\alpha_i)\ N^2\ (1 - \frac{L}{q}$$

$$+ \frac{3}{4} \frac{L^2}{q^2} - \ldots)$$

which shows the progressive deviation from a stiff rod result. Introducing the length $\ell_o = L/N$ and the molecular weight $m_o = M/N$ of one segment, we can express the Cotton-Mouton constant

$$C_m = \frac{2 (n^2 + 2)^2}{135 \ kT \ \lambda \ n} \ \frac{N}{m_o} \ \Delta\chi_i \ \Delta\alpha_i \ \frac{2q}{3\ell_o} \ [1 - \frac{q}{3L} \ (1 - \exp - \frac{3L}{q})] \tag{7}$$

The last bracket becomes equal to 1 when L >> q showing that, as expected, the Cotton-Mouton constant becomes independent of molecular weight with a value $2q/3\ell_o$ that of the segment. Relation (7) is exactly similar to Wilson's expression for magnetic dichroism but extended to arbitrary L. The variation of C_m with L represents therefore an efficient way to measure q, even if $\Delta\alpha_i$ and $\Delta\chi_i$ are not precisely known.

Applications

Synthetic flexible polymers. As quoted above, the value of C_m has been obtained for several long vinyl chains. Comparison with some values of $\Delta\chi_i$ $\Delta\alpha_i$ derived from model compounds is seen to lead, from relation (7) to a value of q/ℓ_o. This has been taken as a "number of elements in the statistical segment." But for consistency one should compare with the statistical length derived from a study of the unperturbed chain dimensions which is 2q. Alkanes have been studied in a systematic way as a function of the number of carbons up to polyethylene chains (18,11). The parallelism between the variation of C_m and that of the optical anisotropy as derived from light-scattering has been used to obtain a reliable value of the magnetic anisotropy increment of a CH_2 unit

$$\Gamma_m = \Delta\chi_{cc} - 2\Delta\chi_{CH} = - 0,21 \ 10^{-28} \ J \ T^{-2}$$

Aromatic polyamides and related polymers. Measurement of the stiffness of these polymers which enter liquid crystalline phases is of much importance in understanding the relation between this molecular flexibility, nematogenic character and the special type of defects encountered in liquid crystalline polymers. Magnetic birefringence may have some advantage over hydrodynamic measurements or optical anisotropy measurements from depolarized light-scattering considering the class of solvent, such as 99% sulfuric acid, in which they are soluble. A comparison of C_m of polybenzamide

(11) in sulfuric acid and in dimethylacetamide/LiCl reveals a large stiffening in the latter solvent. The result obtained in sulfuric acid is consistent with results obtained by a variety of other techniques (17).

The problem of polyelectrolytes and DNA. The study of the conformation of polyelectrolytes and of its change with ionic strength
is very difficult because at low ionic strengths, the scattering of
light, X-rays or neutrons is dominated by long range intermolecular
interactions. Steady state electro-optic measurements fail because
we have no good theory of the ionic polarizability to interpret the
changes of the Kerr constant with molecular weight. Transient measurements have however given reliable values of the relaxation time
and have shown that it was not appreciably dependent upon the polyelectrolyte concentration provided the overall ionic strength is
kept constant. However, reliable formulae on the variation of D
with L and q in the whole range of L/q values seem lacking and the
sensitivity of D to the choice of q in the region $L \sim q$ is doubtful.
It remains however, that despite the existence of strong correlations in position (as for example demonstrated by a peak in the
neutron scattering pattern of low ionic strength polyelectrolyte
solutions (18)) the orientational correlation is very weak. This is
also demonstrated by the concentration independence of the Cotton-
Mouton constant up to high concentrations of DNA (12). One can
therefore use magnetic birefringence to get reliable values of the
persistence length at any overall ionic strength (i.e., taking into
account the contribution to the screening of the osmotically free
counterions originating from the polyion salt) as limited by the
sensitivity of the experiment. This is illustrated by the results
of Maret and Senechal (12,13) on a calf-thymus $(L \sim 3 \ 10^4 \ \overset{o}{A})$ DNA
and a core particle rat liver (143 Base Pairs $L \sim 500 \ \overset{o}{A}$) DNA.

1 M NaCl solutions $C_{DNA} \geq 3$ mgr/ml $\lambda = 6328 \ \overset{o}{A}$

$\quad C_m = 2.8 \mp 0,2 \qquad T^2 \ m^2 \ kg^{-1} \qquad$ calf thymus

$\quad C_m = 2.0 \mp 0,2 \qquad T^2 \ m^2 \ kg^{-1} \qquad$ rat liver

The base pair magnetic anisotropy $\Delta\chi_i$ is taken from the purine and
pyrimidine magnetic anisotropies (19)

$$\Delta\chi_{AT} = 10.4 \ 10^{-28} \ J \ T^{-2}$$

$$\Delta\chi_{GC} = 7.8 \ \ 10^{-28} \ J \ T^{-2}$$

averaged for the DNA compositions (42% GC)

$$\Delta\chi_i = 9.2 \ 10^{-28} \ J \ T^{-2}$$

The corresponding optical anisotropy is taken from the comparison
of electric birefringence and dichroism (20)

$$\Delta\alpha = 1.40 \ 10^{-39} \ F \ m^3 \ (\sim 12.5 \ \overset{o3}{A})$$

with ℓ_o = 3.4 $\overset{o}{A}$ and m_o = 690 (including the two Na$^+$ counterions)
one gets from relation (7)

$q \sim 700 \overset{o}{A}$

in fair agreement with the results of light-scattering at high ionic
strength (21).

A direct calculation using the ratio of the C_m values for the
two DNA's and relation (7), i.e.,

$$\frac{C_m(L = 500 \overset{o}{A})}{C_m(L = 30.000 \overset{o}{A})} = 1 - \frac{q}{1500} \left(1 - \exp - \frac{1500}{q}\right)$$

requires a very high degree of accuracy in the measurements since,
with the precision given above, this ratio is found to be between
0.60 and 0.80, corresponding to $300 \le q \le 600 \overset{o}{A}$. Some results on
the variation of the Cotton-Mouton constant of calf-thymus DNA with
the added salt concentration have been published by Maret (12). C_m
increases below 10^{-1} M NaCl but a quantitative comparison with current theories (22,23) is precluded because the true ionic strength
is in this case dependent on the concentration of DNA which is close
to 10^{-2} mM in phosphate. This calls for an improvement in the sensitivity of the experiment in order to work at lower DNA concentration since the technique seems ideally suited for a direct check of
these theories. It may also be added that the melting of DNA and
other polynucleotides is easily followed by magnetic birefringence
(5) and could be used to study quantitatively the fine structure of
helix-coil transitions.

LARGE COLLOIDAL PARTICLES, AGGREGATES, BIOLOGICAL MATERIALS

It is clear that any correlation in orientation of a large
number of subunits will give rise to an increase in the specific
Cotton-Mouton constant as compared to that of the subunit. This
explains, for example, why silks, keratins, collagen (24) and muscle
fibers (25) have been among the first structures studied by magnetic
birefringence. It also explains why clay minerals can orient to a
high degree in relatively low magnetic fields (7,26). An interesting example is that of kaolinite, for which the sign of the birefringence reverses between low and high fields. This is believed
to be due to magnetic impurities adding a "transverse permanent
moment" orientation mechanism to the magnetic susceptibility anisotropy. Here, we shall only mention in some detail three conceptually
interesting cases.

The Problem of Order in Amorphous Synthetic Polymers

The entropy of amorphous polymer "liquids" as calculated from the enthalpy and temperature of fusion of semi-crystalline polymers has for a long time suggested the existence in the liquid of some orientational order. Electron micrographs seemed to favour a model of fringed micelles. Careful scattering experiments (depolarized light scattering, X-ray and neutron low angle scattering) reveal however the absence of anomalous density fluctuations, orientational correlations, or deviation of one chain scattering factor from that of a Gaussian chain. A simple and good argument has been recently (11) derived from the comparison of the Cotton-Mouton constant of polystyrene in solution and in the bulk.

Transition to a Lyotropic Liquid Crystalline Phase of Semi-Flexible Polymers at High Concentrations

This fast transition is easily followed by an abrupt increase of several orders of magnitude in the specific Cotton-Mouton constant. This has been observed not only on aromatic polyamides (11) and polypeptides (27) in organic solvents, but also on water soluble charged polypeptides (28) and polynucleotides (29,30); of special interest is the case of the high concentration poly(A) poly(U) phase above 70 mg/ml (13). The high orientation achieved in the field remains after the field is turned off in domains ($\sim 20\mu$), allowing the recording of an anisotropic light diffraction pattern corresponding to a superstructure with a periodicity in the range of 12,500 $\overset{o}{A}$. Neutron scattering experiments reveal a short range order with a lattice constant of ~ 50 $\overset{o}{A}$. A tubular spherolitic structure, which can be considered as locally cholesteric has been proposed for this densely packed structure.

Biological Material

Retinal rod outer segments, chloroplasts and photosynthetic algae and bacterial purple membranes have been oriented in an electric field. References can be found in the recent works of Chabre (31) and of Worcester (32) which discuss the origin of the required diamagnetic anisotropy. It has long been considered that diamagnetic anisotropy in biological material was dependent upon aromatic structures. Evaluation of the peptide bond anisotropy from small models and from measurements on polypeptides has led to the prediction of the contribution of one peptide bond to a regular α-helix ($\Delta\chi \sim 0,7 \ 10^{-28}$ J T^{-2} as compared to $\Delta\chi \sim 10 \ 10^{-28}$ J T^{-2} for the benzene ring) and to a β pleated sheet ($\Delta\chi \sim 0,18 \ 10^{-28}$ J T^{-2}).

It is then proposed that orientationally correlated protein molecules, with their α-helical parts directed essentially normal to
the plane of the membrane could play a major role in the diamagnetic anisotropy of these materials.

REFERENCES

1. A. D. Buckingham, and P. J. Stephens, Ann. Rev. Phys. Chem.
    17:399 (1967).
2. I. Tinoco, and A. Bush, Biopol. Symp. I 235 (1964).
3. H. Kaye, and J. S. Hwang, J. Pol. Sc. Al:10,639 (1972).
4. J. V. Champion, D. Downer, G. H. Meeten, and L. F. Gate, J.
    Phys. E 10:1137 (1977).
5. G. Maret, M. V. Schickfus, A. Mayer, and K. Dransfeld, Phys.
    Rev. Letters 35:397 (1975).
6. H. B. Serreze, and R. B. Goldner, Rev. Sci. Instrum. 45:1613
    (1974).
7. J. V. Champion, A. Dandridge, D. Downer, J. C. McGrath, and
    G. H. Meeten, Polymer 17:511 (1976).
8. G. H. Meeten, Polymer 15:187 (1974).
9. J. V. Champion, P. A. Desson, and G. H. Meeten, Polymer 15:301
    (1974).
10. V. N. Tsvetkov, G. I. Kudryavtsev, E. I. Ryumtsev, V. A.
    Nikolaev, V. D. Kalmykova, and A. V. Volokhina, Dokl. Nauk.
    USSR 224:398 (1975).
11. M. Stamm, Ph.D. thesis, Univ. Mainz, July 1979.
12. G. Maret, and K. Dransfeld, Physica 86/88 B:1077 (1977).
13. E. Senechal, G. Maret and K. Dransfeld, submitted to Int. J.
    Biol. Macromol., in press.
14. K. Nagai, and T. Ishikawa, J. Chem. Phys. 43:4508 (1965).
15. P. J. Flory, and R. L. Jernigan, J. Chem. Phys. 48:3823
    (1968).
16. R. L. Jernigan, and D. S. Thomson, in: "Molecular Electro-
    Optics, Part 1," C. O'Konski, ed., Marcel Dekker, Inc.,
    NY, p. 159 (1976).
17. M. Arpin, C. Strazielle, G. Weill, and H. Benoit, Polymer 18:
    262 (1977).
18. M. Nierlich, C. E. Williams, F. Bove, J. P. Cotton, M. Daoud,
    B. Farnoux, G. Jannink, C. Picot, M. Moan, C. Wolff, M. Rinaudo,
    and P. G. DeGennes, J. de Physique 40:701 (1979).
19. A. Veillard, B. Pullman, and G. Berthier, C. R. Acad. Sc. 252:
    2321 (1961).
20. S. Sokerov, and G. Weill, Biophys. Chem. 10:161 (1979).
21. G. Cohen, and H. Eisenberg, Biopolymers 6:1077 (1968).
22. T. Odijk, and A. C. Houwart, J. Pol. Sc. Phys. ed. 16:627
    (1978).
23. T. Skolnick, and M. Fixman, Macromolecules 10:944 (1977).
24. E. Colton-Feytis and E. Fauré-Fremiet, C. R. Acad. Sci. Paris
    D214:996 (1942).

25. W. Arnold, R. Steele, and H. Mueller, <u>Proc. Nat. Acad. Sci. USA</u> 44:1 (1958).
26. R. V. Mehta, <u>J. Coll. Interface Sc.</u> 42:165 (1973).
27. K. Tokyama, and N. Miyata, <u>J. Phys. Soc. Japan</u> 34:1699 (1973).
28. G. Maret, A. Domard, and M. Rinaudo, <u>Biopolymers</u> 18:101 (1979).
29. E. Jizuka, <u>Polymer J.</u> 10:253 and 293 (1978).
30. E. Jizuka, and Y. Kondo, <u>Mol. Cryst. Liq. Cryst.</u> 51:285 (1979).
31. M. Chabre, <u>Proc. Nat. Acad. Sci. USA</u> 75:5471 (1978).
32. D. L. Worcester, <u>Proc. Nat. Acad. Sci. USA</u> 75:5475 (1978).
33. R. W. Wilson, <u>Biopolymers</u> 17:1811 (1978).

INTRODUCTION TO LASER LIGHT SCATTERING

SPECTROSCOPY IN ELECTROPHORESIS

E.E. Uzgiris

General Electric Research and Development Center
P.O. Box 8
Schenectady, N.Y. 12301

INTRODUCTION

The measurement of electrophoretic mobilities by light scattering spectroscopy relies on the detection of small frequency shifts of light scattered from the objects that are migrating in an electric field.[1-4] Even though the velocities involved may be quite small, perhaps 10µm/sec to 100µm/sec, and the frequencies shifts may be only a few tens of cycles per second, the use of rather straightforward optical heterodyne techniques readily allows detection of such small shifts. The difficult problems associated with laser Doppler electrophoresis are: the application of an electric field without causing spectral degradation due to heating, electrode bubbling, or electrode reactions; and the design of a suitable chamber that prevents such degradation and is itself free from systematic velocity offsets due to the phenomena of electroosmosis.[3,5]

Several alternative approaches have been developed. The simplest electrode configuration and the most effective from the point of view of heat dissipation, resistance to turbulent liquid motion, and freedom from electroosmosis is the small gap, parallel rectangular plate electrode assembly immersed in a small rectangular optical cuvette.[1,3,5] This configuration requires extra care in the type and quality of electrodes used since the probed volume is in close proximity to the electrodes. The alternative approaches are to use a capillary or rectangular channel with electrodes far apart[4] or electrodes of cylindrical shape.[6] Although the sensitivity to electrode surface reactions is decreased, these configurations suffer from poor heat dissipation, an increased propensity to turbulence, and velocity offsets

due to electroosmotically driven liquid flow. We shall examine the
resolving power of these approaches as well as some of the more
standard electrophoretic techniques below.

The interesting features of laser Doppler electrophoresis are
speed of measurement, objective data acquisition, and high resolving
power (for suitable electrode and chamber designs). These features
are especially notable in applications involving cell electrophoresis.[1,3] The electrophoretic analysis of biological cells and
other large particles has, heretofore, been chiefly performed by
microscopic manual observations. Not only are these microscopic
methods slow and tedious, but they also suffer from possible operator
bias in the acquisition of particle velocity statistics. Because of
these shortcomings and because of the last problem in particular,
cell electrophoresis methods have largely been without much impact
in immunology. In fact, the macrophage electrophoretic mobility, MEM,
test, perhaps the best known application of cell electrophoresis in
recent years, has now become controversial. It would seem, therefore,
that there are scientific opportunities in the cell electrophoresis
field with the laser Doppler approach.

Protein electrophoretic analysis presents more difficult problems
for the successful implementation of the light scattering mode. It is
inherently more difficult to obtain high resolution Doppler spectra
from small scatterers because of Brownian motion broadening. Furthermore light scattering measurements on such systems are also prone to
interference from contaminating particles such as due to dust or due
to the presence of multiplets or aggragates of a particular species.
Both of these limitations are absent in the case of large particle
systems. For this reason, the applications of laser Doppler Spectroscopy in cell electrophoresis have already yielded interesting measurements that add new information on such topics at T lymphocyte mobility
heterogeneity,[7,8] T lymphocytes interactions with antigen,[9] and the
MEM test.[10] However, laser Doppler studies of macromolecular systems,
although extensive, have not moved beyond the initial demonstration
stage. Nevertheless, there is one interesting and potentially important application, that of blood plasma analysis. This application of
laser Doppler spectroscopy has not been successfully implemented to
date and remains an open problem.

HETERODYNE SPECTROSCOPY

In laser light scattering spectroscopy, particle motion is
detected by the effects on the frequency of scattered light. For
stationary scatterers, the scattered light is unchanged in frequency
and the detected light intensity is constant in time. For moving
scatterers, the scattered light is shifted in frequency by the Doppler
effect given by

$$\delta\nu = \frac{2\nu_0}{C} \ v \ \sin\theta/2 \tag{1}$$

where $\delta\nu$ is the Doppler shift, C is the velocity of light in the medium, ν_0 is the incident laser light frequency, v is the velocity and θ is the scattering angle. Note that for $\theta \to o$, the Doppler shift vanishes: the red shift of an object moving away from the laser source is compensated by the blue shift of an object moving toward the detector.

We now consider what sort of magnitudes are involved for the directed velocities and the associated Doppler shifts. The directed motion we are considering is due to particle charge. In an electric field, the steady state velocity of a particle per unit electric field is the electrophoretic mobility. For particles large compared to the Debye shielding distance, the mobility, μ, is given by

$$\mu = \frac{\epsilon\zeta}{4\pi\eta} \tag{2}$$

for ϵ the dielectric constant, η the viscosity, and ζ the electric potential at the slipping plane. For particles small compared to the counter-ion shielding distance, the factor 4π is replaced by 6π. A reasonable value for ζ is $\zeta \approx kT/e \approx 25mV$. Hence $\mu \approx 1\mu m \ sec^{-1}V^{-1}cm$, as an order of magnitude. And for $E \approx 50V/cm$, a reasonable upper value for applied electric field levels, and $\theta=10°$, $\delta\nu=18$ Hz.

The detection of such a small frequency shift, 10^{-14}, of the incident light frequency, appears to be a formidable task. However, it is not as hard as it seems. What is required is a laser source of reasonably constant amplitude and a detection technique known as optical heterodyning. In this technique, two light signals, frequency shifted signal light and frequency unshifted reference light, both derived from the same laser source, are combined or, more accurately, multiplied together in a square law detector such as a photomultiplier, the output of which is proportional to the square of the total light field at the photocathode. As a result of this multiplication, sum and difference frequency photocurrents are generated and only the latter can pass through the subsequent electronic circuits

It is important to note that in the more common light scattering studies of Brownian motion either homodyne (no reference beam) or heterodyne modes of detection can be used; however, for measurements of Doppler frequency shifts arising from uniform directed motion, it is necessary to use heterodyne detection. Otherwise, no net Doppler frequency shift will result in the signal current and the signal will be merely broadened into a Lorentzian band of width given by the diffusion constant in the manner indicated below.

The power spectrum of a heterodyne signal is the power spectrum of the signal light field and has the form

$$P(\nu) \; \alpha \; \frac{(DK^2)}{(DK^2)^2 + (\omega-\omega')^2} \tag{3}$$

where $\omega'=2\pi\delta\nu$, ω is the radial frequency, D is the diffusion constant, and K the scattering vector ($K=k_0-k_s \cong 4\pi/\lambda \; \sin\theta/2$ for θ the scattering angle, λ the wavelength of laser light in the scattering medium). This line profile assumes no other source of line broadening other than diffusion. The full width at half maximum of Eqn. 3 is

$$\Delta\nu_D = \frac{DK^2}{\pi} \tag{4}$$

in Hz.

Thus the shift to width ratio, or quality parameter, Q, for a given velocity, is

$$\frac{\delta\nu}{\Delta\nu_D} = \frac{vK}{2DK^2} \tag{5}$$

or

$$\frac{\delta\dot{\nu}}{\Delta\nu} \; \alpha \; \frac{1}{K} \; \alpha \; \frac{1}{\sin\theta/2} \tag{6}$$

This indicates that the smaller the angle the better for any size particle. This is an erroneous conclusion since this analysis has ignored line broadening from other sources such as spectrum analyzer resolution, transit time effects, electric field inhomogeneity broadening and the like. In fact, it can be shown that the optimum angle is in fact set by the following criterion:

$$DK^2 = \Delta\nu_{res} \tag{7}$$

where $\Delta\nu_{res}$ is the residual linewidth even if $D\rightarrow o$ in Eqn. 3. Since by the Stoke Einstein relation, we can write,

$$D = \frac{kT}{6\pi\eta r} \tag{8}$$

for k the Boltzmann constant, T absolute temperature, η the viscosity, and r the particle radius, then we deduce from Eqn. 7 that the optimum angle or optimum scattering vector as the relation specifies is

$$K^2_{opt} = \Delta\nu_{res} \frac{6\pi\eta}{kT} \, r \, . \tag{9}$$

The larger the particle the larger K and the larger the scattering angle should be in order to achieve the highest quality parameter as shown in Fig. 1. Assuming a 1 Hz residual broadening and blood cells of 10μm diameter, θ_{opt} is approximately 30°, whereas for 1μm particles $\theta_{opt} \sim 7°$. Hence depending on the scattering system to be studied one might desire small scattering angles in the case of proteins or rather large ones in the case of blood cells.

The generation of a reference beam can in many applications be done rather trivially through a single beam approach, as shown in Fig. 2a. The reference light is generated from the scattering centers on the cuvette walls and this light as well as the signal light from the moving particles is collected along a certain chosen angle. This simple approach works well for angles that are less than about 15°. Beyond this value the reference light intensity decreases rather severely and it becomes difficult to generate the difference frequency signal in the photocurrent. To achieve wide angles, a crossed beam, reference beam mode is used as shown in Fig. 2b. This approach is also easy to align and implement.[11]

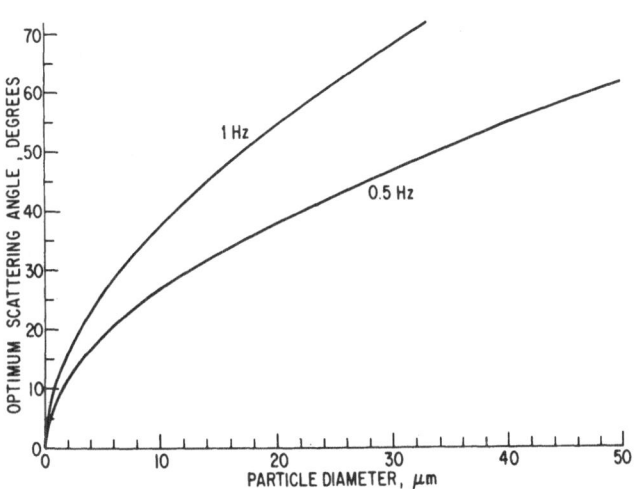

Fig. 1.   Optimum scattering angle for various size particles and for two values of residual line broadening.

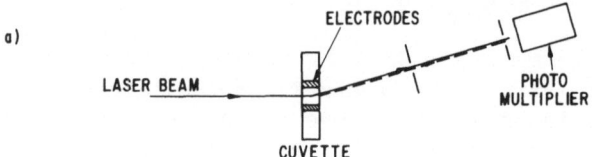

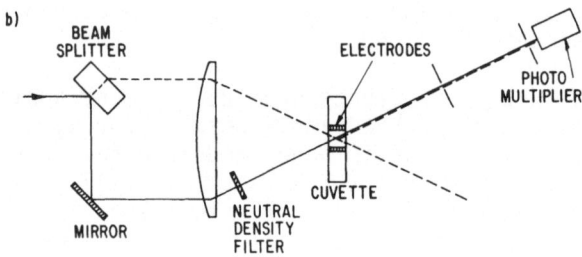

Fig. 2.   Schematic of small angle and wide angle optics.

CHAMBER DESIGN

   Of central importance in laser Doppler electrophoresis is the
electrophoresis chamber itself and the electrodes. Most experimental
difficulties can be traced to shortcomings in either of these components. Therefore, careful attention must be given to the problem of
how to apply the field and how to contain the sample of interest.
Also, we must realize that probing of instantaneous velocity by light
scattering spectroscopy is a localized measurement requiring small
volumes and short measurement times. Thus the chamber design can
depart radically from previous chamber types that depended on migration distance measurements over prolonged times. We shall consider
several current approaches to this problem in detail in the lecture.
Here I shall outline only the essential requirements and briefly
describe the simplest· of the laser Doppler chambers.

   We must first realize that fairly sizable electric currents are
involved in electrophoresis experiments with the concominant generation of heating, temperature gradients, and ultimately, in high
enough fields, turbulence. Thus the central problem is how to apply

a large electric field to a rather conducting solution (physiological saline, 0.15M NaCl with a specific conductivity of 0.015 mhos) without destroying the uniformity of the driven electrophoretic motion. Another important consideration is whether electroosmotic liquid flow[12] that distorts the observed particle velocity profiles can be minimized or all together eliminated by a suitable chamber configuration. Somewhat less fundamental are questions of generating satisfactory particle statistics, of minimizing measurement time (full duty cycle electric field applications vs. partial duty cycle), and of chamber simplicity and ease of system manipulation.

I restate the problem by considering some relevant parameters. For blood cells $\theta/2 \sim 15°$ for optimum resolution. Thus $\delta\nu$=11, 22, 33Hz for 10, 20, and 30V/cm for μ=1μm sec^{-1}V^{-1}cm. Clearly, for a constant residual line width of $\sim$1Hz, the best resolution is for the highest fields that do not induce extra line broadening. Yet fields of 30V/cm in 0.15M NaCl are not trivial to implement: the rise in temperature of 1 mℓ of physiological saline for such an applied field would be 3°C in one second if there was no heat dissipation.

The key to chamber optimization is the Kohlrausch model and the Rayleigh criterion.[5] Let us consider the former. The temperature profile for a volume of liquid heated uniformly between infinite plane boundaries separated by a distance d and held at constant temperature is parabolic with the maximum temperature rise ΔT at d/2 given by

$$\Delta T = \frac{0.24V^2}{8\rho\ k} \quad , \tag{10}$$

where k is the thermal conductivity, ρ the electrical resistivity, and V the applied voltage. The remarkable point is that ΔT is independent of the distance between the isothermal boundaries. To maximize the electric field, E=V/d, for the case of a field applied by parallel plane electrodes one needs to minimize the gap spacing d since V will be limited to some maximum value given by Eqn. 10 and some maximum ΔT_{MAX} (a few degrees in practice).[13] This is the approach I have taken in designing a parallel plate, narrow-gap electrode assembly for use in a small optical cuvette.

In addition to minimizing the temperature rise in a given field, the use of a narrow gap also helps in resisting convective instabilities. It is known that if the Rayleigh number is less than 1700 convective turbulent motion will not take place for a horizontal layer of water heated from below. The Rayleigh number, R, is given by

$$R \propto \Delta T\ d^3 \quad ,$$

where d is the layer thickness. Again the conclusion is that to maintain uniform electrophoretic motion d should be small. For R<1700,
and d=5mm, 3mm, and 1mm, we have ΔT limits of 1°C, 4°C, and 5°C respectively before the onset of a Bernard type of instability.

These conclusions about gap spacing have been verified in practice although the limits set by the Rayleigh criterion have not been
attained. An example of high field spectra from red blood cells in
physiological saline is shown in Fig. 3. The solid curve is for a
field of ∿50V/cm and a high quality factor for the red blood cell peak
is still observed. Beyond 50V/cm there is a degradation in the shift
to width ratio as indicated by the dashed spectrum.

The other important point is that in the narrow gap configuration
electroosmotic effects are not present. This has not been treated
theoretically since the solutions to the Navier-Stokes equations have
not been obtained. We can appeal to a limiting case and to experimental observation. In the limit that the electrode edges are far
from the cuvette walls, there will be no driving force on the ions
near the cuvette surface. In practice the velocity profile is observed
to be flat and uniform as a function of position inside the space
spanned by the electrodes even when the cuvette walls are close to
the electrode edges.[13] In contrast, all channel and capillary systems
for carrying electrical current suffer from the parabolic distortion
of the liquid flow induced by the charges near the channel walls.
There is a body force on the liquid near the channel walls as a field
is applied. This results in a plug liquid flow which, within perhaps
milliseconds, becomes a parabolic liquid flow as a hydrostatic head
forces liquid back down the middle of the tube. This phenomena is
well known in the microscopic electrophoresis literature.[12] Recently,
attempts have been made to reduce the magnitude of the liquid flow in
channels by using coatings on the channel walls with low electrical
charge.[6,14] The performance of various coatings is not well established: There is the problem of coating stability itself and of ion
adsorption and desorption during the course of time.[14]

In summary, there are now two principal chamber designs in use
with concominant advantages and disadvantages. The narrow gap rectangular electrode system is free of electroosmotically driven
velocity distortion and has good heat dissipation properties as well
as high resistance to the onset of turbulent liquid motion; however,
it suffers from a sensitivity to electrode surface reactions. The
capillary system is capable of high current density operation as well
and is rather insensitive to electrode reactions but requires capillary
coating procedures to reduce electroosmotic liquid flow and also a
restricted sampling volume to minimize the effects of velocity
gradients.

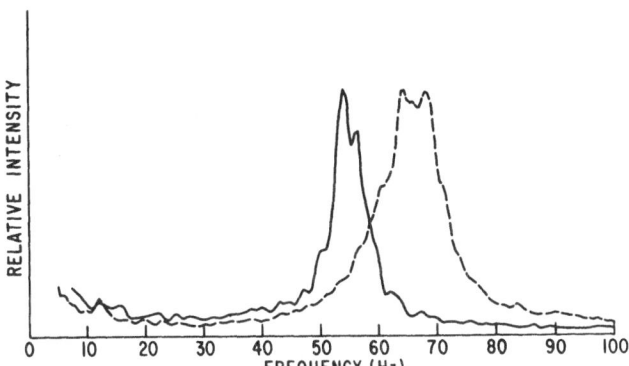

Fig. 3. Doppler spectra of human red blood cells. Solid curve,
         a large peak shift to peak width ratio obtained at 50 V/cm
         for cells in physiological saline. Dashed curve, spectrum
         from same sample but at a higher field, ∿ 60 V/cm, showing
         a degradation of the quality parameter, i.e., the shift
         to width ratio.

I will not consider any further details of chamber design or of
light scattering theory. These can be found in several recent
reviews.[13,15] Rather I will consider in the following section some
practical issues in Doppler measurements.

RECENT EXPERIMENTAL DEVELOPMENTS

Narrow gap electrodes suitable for high current density and high
salt conditions have been described.[16] One type is platinized silver
electrodes and the other is platinized silver/silver chloride electrodes. These are capable of up to 500mA/cm^2 and 50V/cm in physiological saline without causing degradation in the Doppler spectra.
By contrast the familiar reversible electrodes, Ag/AgCl, tend to
emit particles into solution while another common electrode type,
platinized platinum, is incapable of high current density without
bubbling.

The first type is the simplest and the least sensitive to abuse
in operation.   Platinum black is electrodeposited from a solution of
chloroplatinic acid at 10mA for 3 minutes. The electrodes are a pair
of rectangular Ag electrodes 2mm wide and are typically 2mm high.
The electrodes are then soaked in water and after about 30 minutes
with an applied square wave of about 25V/cm in 0.15N NaCl, the electrodes become sufficiently equilibrated to give a spectrum from red
blood cells such as shown in Fig. 4. With further aging and running,
the field homogeneity becomes remarkably good and can yield spectra
such as shown in Fig. 5.

There are two problems that may emerge in practice: 1) the
electrodes may, after an extended period of several days, induce a
local pH drift to around pH 8.5 in the immediate region of the electrodes; and 2) there may develop a polarization offset even for a
square wave voltage source that is symmetrical to 1% or better. In
the first case buffering with a reasonably strong buffer in 0.15N NaCl
can clamp the pH at a lower value than pH 8.5, if that is desired, and
in the second, reversal of electrode leads to the square wave source
may decrease the "polarization" effect (it is surmised that the
electrode polarization is the reason for the observed symptom of a
broadening of a spectral peak for red blood cells and ultimately of
the splitting of a peak into two peaks of equal amplitude) and give
spectra of the quality shown in Fig. 3 and Fig. 5. This electrode
system is simple and easy to prepare and in the event that aging and
prolonged operation or surface contamination preclude the attainment
of high resolution, the electrodes can be scraped clean and replati-
nized easily and quickly.

A detailed comparison of quality parameters of spectra and
mobility histograms obtained for homogeneous cell systems by a
variety of techniques has been given.[11] The remarkable point is
that in physiological saline spectra for red blood cells obtained
with the simple narrow gap electrode system is as good, if not better
than by any other technique (see for example Fig. 5, $Q \sim 11$ in that
case). Thus the inherent instrumental broadening by field inhomogeneities is not an important factor in practice. The mobility heterogeneity of even homogeneous cell systems such as red blood cells
exceeds the instrumental broadening. Nevertheless, it is striking
that a simple parallel plate electrode system can achieve such high
resolution. The high homogeneity must be the consequence of the
high conductivity of the solution spanning the space between the
electrodes.

The electrode geometry as depicted in Fig. 6 is typical and departures from these dimensions to a wider gap spacing leads to degradation in the maximum electric field and in the maximum resolution
that can be attained. Wide electrode gaps do not allow the attainment of high fields for the reasons given in Section III.

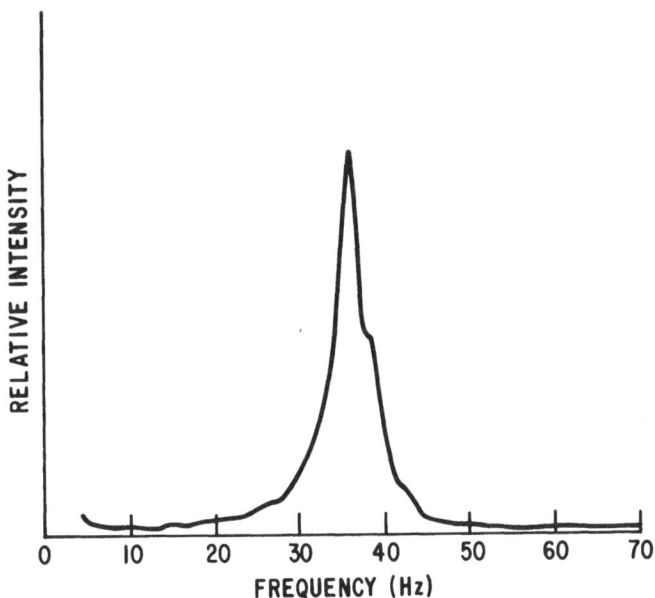

Fig. 4. Typical spectrum from red blood cells in physiological
         saline shortly after electrode preparation.

        The optical configurations used in this laboratory have been
described in detail[11] and are represented schematically in Fig. 2.
Wide angle scattering, i.e. at an angle of approximately 30° in water,
is the preferred optical configuration for blood cell studies. Because it is possible to observe a very large volume by virtue of
the absence of electroosmotically driven liquid flow in the narrow
gap parallel plate electrode configuration, the collecting optics
for the scattered light is extremely simple, rugged, and requires a
minimal of alignment as no collecting lens is used. Electrodes,
samples, and cuvettes can be readily and rapidly interchanged. The
sensitivity to microphonic vibration is greater than for the simple
single beam case shown in Fig. 2a. Thus, for best results, it is
desirable to have the optics in a quiet environment on a damped
optical table. Also it is beneficial to mount the laser and the
optics platform on sturdy mounts without introducing long pivot arms
anywhere in the optical system.

        Another optics approach is the differential Doppler mode in
which two beams of equal amplitude are made to intersect and the
scattered light from each beam is collected at any convenient position relative to the point of intersection. The Doppler shift is

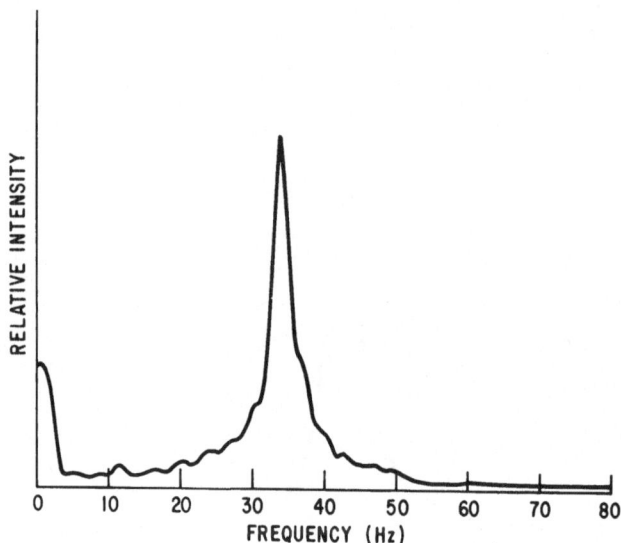

Fig. 5.   Spectrum from red blood cells in saline showing a peak
           shift to peak width which is in the range of the highest
           values that have been observed with the narrow gap rec-
           tangular electrodes. This high resolution depends on
           quality of electrodes. Operationally this means an ab-
           sence of electrode polarization that give a small asym-
           metry to the applied electric field in solution as well
           as an absence of any local pH drift in the liquid volume
           that is probed.

given by the angle of intersection. Because a strong reference beam
is lacking, a collecting lens must generally be used in this approach.
In our hands we found this approach to require rather sensitive alignment. If disturbed, both the cuvette and electrode position required
careful manipulation to regain an optimum signal. The sensitivity to
microphonics was about the same as in the reference beam mode. Recently workers in another laboratory have reported a very successful
operation of a differential Doppler spectrometer in a noisy environment.[17] Evidently, careful optics design can significantly decrease
sensitivity to microphonic disturbance.

     Yet another configuration involves the scattering of a second
beam from a diffuse scatterer and the reflection of this light from
the cuvette face itself into the photomultiplier collecting lens.[14]
Scattering angles of up to 57° have been used with this configuration.
This approach appears to be more cumbersome and less easily control-
lable than either of the two standard Doppler configurations described
above. I know of no special advantages of this approach. The use of
the cuvette itself as an integral component of the optics would make

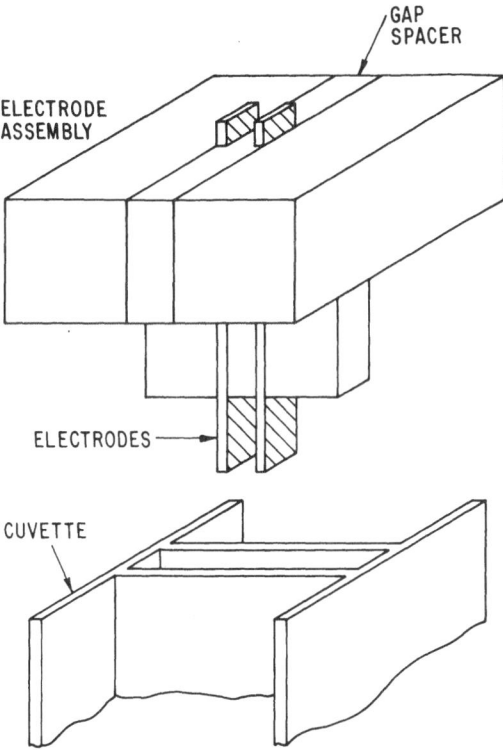

Fig. 6. Schematic of electrode assembly. The gap is typically
         1 mm, the electrodes are 1.8 mm wide and about 2 mm high
         in the active region.

cuvette interchanges more difficult requiring careful realignment
after reinsertion.

        The voltage to the electrodes is applied as a square wave in
order that the velocity information be obtained with 100% efficiency.
The polarity of motion is not important for the production of a given
Doppler shift. Of course, the simplest situation, that of a dc
electric field, is not possible because of the development of electrode polarization, gas bubbling, and surface reactions that lead to
a pH drift in the sample volume. In general, a square wave voltage
will introduce harmonic structure in the spectra at intervals of the
square wave frequency. A significant spectral smoothing can occur
if the diffusion width and the harmonic structure spacing are equivalent, and if the spectral bandwidth resolution $\Delta\nu_{Anal}$ is approximately
the same as $\Delta\nu_D$. Thus, the condition for spectra that are substantially free of harmonic structure is

$$\Delta\nu_D \sim \Delta\nu_{Anal} \sim 1/T$$

where T is the period of the square wave. In fact, Figs. 3-5 demonstrate this to be the case for red blood cells with $\Delta\nu_{Anal}$ = 1 Hz, $\Delta\nu_D$ = 1 Hz, and T = 0.56 sec.

Both for capillaries at high angles[17] (90°) and for channel systems[10] (57°) some spectra from red blood cell[17] and macrophage[10] suspensions contain amplitude modulation that is not of harmonic nature. In the first case the square wave frequency is quite low, 1/2 Hz, and the amplitude modulation is spaced at several Hz or more. In the second, the electric field is applied in a series of alternating polarity pulses of 2 sec duration and 20 sec spacing; hence, there should not arise harmonic structure in the power spectra at all. Some spectra (for example, Fig. 1 and Fig. 2 in Ref. 10), however, contain intensities modulated at a few Hz. The modulated power spectra must arise due to statistical effects. These effects have not been treated in the literature to date.

The question of particle statistics in a given measurement time is an important consideration. In capillary and channel systems, the scattering volume tends to be restricted by design in order to observe particles at a stationary layer of liquid flow and avoid velocity broadening. In alternating polarity, the same cells are viewed in one measurement cycle and the statistics for a single measurement may be poor since no new cells are sampled. New cells can be analyzed only by stopping the spectral measurement and injecting new sample into the probe volume or moving the probe volume along the capillary axis. Because the electrophoretic velocity is along the axial direction, the measurement cycle must stop or extraneous Doppler shifts would result. The situation is quite different for the narrow gap system: The scattering plane is perpendicular to the "long" axis of the electrodes and it is possible to introduce a continuous stream of new cells into the probe volume. In practice, for the conditions and electrode geometries typically used and for the reference beam mode, we estimate that as a minimum 10 cells are instantaneously observed for a cell density of 2×10^6/ml. And, in a one minute spectral measurement, a few hundred cells are observed by the spectrometer. By altering the operating parameters and the optics, even higher sampling rates are possible with this reference beam system.

An example of spectra from mixed cell systems is shown in Fig. 7 and Fig. 8. The statistical sampling of the spectrometer can be vertical by comparing the spectral peak areas for a sequence of spectral measurements. It should be noted that the spectrum analyzer sums data in discreet steps. In one minute, there are 64 summations to the integrated power spectrum for 200 Hz full scale analysis range that is typically used in this laboratory.

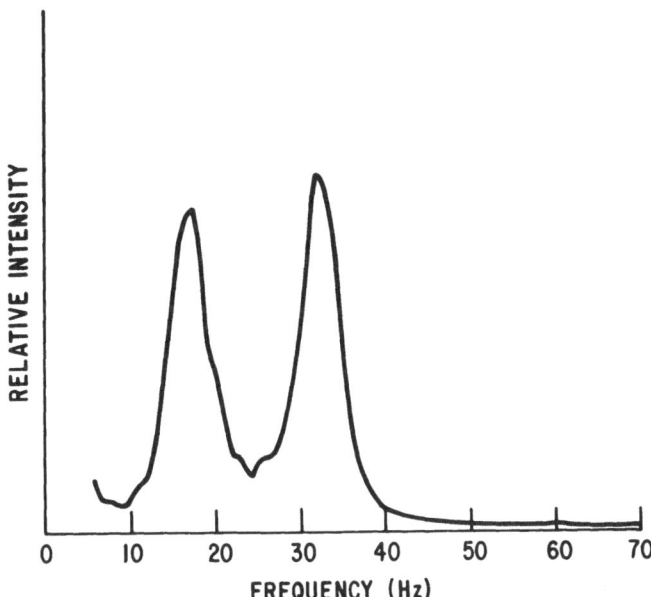

Fig. 7.   Spectrum from a mixture of rabbit and human red blood cells
          in physiological saline. The lower peak is associated with
          the rabbit cells and the higher peak with the human cells.

     Doppler spectra of red blood cells in physiological saline have
been obtained with a capillary system and wide angle differential
Doppler optics.[17] A quality factor of about 7 was obtained for the
red blood cell peaks. These peaks are modulated in amplitude at a
few Hz spacing presumably due to statistical limitations in individual
measurements. A review of earlier work with capillaries has been
given.[11,13,15] In low salt conditions, high mobility components for
fresh human red blood cells have been observed in the Doppler spectra
when measured with a capillary system;[18] however, red blood cells are
found to be homogeneous by other Doppler measurements, by electrophoretic cell separation techniques, and by microelectrophoresis.

DISCUSSION

     In the lecture, I will discuss various applications of laser
Doppler spectroscopy including blood plasma analysis, detection of
immune reactions, lymphocyte analysis, and lymphocyte-antigen interactions. I present here a very brief overview of recent blood cell
studies.

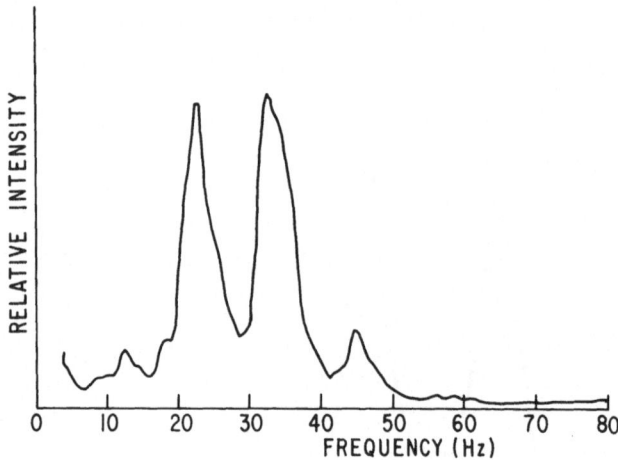

Fig. 8.   Spectrum from a mixture of human red blood cells that
          have and have not been exposed to neuraminidase. The
          lower frequency peak is from the neuraminidase treated
          cells. Smaller amplitude peaks at 12Hz and 45Hz are due to
          effects of vibratton.

        Before discussing the applications, it is necessary to consider
the performance of the method. That the spectral measurements are
fast and bias free is reasonably clear, but not so clear is whether
the resolving power is as good as achieved by other cell electrophoresis methods. A comparison of reported data on well defined homogeneous cell populations was made recently and indicated that the
resolution of red blood cells in physiological saline was as good or
better than that of many reports in the literature using microelectrophoresis techniques.[11]   In terms of quality parameter, Q, defined as
the peak shift frequency divided by the full width at half maximum,
or the equivalent definition for mobility histograms, the laser Doppler narrow electrode gap spectrometer yielded Q~8 for human red blood
cells in 0.15N NaCl and sometimes Q~10 or higher as indicated in this
paper whereas microelectrophoresis measurements spanned the range of
6 to 9. In more dilute salt, the reported Q values increased for all
measurements and remained comparable - of course the higher the Q
value the sharper the spectral peak or mobility histogram. For fixed
cells and certain dilute salt buffers, the density gradient method
or the free flow analytic method may have somewhat higher resolving

power than either laser Doppler or microelectrophoresis. However, either of the former methods are not suited for measurements in physiological saline which is the condition of most interest for general clinical and biological applications.

The first lymphocyte studies involving laser Doppler spectroscopy were on T and B cell mobility differences.[19] The finding that B cells are of lower mobility than T cells is in agreement with earlier reports and is now well established by subsequent work as well. Laser Doppler analysis gave one of the first indications that T cells in human peripheral blood are heterogeneous in mobility.[20] The faster T cells exhibit the property of "high-affinity" binding to sheep red blood cells and possess IgM receptors, whereas the slower T cells are "low-affinity" binding to sheep red blood cells and possess receptors for IgG.[7,21] Figures 9 and 10 indicate the T cell mobility patterns obtained before and after enrichment for IgM receptor bearing cells and IgG receptor bearing cells respectively. These findings do not yet suggest an important breakthrough for clinical application but

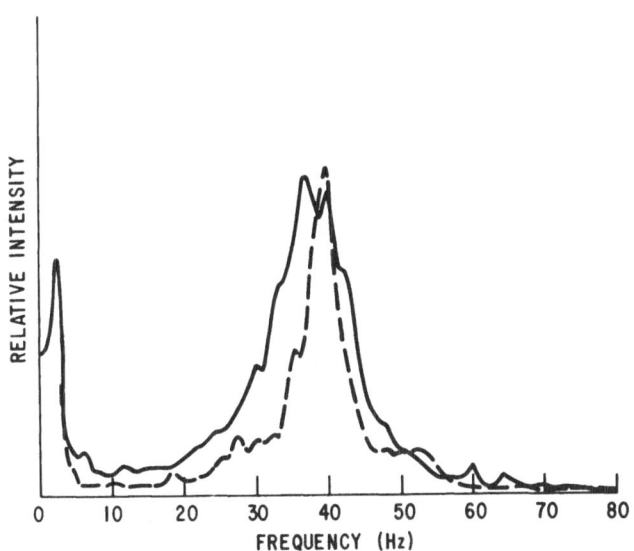

Fig. 9.   Doppler spectrum from lymphocytes purified on Ficoll-Hypaque, solid curve. Dashed curve, lymphocytes that are nonadherent to an IgG coated column. Note the heterogeneity is greatly reduced after passage through column. These cells are all of high mobility and are inferred to be T_{μ} type cells.

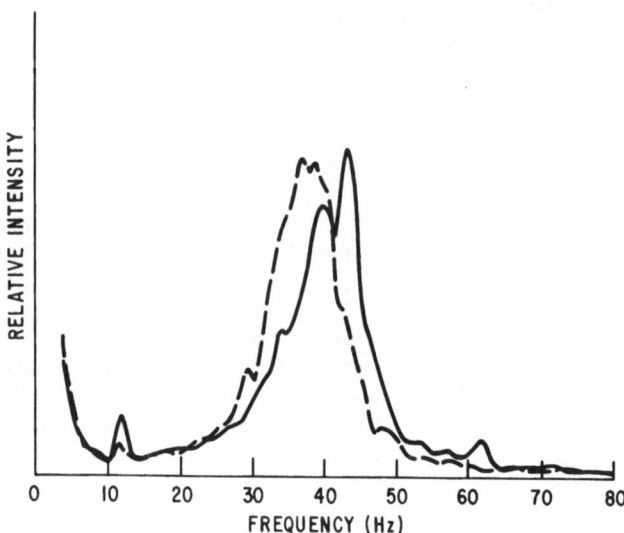

Fig. 10.   Ficoll-Hypaque purified lymphocytes solid, and after
           depletion of Tμ cells by passage through an IgM coated
           column , dashed.

certain differences between T cell mobility populations between normal
and diseased donors have been demonstrated.[8]

Figure 11 shows a comparison of spectra from preoperative cancer
patient lymphocytes and spectra from normal donor lymphocytes.   Clear
enhancement of slow type T lymphocytes in cancer patients is observed.
It may be that this heterogeneity and its variation among individuals
may be easier to analyze by laser Doppler spectroscopy than by other
electrophoretic techniques because of measurement speed and accuracy,
latitude of ionic conditions suitable for measurement, and the objective nature of the data acquisition.   Furthermore, it appears that
analysis of T cell populations by immunological markers such as by
sheep red blood cell binding differences is more difficult, more
cumbersome, and more resistant to standardization among different
laboratories.[8]

With regard to immune reactions there has been recent progress
on antigen-lymphocyte interactions.   This reaction can be observed
directly by analyzing lymphocyte populations before and after antigen
application.[9]   The first such measurements gave rise to rather weak

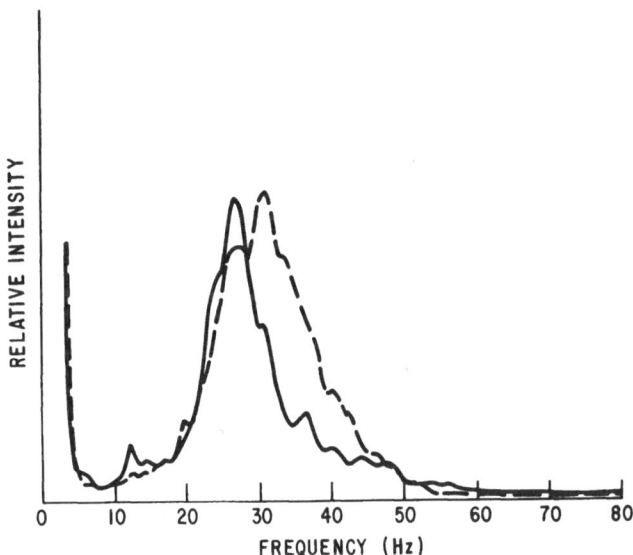

Fig. 11.   Comparison of Doppler spectra from normal donor (dashed)
           and preoperative cancer donor lymphocytes (solid).

signals that were difficult to discern - small mobility changes in
a small fraction of the cells observed. This lack of a strong signal
in the mobility analysis was in part due to measurement in low salt
conditions and in part due to improper protocol. With improved cell
handling procedures and protocol and higher resolution measurements
in physiological saline, the mobility changes upon antigen application are readily observed. This is in contrast to the indirect electrophoretic tests which involve the mobility measurement of another
type of cell, the macrophage, in the presence of reacting lymphocytes.
There is now controversy regarding this so-called MEM test.[10] It
would appear that a direct assay for antigen lymphocyte interaction
such as I describe will be far less cumbersome and quicker than the
indirect MEM test. If the specificity of the direct lymphocyte assay
holds up, it may be a valuable tool for studying cell mediated immunity and, perhaps, in clinical diagnosis.

ACKNOWLEDGMENTS

     I would like to thank my collaborators J.H. Kaplan and S.H.
Lockwood for kindly providing some samples and D.H. Cluxton for valuable experimental assistance.

REFERENCES

1. E.E. Uzgiris, Biophys. J. 12, 143a (1972).
2. B.R. Ware and W.H. Flygare, Chem. Phys. Lett. 12, 81 (1971).
3. E.E. Uzgiris, Optics Commun. 6, 55 (1972).
4. T. Yoshimura, A. Kikkawa, and N. Suzuki, Jpn. J. Appl. Phys. 11, 1797 (1972).
5. E.E. Uzgiris, Rev. Sci. Inst. 45, 74 (1974).
6. D.D. Haas and B.R. Ware, Anal. Biochem. 74, 175 (1976).
7. J.H. Kaplan, E.E. Uzgiris, S.H. Lockwood, J. Immunol. Methods 27, 241 (1979).
8. E.E. Uzgiris, J.H. Kaplan, T.J. Cunningham, S.H. Lockwood, and D. Steiner, European J. of Cancer 15, 1275 (1979).
9. E.E. Uzgiris and J.H. Kaplan, J. Immunology 117, 2165 (1976).
10. H.R. Petty, B.R. Ware, H.G. Remold, and R.E. Rocklin, J. Immunology 124, 381 (1980).
11. E.E. Uzgiris and D.H. Cluxton, Rev. Sci. Instruments 51, 44 (1980).
12. C.C. Brinton, Jr. and M.A. Lauffer in Electrophoresis, Vol. I, M. Bier, Ed., Academic Press, London (1959) p. 427.
13. E.E. Uzgiris, Prog. in Surface Science (in the press).
14. B.A. Smith and B.R. Ware, in Contemporary Topics in Analytical and Clinical Chemistry, Vol. 2, Hercules et al., Ed., Plenum Press, New York, 1978, p. 29.
15. E.E. Uzgiris, Adv. in Colloid and Interface Science (in the press).
16. E.E. Uzgiris, Rev. Sci. Instruments 51, 1004 (1980).
17. A.W. Preece, Private Communication.
18. R. Kaufmann and R. Steiner in Cell Electrophoresis, A. Preece and D. Sabolovic, Ed., Elsevier/North Holland, Amsterdam 1979, p. 435.
19. J.H. Kaplan and E.E. Uzgiris, J. Immunology 117, 1732 (1976).
20. E.E. Uzgiris and J.H. Kaplan, J. of Colloid and Interface Science 55, 148 (1976).
21. C.D. Platsoucas, R.A. Good, and S. Gupta, Proc. Nat. Acad. Sci., USA 76, 1972 (1979).

LECTURERS

F. S. Allen	Department of Chemistry The University of New Mexico
J. Bernengo	Laboratoire de Biophysique Université de Nice
A. D. Buckingham	Department of Theoretical Chemistry University of Cambridge
E. Charney	Laboratory of Chemical Physics National Institutes of Health
D. A. Dunmur	Department of Chemistry The University of Sheffield
J. G. de la Torre	Departmento de Quimica Fisica Universidad de Extremadura
C. Houssier	Laboratoire de Chimie Physique Université de Liège
B. R. Jennings	Department of Physics Brunel University
R. L. Jernigan	Laboratory of Theoretical Biology National Institutes of Health
M. Mandel	Department of Physical Chemistry University of Leyden
E. G. Neumann	Department of Physical Biochemistry Max Planck Institut für Biochemie
C. T. O'Konski	Department of Chemistry University of California

D. Pörschke   Max Planck Institute für Biophysikalische
                       Chemie

S. P. Stoylov   Institute of Physical Chemistry
                       Bulgarian Academy of Sciences

E. E. Uzgiris   Research and Development Center
                       General Electric Company

G. E. Weill   Center de Recherches sur les Macromolecules

INDEX